Aruba Certified Design Professional
OFFICIAL CERTIFICATION STUDY GUIDE (EXAM HPE6-A47)

First Edition

Aruba Education Development Team

HPE Press
660 4th Street, #802
San Francisco, CA 94107

Aruba Certified Design Professional (ACDP)
Official Certification Study Guide (Exam HPE6-A47)
Author Name: Aruba Education Development Team

Published by:

Hewlett Packard Enterprise Press
660 4th Street, #802
San Francisco, CA 94107

ISBN: **978**-1-942741-94-7

Printed in Mexico

WARNING AND DISCLAIMER
This book provides information about the topics covered in the Designing Aruba Solutions (HPE6-A47) certification exam. Every effort has been made to make this book as complete and as accurate as possible, but no warranty or fitness is implied.

The information is provided on an "as is" basis. The author, and Hewlett Packard Enterprise Press, shall have neither liability nor responsibility to any person or entity with respect to any loss or damages arising from the information contained in this book or from the use of the discs or programs that may accompany it.

The opinions expressed in this book belong to the author and are not necessarily those of Hewlett Packard Enterprise Press.

Feedback Information

At HPE Press, our goal is to create in-depth reference books of the best quality and value. Each book is crafted with care and precision, undergoing rigorous development that involves the expertise of members from the professional technical community.

Readers' feedback is a continuation of the process. If you have any comments regarding howwe could improve the quality of this book, or otherwise alter it to better suit your needs, you can contact us through email at hpepress@epac.com. Please make sure to include the book title and ISBN in your message.

We appreciate your feedback.

Publisher: Hewlett Packard Enterprise Press

HPE Aruba Contributors: Kimberly Graves, Leo Banville, Wim Groeneveld

HPE Press Program Manager: Michael Bishop

About the Authors

The material in this Study Guide was developed by Kimberly Graves and Leo Banville from the Aruba Education Development team and by Miriam Allred.

Introduction

This book is based on the Aruba Certified Design Professional (ACDP) course, which teaches networking professionals how to design best practice Aruba networking solutions, as well as how to explain the benefits of the solution to customers. The book explains how to gather information about customer requirements and how to translate those requirements into technical needs.

The guide covers best practices for security and high availability, as well as strategies for designing Quality of Service (QoS) solutions that protect voice and other time-sensitive traffic. Design topics include high density deployments for auditoriums and also solutions for telecommuters and traveling employees. Finally, the guide discusses Aruba AirWave and Central and explains when and how to add these solutions to a proposal.

As the guide progresses, readers can apply what they learn in a detailed scenario using practical skills. For example, there are activities to create VisualRF plans for AP locations and to design a logical network topology based on a set of requirements. At the end of the guide, you can put everything together, creating a complete Bill of Materials (BOM) and network solution design for a new scenario.

Certification and Learning

Hewlett Packard Enterprise Partner Ready Certification and Learning provides end-to-end continuous learning programs and professional certifications that can help you open doors and accelerate your career.

We provide

- **Professional sales and technical training and certifications** to give you the critical skills needed to design, manage and implement the most sought-after IT disciplines; and
- **Continuous learning activities and job-role based learning plans** to help you keep pace with the demands of the dynamic, fast paced IT industry
- **Advanced training** to help you navigate and seize opportunities within the top IT transformation areas that enable business advantage today.

As a Partner Ready Certification and Learning certified member, your skills, knowledge, and real-world experience are recognized and valued in the marketplace. To continue your professional and career growth, you have access to our large HPE community of world-class IT professionals, trend-makers and decision-makers. Share ideas, best practices, business insights, and challenges as you gain professional connections globally.

To learn more about HPE Partner Ready Certification and Learning certifications and continuous learning programs, please visit
http://certification-learning.hpe.com

Audience

This book is designed for presales solution architects involved in supporting the sale of Aruba solutions.

It is assumed that you have an understanding of basic wireless and wired technology and familiarity with the Aruba 8 architecture as well as interest in learning about design best practices to help you understand customers' business issues and to propose appropriate solutions.

Assumed Knowledge

Aruba Certified Design Professional Official Certification Study guide is a professional level book, which assumes some prior knowledge of Aruba solutions.

Minimum Qualifications

Typical candidates for this exam are networking IT professionals who have architect experience with Aruba wireless and wired switching solutions. They have relevant field experience focused on interpreting architectures and customer requirements to design Aruba subsystems or single campus network solutions.

Relevant Certifications

After you pass the exam, your achievement may be applicable toward more than one certification. To determine which certifications can be credited with this achievement, log in to The Learning Center and view the certifications listed on the exam's More Details tab. You might be on your way to achieving additional certifications.

Preparing for Exam HPE6-A47

This self-study guide does not guarantee that you will have all the knowledge you need to pass the exam. It is expected that you will also draw on real-world experience and would benefit from completing the hands-on lab activities provided in the instructor-led training. To pass the certification exam, you should be able to demonstrate that you can collect the information relevant to designing the solution, analyze this information to determine customer requirements, and create a plan to meet those requirements. The exam tests you on all components of the design from selecting the wired and wireless products; designing the physical and logical topology; creating the security, QoS, and management plans; and delivering the BOM.

Recommended HPE Training

Recommended training to prepare for each exam is accessible from the exam's page in The Learning Center. See the exam attachment, "Supporting courses," to view and register for the courses.

Obtain Hands-on Experience

You are not required to take the recommended, supported courses, and completion of training does not guarantee that you will pass the exams. Hewlett Packard Enterprise strongly recommends a combination of training, thorough review of courseware and additional study references, and sufficient on-the-job experience prior to taking an exam.

Exam Registration

To register for an exam, go to https://certification-learning.hpe.com/tr/learn_more_about_exams.html

CONTENTS

1 Information Gathering

EXAM OBJECTIVES

✓ Given a customer's needs for a single site, determine the information required to create a solution.

✓ Evaluate a customer's needs for a single-site campus, identify gaps, and recommend components.

Assumed knowledge

- Aruba Mobility Master and Controller architecture
- Basic knowledge of 802.11 and radio frequency (RF) concepts
- Basic knowledge of switching and routing

Introduction

In this chapter you will determine and document the existing network environment and identify key stakeholders. The scope and size of the project will be defined using the physical site information and the current network usage.

Key stakeholders

You will first consider how to identify and interview an organization's key stakeholders.

Key stakeholders of a deployment project

Information gathering for network design projects requires information from many different sources within the organization. Usually, someone at the executive level within the organization initiates a new or network upgrade project. Other C-level executives such as Chief Technology Officer (CTO), Chief Executive Officer (CEO), or Chief Financial Officer (CFO) may also need to "buy-in" and support the project, depending on the level of complexity and cost of the project. Most obviously technical employees such as network administrators or managers will need to be involved in the technical aspects of the project. Lastly, if the project will affect end-users, then appropriate training and information will need to be provided to the ultimate end-users of the system. It is good to understand what the customer wants to do with the network in terms of the business reasoning behind it. This information will enable you to discuss the appropriate Aruba solutions.

Many large and complex network design projects will require a long sales cycle. A proof of concept (POC) and possibly a pilot of the new network systems might be required to show the value they can bring to an organization. In many cases, the scope of the project may be defined in a Request for Proposal (RFP) or Request for Quote (RFQ), and the sales process will be highly competitive more often than not. Price and benefits of the new network design as well as solving the business needs are of utmost importance when discussing deployment projects.

Interview stakeholders

Once a network design project is underway, the information gathering process starts with talking to the organization's stakeholders and clarifying their expectations. It is important to define the scope of the project by discussing the current network deployment and issues or problems with the current systems. During the interview process, the critical needs and pain points of the current network and systems should be discussed. The anticipated growth of the organization, including additional expanded locations and additional users or devices, should also be discussed.

Stakeholder interview checklist

1. What is the scope of the project?
2. What is the primary purpose of the new network deployment?
3. What business problems are you trying to solve?
4. What are your most critical needs?
5. What are your pain points or most important issues?
6. What is your anticipated growth of the environment?

Wireless RF/network questionnaire

A common way to gather information is to use a questionnaire to ask about the network environment and customer needs. Figure 1-1 outlines the topics you would cover in a typical questionnaire.

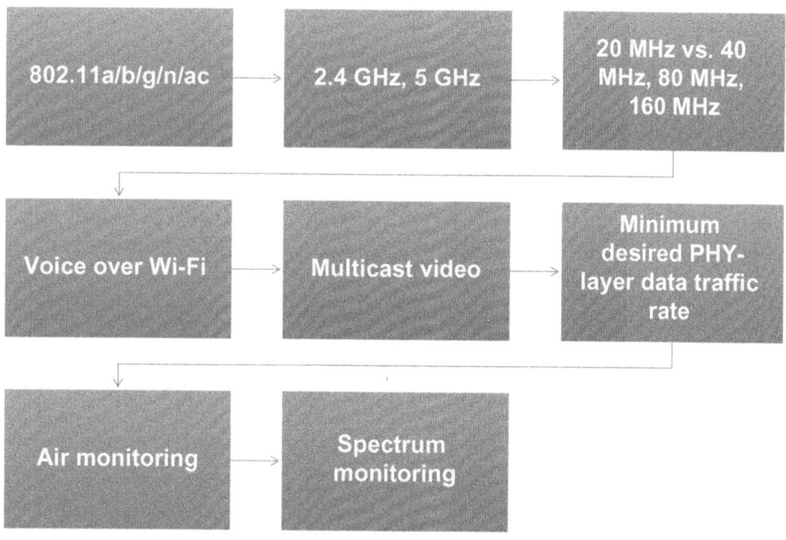

Figure 1-1 Wireless RF/network questionnaire

Prior to creating a network design, you should determine which 802.11 standards the customer devices currently support. You should also determine if client devices will be upgraded in the near future.

Almost all new enterprise deployments will be based upon 5 GHz radio frequency channels unless there is a compelling need for 2.4 GHz coverage. It may also be necessary to perform an on-premises walkthrough and spectrum analysis to determine if it is feasible to use 40 MHz or wider channels.

The primary use of the network such as voice, video, or data is critical when designing an RF network. Knowing a targeted minimum data rate is also helpful prior to starting the design process. Lastly, It is important to ask the customer if air monitoring/WIPS with containment or spectrum-monitoring services are necessary for the network.

Keep in mind that many customers might not know answers for all of these questions. Many environments feature clients introduced by the users themselves, and some customers lack insight into all the applications in use. It is generally best practice to design to support 802.11ac, as well as voice and video applications, in order to future proof.

Wireless RF/network questionnaire

1. What 802.11 PHY types are required over the course of the WLAN lifecycle (802.11a/b/g/n)?
2. Which RF bands (2.4 GHz, 5 GHz) will be used? Plan to use both bands due to increases in client density.
3. What channel width (20 MHz vs. 40 MHz) will be used in each band? Typically, 20 MHz channels are used in 2.4 GHz, and 40 MHz channels are used in the 5 GHz band.
4. Will voice over Wi-Fi be used? This answer will affect your planning for roaming and access point (AP) signal strength calculations.
5. Will multicast video over Wi-Fi be used? Use of roaming video has a similar effect as voice.
6. What is the minimum desired PHY-layer data traffic rate that must be available throughout the coverage area? Do some areas have different minimum data rate needs?
7. What are the desired air monitoring rates? Are dedicated air monitors required for security or compliance purposes?

Wired network questionnaire

You should also investigate the customers' wired network needs by asking, how many wired ports does the customer require for users and devices? What types of applications run over the network? It is also important to ask the customer about device capabilities (including Power over Ethernet [PoE]), logical topology, physical wiring closet locations, and site cabling.

Typical questions to ask the customer about the wired network include how many wired ports the customer needs to support, including ports for wired users and other devices such as printers and Internet of Things (IoT) devices. You should also ask what applications users run on the wired network. Ask whether the customer can share information about the current topology and device capabilities. You will also need to assess the physical requirements. In addition to closet locations, you need to gather information about the site's cabling. This will help you to select the correct switches and transceivers, and it might also affect whether you propose switches that support Smart Rate ports.

You should also keep in mind that information that the customer provides about the topology might not be update. If the customer is open to you passively collecting some information about the network, you could install an evaluation copy of AirWave or use a tool such as SolarWinds to map the network.

Wired network questionnaire

1. How many wired ports for users and devices?
2. What types of applications are used on the network?
3. What are the current device capabilities and topology?
4. Where are closets located?
5. How is the site cabled?

Device needs

Determining the user device needs is critical for accurately designing a wireless network. To better refine customer needs around AP model and number of APs, begin by asking, how many devices will each user have? Today, Aruba recommends you plan for at least three devices per user: a laptop, a tablet, and a smartphone.

As Figure 1-2 shows, other considerations about device capabilities such as max transmit power, number of supported spatial streams, and support for DFS channels are all important.

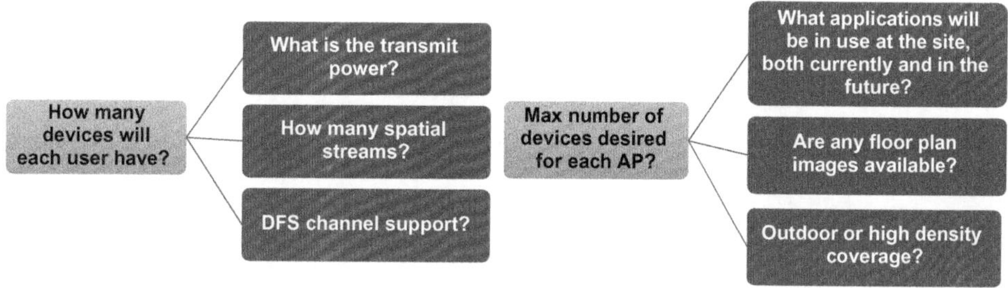

Figure 1-2 Device needs

Next, it is important to determine the maximum number of devices desired for each AP. Typically, Aruba recommends 20–30 devices per radio (40–60 per dual-radio AP). This number may be more or less, depending on traffic type (voice or data), expected load, and connection type (802.11a, b, g, or n).

Prior to starting an RF plan, you should try to gather any floorplans or image files from the customer and determine the scale of the floor plans.

You should ask the customer what their requirements are in terms of security—not only technical requirements but also regulatory. This is true for both wired and wireless. You should also gather all information they have in terms of regulatory requirements for their environment.

Customer user device needs job aid

1. How many devices will each user have? Today, Aruba recommends that you plan for at least three devices per user: a laptop, a tablet, and a smartphone. The number of devices per user also has ramifications in the design of VLANs and subnets. Consider if all devices will be active simultaneously, which also impacts AP density.
2. What is the maximum number of devices desired for each AP? Typically, Aruba recommends 20–30 devices per radio (40–60 per dual-radio AP). This number may be more or less depending on traffic type (voice or data), offered load, and connection type (802.11a, b, g, or n).
3. What applications will be in use at the site, both presently and in the future? Bandwidth requirements help determine coverage versus capacity requirements.
4. Are any floor plan images available? VisualRF Plan supports direct importation of JPEG, GIF, PNG, PDF, and CAD (.dwg and .dwf) files for floor plan formats.
5. What is the maximum transmit power of the least-capable common device in the network?
6. How many transmit, receive, and special streams do the most common devices support?
7. If DFS channels are being considered, do the devices most commonly used in the network support DFS channels?

Current physical environment

You will now focus on gathering information about the customer's current physical environment.

Physical sites

You also need to document the number of physical sites or locations that are within the scope of the project. It is a good idea to identify the purpose of the network in each location and the types of users the network will serve such as main campus or HQ (headquarters) location, branch office or remote workers. It is also important to find out the size in square feet or square meters of each site or location and the number of buildings and number of floors in each building. You should also document the number of users and types of devices for each site/building/floor where coverage will be needed.

Physical environment

Prior to creating an RF plan and determining the number of APs in each location or floor, you should understand the current RF environment. Ideally, this would include a current site survey, if one is available.

You will need a building floorplan file. This file can be in CAD, JPG, PNG, GIF, or other file format but needs to be an electronic version for each building or floor needing wireless coverage. You also need to know the scale of the floorplan and the building materials, so you can begin to plan wireless coverage. In addition, you need to know the type of space (such as retail, hospital, or education).

To complete the wired network design, you should obtain a floorplan or indoor/outdoor schematic of the building with locations of wiring closets and power and cooling capabilities. The customer should also provide an existing wired network diagram.

Devices

Figure 1-3 outlines the information you should gather about devices.

The types of devices accessing the network will be in large part determined by organizational and application needs. The mobility of the devices will dictate the level of roaming support needed on the network.

Some devices such as printers, copy machines, or other specialized equipment may connect to the network wirelessly, but remain stationary. Other devices such as laptops are used from different locations, but not usually used to access the network while in motion. These are sometimes called nomadic devices. Mobile phones and tables are most definitely mobile devices and will require the wireless infrastructure to enable the devices to access the network while in motion.

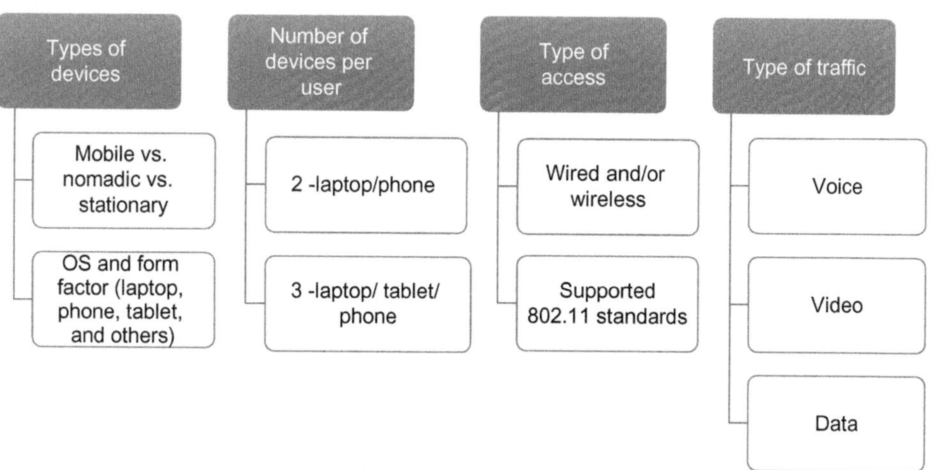

Figure 1-3 Devices

The number of devices per user is another consideration and must be factored into the entire device number that the network must support. You must also consider the mix of wired and wireless devices and the 802.11 standards wireless devices will support. The type of traffic is also critical for designing a well-functioning network that will support the number of devices and applications.

Mobile device types—portability

A WLAN must be flexible enough to accommodate wireless devices that have varying degrees of mobility. Devices can be grouped into three mobility categories, based on their usage characteristics and roaming frequency:

1. Stationary devices (SDs)

   If a wireless network uses only stationary devices, such as wireless printers, scales, or fixed Point of Sale (POS) terminals, the network planning is relatively simple.

2. Somewhat mobile devices (SMDs)

   In some settings, wireless devices move infrequently. Or they may move regularly, but they are only used while stationary. Examples include laptops used in auditoriums or conference rooms, or mobile barcode scanners or mobile PoS terminals.

3. Highly mobile devices (s) or roaming devices

   Handheld scanners or voice handsets present particular challenges to a WLAN (see Figure 1-4).

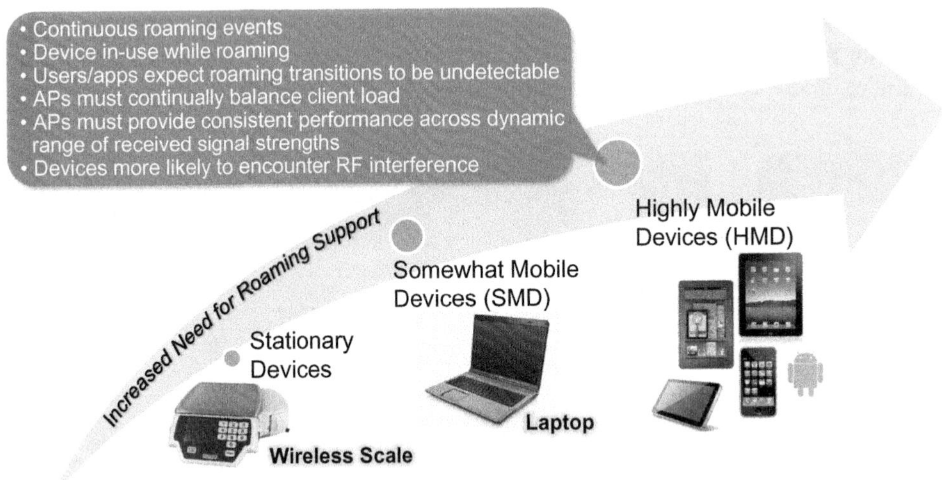

Figure 1-4 Mobile device types—portability

If a wireless network includes only stationary devices, such as wireless printers, scales, or fixed POS terminals, the network planning is relatively simple. Many wireless Internet of Things (IoT) devices are also stationary.

The most difficult WLAN to plan and implement is the one in which there are many highly mobile devices. As Figure 1-4 shows, highly mobile devices such as handheld scanners or voice handsets present the following challenges to a WLAN:

- Continuous roaming events
- Device is in use while roaming
- Users and applications expect roaming transitions to be undetectable
- APs must continually balance client load
- APs must provide consistent performance across a dynamic range of received signal strengths
- Devices are more likely to encounter RF interference

Users

The number of users and the types of users need to be documented for each location in the scope of the project. For example, the application needs of doctors needing wireless coverage at the bedside will differ from the wireless needs of nurses who are using voice devices to communicate. The characteristics of mobile devices connecting to a WLAN affect the configuration of the network infrastructure. Mobile devices are by nature more portable than nonmobile devices. When a WLAN has to support portable devices it requires robust roaming support as the devices move through the WLAN. Roaming support allows for seamless transition from one AP to another. The user does not have to reconnect, reassociate, or reauthenticate.

Assess roaming requirements

When users access the network through a wireless connection, they expect to be able to roam. First, you need to clarify what users mean by "roaming." A user working on a laptop and then taking that laptop elsewhere introduces less rigorous roaming requirements than users talking over a wireless voice app on the go. Voice over WLAN requires roaming to occur under 50 ms to remain undetectable to the user.

The customers' roaming requirements will affect both your RF plan and the overall solution architecture. You need to determine the extent of roaming domains for your customer. For example, should users be able to roam seamlessly throughout a building without a session interruption? Should coverage and roaming extend between buildings? If the latter, you will need to plan the outdoor coverage. If roaming domains extend far, you will need to consider how to cluster controllers to meet the needs.

You might also need to meet special requirements for manufacturing or warehouse environments.

Anticipated growth

During the customer interview, you need to factor in the anticipated growth of the organization as well. If the organization is planning a new location or site, or a large increase in the number of users or a change to a new device type, you should factor these changes into the design plan from the beginning. In addition, you should discuss any anticipated changes to the network in the future.

Basic wired connectivity requirements

The wired network will almost always function as the backhaul for wireless network, connecting wireless users and APs to their controllers. As Figure 1-5 shows, you will need to consider the locations for APs, as well as the requirements, such as PoE+, for their wired connections. You must also determine the type of cable because PoE+ should have at least CAT5e cable. CAT6a cable and above is preferable in modern networks.

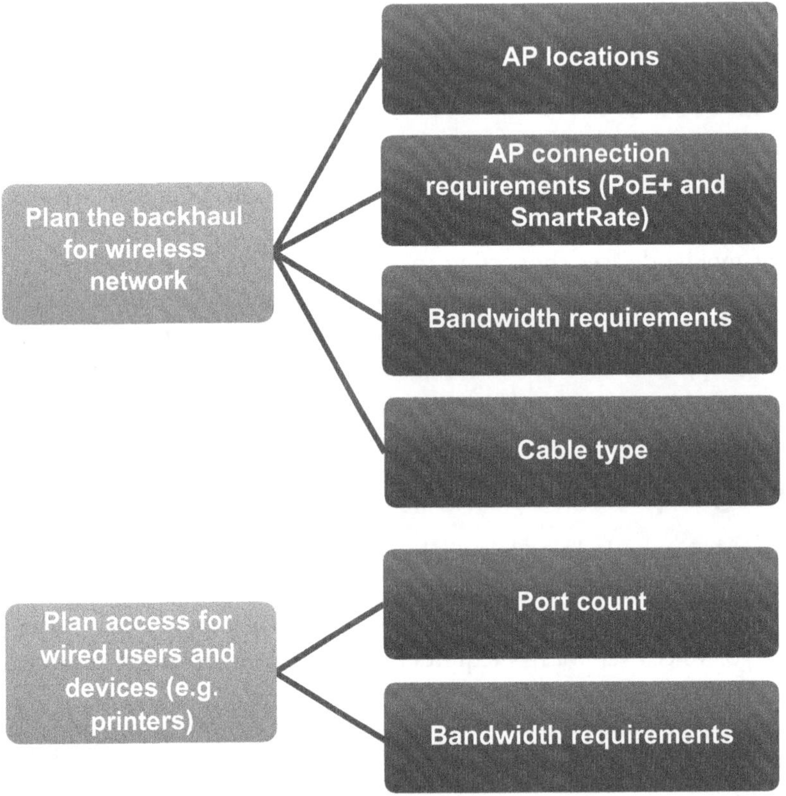

Figure 1-5 Basic wired connectivity requirements

Information you collect about user density and application requirements will help you to assess how much bandwidth you will need to plan on the APs' switches' uplinks and across the network. Customers might require wired connectivity for traditional workstations or devices such as printers.

As customers are beginning to use wireless connectivity more for their employees, many are repurposing Ethernet ports for IoT devices such as smart lights, heating, ventilation, and air conditioning (HVAC) systems, and more.

Again, port count and application requirements can help you to plan the uplink bandwidth.

Assess application requirements

Different applications have different requirements. In today's networks, a mix of application types is normal. Networks must support voice traffic, video traffic, critical transactions, routine database queries, requests for Web sites, file transfers, and more. Each of these applications must compete for network bandwidth. Ensuring availability and good performance for mission-critical and business-critical applications can present significant challenges, particularly in an environment that extends across wireless and wired media.

The first key to ensuring that users have a good experience running all applications is to provision adequate bandwidth from end to end.

Understanding applications' bandwidth requirements helps you plan for capacity. You should also consider whether applications tend to send bursts of traffic or a constant flow. You also need to know:

- What are the throughput requirements across multiple users over the long term?
- Does traffic flow more heavily from the core to the edge than the reverse? Also, does traffic flow east to west or north to south or vice versa?

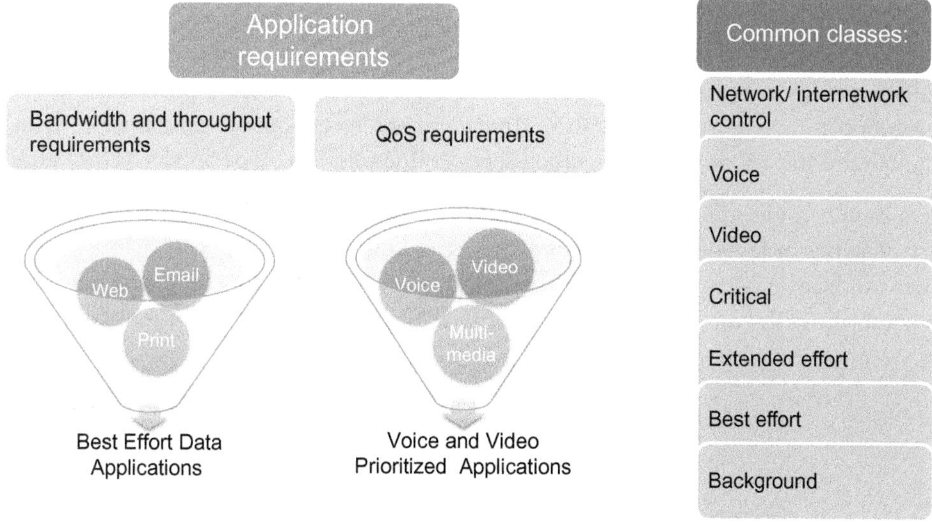

Figure 1-6 Assess application requirements

Only thorough examination of the throughput requirements can ensure that your plan will deliver the customers the experience they require. In the Ethernet network, you must scope the expected traffic loads on each physical link and plan appropriately. Planning capacity for the wireless network, though, can introduce the true challenges because multiple users share the wireless media (although multiuser MIMO [MU-MIMO] introduces another wrinkle). Other chapters in this study guide will help you plan wireless cells for typical use cases and high-capacity ones.

Applications' special requirements can extend beyond bandwidth requirements. Real-time applications such as voice and video also need special handing to ensure a good quality of service (QoS). These applications tend to be sensitive to excessive packet loss and latency. Even if the application can tolerate latency, jitter, which is uneven and unpredictable latency, can wreak havoc on the user experience. Jitter can occur even in networks with adequate provisioning.

To protect these applications, you should plan to implement QoS technologies that can prioritize traffic and give it different handling based on its class. A list of common classes is provided in Figure 1-6.

Work with the customer to define traffic classes, which designate sets of traffic that require the same sort of handling. Traffic classes suggested by IEEE 802.1Q-2005 include network and Internet work control traffic (the packets exchanged between network infrastructure devices for Layer 2 and 3 protocols), voice, video, mission-critical applications, important applications that needs extended effort, best effort (which normal traffic that needs typical handling), and background traffic (which is less important traffic or traffic that can tolerate delays). Later in this study guide, you will learn more how to create a plan to provide the correct service to each class, using the appropriate Aruba network technologies.

Assess security/regulatory requirements

You should discuss security needs with C-level executives so you can plan to provide the proper access security solution for the customer. Unauthorized network access opens companies to risks of data breaches, malware infections, and interrupted services due to a Denial of Service (DoS) attack.

Customers should understand what they stand to lose (see Figure 1-7). Data breaches can lead to regulatory penalties or loss of intellectual property. Unplanned downtime causes lost revenue. All of these issues can lead to costs to remediate the issue and possibly to provide restitution if customers' data was compromised. Less tangible, but also critical effects, include a loss of reputation and future business.

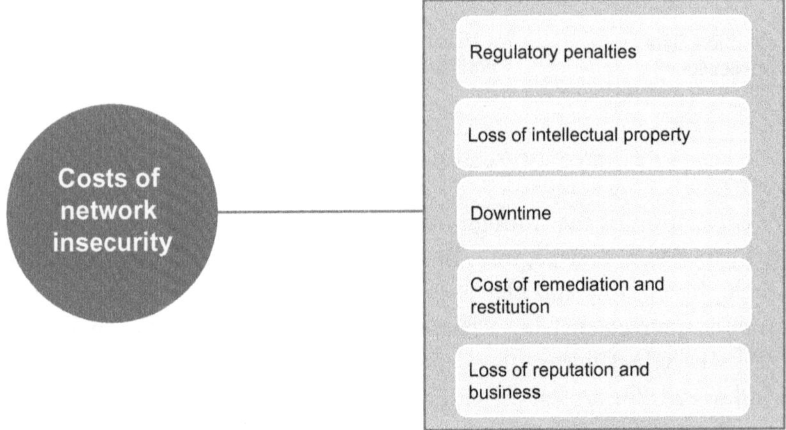

Figure 1-7 Assess security/regulatory requirements

Customers should understand that an enterprise wireless network should not use a single password but instead the true user-based authentication provided by 802.1X. This approach not only provides better protection against hackers but also enables the Aruba solution to control authorized users in different ways with a role-based and stateful firewall. Recent attacks should show customers that internal threats can pose real risks to their operations and bottom line.

In more detail, risks include regulatory penalties. The Health Insurance Portability and Accountability Act (HIPAA), the Payment Card Industry Data Security Standard (PCI DSS 2.0), and other federal and private regulations require particular security measures. These might extend well beyond personal firewalls and local data encryption into extensive network security requirements. While regulations may not specifically state network security requirements, companies will want to add extra layers of protection for applicable servers and data in order to protect themselves from security breaches that may result in fines or other types of penalties. These fines can add up quickly since each piece of data that is compromised can incur a fine. One security breach may compromise thousands of records and incur fines that can place an organization's financial stability at risk.

Legal protections aside, most companies are interested in making sure their proprietary data and company plans do not fall into the wrong hands. While legal measures for restitution exist, most companies would rather avoid the legal hassle by ensuring that their intellectual property is free from snoopers.

Security breaches, DoS attacks, and viruses can cause disruptions or failures in the network that result in downtime. And downtime represents revenue loss to most companies.

Other costs might include hiring IT security professionals to assess a breach, estimate its scope, and plan a response, as well as a way to avoid problems in the future. When regulated data is breached, the company must also spend money notifying those whose records have been compromised and receiving legal advisement about the implications. The company might need to offer restitution to those whose records were compromised in the form of identity protection.

Security breaches can cause companies to lose status or reputation with their customers and partners. A loss in reputation can impact a company's bottom line and result in revenue loss that can continue for years after the security breach.

Assess security requirements

A company's vulnerability to these risks depends on a variety of factors, including the types of network access offered, the types of users at the site, and the types of devices in use. Wireless access is inherently more open than wired access, but you can plan solutions that control both wireless and wired access appropriately. In fact, because many companies implement 802.1X on wireless connections as a matter of course, but not on wired ones, in a campus without physical monitoring, wireless connectivity could be more secure than Ethernet. You should discuss with customers whether they would like to integrate their wired connections with the security options available for wireless connectivity.

Aruba solutions are designed specially to offer role-based control at a deep level. You should work with the customers to understand which categories of users they need to control. You should also discuss whether they need a solution for handling Bring Your Own Device (BYOD), which you can offer with Aruba ClearPass Onboard.

Some customers simply want to classify users as employees, contractors, and guests or, in an education setting, students and faculty. Other customers will want to take full advantage of the role-based firewall and classify their users in more granular ways.

As you discuss the types of devices that the solution needs to control, assess whether the customer has IoT devices, which can be vulnerable to attack. This can affect whether you recommend User and Entity Behavior Analytics (UEBA) in addition to an access control solution such as ClearPass. An Aruba product that can provide UEBA is IntroSpect, an AI machine-learning solution that detects changes in behavior that could be indicative of an attack. It combines more than 100 supervised and unsupervised machine learning models focused on detecting targeted attacks with forensic insight into the network.

Assess availability requirements

Because customers rely on their network services, they need those services to be consistently available. But a wide variety of issues—from a cut cable to a configuration error to a power failure—can interrupt services. As you will learn in a later chapter, you can plan Aruba solutions to provide varying levels of availability. Higher degrees of availability require more equipment—for example, an additional controller for redundancy—as well as technologies for detecting and responding to failures.

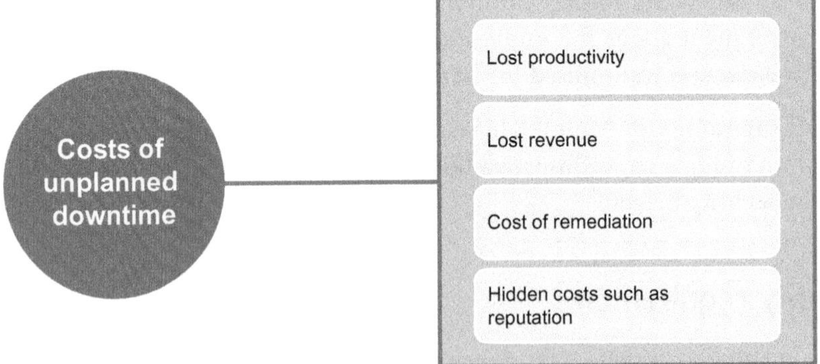

Figure 1-8 Assess availability requirements

Because these requirements translate to additional costs and complexity, you need to assess the availability level this particular customer requires based on the costs of unplanned downtime. The required level typically depends on how vital a service is to the customer's mission and the risks they run from loss of that service. As Figure 1-8 shows, these risks include lost productivity, lost revenue, and the cost of remediating the issue, as well as hidden costs such as damage to the customer's reputation.

For example, if the customer relies on wireless voice applications, the wireless network will need to provide higher availability than if the customer uses wireless as a convenience for guests. And a hospital that uses the network for supporting critical services during surgery will have much higher-availability requirements than a typical office.

Customer checklist

The following checklist can help verify all the needed information has been gathered during initial customer meetings prior to starting design plan construction.

Checklist

- ✓ Building floorplan/blueprint
- ✓ Existing wired environment
- ✓ User information
- ✓ Site information
- ✓ Device information
- ✓ Mobility requirements
- ✓ Roaming capabilities
- ✓ Application requirements
- ✓ Security requirements
- ✓ Growth expected

Activity 1: Determine customer network and requirements

Throughout this study guide, you will be given opportunities to practice designing a network by completing activities. In Chapters 1–11, these activities will be based on the same customer scenario. Each activity in these chapters will build on the previous chapters, guiding you through the process of designing a network upgrade and creating a bill of materials (BOM) and network design for a fictitious customer.

"Appendix: Activities and Learning Check Answers" includes the answers for the questions outlined in each activity.

In this first activity, you will read a customer scenario and determine the customer's existing network in preparation for designing a new network.

Customer scenario

Corp1 is a reseller of residential products. This company sells and distributes to more than 200 customers, each of which have a multitude of stores, including Walcheap, Costalot, and Northofme, around the United States.

Corp1's main corporate campus is in Big City. The company also has 10 warehouses around the country.

Corp1 goals

Corp1 is looking for a total upgrade of its Wi-Fi and campus wired network. Phase 1 will be the corporate offices in Big City. If the upgrade is successful, Phase 2 will be completed next year and will include the warehouses.

Corp1 has not had a campus network upgrade for approximately seven years. Employee surveys have shown that while some employees still use wired connections, many would like to connect on wireless alone. But the existing Wi-Fi coverage is spotty. Even though employees do not use highly demanding applications, performance is poor. Even on wired connections, some of the more modern applications are running slowly.

After the upgrade, the corporate office should have 100% Wi-Fi coverage and, if possible, seamless roaming between the main office buildings. To give the network longevity, the design should provide 802.11ac Wave 2 coverage. The initial bid should provide a signal level of –75dBm across the campus.

The customer also wants to upgrade the aging campus wired network to support the new 802.11ac APs and to improve performance. Although the customer is interested to shifting toward greater reliance on the wireless network, key decision-makers want to ensure that employees can still obtain wired connections through their docking stations. They also want to continue to support all existing wired drops for potential repurposing.

The customer has expressed interest in a three-year support contract.

The installation and training will be negotiated with the local partner.

Main site information

Corp1 has a main corporate campus in a business park where they lease two adjacent buildings that are 50 feet (15.2 m) apart. This is an open campus with no obstructions between the buildings.

- Each building has three floors.
- Each floor is 310 x 173 feet (94 x 53 m) for 53,630 square feet total (4,982 sq m).
- The ceiling for each floor is 15 feet high (4.5 m) with a drop-down ceiling at 10 feet (3 m).

The sections below include information about each floor and the associated floor plans.

Building 1, Floor 1

- Building 1, Floor 1 has a front door reception area with offices and a mailroom that takes up 25% of the southeast side of the floor (see Figure 1-9).
- The data center is located in building 1, floor 1 on the west side.
- The north side of this floor is open space.
- The main corridor is at the center of the building and includes washrooms, stairs, elevators, and supply and service cabinets.

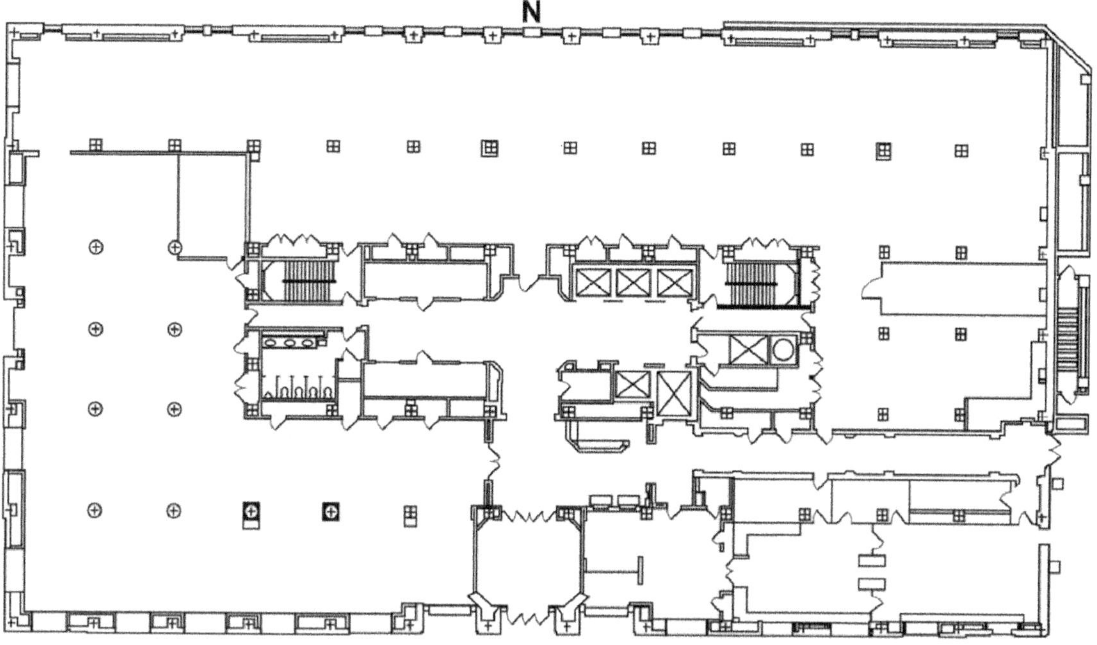

Figure 1-9 Bulding 1, Floor 1

Building 1, Floor 2

- Building 1, Floor 2 has a central main corridor with washrooms, stairs, elevators, and supply and network cabinets (see Figure 1-10).
- There are cubicles around the perimeter of the floor.
- The northeast and northwest sides of the floor have conference rooms with wired devices for video conferencing.

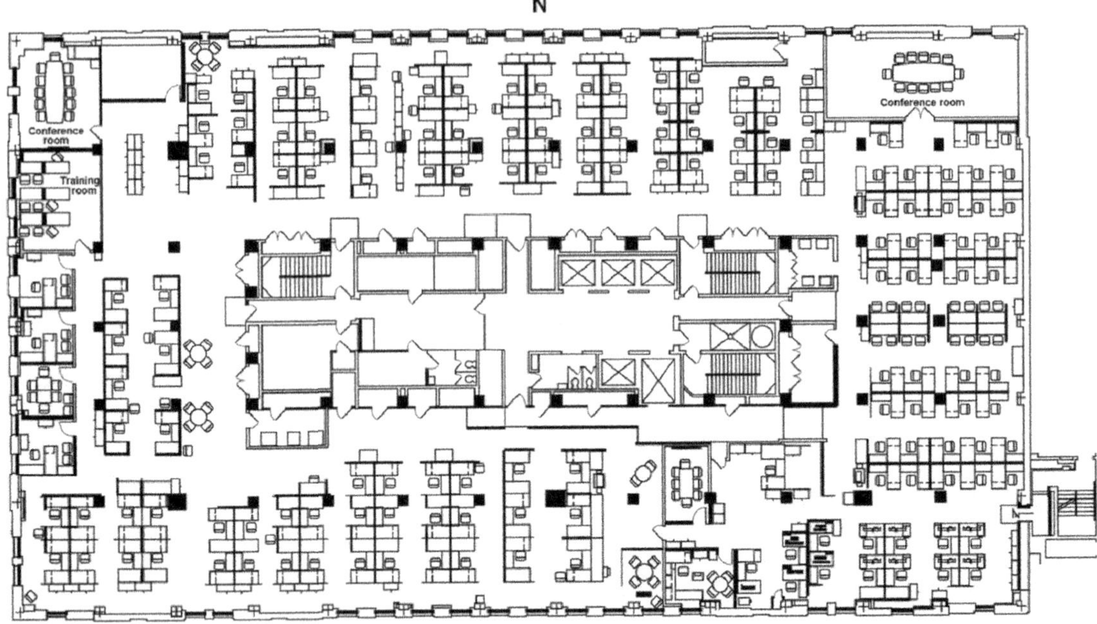

Figure 1-10 Bulding 1, Floor 2

Building 1, Floor 3

- Building 1, Floor 3 has more office space and fewer cubicles (see Figure 1-11).
- This space includes the president's office, vice president's office, as well as the HR and Finance departments.
- The north side of the floor has two boardrooms, which have wired video conferencing devices.
- The center corridor is a repeat of the second and first floor.

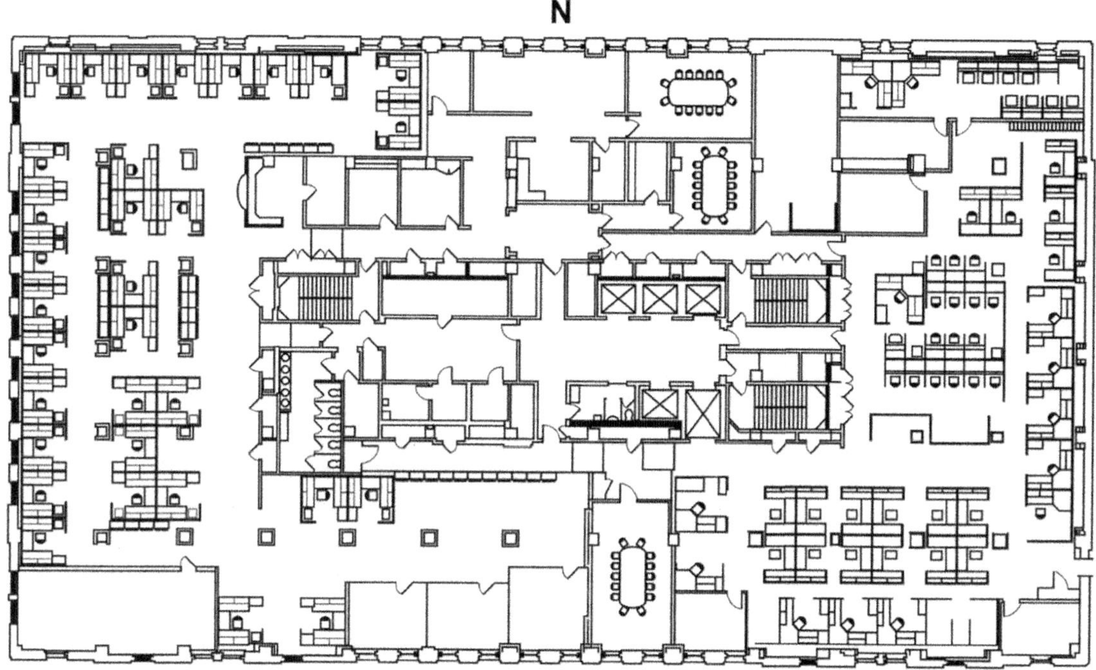

Figure 1-11 Bulding 1, Floor 3

Building 2 Floors 1-3

In the second building all floors have a central main corridor with washrooms, stairs, elevators, supply cabinets and network cabinets (see Figure 1-12). The perimeter of each floor includes cubicles and conference rooms, which might need to support video conferencing.

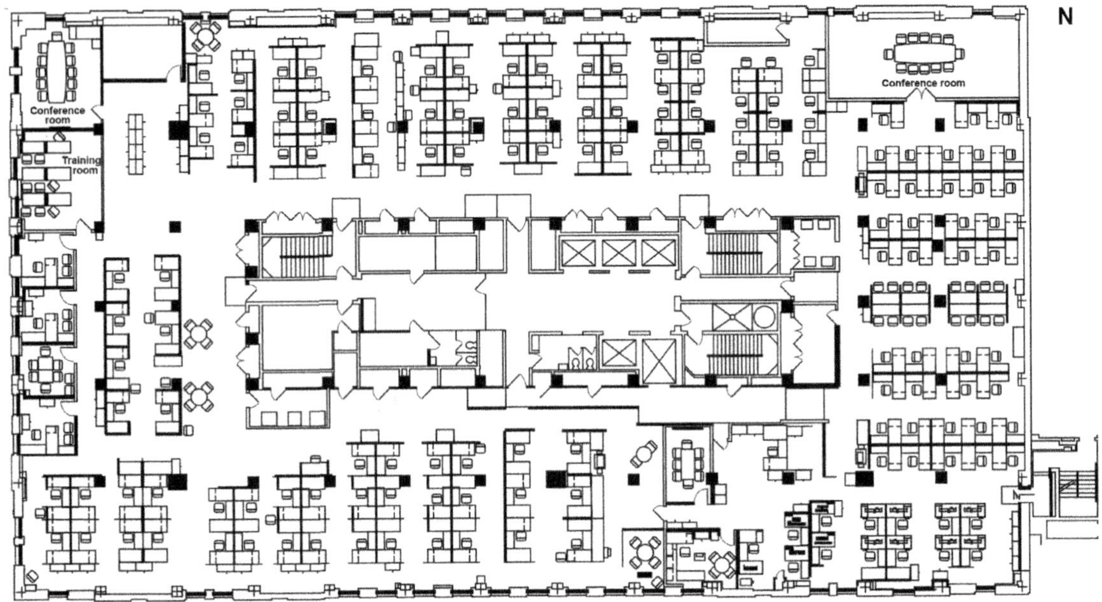

Figure 1-12 Bulding 2 Floors 1-3

Warehouse information

There are 10 warehouses, which are 40,000 sq. feet each (3716 sq m). The dimensions for each building are 100 x 400 feet (30.5 x 122 m) and 50 feet (15.3 m) high. Each warehouse has a truck loading bay as well as numerous racks, cabinets, and distribution lines. At on one end of the building, a raised administrative section includes 10 offices that measure 30 x 100 feet (9 x 30.5 m) with a 15-foot (4.5 m) ceiling.

Corp1 existing network

Seven years ago, Corp1 installed Wi-Fi and upgraded the access and aggregation layer of the main corporate office network. At the time, the Wi-Fi was just an add-on and not of critical importance. The customer has supplied Figures 1-13 and 1-14, illustrating the logical and physical topology, and described the network.

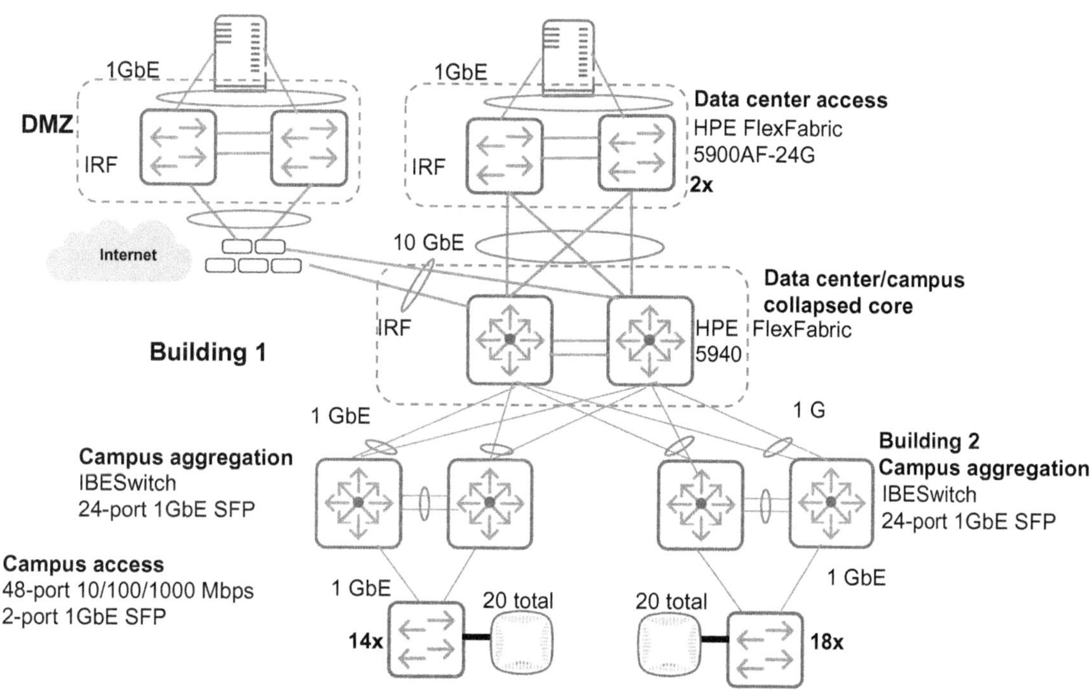

Figure 1-13 Logical topology of Corp1's existing network

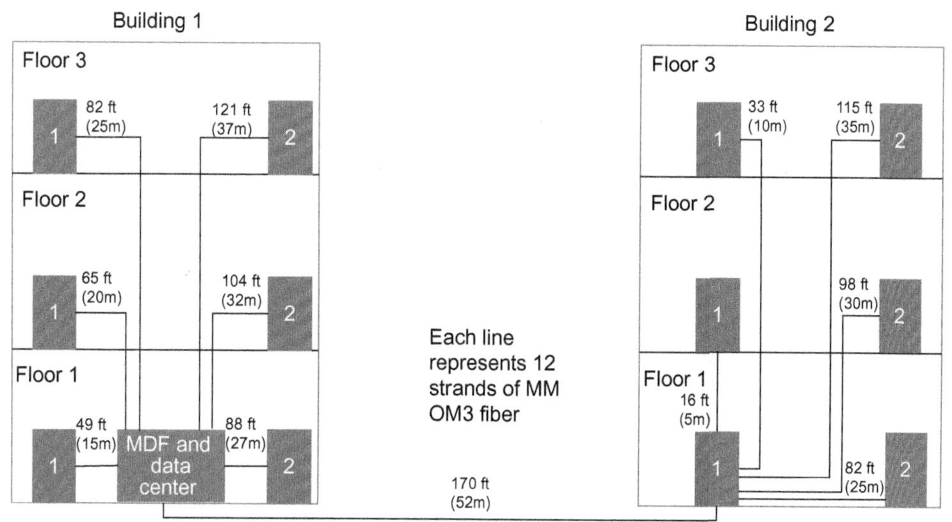

Figure 1-14 Physical topology of Corp1's existing network

Wireless network

The IBEWIFI vendor supplied the current a/bg Wi-Fi network equipment. The Wi-Fi deployment is sparse and does not cover every corner of the building. Each building has about 6 to 8 APs per floor.

Campus wired access layer

There are two network closets in the main corridor of every floor. Figure 1-15 shows the closet location on Building 1, Floor 2 as an example. The cable drops use CAT5e cable. The maximum distance from a far corner of the floor to any wiring closet is 294 feet (89.6 m).

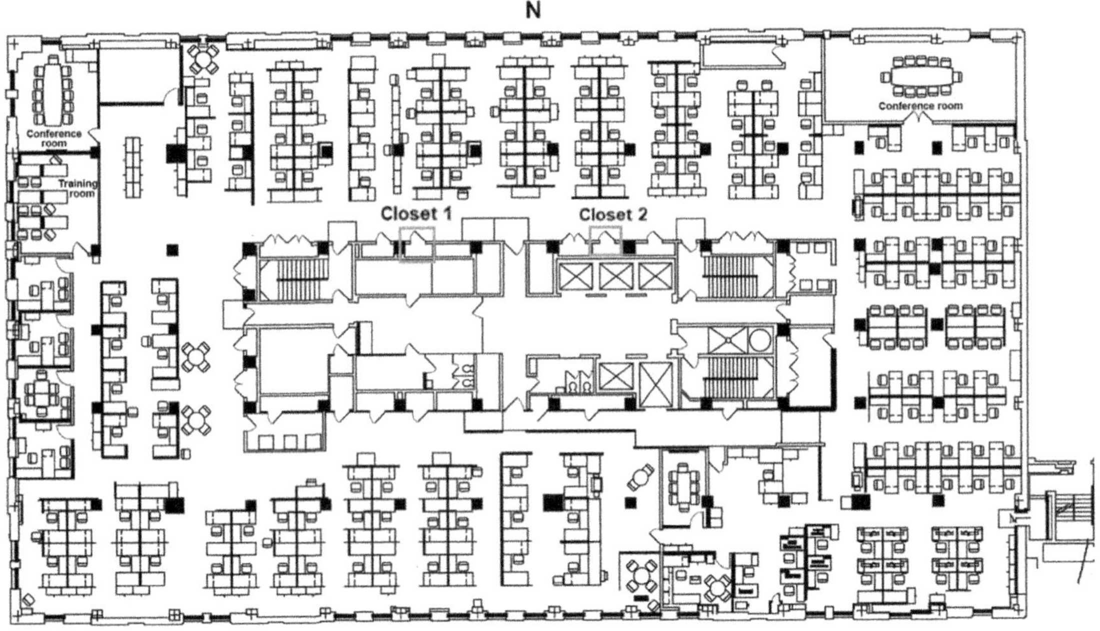

Figure 1-15 Closet locations on Building 1, Floor 2

Each closet has 20U of rack space available. The closets are equipped with switches from IBESwitch. Each switch has these capabilities:

- 48 10/100/1000 Mbps ports
- PoE (802.3af) on 8 ports
- Two 1 GbE SFP SX transceivers
- Layer 2 forwarding
- No authentication capabilities

Each wiring closet has three switches and CAT5e cable extended to 125 wired jacks, as well as drops to three or four APs. The exceptions are the two wiring closets in Building 1, Floor 1. Each of these closets has one switch, drops to 35 jacks, and drops to three APs.

The aggregation layer of the network only supports 1 GbE ports. The switches also have a limitation on IP-to-VLAN mappings and have limited ARP and MAC tables.

There are four active fiber links between the two buildings that currently use 1 GbE. The fiber can also support 10 GbE or 40 GbE.

The aggregation switches for Building 1 are located on Floor 1 in the data center. The aggregation switches for Building 2 are located in Floor 1, Closet 1.

Corp1 has its own data center with several server racks and a DMZ. The company updated the data center network two years ago. The data center uses HPE FlexFabric switches with 10 GbE/40 GbE interfaces for routing and switching. The server and networking teams are quite isolated from each other; even the data center networking and campus networking administration is separate.

The admin section of the warehouse has a wired cabinet with a single switch and gateway equipment for MPLS and corporate access. The switch is also wired to four other wired cabinets on the warehouse floor. Each cabinet has one 48 port switch that supports BASE10/100/1000. The site is wired with CAT5e cabling.

Users

The main site has 900 employees:

- Building 1 Floor 130 employees
- Building 1 Floor 2200 employees
- Building 1 Floor 370 employees
- Building 2 Floor 1200 employees
- Building 2 Floor 2200 employees
- Building 2 Floor 3200 employees

The daily guest average is 20 per day for the main site.

All warehouse office spaces house approximately 10 employees while the floor of the warehouse has approximately 25 employees. With 10 warehouses, Corp1 has 350 warehouse employees total.

Existing client equipment

Corp1 has shared this information about the devices that require connectivity.

The employees' desk workstations have been swapped for HP EliteBook Folio G1 laptops with 802.11ac capabilities and Windows 10. Each employee at the main site has a docking station for the laptop, has a wired connection. Many employees also connect the laptops wirelessly.

All employees have their own smartphones and tablets that are allowed on the network.

Employees typically use the network for Web browsing, accessing shared files, sending email, and printing files, as well as to look up sales and inventory information.

Each building has two large color printers in the center corridor room of the first floor. The large printers require 1 GbE wired access.

Many departments have also purchased their own small printers that have both wired and Wi-Fi capabilities. We estimate approximately ten of these types of printers per floor. Some of the older printers are 802.11ag capable, and others are 802.11n capable.

The company has video conferencing equipment in the conference rooms on Building 1, Floor 2 and all three floors of Building 2. Two boardrooms on Building 1, Floor 3 also have video conferencing equipment.

Activity 1.1: Determine the customer's existing network

Activity 1.1 objective

You will review the customer scenario and determine important information about the existing network.

Activity 1.1 steps

1. Briefly describe the scope of your project. For which sites and buildings do you need to propose a network solution?
2. Determine the customer's present Wi-Fi network structure at this site:
    a. Existing Wi-Fi network: Y/N ____________
    b. Wi-Fi vendor: ____________
    c. Radio capabilities: ____________
    d. Give a brief explanation of the customer's current Wi-Fi issues:

3. Determine the customer's existing wired network(s) at this site:
   a. Wired network to be replaced: Y/N ___________
   b. If yes, which segments ___________
   c. Data center network vendor: ___________
   d. Campus network vendor: ___________
   e. Number of wired closets: ___________
   f. Rack space in each closet: ___________
   g. Number of access layer switches per closet: ___________
   h. Number and speed of ports on access layer switches: ___________
   i. Current access layer switch capabilities such as PoE and PoE+:
   j. Is every desk wired (Y/N): ___________
   k. Wired network diagram supplied (Y/N): ___________
   l. Give a brief explanation of the customer's present wired issues. Also note features that might be required that the current switches do not support:
   m. Does the existing wired network support enough drops for an upgraded Wi-Fi network?

Activity 1.2: Determine site specifics

Activity 1.2: Objective

You will begin to assess the physical environment.

Activity 1.2: Steps

1. You have determined that the customer needs a Wi-Fi upgrade for two buildings at the main campus. Fill in more details in Table 1-1. If the information is the same for both buildings, you can simply draw an arrow to the other cell.

Table 1-1 Details of Buildings 1 and 2 at the Main Campus

	Building 1	Building 2
Dimensions:		
Length		
Width		
Area per floor		
Number of floors		
Ceiling height		
Drop-ceiling height		
What obstruction would you typically find above the drop-ceiling?		
Floorplans supplied (Y/N)		
General description of the physical environment and list of features relevant to RF		

2. What is the distance between Building 1 and Building 2: ____________

Activity 1.3: Collect information on users and devices

Activity 1.3: Objectives

Review the customer scenario and determine how many users are at each location, as well as what devices and applications they are using.

Activity 1.3: Steps

1. Determine the customer's wireless user base:
   a. Number of users in main campus: ____________
   b. User types: ____________, ____________, ____________
   c. Device types: ____________, ____________, ____________
   d. Types of applications in use: ____________, ____________, ____________
2. What are the requirements for roaming on these devices?
3. Use the Internet to research the wireless capabilities of some smartphones. You can use this site http://clients.mikealbano.com/. Or you can search for the official specifications on the providers' sites. Note features such as the Wi-Fi bands supported, 802.11ac support, and number of spatial streams. Look for IPhone 7, Samsung Galaxy S8 plus, and the Sony Z4. (If you want to look up the EliteBook Folio G1 specifications, search for the HP QuickSpecs because the clients.mikealbano.com website does not list this specific model.)

4. Determine the customer's wired user and device base:
    a. Excluding APs, how many drops per closet must the new switches support? If the answer is the same for multiple closets, you can draw an arrow down to indicate that.
        i. Building 1
            1. Floor 1 Closet 1: __________
            2. Floor 1 Closet 2: __________
            3. Floor 2 Closet 1: __________
            4. Floor 2 Closet 2: __________
            5. Floor 3 Closet 1: __________
            6. Floor 3 Closet 2: __________
        ii. Building 2
            1. Floor 1 Closet 1: __________
            2. Floor 1 Closet 2: __________
            3. Floor 2 Closet 1: __________
            4. Floor 2 Closet 2: __________
            5. Floor 3 Closet 1: __________
            6. Floor 3 Closet 2: __________
    b. User types: __________, __________, __________
    c. Device types: __________, __________, __________
    d. Types of applications in use: __________, __________, __________

Answers to the questions included in activities are listed in "Appendix: Activities and Learning Check Answers."

Summary

The first step in a network design is to identify stakeholders and determine the customers needs and expectations. The physical environment, including site information and current network usage, and the scope of the project need to be determined prior to the design process.

Learning check

1. What information do you need to gather about devices to create an RF plan? (Select two.)
   a. Type of applications running on the devices
   b. Size of the hard drive and available RAM
   c. Security features supported
   d. Degree of mobility
   e. Whether multiple users share laptops or workstations
2. What challenges do highly mobile devices such as voice handsets pose for architects planning a WLAN?
   a. These devices are in use while roaming, and users expect roaming transitions to be seamless.
   b. These devices require APs operating in the 5.0 GHz range, which limits the ability to implement 40 MHz or wider channels.
   c. These devices typically do not support QoS, making it more difficult to ensure traffic is handled appropriately across the network.
   d. These devices do not support 802.11ac, forcing network designers to deploy APs that support 802.11 a/g.

Answers to the learning checks in this study guide are in "Appendix: Activities and Learning Check Answers."

2 RF Planning

EXAM OBJECTIVES

- ✓ Given a scenario, translate the business needs of a single-site campus environment or subsystems of an enterprise-wide environment into technical customer requirements.
- ✓ Given a customer's requirements for a single-site campus, design the high-level architecture.
- ✓ Select the appropriate products based on a customer's needs for a single-site campus.
- ✓ Given a scenario, explain how a specific technology or solution would meet the customer's requirements.
- ✓ Given a customer scenario for a single-site campus environment, design and document the logical and physical network solutions.

Assumed knowledge

- Basic wireless network concepts, such as the 802.11 standards
- Frequencies and channels used by 802.11 and the basics of co-channel interference
- Familiarity with the Aruba Mobile First Architecture
- Basic network design principles

Introduction

Understanding Radio Frequency (RF) behavior and RF planning is critical to properly designing a highly available and responsive wireless LAN (WLAN). The first step in RF planning is validating the physical RF environment at the network premises. After you perform that step, selecting the AP and antenna is the next step of the design process followed by AP placement and an RF channel plan.

Review RF fundamentals

You will begin by reviewing RF fundamental concepts.

Review IEEE 802.11 amendments

The 802.11 standard was ratified in 1997. It was capable of transmitting at 1 and 2 Mbps, using either Frequency Hopping Spread Spectrum (FHSS) or Direct Sequence Spread Spectrum (DSSS) with the 2.4 GHz Industrial Scientific and Medical (ISM) band. Most early 802.11 installations used FHSS; the early 802.11 is rarely used anymore.

The 802.11b and 802.11a amendments were both ratified at the same time in 1999. Products that supported 802.11b shipped immediately, whereas products that supported 802.11a did not ship until almost a year later. Providing an upgrade to 802.11, 802.11b supported transmission speeds of up to 11 Mbps while using the same frequency band as 802.11. The 802.11b amendment also provided backward compatibility with 802.11 DSSS and was the first of the wireless technologies to gain widespread consumer acceptance. This was mainly due to the decrease in prices for the equipment.

The 802.11n amendment was ratified in 2009 and provided data rates up to 600 Mbps, even though 300 Mbps is the highest commonly supported rate. In addition, 802.11n operates in the 2.4 GHz and 5 GHz bands.

The 802.11ac amendment, which was ratified in 2013, provided wider channels and better modulation with up to eight spatial streams.

Table 2-1 summarizes the 802.11 standards.

Table 2-1 The 802.11 Standards

IEEE Standard	Transmission Speed	Frequency and Band	Comments
802.11 (1997)	1, 2 Mbps	2.4 GHz ISM	Original standard. Rarely used anymore. FHSS and DSSS.
802.11b (1999)	1, 2, 5.5, 11 Mbps	2.4 GHz ISM	First standard to gain consumer popularity. Backward compatible with 802.11 DSSS.
802.11a (1999)	6, 9, 12, 18, 24, 36, 48, 54 Mbps	5 GHz UNII	Slowly gained popularity due to less interference in the 5 GHz frequency range. OFDM.
802.11g (2003)	1, 2, 5.5, 6, 9, 11, 12, 18, 24, 36, 48, 54 Mbps	2.4 GHz ISM	Popular standard, quickly being replaced by 802.11n. Backward compatible with 802.11 DSSS and 802.11b. OFDM.
802.11n (2009)	70+ different rates, from 6.5 to 600 Mbps	2.4 GHz ISM and 5 GHz UNII	Offers high performance along with backward compatibility.
802.11ac (2013)	MCS 0-9 with data rate from 6 Mbps to 6 Gbps	5 GHz UNII	Wider channels (up to 160 MHz), better modulation (256 QAM), additional spatial streams (up to 8). Only backward compatible with 802.11a and 802.11n.

The 802.11ac amendment to the 802.11 standard

The 802.11ac amendment to the 802.11 standard was ratified in 2013 and was incorporated into the new 802.11 standard in 2016.

The 802.11ac amendment provided features that essentially increase the speed and distance of an 802.11 WLAN signal. These features include additional spatial streams, wider bandwidths, and Multi-User MIMO (MU-MIMO), which provide an improved user experience for clients that support these features.

Beamforming is a technique implemented in Digital Signal Processing (DSP) logic to improve range and data rates for a given client. In a basic system (single stream), beamforming works on the principle that signals sent on separate antennas can be coordinated to combine constructively at the receiving antenna.

Specifically, the phases of the transmit signals are manipulated to improve directivity. Transmit beamforming is specified in the IEEE 802.11n specification and takes advantage of the multiple transmit antennas available in a multiple-input multiple-output (MIMO) system. Efficient steering of individual streams in a system provides overall gain. This can be achieved through knowledge of the channel between the transmitter and receiver and viewed as a form of transmit diversity with a known channel.

Multi-User MIMO (MU-MIMO)

The 802.11ac standard also introduced MU-MIMO. In the 802.11n and 802.11ac Wave 1 standards, all the wireless communication is either point-to-point (one-to-one) or broadcast (one-to-all). However, 802.11ac Wave 2 supports MU-MIMO (see Figure 2-1).

With MU-MIMO an AP, by using its different streams, can transmit data to multiple clients at the same time (see Figure 2-1). Instead of using multiple spatial streams between a given pair of devices, spatial diversity is able to send multiple data streams between several devices at a given instant.

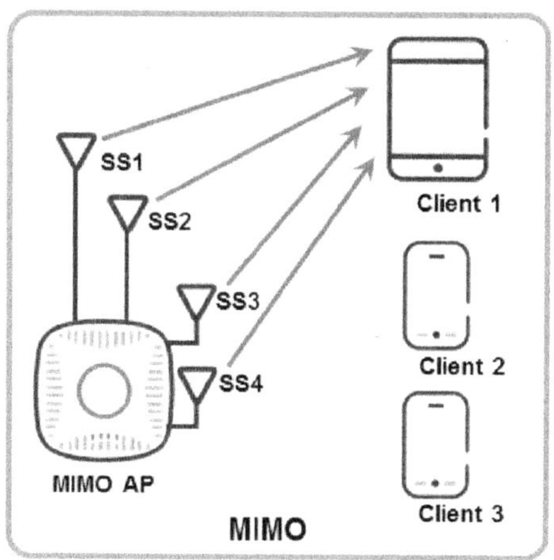

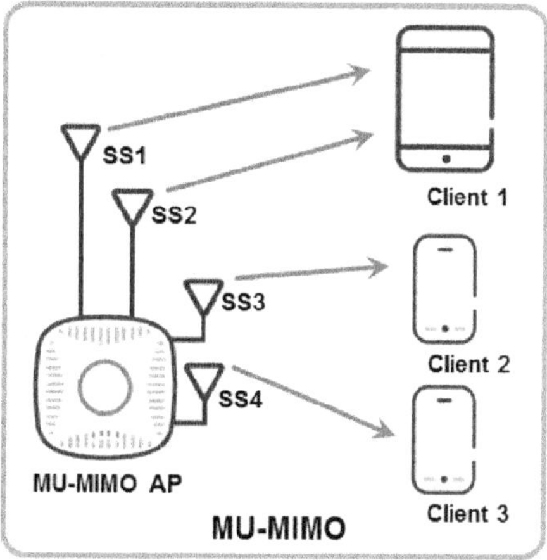

Figure 2-1 Multi-User MIMO (MU-MIMO)

802.11ac Wave 2—MU-MIMO Spatial Streams

The 802.11ac Wave 2 APs with MU-MIMO have two variables on the 5 GHz radio that you should be aware of when creating a MU-MIMO design. In addition to the number of transmit chains, receive chains and total spatial streams, 802.11ac Wave 2 AP radios can be described by the number of Multi-User (MU) spatial streams and the total number of MU clients supported per MU-MIMO group. A MU-MIMO group is a group of MU-MIMO capable clients to which an AP can transmit simultaneously.

There is no industry standard for reporting number of MU spatial streams and MU clients. Aruba uses the convention shown in Figure 2-2. The first three numbers indicate the number of transmit chains, receive chains, and total spatial streams. The fourth number is the maximum number of MU spatial streams, which might be the same or less than the number of total spatial streams. In the Figure 2-2 example, the number is the same, so the AP-315 can transmit up four spatial streams to a single client or four spatial streams to different clients in a MU-MIMO group. The fifth number indicates the number of MU clients per MU-MIMO group. It can be the same or less than the number of MU spatial streams. In this example, it is less, which means that the AP-315 can transmit two streams to a 2x2 MU-MIMO client, for example, one stream to another client, and one stream to a third client. However, it cannot transmit one stream each to four clients.

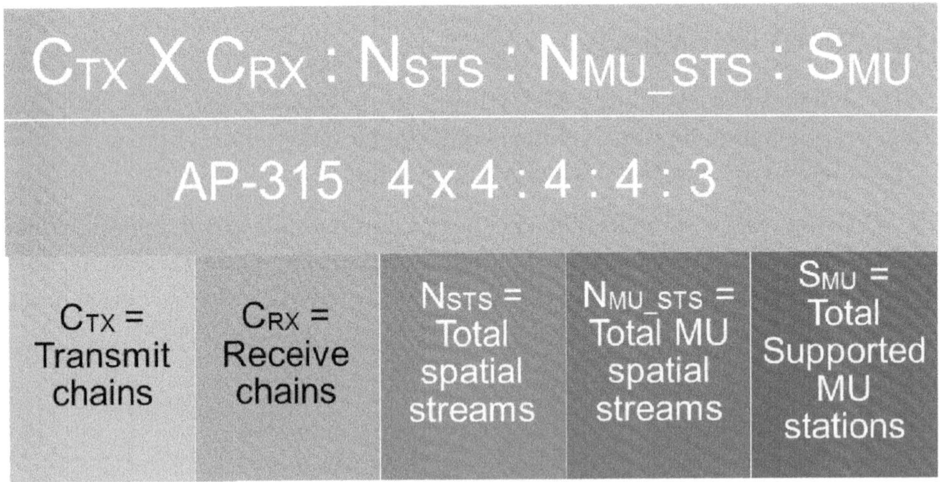

Figure 2-2 802.11ac Wave 2—MU-MIMO spatial streams

Figure 2-2 shows the Aruba syntax. For example, on the Aruba AP models these are the syntax that denotes the number of transmit chains, receive chains, spatial streams, MU spatial streams, and MU clients:

AP-335 = 4x4:4:4:3

AP-325 = 4x4:4:3:3

AP-315 = 4x4:4:4:3

AP-305 = 3x3:3:2:2

AP-303H = 2x2:2:2:2

AP-365 = 2x2:2:2:2

AP-367 = 2x2:2:2:2

802.11ac data rates by client capability

The key determinants of the physical layer (PHY) data rate for 802.11ac clients are channel width, modulation, and coding—otherwise known as the Modulation and Coding Scheme (MCS)—and the guard interval. As the channel width increases, the PHY layer data rates increase for that client (see Table 2-2).

You can look up details about what features clients support at this site: http://clients.mikealbano.com/

Table 2-2 Client capabilities and 802.11n and 802.11ac data rates

Client capability	802.11n–40 MHz channel	802.11ac–40 MHz channel	802.11ac–80 MHz channel
1x1 (Smart phone)	150 Mbps	200 Mbps	433 Mbps
2x2 (Tablet, PC)	300 Mbps	400 Mbps	867 Mbps
3x3 (PC)	450 Mbps	600 Mbps	1300 Mbps

- Channel width—You can configure an 802.11ac AP to use 20 MHz, 40 MHz, or 80 MHz channel width; 802.11ac Wave 2 products also support 160 MHz channel width or noncontiguous 80 + 80 MHz channel width.
- Modulation and coding—All the earlier options are still available and are used if signal-to-noise ratio (SNR) is too low to sustain the highest rate. In the modulation and coding scheme (MCS) table, however, the 802.11n modulation is extended to add 256-QAM options with coding of $^{3}/_{4}$ and $^{5}/_{6}$.
- Guard interval—Unchanged from 802.11n, the long guard interval of 800 ns is mandatory, while the short guard interval of 400 ns is an available option. The guard interval is the pause between transmitted RF symbols. It is necessary to avoid multipath reflections of one symbol from arriving late and interfering with the next symbol.

Minimum RSSI requirements by data rate

When designing an 802.11ac WLAN, you should understand that increased MCS coding, in terms of bits/sec per hertz spectrum, comes at a price. The required signal level for good reception increases with the complexity of modulation and the channel bandwidth. For example, the graph in Figure 2-3 shows that while –64 dBm was sufficient for the top rate (72 Mbps) of 802.11n in a 20 MHz channel, the requirement increases to –59 dBm for the top rate (86 Mbps) of 802.11ac, single stream in a 20 MHz channel. It increases to –49 dBm for the top rate (866 Mbps) in a 160 MHz channel. As the client distance from the AP increases, the RSSI will decrease, and the data rate will decrease as well.

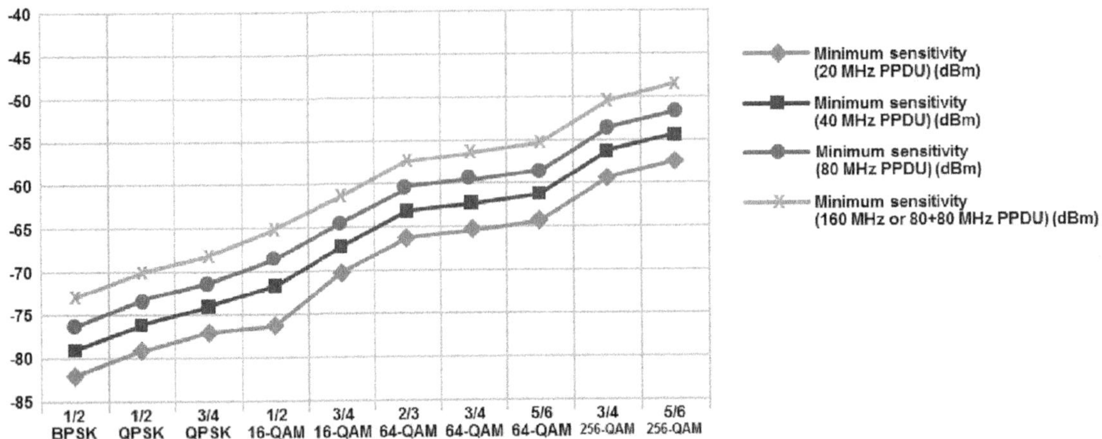

Required receive sensitivity for different modulation and coding rates channel, and to -49 dBm for the top rate (866 Mbps) in a 160-MHz channel

Figure 2-3 Minimum RSSI requirements by data rate

Deployment models

You will now consider models for deploying APs.

Review common WLAN types

Figure 2-4 summarizes common WLAN types. The first and most common type of WLAN is basic coverage in which the space is a typical, carpeted office space. This is a fairly simple network to design for full coverage of the area.

A high-density WLAN is typically seen in an auditorium or classroom space where the density of users is higher than usual. High-density networks require additional planning and design than basic coverage. Certain applications such as voice or video require even more planning and design considerations such as Quality of Service (QoS) and firewall configuration.

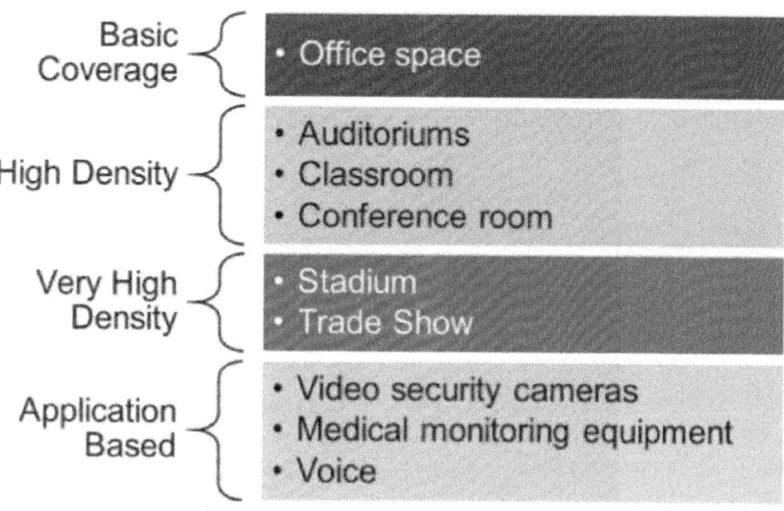
Basic Coverage
• Office space
High Density
• Auditoriums
• Classroom
• Conference room
Very High Density
• Stadium
• Trade Show
Application Based
• Video security cameras
• Medical monitoring equipment
• Voice

Figure 2-4 Review common WLAN types

Review coverage versus capacity

RF coverage is a critical consideration when you are designing and planning a WLAN deployment. Without RF coverage of the physical space, the clients will not be able to "hear" an AP.

Planning and designing the appropriate amount of overlap between APs' RF coverage areas is a critical process. Too much overlap and APs' coverage areas will be on the same channel in the same physical area, resulting in Co-Channel Interference (CCI). Too little coverage and clients may not be able to reliably connect to the WLAN and transmit and receive data to and from the AP.

As Figure 2-5 shows, there are two design models for providing adequate RF signal for the clients. The first is the coverage-based design suitable for low-bandwidth and sparse client deployments, where only enough APs are deployed to provide RF coverage at a basic dBM level.

The second is the capacity-based design used for modern enterprise, high-bandwidth and high-density of users, where the number of users and the applications' bandwidth requirements are calculated in addition to basic RF coverage.

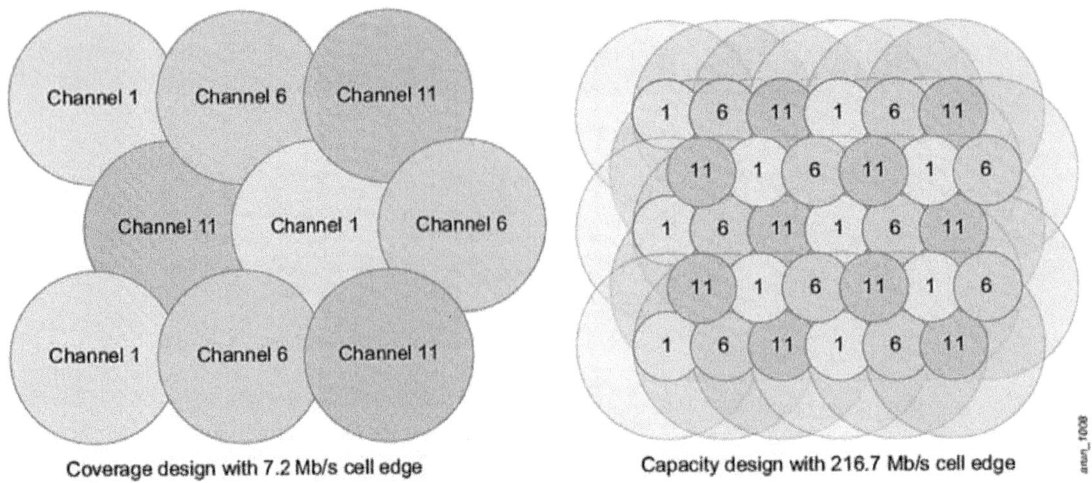

Figure 2-5 Review coverage versus capacity

Coverage model (low bandwidth)

The coverage model is for low-bandwidth deployments where coverage is required for applications such as a scanner solution or limited guest access. A coverage-based deployment might consist of APs placed roughly 70 to as much as 200 feet (approx. 30.5 m to 61 m) apart in an open space, running at 50%–75% of power. If redundancy is not required, APs can run at 100% power, but Aruba does not recommend this solution.

As an example, if the determined application is a scanning solution with minimal traffic, the site might be a good candidate for a coverage-based model. This deployment would consist of an AP installation base with clients that associate at greater distances and at lower traffic rates. This coverage model would mandate a ceiling deployment. Aruba no longer recommends coverage-based deployments as enterprise networks that had very few clients previously are now seeing new services and applications deployed.

Capacity model (high bandwidth)

The capacity model is for dense deployments with high device counts and traffic rates. A capacity-based deployment might consist of APs placed roughly 45 feet to 60 feet (approx. 13.75 m–18.25 m) apart running at 25%–50% or 50%–75% of power. In general, you should set the transmit power of the AP to match that of the least-capable device in the network. If the requirement is a "desktop like" experience for employee laptops, where the employee can run multiple applications simultaneously, the site requires a capacity-based deployment. Aruba recommends capacity-based deployments for all office and education settings.

High-density and outdoor

Two other models exist: high-density and outdoor deployments.

High-density deployments include large spaces (such as lecture halls, libraries, and stadiums), where many devices will be present. Outdoor deployments cover a range of deployments, including metro mesh and point-to-point bridging. For both of these models, Aruba has published Validated Reference Designs (VRDs), which are available at http://www.arubanetworks.com/vrd.

5 GHz versus 2.4 GHz coverage

As Figure 2-6 shows, 802.11 b/g/n WLANs provide a larger coverage area in the 2.4 GHz frequency range than 802.11a/n WLANs in the 5 GHz frequency range. This is due to many factors, including the higher attenuation at 5 GHz and the lower available data rates at 2.4 GHz.

Why would you want to use 5 GHz if it has a smaller coverage area? Using 5 GHz can actually provide a WLAN designed with many benefits, including less noise because fewer devices use the 5 GHz range. Many home automation and small Internet of Things (IoT) devices operate in the 2.4 GHz range as fewer devices are 5 GHz capable. Next, there is more available bandwidth in the 5 GHz range and more available channels.

Lastly, there is minimal adjacent channel interference because 5 GHz channels do not directly overlap as 2.4 GHz channels do.

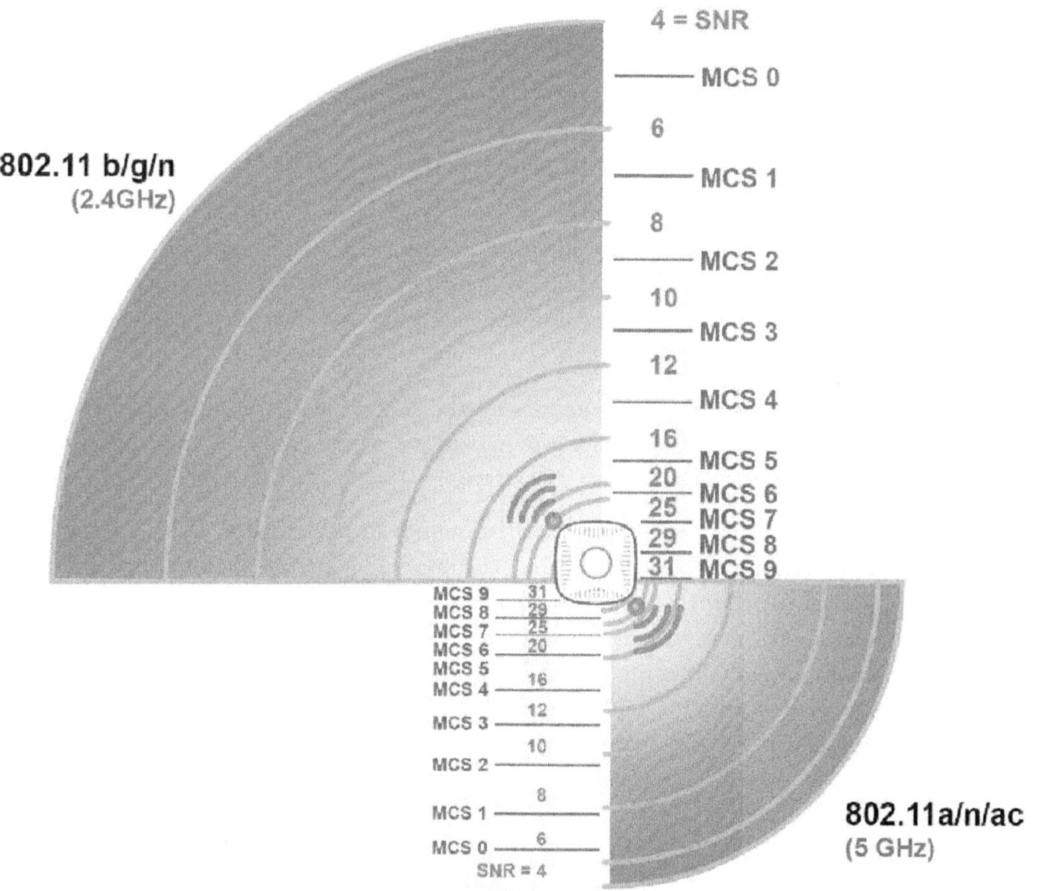

Figure 2-6 5 GHz versus 2.4 GHz coverage

Types of 802.11ac deployments

In a Greenfield design, the network is being designed from scratch, which means there is no legacy network support and all clients will be 802.11ac. Because there is no legacy network in place, you can design a network with support for future technology in mind. A campus that does not have a wireless network today would be considered a Greenfield environment.

A Brownfield deployment is a scenario where a legacy wireless network already exists, and there is plan to deploy 802.11ac. Because there is a dependency on how the network can be designed, you should adopt a phased approach to roll out the new technology. A campus that already has an 802.11 a/b/g or 802.11n-based wireless network and requires you to implement 802.11ac is an example of a mixed environment.

Your deployment recommendations might vary slightly, based on the environment you are considering.

RF planning

You will now turn your attention to RF planning.

WLAN RF planning process

As Figure 2-7 shows, there are seven steps to plan a wireless network deployment. These steps are described below:

1. First, perform an initial environment evaluation. You must know what to look for and the questions to ask to effectively determine the environment type and the appropriate deployment type. The Aruba environment evaluation questionnaire and other tools provided in Chapter 1 can help you to obtain this information.
2. Next, select the appropriate APs and antennas for the deployment. You must understand Aruba APs and antenna types to determine the products that are best suited for the environment and will provide optimal performance and RF coverage.
3. Then enter the collected and determined information into Aruba VisualRF Plan. VisualRF Plan is the Aruba pre-deployment site-planning tool. (For more information about VisualRF Plan, visit https://support.arubanetworks.com/Documentation/tabid/77/DMXModule/512/EntryId/4830/Default.aspx.) In most instances, you can perform a standard deployment based on the VisualRF Plan output without completing a physical site survey. This is called a "virtual" or "predictive" site survey. For complex deployments, you can use a VisualRF Plan to generate a basic foundation for planning. You should visit the site to verify AP location and signal coverage

4. Next, conduct a physical site survey (optional). To properly characterize the RF propagation of a given facility, conduct a passive and/or active physical site survey. When you select AP locations, you must identify the worst-case challenges in the installation environment. A walk-through is crucial to effectively plan a WLAN deployment in a complex environment

5. Next, make adjustments to VisualRF Plan (optional). After you have conducted a physical site survey, the RF propagation assessment affects the best choice of actual AP locations. Aside from the general environment, you should consider specific physical obstructions such as poles, lights, ventilation, and cable runs. Change the floor plan to adjust for these findings.

6. Then, install the selected APs and external antennas (if applicable). Installation guides are available with the products and on the Aruba support site at https://support.arubanetworks.com.

7. Lastly, configure the APs. Perform AP configuration following the best practices outlined in the Validated Reference Design (VRD) series available at http://www.arubanetworks.com/vrd.

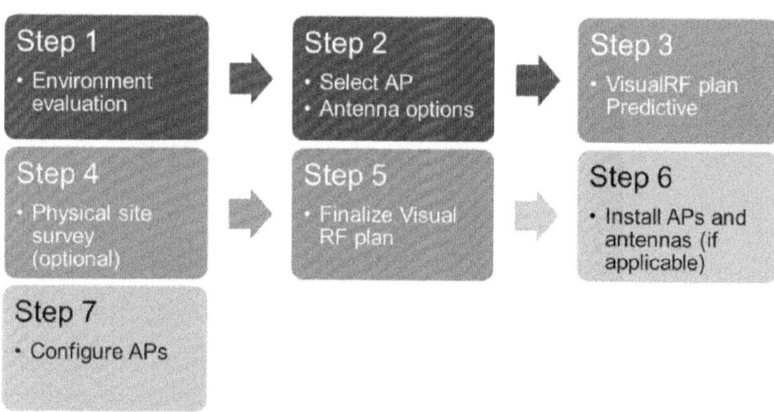

Figure 2-7 WLAN RF planning process

Survey methods

The simple goal of an RF site survey is to accurately determine how many APs are necessary to provide a targeted minimum data rate in a given area. The survey also helps to identify where to place the APs to enable optimum performance. AP coverage can be modeled in a virtual site survey in many open office environments.

As Figure 2-8 shows, RF coverage in the actual world differs from that of theoretical coverage due to factors such as environmental conditions, obstructions, and interference, which all affect RF energy propagation. RF behavior is notoriously difficult to accurately predict in challenging environments. Environments that are more complex and have many RF obstructions are more likely to require a physical site survey for optimal performance.

Theoretical RF coverage
–Virtual site survey – open office environments

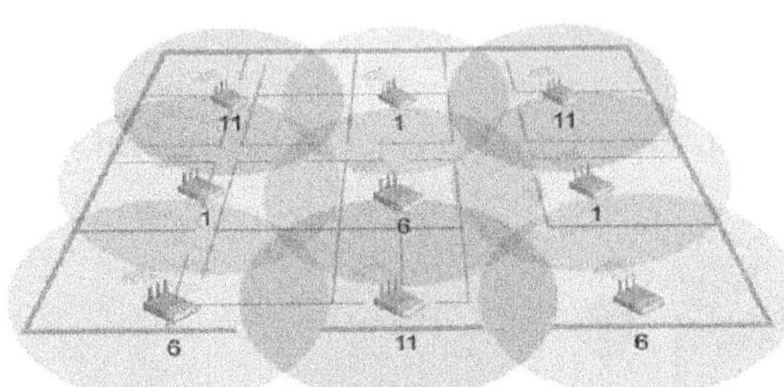

Realistic RF coverage
–Physical site survey – complex environments, (i.e. environmental conditions, obstructions, interference)

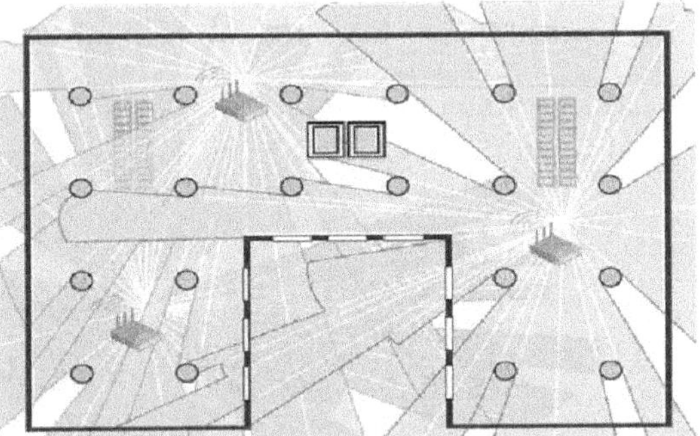

Figure 2-8 Survey methods

Selecting a survey type

The term "site survey" is really a category of activities that means different things to different people. Consulting firms and wireless integrators that provide engineering services generally offer four different types of RF site surveys. Table 2-3 summarizes these survey types.

You will typically perform a virtual survey in low-complexity environments, such as relatively new constructions with primarily cubicles for employees. In a virtual site survey, Aruba VisualRF Plan is used to plan and model a deployment and to create a Bill of Materials (BOM). In higher-complexity deployments, you can use passive and active surveys typically. You can do spectrum-clearing surveys to evaluate the level of non-Wi-Fi interference that may be encountered in a location. In many cases, a physical site survey is necessary to validate the virtual survey for coverage or capacity design.

Table 2-3 Survey types

	Virtual survey (or predictive survey)	Passive survey	Active survey	Spectrum clearing
Description	Uses customer-supplied building drawings in JPG, PDF, or DWG format to place APs	Involves passive data collection of the ambient RF environment (no active testing) based on actual RF data	Involves active testing of real APs throughout a facility (indoor or outdoor) to determine the actual AP coverage footprint and RF hazards	Same as active RF survey, but also includes a spectrum analysis at each active test location
Location	Remote	Onsite (typically indoors)	Onsite	Onsite (typically indoors)
Deliverables	• Marked-up JPG file, indicating AP locations and controller location codes • Site bill of materials	• Heat maps of existing 2.4 GHz and 5 GHz RF environment • Marked-up JPG showing AP locations • Summary narrative analysis	• Heat maps of test APs with actual measured coverage • Marked-up JPG showing AP locations • Detailed data analysis	• Pinpoint locations of the 2.4 GHz and 5 GHz interference sources
Accuracy	*	**	***	****
Cost	$	$$	$$$	$$$$

Virtual survey

VisualRF Plan is the Aruba AP planning software package. VisualRF Plan is available as part of the AirWave server product and as a free standalone planning tool for Windows®.

To perform a virtual survey:

1. Complete environmental assessment
2. Obtain current facility electronic floor plan
3. Facility walk through or obtain images of site to compare to floor plan. Note anomalies.
4. Gather information about building age, materials used and any specially shielded areas
5. Place APs in VisualRF to complete predictive plan

Onsite passive survey (Ekahau)

Physical site surveys are the traditional method for planning wireless deployments of all kinds. Prior to technology advancements such as Aruba AirMatch and ARM, APs were planned to use static power and channel settings as determined by the physical site survey.

Although many of the same principals in planning apply today, AirMatch and ARM eliminate the need to carefully select exact channel and power settings. The three types of physical surveys are passive, active, and spectrum clearing.

To complete a passive survey, follow these steps:

1. Obtain a current electronic floor plan of the facility.
2. Walk through the coverage area with a site-survey software application such as Ekahau. Using this application, sample the RF path every few feet.
3. Analyze the data to produce heat maps of the existing coverage such as the map in Figure 2-9. Also look for sources of external interference.
4. Use this survey data to validate your choice of AP locations.

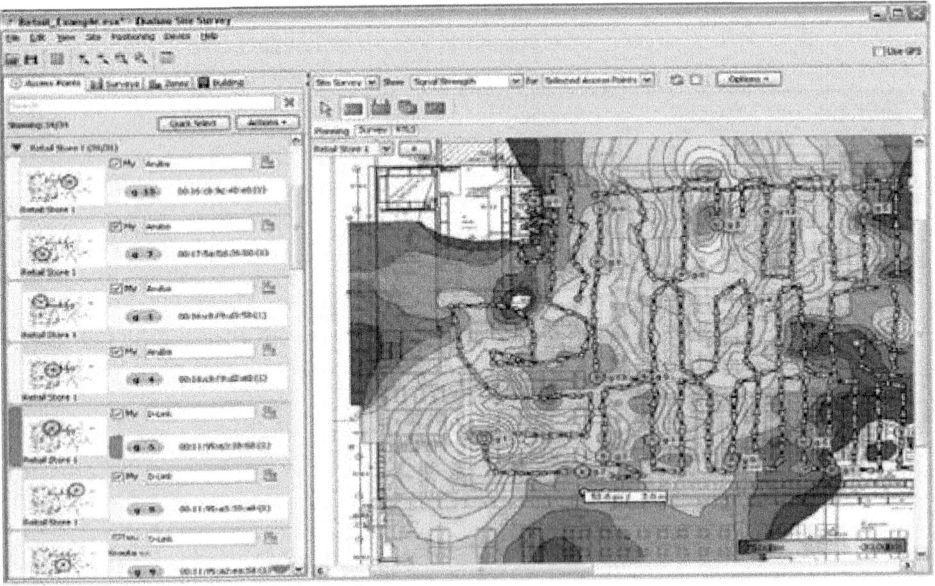

Figure 2-9 On-site passive survey (Ekahau)

Physical building materials

The building materials in an environment will affect the RF signal propagation and consequently may impact the client throughput. Most enterprises have carpeted spaces that have some materials that may cause reflection or absorption of the RF signals.

When performing a walkthrough of the environment, you should look for cubicles, stairwells, and elevators because they may cause a significant loss of the RF signal propagation. Concrete walls and floors and metal firewalls between office suites are also a consideration because of their effect on RF signals (see Figure 2-10). Concrete walls and floors and metal firewalls may be difficult to see during a visual walkthrough inspection and may require examination of a floorplan or blueprint of the building.

Figure 2-10 Physical building materials

Active survey

An active site survey uses operational APs to test WLAN signals by connecting actual clients to the survey AP. For each deployment area, you can use a series of seven steps, which are listed below.

Active site surveys are very time-consuming but provide the most true and validated testing possible for designing a WLAN. The active survey will most accurately determine the best placement of the APs and the proper AP density. It will also ascertain any pre-existing RF conditions that may influence the outcome of the implementation.

You will perform these steps during an active site survey.

1. Obtain current facility electronic floor plan—mark locations where you will perform active tests.
2. Provision radio minimum transmit power for the deployment.
3. Mount AP to portable tripod, speaker mount, or other stable movable platform—or, if available, mount directly to ceiling rails.
4. Position AP at a test location and connect to power source and data link.
5. Use professional site survey application to complete a passive survey of area immediately surrounding test AP.
6. Repeat steps 4 and 5 for all identified test locations.
7. Have an experienced WLAN engineer analyze active survey data to determine proper AP density for coverage area.

Spectrum clearing methodology

By the very nature of RF, multiple devices that operate in the same frequency space will share the unlicensed 2.4 GHz and 5 GHz spectrum and create interference for one another. This situation can result in poor 802.11 network performance.

Common examples of such devices include APs in neighboring stores or warehouses, cordless phones, analog and digital video cameras, Bluetooth devices, and microwave ovens in break areas. When you design a wireless network, it is important to understand the overall RF environment typical of the facility types where you will deploy the network to mitigate any interference problems.

Spectrum clearing refers to the use of a portable spectrum analyzer to discover and pinpoint interference sources before the network is deployed.

The methodology is outlined below:

- Configure the spectrum analyzer to record peak, average, and maximum hold for 2.4 and 5 GHz bands – if supported, also enable swept spectrogram for both bands
- Walk a carefully planned route for each selected location looking for active devices
- If strong interfering signals observed, pause in that location and record spectrum trace for 60-90 seconds
- If interferers are found, pinpoint using a directional antenna

After the interferers have been identified, you should remove or migrate any devices you can to lower the interference effects.

RF design

The majority of modern WLAN deployments are at the ceiling level. A ceiling deployment can occur at or below the level of the ceiling material. In general, Aruba does not recommend mounting APs above any type of ceiling material, especially suspended or "false" ceilings.

There are two reasons for this: first, many ceiling tiles contain materials or metallic backing that can greatly reduce signal quality. Second, the space above the ceiling is full of fixtures, air conditioning ducts, pipes, conduits, and other normal mechanical items. These items directly obstruct signal and can harm the user experience.

Wall deployments are not as common as ceiling deployments but are often found in hotels and dormitory rooms. Walls are a common deployment location for large spaces such as lecture halls because reaching the ceiling is difficult. Wall deployments may also be preferable in areas with a hard ceiling where you cannot run cabling (see Figure 2-11).

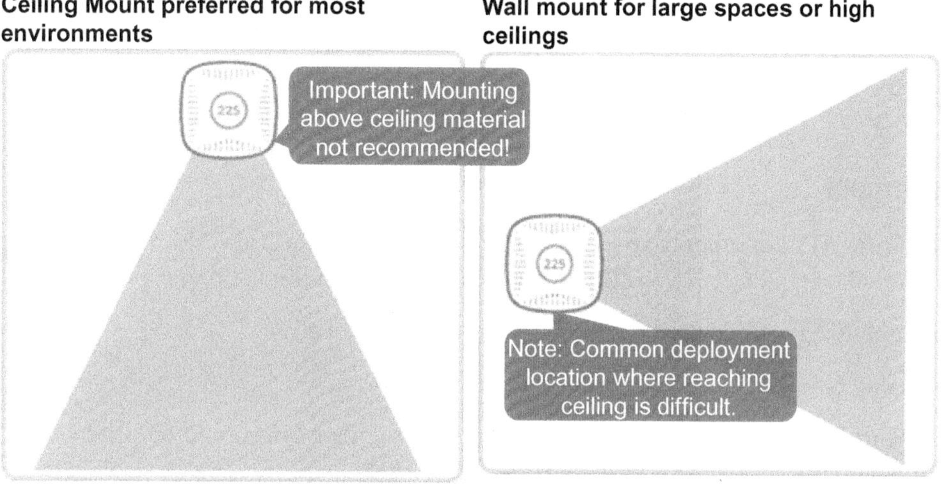

Figure 2-11 RF design

Deploy APs in rooms—not in hallways

Instead of trying to use hallway APs at high power to cover users through a wall, Aruba recommends using many room APs at low power to cover the user space directly.

Depending on how much attenuation the building presents, you will place an AP in every 1, 2, or 3 rooms. As Figure 2-12 shows, you will stagger APs inside of rooms on either side of the hallway and vertically. On one floor, install an AP in every other room, and you should not place APs in rooms directly across the hall from one another. On adjacent floors, vary the deployment so that APs are in different rooms. However, staggering the APs across floors can be less important when the floor has a highly dampening material such as metal shielding.

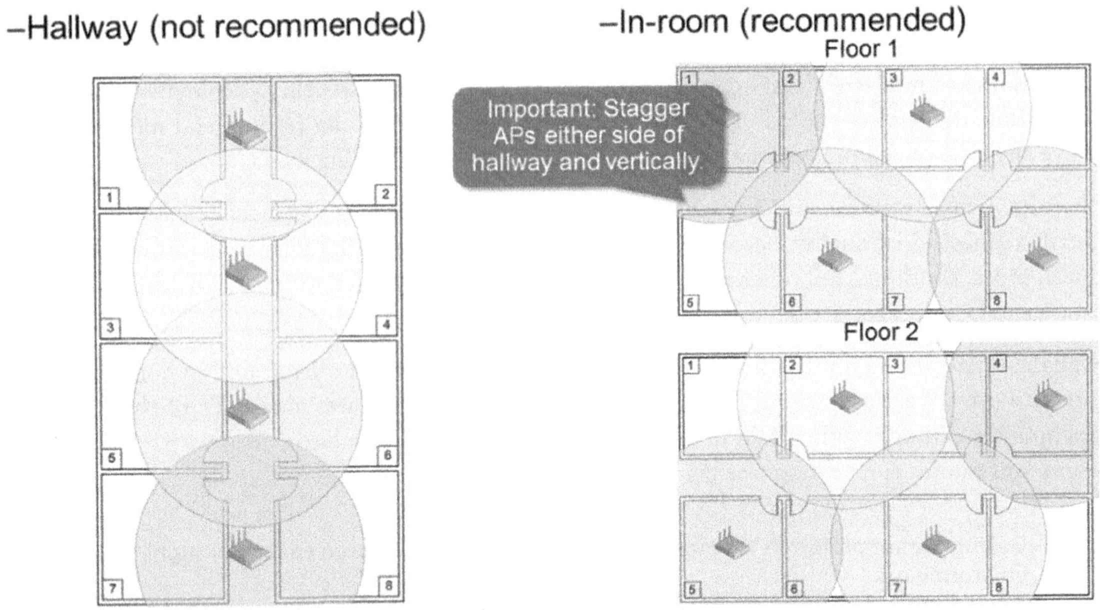

Figure 2-12 Deploy APs in rooms—not in hallways

AP placement

Although Aruba APs can support up to 255 client devices per radio, Aruba recommends you plan for a distance of 40 feet–60 feet (16 m–20 m) between APs to provide a good user experience.

The minimum RSSI value should be –65 dBm for the coverage area.

SNR (Signal to Noise ratio) should be greater than 25 dB.

AP placement should be in a honeycomb pattern, and the network should be planned for 5 GHz instead of 2.4 GHz.

These guidelines for AP placement are summarized in Figure 2-13.

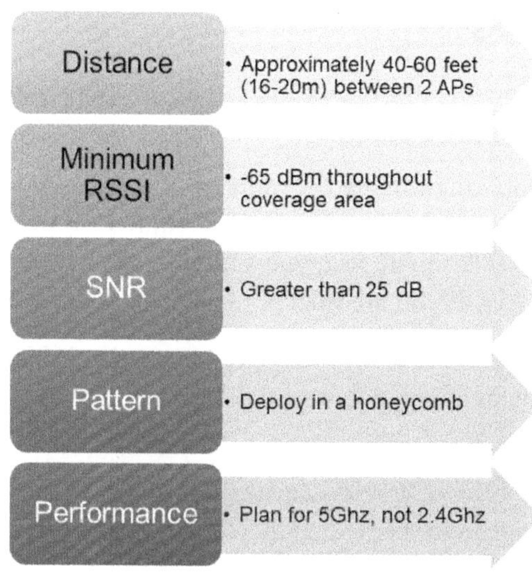

Distance
• Approximately 40-60 feet (16-20m) between 2 APs
Minimum RSSI
• -65 dBm throughout coverage area
SNR
• Greater than 25 dB
Pattern
• Deploy in a honeycomb
Performance
• Plan for 5Ghz, not 2.4Ghz

Figure 2-13 AP placement

RF planning general recommendations

Aruba provides some RF general planning recommendations in the Validated Reference Design (VRD) Guides. These recommendations have been tested in many Aruba deployments to create the best user experience. The VRDs are based upon documented Aruba deployments and the best configuration settings for given deployments such as university, hospital, or enterprise deployments. Table 2-4 lists some general recommendations.

Table 2-4 RF Planning general recommendations

Feature	Recommendation
AP Mounting Recommendation	Ceiling mount
AP Placement	• Place APs approximately 40–60 feet (16–20 m) apart. • Minimum Received Signal Strength Indicator (RSSI) should be -65 dB throughout your coverage area. • SNR should always be greater than 25 dB. • APs should be deployed in honeycomb pattern. • Plan your network for 5 GHz performance.
AP Forwarding Mode	• Tunnel mode is preferred. • Decrypt-tunnel mode.
Channel Width Selection	• In Greenfield deployments, deploy 80 MHz channels, including the use of Dynamic Frequency Selection (DFS) channels if no radar signal interference is detected near your facility; otherwise, deploy 40 MHz channels. Consider 40 MHz or 20 MHz channel width for better channel separation. • In Brownfield deployments, 80 MHz channel is recommended with DFS channels only if no radar signal interference is detected near your facility. Also make sure that your legacy clients are not having wireless issues with 80 MHz channels. If they are experiencing issues, deploy 40 MHz channels.

Estimating number of APs

To help you estimate the number of APs, Figure 2-14 outlines best practices for both coverage-based and capacity-based designs.

For coverage-based designs, you can estimate the number of APs needed by dividing the total square footage by 5000. This will give you the number of APs required to provide adequate coverage. For example, if the building is 100,000 square feet, you would divide 100,000 by 5000. The result if 20, meaning you would need 20 APs. If you are working in square meters, divide by about 465.

For capacity-based designs, you divide the total square footage by 2500 (or area in square meters by 232. For example, for the same building, you would divide 100,000 by 2500. In this case, you would need 40 APs.

In some cases, with a higher density of wireless devices, or devices that run more demanding applications, you should divide the square footage by 1500 (or the area in square meters by 139).

Coverage-based Design (Estimate)	Capacity-based Design—High Density (Estimate)
• Total Sq footage divided by 5000 equals number of APs to provide adequate coverage	• Total Sq footage divided by 2500 equals number of APs to provide adequate coverage
• Example Calculation • 100,000 sq foot building • 100,000/5000 – 20 APs	• Example Calculation • 100,000 sq foot building • 100,000/2500 – 40 APs

Figure 2-14 Estimating number of APs

Considerations in AP design

The coverage threshold must be planned. A Wi-Fi handset requires a continuous, reliable connection as a user moves throughout the coverage area. In addition, voice applications have a low tolerance for network errors and delays. Using a Wi-Fi network for voice can be complex, without proper planning. A typical voice application needs a minimum of –65 db to an AP at all times.

Signal overlap between AP cells is important for continuous connections. 15% to 20% of signal overlap between AP cells generally works well with the typical walking speed of a user. If the speed of the moving user is greater, such as a fork lift or golf carts, or if the cell size is smaller, then a different overlap strategy may be necessary for successful handoff between APs. The amount of time needed to find a new AP is a fixed constant. Smaller cells or faster roaming speeds require larger overlap percentages due to the need to maintain an overlap area that still allows time to find the next AP.

Minimum throughput and association rates, for the clients, should be taken into consideration. Minimum throughput per client is the minimum acceptable rate of successful message delivery over a communication channel to an AP. Minimum association rate is the rate at which a client can associate with the WLAN. The supported data rates are configured as part of the network policy. Ensure that the clients can properly decode and receive management frames like beacons as well as broadcast and multicast traffic, which is essential to successful network connectivity.

If the network has voice implemented, then you must consider the number of voice associations per AP. Because voice is affected by jitter or loss of packets, voice traffic needs assured bandwidth and to avoid Interference. Therefore, total clients and concurrent calls per AP must be properly planned. The number of total voice clients and concurrent calls varies, depending on the type of AP and the capabilities of the clients.

In addition, high-priority areas, such as meeting rooms, may need their own APs. Many of these rooms are closed in, limiting WI-FI signals. The number of users in the room may also call for an AP dedicated to that room.

Sources of attenuation/concentration

Attenuation is a decreased signal strength or amplitude that occurs when a signal passes through an object. Elevator shafts and stairwell shafts are normally heavily walled and will attenuate radio signals. Electric rooms will also cause radio attenuation. Furthermore, open office spaces are not necessarily totally open (see Figure 2-15).

The wave weakens as it moves through an object. The denser material absorbs more signal. Any signal that travels through walls is partially absorbed or reflected. This also occurs naturally due to the broadening of the waves as it moves further from the source. This is referred to as Free Space Path Loss.

Therefore, a wireless signal is strong in some areas and weak in others. It is not only a matter of distance between your wireless APs and your wireless clients. It is also a matter of what objects (walls, doors, furniture, and electrical equipment/outlets) interfere with a good signal.

You should also examine the site for locations that tend to create a concentration of users. For example, classrooms and conference rooms have a large concentration of users. Aruba recommends locating an AP in these rooms. Entrances and lobbies, typically have large amounts of users at specific times, all authenticating to the first AP in the building.

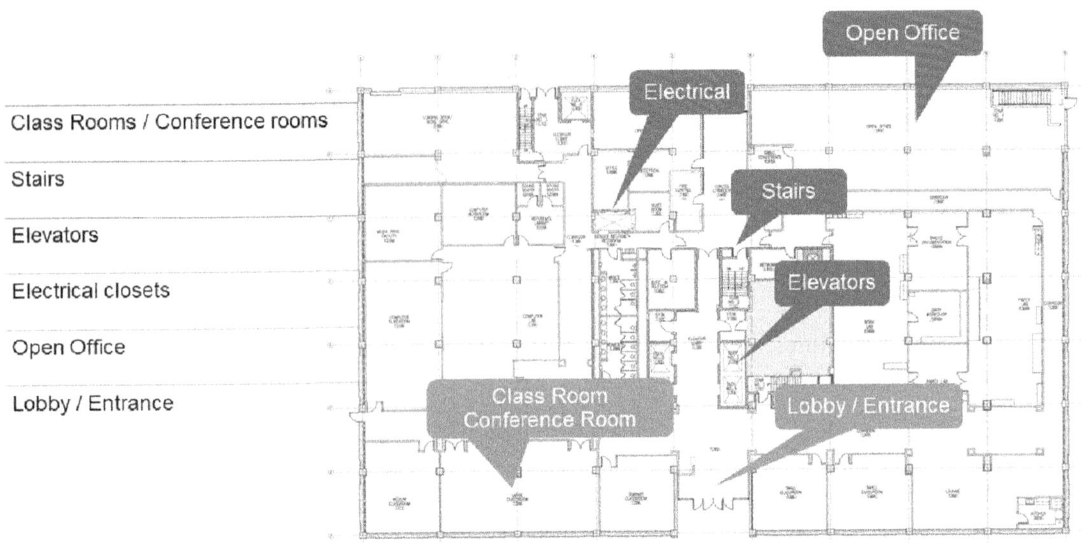

Figure 2-15 Sources of attenuation/concentration

Open office environments

An open office space is often considered open free space where you would need few APs to cover the area. However, not all offices are the same. The picture on the left in Figure 2-16 shows low cubicles where Wi-Fi signal could travel a long distance.

The picture on the right in Figure 2-16, on the other hand, shows high-wall cubicles with metal cabinets. These would attenuate the signal. This environment requires you to use a denser deployment of APs.

You would also need to consider the quantity of employees in the area.

How would you plan APs in these two open offices ?

Figure 2-16 Open office environments

Selecting APs and antennas

You will now think about how to select the appropriate APs and antennas.

AP replacement from 802.11a/b/g/n to 802.11ac

If you are designing an 802.11ac deployment to replace a legacy Aruba deployment of 802.11a/b/g/n, you will need to ask the customer a standard set of questions.

- Is your current network planned for coverage as opposed to capacity?
- Are there any coverage holes in your current environment where users do not receive RF signals?
- Are there any known RF-related issues such as poor connectivity or poor performance?
- Do you want to provide a seamless roaming experience for users?
- Do you need location-based services?
- Are there any architectural changes in your facility, such as new walls or metal cabinets, which are resulting in poor Wi-Fi connection?

If the answer to any question is yes, then one-to-one replacement may not be a recommended best practice.

AP design criteria

In the 802.11ac capacity-based design, Aruba recommends the distance between the center of two APs be approximately 40 feet–60 feet (16 m–20 m). AP placement also depends on client density.

In all wireless offices where APs are deployed every 50 feet (15.24 m), the expected client count on an AP's radio is approximately 40 to 60 clients. If the client density is higher than this, you should deploy APs closer.

Omnidirectional antenna review

Omnidirectional antennas are like a light bulb, emitting energy in all directions. Figure 2-17 shows the radiation pattern of an omnidirectional antenna.

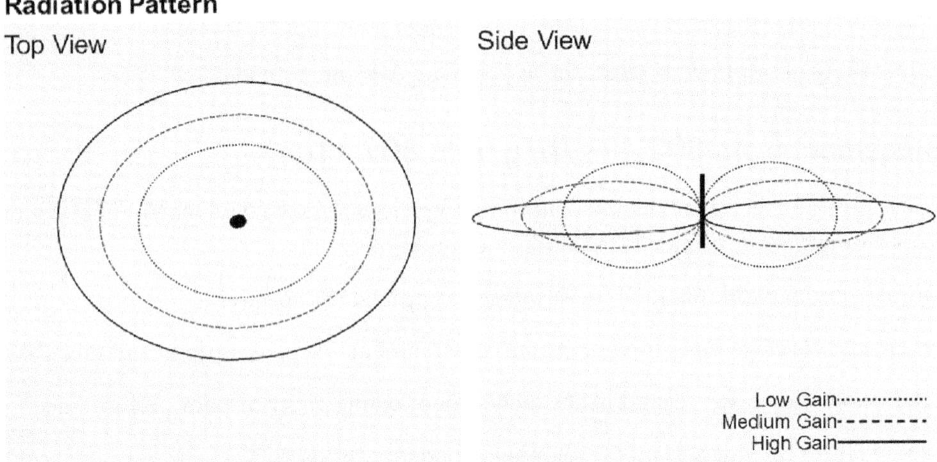

Figure 2-17 Omnidirectional antenna review

Directional antennas

The gain of directional antennas will determine the radiation patterns. As Figure 2-18 shows, high-gain directional antennas will have a more narrow beam-width and more coverage in a specific direction.

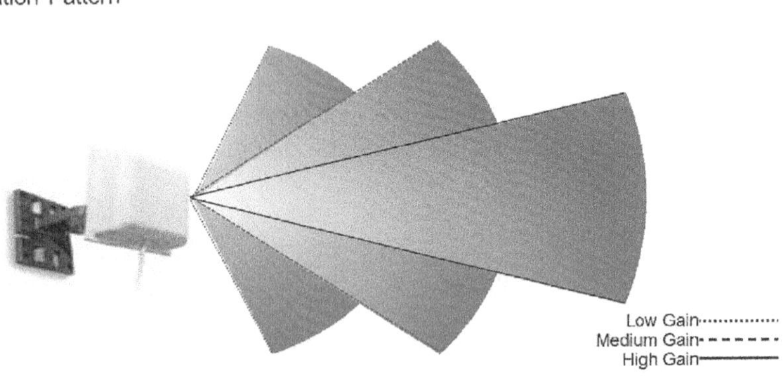

Figure 2-18 Directional antennas

Maximum range and coverage

To determine the maximum range from an AP, you must know some basic information, including the AP power and antenna gain and any additional cable loss at the AP. You also need to know if the customer uses any extension cables. In addition, you need to know the desired data rate(s) and services (802.11a/b/g/n/ag), the client power, client antenna gain, and any additional cable loss at client station.

AP coverage is determined by locations where a connection is possible. The AP must be within maximum range (including client range) and within the antenna pattern. The signal cannot be obstructed or absorbed in an unusual way such as to create a coverage "hole."

Review link budget

The link budget calculation is a predicted maximum value of the power generated at the transmitter, lightning arrestor attenuation, antenna gain, Free Space Path Loss (FSPL), and client antenna gain (see Figure 2-19). The actual signal level may be less due to the outside variables that you may not have taken into consideration. The link budget differs in the opposite direction.

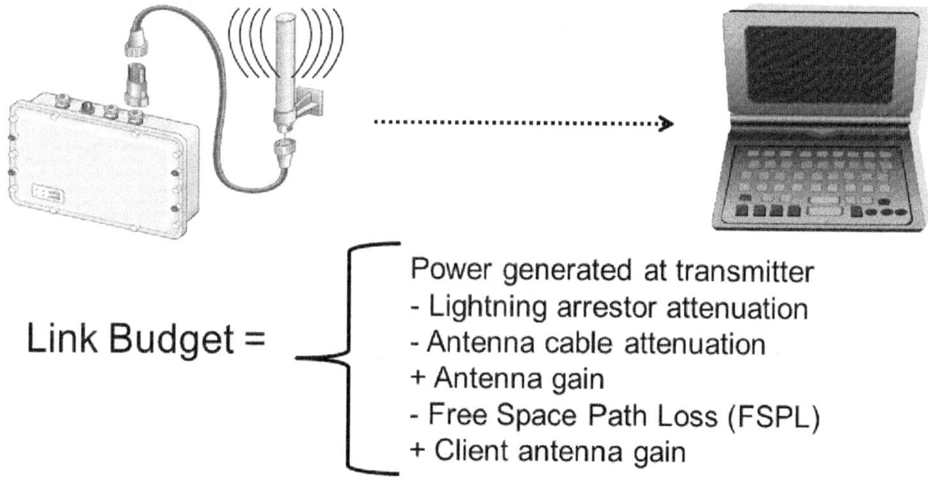

Figure 2-19 Review link budget

Beam-width and patterns

The beam-width and patterns measure how an antenna directs the energy it radiates by using the half-power or –3 dB beam-width in degrees for measurement (see Figure 2-20). The measurements are represented as a polar plot for horizontal (azimuth) and vertical (elevation).

–Half-power or 3 dB beam-width in degrees

–Horizontal (azimuth)

–Vertical (elevation)

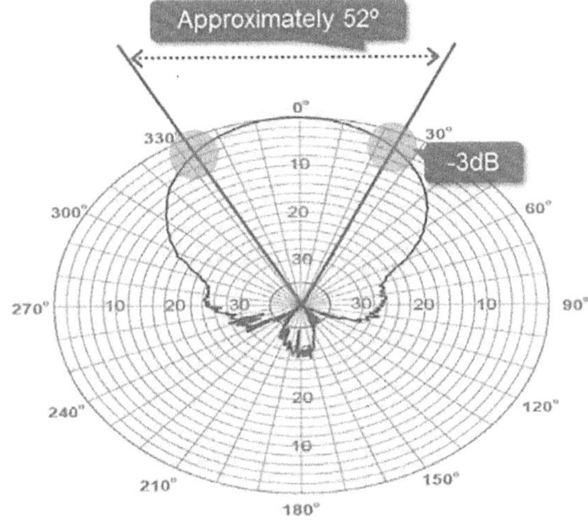

Figure 2-20 Beam-width and patterns

Antenna frequency and gain

The antenna frequency and gain have certain requirements in order to work properly. The antenna size is related to the wavelength for each discrete frequency. To avoid issues, you should not use antennas designed for one band for another band.

The gain of an antenna only improves the received power if you are in the direction of the gain. If you are somewhere else, the gain is actually "loss." The higher the gain of an antenna, the greater range in the direction of the gain and consequently the less coverage in other areas.

Think of a round balloon, and now imagine pressing your fingers on both ends of the balloon to turn the balloon into the shape of a donut. Now press the donut shape even flatter into the shape of a pancake. This is the same as shaping the coverage area of an antenna. The round balloon shape is an omnidirectional antenna. The donut is a directional antenna, and the pancake shape is a highly directional antenna. They are all covering the same area, but as coverage is gained in one direction, it is lost in another direction.

Reading antenna pattern plots—omni

In order to read the antenna pattern plot, it is important to understand that antenna pattern plots are "slices" of a 3D pattern. The Azimuth is the horizontal plane, or top view. Elevation is the vertical plane, or side view (see Figure 2-21).

Antenna plots are scaled in either dBi (dB above isotropic), dBd (dB above dipole), or dBr (dB relative to maximum) values. Antenna data sheets often show dBr; therefore, the 0 dB on the scale corresponds to the direction of maximum specified gain.

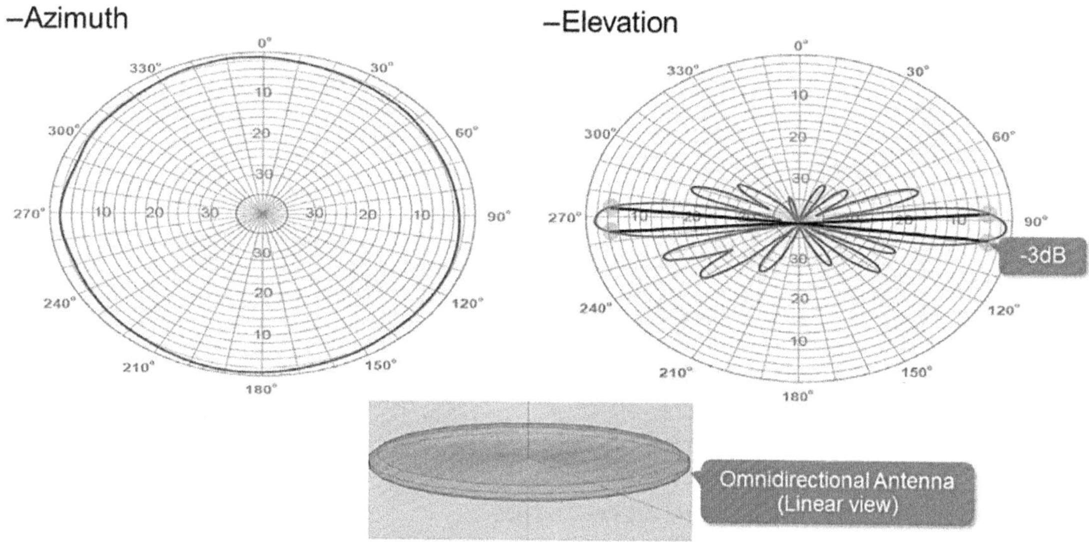

Figure 2-21 Reading antenna pattern plots—omni

Current AP products

Aruba's WLAN products include indoor and outdoor APs as well as different types of antennas, including omnidirectional and directional antennas. When you are designing a WLAN deployment, it is important to use the appropriate AP and antenna to meet the customer requirements.

Considerations for selecting Aruba 802.11ac indoor APs

For new deployments, you should typically recommend APs in the latest series. As of the publication of this study guide, those are 802.11ac Wave 2 APs with model numbers that begin with 3. As Figure 2-22 shows, the second number in the model name helps you to determine the relative performance level and the density of deployment for which that AP series is targeted. For example, AP 330 and 340 Series APs are 802.11ac Wave 2 APs designed to deliver the highest performance

Traditionally dual-radio APs have offered a 2.4 GHz radio and a 5 GHz radio to offer coverage in both bands. These APs remain the best choice for environments in which support for 5 GHz is uncertain. However, more and more devices do support 5 GHz, and many deployments are shifting to use primarily 5 GHz. With the 2.4 GHz coverage less useful, a dual-radio AP does not offer the full capacity that it could. An AP with two 5 GHz radios, such as an Aruba AP 340 Series, essentially acts as two APs deployed directly on top of each other. Each radio operates on a different channel, doubling the capacity.

For a deployment with dual 5 GHz APs, you should keep in mind factors (such as nearby radar or the use of 80 MHz channels) that could limit the number of available channels available.

Performance and density
- x0x = low
- x1x and x2x = medium
- x3x and x4x = high (such as AP 330 and 340)

Type of dual-radio
- 2.4/5GHz for guest/legacy support
- Dual 5GHz for high capacity

Figure 2-22 Considerations for selecting Aruba 802.11ac indoor APs

Aruba antennas, mounts, and accessories

In most cases, Aruba has similar antennas (types and patterns) for indoor (RP-SMA) and outdoor (N-connectors). Indoor antennas do not usually come with mount kits. However, outdoor antennas do include mounting kits.

There are multiple mounting options to cover nearly any outdoor installation requirement. They are designed to require minimal installation effort and time.

Aruba sells lightning arrestors, antenna cables, power injectors, and so on (see Figure 2-23). In some special installation cases, partners and customers may have to source special cables or hardware to support their specific installation.

Note

Lightning arrestors for the 274 are only required when the cables are longer than 6.6 feet (2 m).

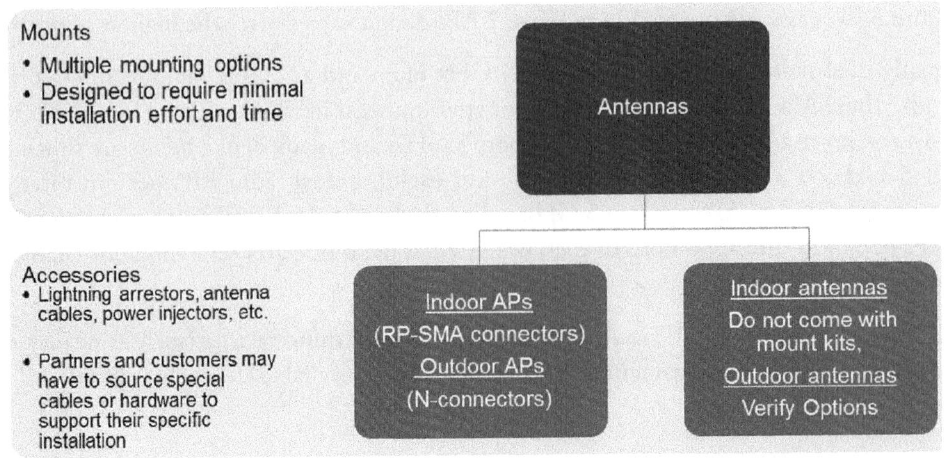

Figure 2-23 Aruba antennas, mounts, and accessories

External antenna considerations

Aruba has a wide variety of antennas for indoor and outdoor use. When selecting an antenna, consider the coverage pattern that you want to provide—whether omnidirectional or sector (another term for directional). For sector antennas, you can select from wide sector (90 deg–100 deg), narrow sector, and high-gain sector. You must also choose an antenna for the correct number of spatial streams supported on your AP radio.

The antenna must also be for the correct frequency band. Aruba offers dual-band antennas.

Use tools such as Iris and the HPE Networking Online Configurator to find valid and up-to-date selections for your AP and attach them to the BOM.

Wired network considerations PoE

The most important consideration while choosing a wired access switch for powering 802.11ac Wave 2 APs is the PoE support provided.

In normal conditions, without a USB plugged in, 802.11ac APs consume near or above the 802.3af maximums to power all supported streams on both the radios. For some APs, some 802.af devices can supply the necessary power. In some cases, however, they cannot—particularly since the power actually delivered to the AP depends on the cable quality and cable length. The AP-225 measures the PoE voltage. If it is at least 51V, it will operate in full functional/performance mode even if the source is identified as an 802.3af source. With ArubaOS 6.3.1 and above, if these APs connect to an 802.3af source and cannot receive enough power, they may operate with slightly reduced capabilities. For example, they cannot support the full number of streams on the 2.4 GHz radio (Table 2-5 provides an example). For other AP models, the capabilities might be even more limited.

Therefore, Aruba recommends that access switches support PoE+ on every port that connects to an AP to avoid any performance degradation on the AP.

Table 2-5 Example of capabilities on an Aruba 220 series AP based on power source

	802.3af	802.3at*
2.4 GHz radio	1x3:1	3x3:3
5 GHz radio	3x3:3	3x3:3
Ethernet ports	1	2
USB	Disabled	Enabled

*Ensure PoE+ (802.3at) is provided to prevent loss of capabilities.

Aruba VisualRF Plan

In the activity for this chapter, you will be instructed to use Aruba VisualRF Plan to plan the APs for the customer scenario that was introduced in "Activity 1: Determine Customer Network and Requirements" (see Chapter 1). This section provides some tips for using Aruba VisualRF.

Add calculated APs/remove APs in VisualRF Plan

When planning APs, you must slide the Density bar to see the AP count.

After you have added the APs to the floor plan, you might want to start over with a new AP count. When you want to delete all the planned APs, you select the floor. Then click **Edit** and then click **Delete All Planned Devices** (see Figure 2-24).

You may need to back out to the building level and then return to the floor to complete the change and see the heat map update.

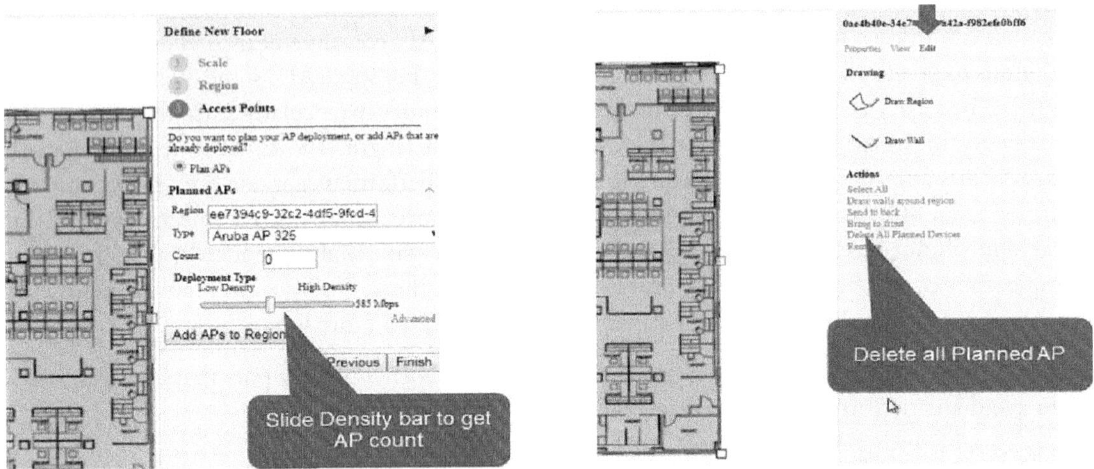

Figure 2-24 Add calculated APs/Remove APs in VisualRF plan

Recalculate AP count in VisualRF plan

To recalculate the AP count, you click **Properties** and then click on the background. This will change the menu display. Select the type of AP and then click **Coverage Calculator**. Include your service level and click **Calculate AP Count** (see Figure 2-25).

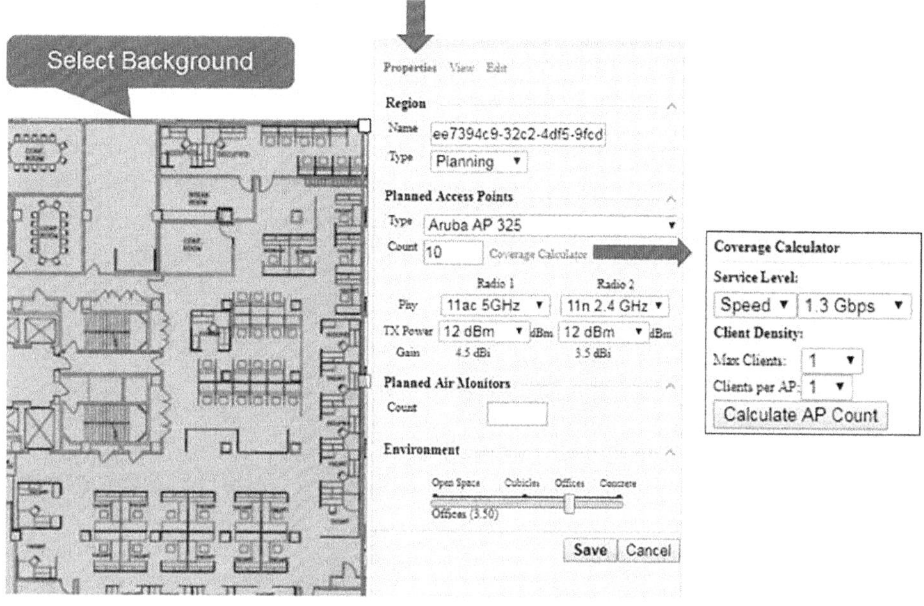

Figure 2-25 Recalculate AP count in VisualRF plan

Iris or HPE networking online configurator

The activities in this study guide are designed to help you plan the network for a customer scenario. As part of these activities, you will create a BOM. The activities were designed to use Intagi Iris, a tool that helps architects create proposals for customer networks.

If you are an HPE partner, you have access to Iris.

If you do not have access to Iris, you can use the HPE Networking Online Configurator to create a BOM. The activities in this study guide do not include step-by-step instructions for the HPE Networking Online Configurator, but the tool is easy to use. You can use this tool to create a BOM; however, you will need to skip the steps that relate to building a topology. (Visit http://hpe.com/networking/configurator to access this tool.)

Iris

To use Iris, you begin by creating a site. You can then select many different HPE and Aruba solutions from a catalog. Once you have added a product to the work area, you can click it; the Properties window is displayed. This window lets you choose a precise model from a series, select modules and transceivers for switches, add licenses for MCs, select accessories, and more.

In the Attributes tab, you can set a Quantity, which multiplies out the same product configuration, allowing you to quickly fill out a BOM.

If you ever want to find out more about a product, right click it and select **Documentation**. You can then choose to see product information or link to a datasheet or Quick Specs on the Internet.

To get help selecting a product or comparing similar products, click the **Advisors** tab.

And when you're ready to check out the BOM and price quotation, click the dollar icon highlighted in Figure 2-26.

Iris offers many more capabilities than the ones highlighted in Figure 2-26 or even those that you explore in the activities for this study guide. You can use Iris to create complete logical topologies for multi-site networks. It supports design groups and templates to help you work more efficiently. To learn more about using Iris, select **Help** and open the User Manual, watch training, or visit an online forum.

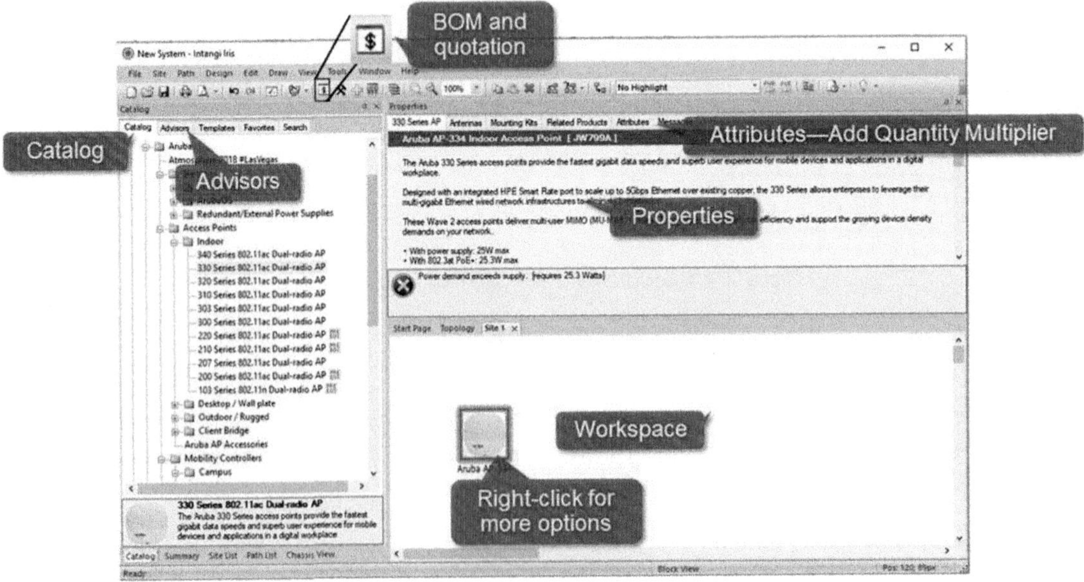

Figure 2-26 Iris

HPE Networking Online Configurator

You will need to run Microsoft Internet Explorer to use the HPE Networking Online Configurator. (Visit http://hpe.com/networking/configurator to access this tool.)

As you can see in Figure 2-27, you begin by selecting the appropriate country for the price list. You can click **Product Selectors** for help finding a product that meets specific criteria. You can then select the HPE Networking Switch Selector, HPE Networking Router Selector, or HPE Networking Wireless Selector. If you already know the product that you want to add, you can also click **Add** next to the **Product List**.

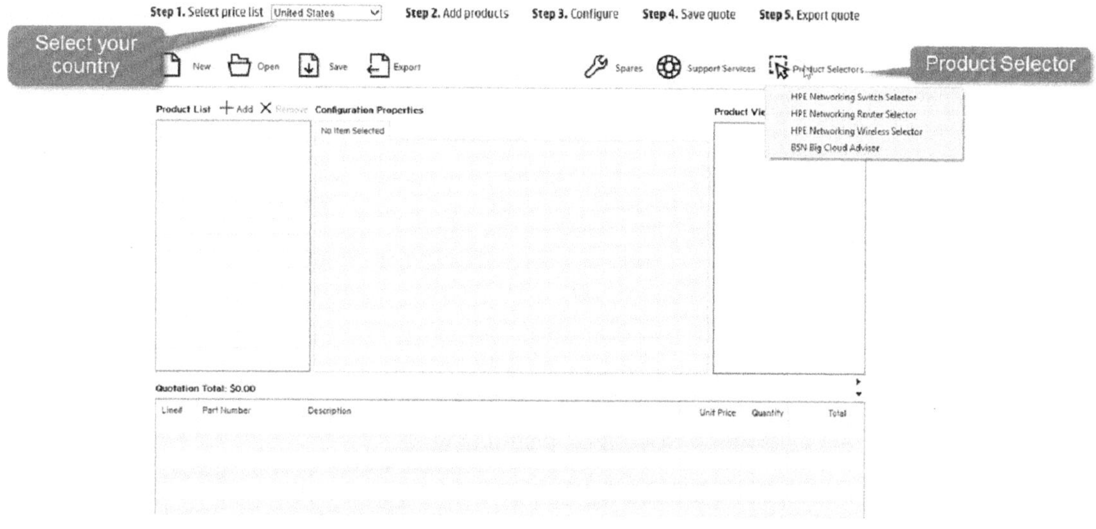

Figure 2-27 HPE Networking Online Configurator

Activity 2: RF planning

In this activity, you will review the information that you collected in Activity 1, select AP models, and add the APs to Iris (or the HPE Networking Online Configurator). You will then use Aruba VisualRF Plan to plan the number of APs for your customer buildings. The section below summarizes what you learned in Activity 1 and provides additional details.

Customer scenario: Corp1 requirements

Corp1 is replacing all of its old APs for the new Aruba installation. To give the network longevity, the design should provide 802.11ac wave 2 coverage. Corp1 requires complete coverage across every floor and between the buildings. Due to a relatively low number of users and nondemanding applications, the customer does not require high density. However, the customer does want to avoid an upgrade for several years.

Table 2-6 Information collected in Activity 1

	Building 1	Building 2
Dimensions:		
Length	310 feet (94 m)	310 feet (94 m)
Width	173 feet (53 m)	173 feet (53 m)
Area per floor	53,620 square feet (4982 square m)	53,620 square feet (4982 square m)
Number of floors	3	3
Ceiling height	15	15
Drop-ceiling height	10	10

(Continued)

Table 2-6 Information collected in Activity 1—cont'd

	Building 1	Building 2
Obstruction within drop-ceiling	yes	yes
Floor plans supplied (Y/N)	Y	Y
General description of the physical environment and list of features relevant to RF	This is an office building with mostly closed offices. Center corridor on each floor has restrooms, elevators, stairs and many closets. Floor 1 has wide open spaces, including a data center.	This is an office building with mostly closed offices. Center corridor on each floor has restrooms, elevators, stairs and many closets.

You collected this information about the wireless user base:

a. Number of users in main campus: 900–920, including guests

   About 200 per floor except 30 on Building 1 Floor 1 and 70 on Building 1 Floor 3

b. User types: Employees and a few guests

c. Device types: HP Elitebook Folio G1 laptops, printers, smartphones (BYOD), and tablets (BYOD)

d. Types of applications in use: Email, Web, print, accessing files on servers

You also determined that wireless devices are rather mobile, and the customer wants seamless roaming.

You researched device capabilities and discovered that most in this environment support 802.11a/gb/n/ac and 2 spatial streams.

Current infrastructure

The existing Wi-Fi structure is a/b/g only and has six to eight APs per floor. There is a small controller per floor.

Each floor has two wiring closets with three 48-port PoE switches each; you will be upgrading these closets as part of the proposal. In addition, the first floor has a datacenter with several HPE FlexFabric 5900 switches. The datacenter primarily uses fiber. The datacenter core switches are HPE FlexFabric 5940 switches, which have an interface module with 10 GbE SFP+ and 40 GbE QSFP+ ports. These switches have free slots that could support different types of interfaces, including SFP+, QSFP+, or even 1000BASE-T.

Activity 2.1: Select APs and begin creating a BOM

Activity 2.1 objectives

In this activity, you will select AP models and add them to Iris to start creating a BOM. (If you do not have access to Iris, you can use the HPE Networking Online Configurator. Visit http://hpe.com/networking/configurator to access this tool.)

You will also make a rough estimate at the number of APs that you require.

Activity 2.1 steps

1. Based on the customer requirements, which AP model will you propose? More than one option can be correct, but give your reasons for your selection.
2. Make a rough estimate for the number of APs required per floor. For a deployment such as this, divide the square footage by 2500 (or the area in square meters by 232). This is only an estimate; make adjustments for conference rooms and the central area.

Now you will add the AP model to Iris to begin creating a BOM.

1. Launch Iris on your device.

Note

There should be no errors in your Iris BOM when completed. However, you will have one error in this activity, which is that the AP is not yet powered. You can ignore that error for now.

2. Create a new site with **File > New Express Mode** (see Figure 2-28).

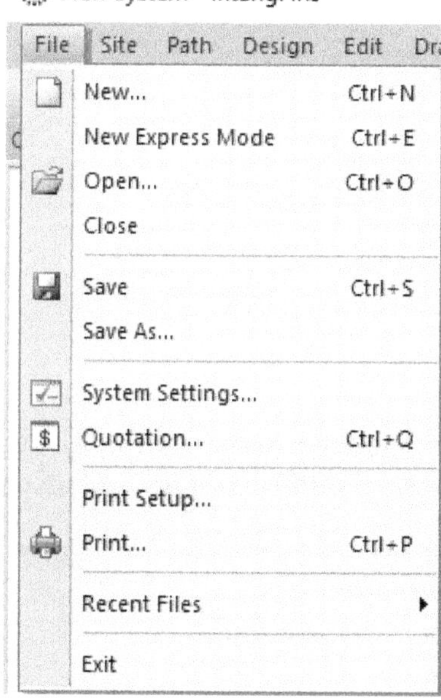

Figure 2-28 Create a new site in Iris

3. In the Catalog, expand **HPE Networking > Aruba > Access Points**.
4. Select the AP series that you chose from the product catalog and double-click to add it to the topology (see Figure 2-29).

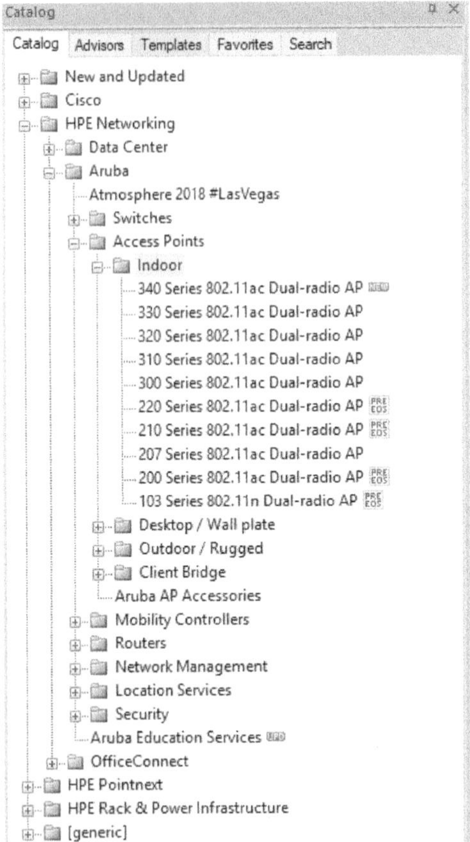

Figure 2-29 Select the AP series

5. Make sure that you select the exact model that you want from the Properties window. Figure 2-30 shows an example. You might select a different series and model.

   Also note that for AP series lower than 340, the APs are purchased as controller managed or Instant. Make sure to select the correct type.

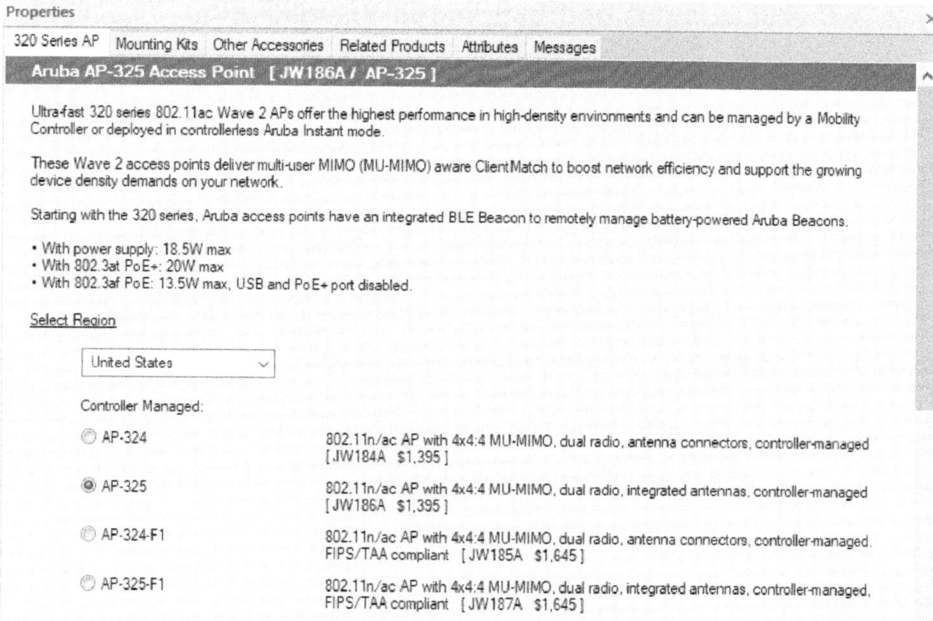

Figure 2-30 Select the Exact AP model number

6. The AP will have a red outline in the workspace because it does not have any power (see Figure 2-31). Do not worry about this. Later you will add a PoE+ switch to the plan and connect the AP to it to resolve the issue.

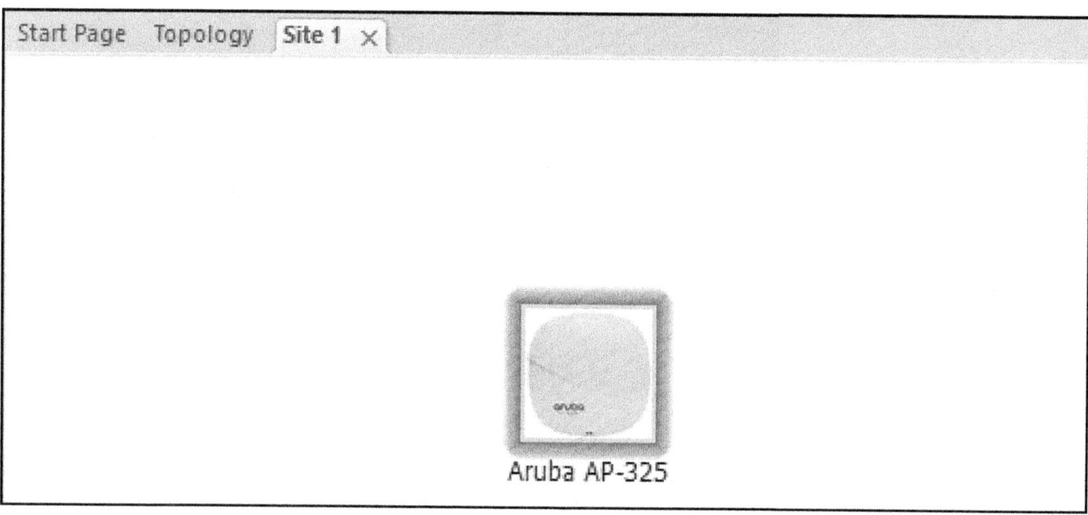

Figure 2-31 Selected AP in the workspace

7. Keep Iris open; you will return to it later. For safety, though, save your plan (by selecting **File > Save** or click the disk icon.)

Activity 2.2: Create campus and building in VisualRF Plan

Activity 2.2 objective

The purpose of this activity is to create a structure in VisualRF Plan. You will use this structure to introduce the building plans for the customer scenario.

Activity 2.2 steps

If you need to download and install VisualRF Plan, start with step 1.

If you already have VisualRF Plan running on your device, start with step 9.

Install VisualRF Plan

1. Download a copy of VisualRF Plan from the Aruba support site.
    a. Go to http://support.arubanetworks.com.
    b. Log in.
    c. Select the Download Software tab.
    d. Expand AirWave and select VisualRF Plan.

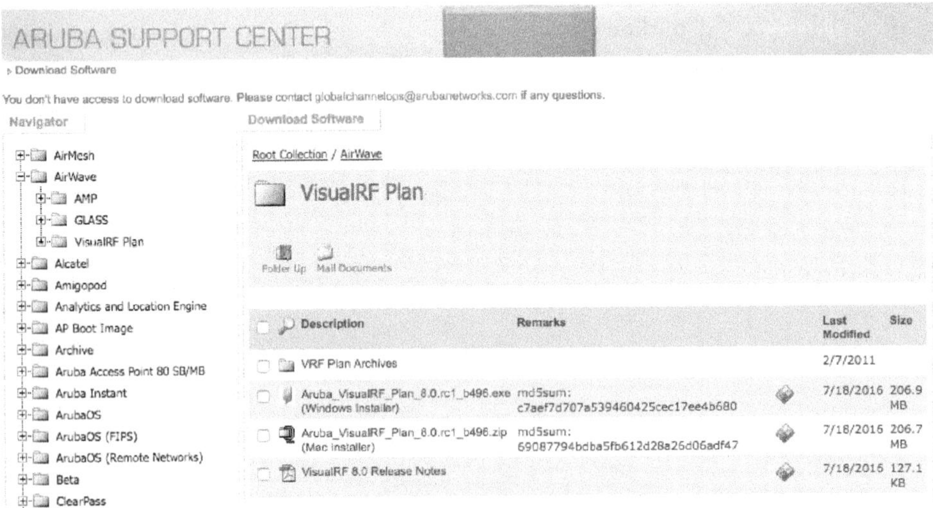

Figure 2-32 Download VisualRF plan

2. Choose either the Windows installer or MAC installer and download it.
3. Run the application file you just downloaded.
4. Click **Next** and follow the installation instructions.

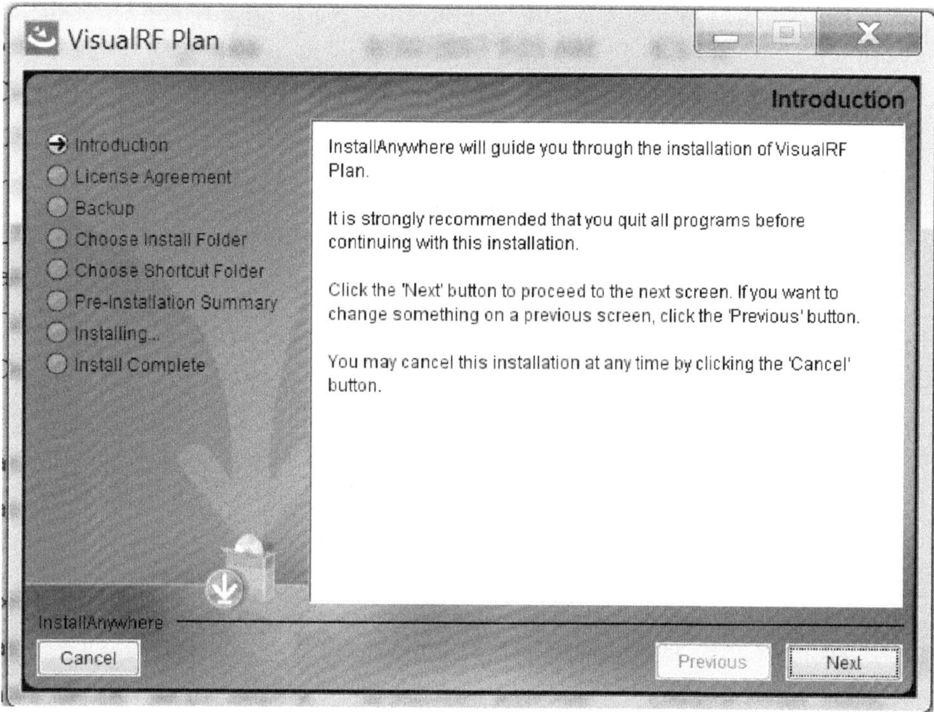

Figure 2-33 VisualRF plan installation

Note

When you get to the Choose Shortcut Folder option, we recommend putting the shortcut on your desktop. The shortcut will be easy to find when you need it.

5. Once you have completed the installation, double-click on the VisualRF Plan shortcut on your desktop.

Figure 2-34 Shortcut on Windows

Figure 2-35 Shortcut on MAC

6. You might be prompted to specify a data root directory. Select **Specify** and select the directory on your laptop.

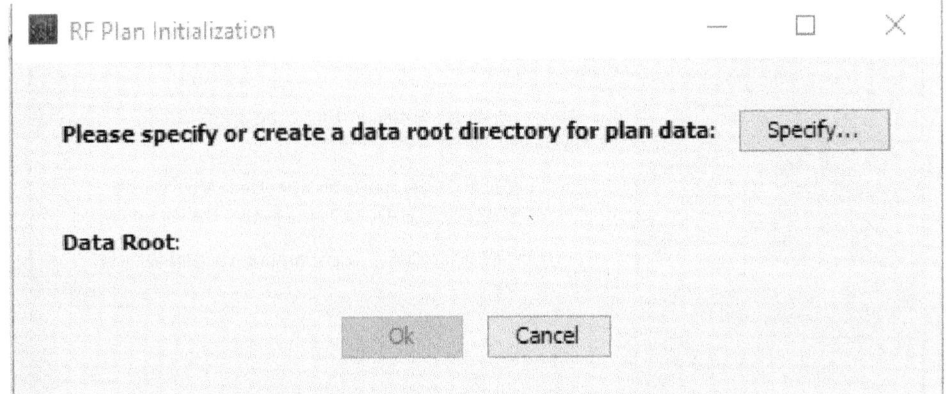

Figure 2-36 Specify a data root directory

7. Select **OK**.
8. When you are prompted, Select **Launch Default browser** (see Figure 2-37).

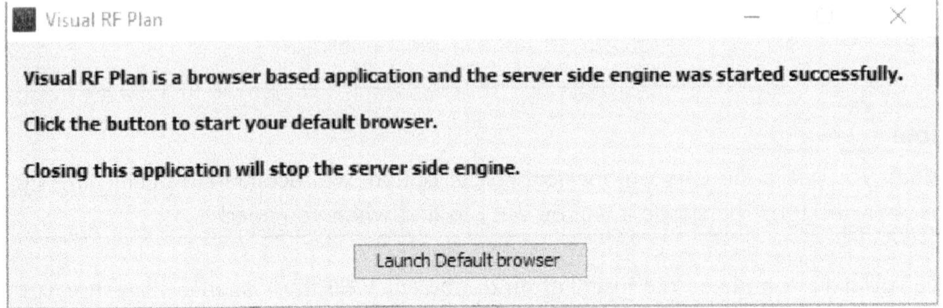

Figure 2-37 Launch default browser

Begin to Create a Structure in VisualRF Plan

9. If have not already done so, launch VisualRF Plan is launched. Then right-click on the background and select **New Campus** in the pop up menu (see Figure 2-38).

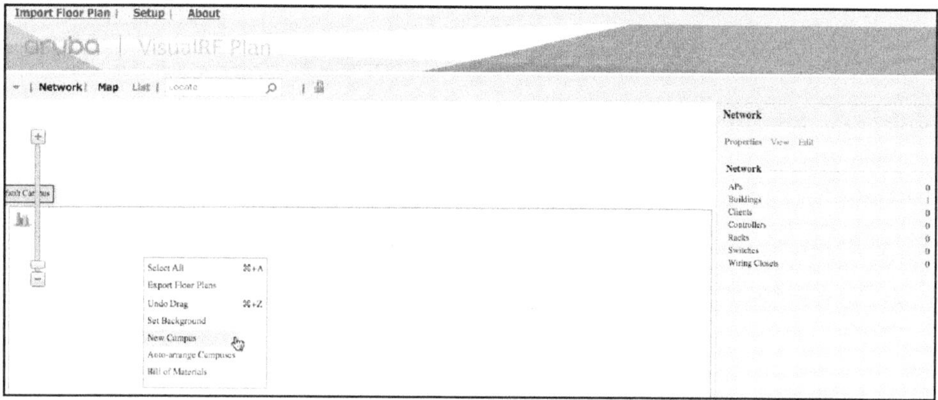

Figure 2-38 Select new campus

10. Give the campus the name **Corp1** and select **Save**.
11. Double click **Corp1**.
12. Right-click on the background and select **New Building** (see Figure 2-39).

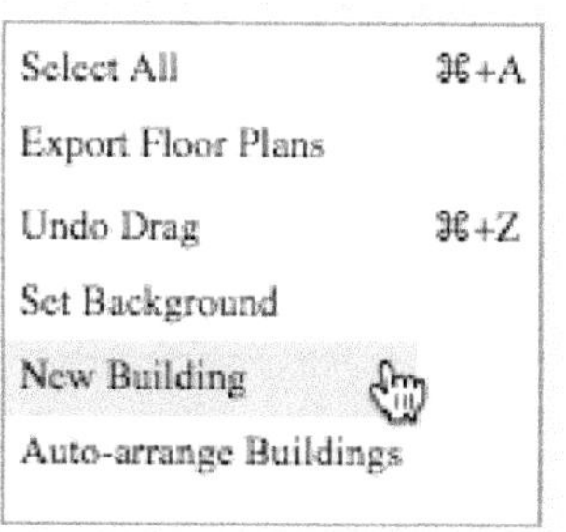

Figure 2-39 Select new building

13. Use the name **Building 1** and select **Save**.
14. Right-click on the background and select **New Building.**
15. Use the name **Building 2** and select **Save**.

Activity 2.3: Add floor plans to the buildings and APs

Activity 2.3 objective

Now that you have created your campus and building, you will add the floorplans.

Activity 2.3 steps

1. Floorplans for this customer scenario were included in the email message you received from HPE Press when you purchased this study guide. You can use these floorplans to plan coverage in Aruba VisualRF Plan.
2. In VisualRF, double-click on **Building 1**.
3. Right-click on the background and select **New Floorplan** (see Figure 2-40).

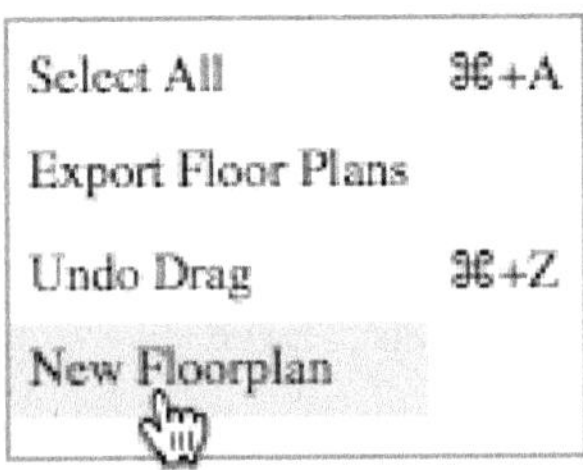

Figure 2-40 Select new floorplan

4. Now browse to the **ACDP office 1** JPEG file.
   a. Set the floor name to **Floor 1.**
   b. Set the number of the floor to **1** and select **Save** (see Figure 2-41.)

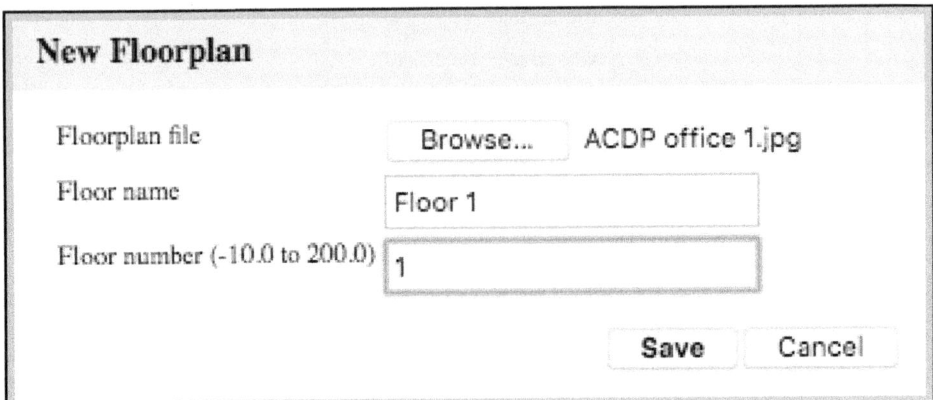

Figure 2-41 Select new floorplan

5. Now you will set the dimensions of the building.
    a. Select the **Measure** stick in the right pane.
    b. Select one side of the building and slide your cursor to the other side of the building (see Figure 2-42).

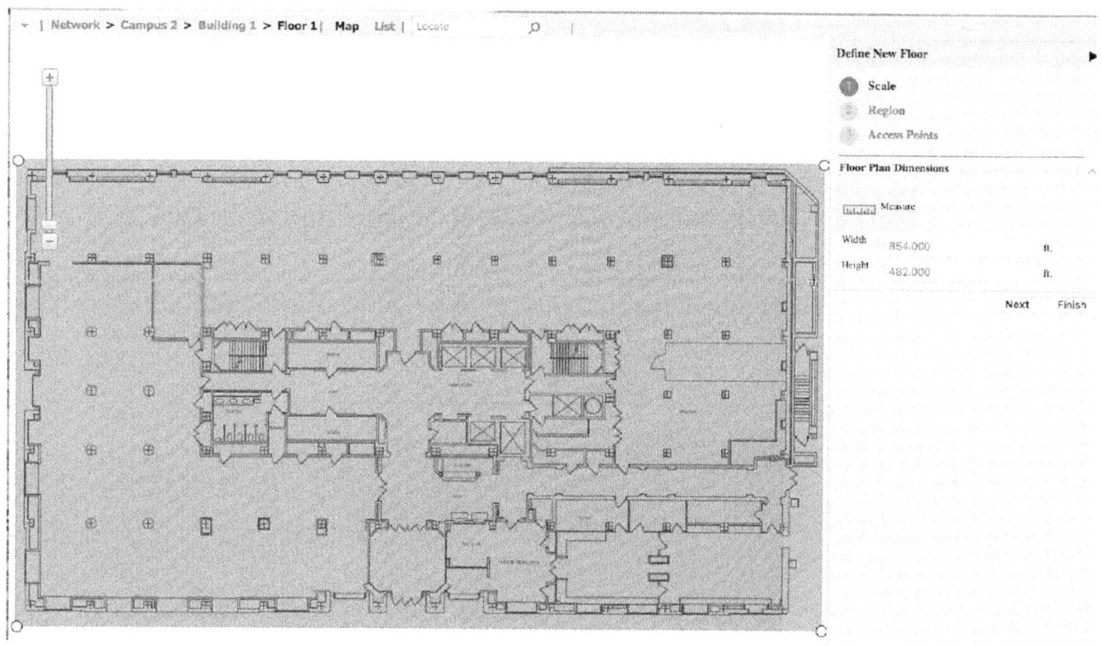

Figure 2-42 Set measurement on floorplan

    c. In the pop up window enter 310 for the distance and select **OK** (see Figure 2-43).

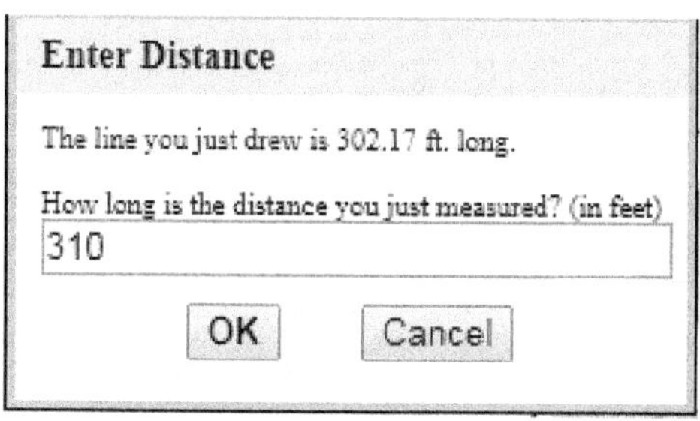

Figure 2-43 Specify distance for measurement

Note

VisualRF Plan will make some adjustments to the building dimensions to match length and width of the total image.

6. In the right pane, select **Next** for the floorplan boundary.
7. Select **Define Planning Region(s)** (see Figure 2-44).

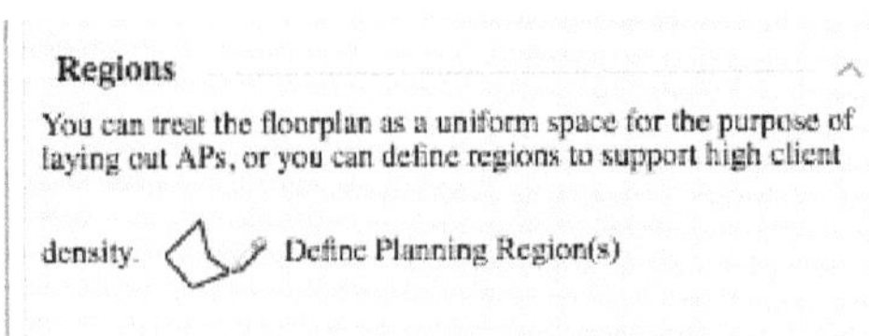

Figure 2-44 Define planning region

8. Select a corner of the building then slide your mouse to the next corner of the building and click again.
9. Make your way all around the building and return to your starting point and click. This will close the shape and define the planning region (see Figure 2-45).

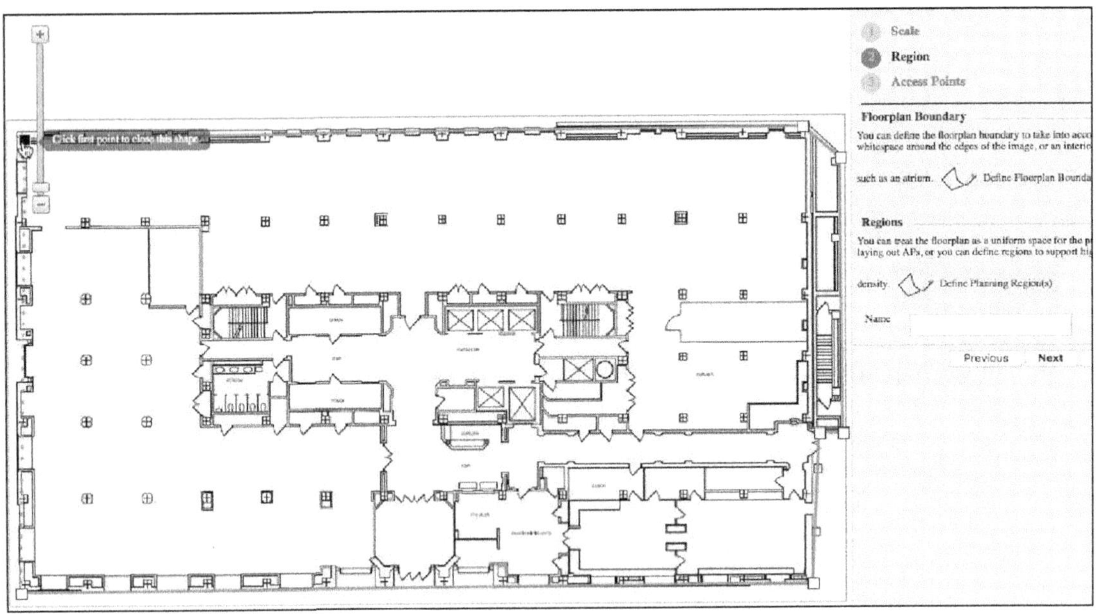

Figure 2-45 Finish defining planning regions

 **Note**

You can click at each corner in the building and follow its contour. Note the diagram above.

10. Now you have defined the planning region. You'll use this region to plan the number of APs you need. Select **Next.**

11. Now you can plan APs.

12. Select the type of APs that you chose at the beginning of the activity.

    If the AP model that you selected is not available, choose another AP for the purposes of the exercise.

13. As you see, a slider lets you plan the deployment from low density to high density. Next to the slider you see the bandwidth associated with that density.

14. Earlier, you made an estimate for the number of APs by dividing the square footage by 2500. Drag the slider down and see the number that appears for **Count** (see Figure 2-46). Adjust the slider until you see an AP count roughly equivalent to your estimate. (Note that this floor could have a bit lower count because it has open spaces.)

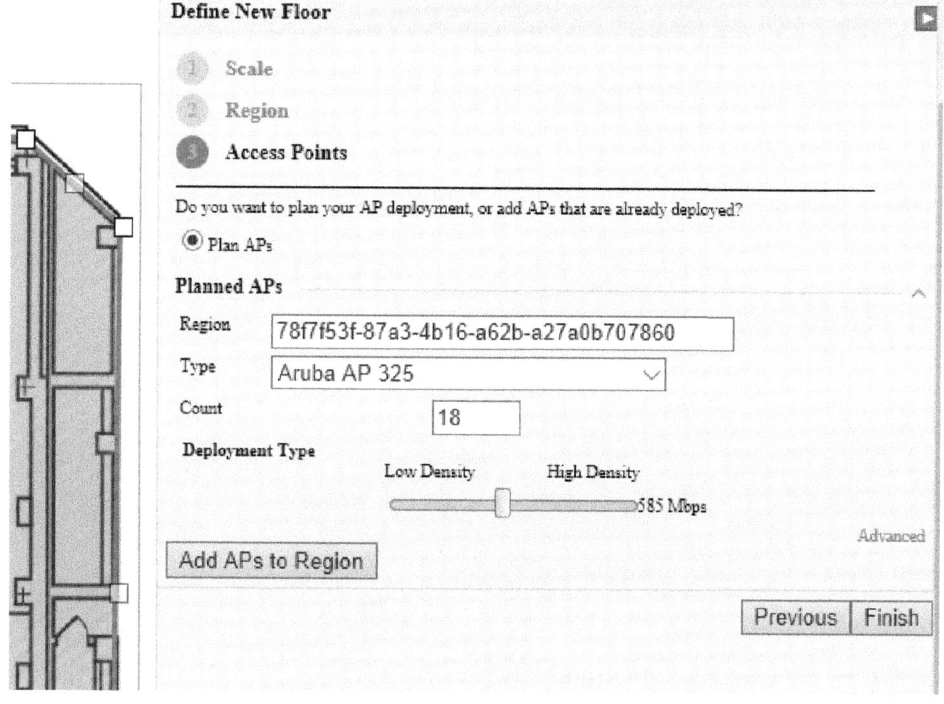

Figure 2-46 Select the AP count based on your estimates

15. How many APs are required on this floor? ___________

    Begin to record the number of APs that you need on each floor. Activity 2.6 provides space for documenting the AP number.

16. Click **Add APs to Region**.
17. Select **Finish**.

You will now repeat these steps to plan the other five floors.

18. In the top bar select **Building 1** (see Figure 2-47).

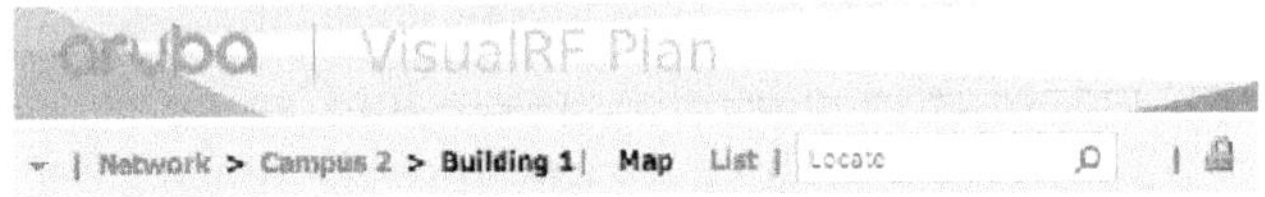

Figure 2-47 Select Building 1 in top bar

19. In the Building 1 area right-click on the background and select **New Floorplan**.
20. Browse to the second floor **ACDP office 2.jpg** file.
    a. Specify **Floor 2**.
    b. For the floor number select **2** (see Figure 2-48).

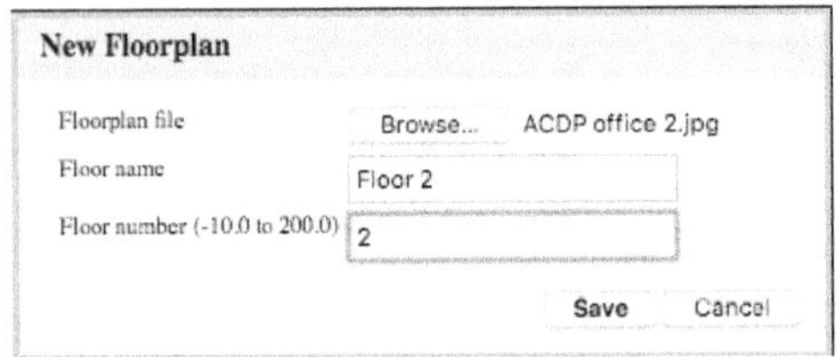

Figure 2-48 Select new floorplan

21. Proceed to measure and Define Planning Region(s).
22. Then plan your APs.
23. Complete the same steps for your third floor.
24. Once Building 1 is completed, select Corp-1 in the top bar.
25. Here you can see the two buildings. Double click on **Building-2**
26. Add the three floors of Building 2 using the same floor plan for each. Measure and Define Planning Region(s) and then plan your APs.

Activity 2.4: Manually add in APs

Activity 2.4 Site survey results

After a physical site survey was completed, it was reported that conference rooms on the third floor have metallic vertical blinds that reduce or cut off the AP signals.

It was also reported that APs in the main employee areas did not propagate very well in the central corridor.

Activity 2.4 objectives

Determine if you need to add more APs in the buildings

Activity 2.4 steps

1. In the top bar select **Building 1** (see Figure 2-49).

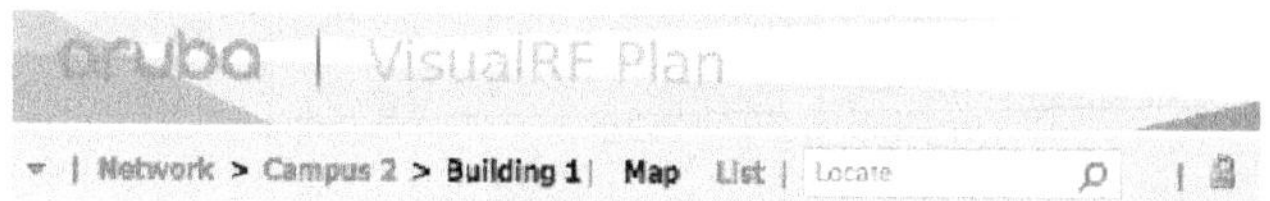

Figure 2-49 Select Building 1 in top bar

2. Select the third floor of the building.
3. Click the **Edit** link to enter Edit mode. Select **Add Planned Devices** (see Figure 2-50).

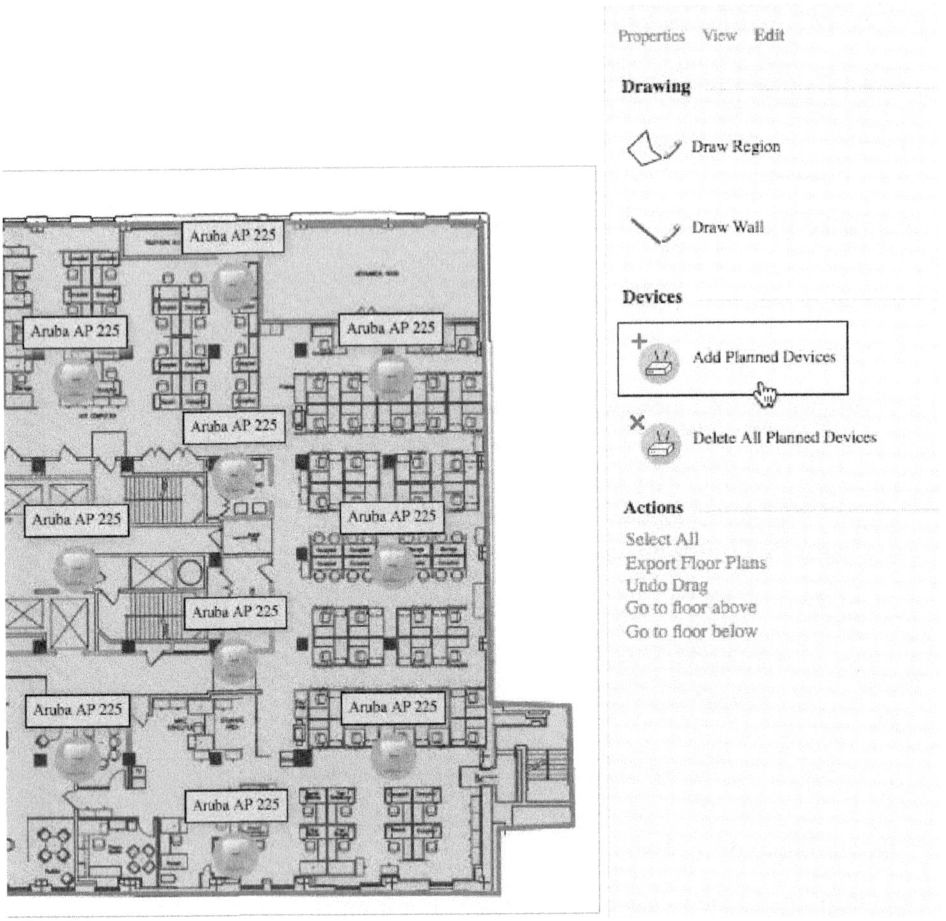

Figure 2-50 Begin to plan the third floor

4. In the window, full of APs, find the AP you want. Then select, hold, and move it into the floor plan where you think you may need more APs (see Figure 2-51).

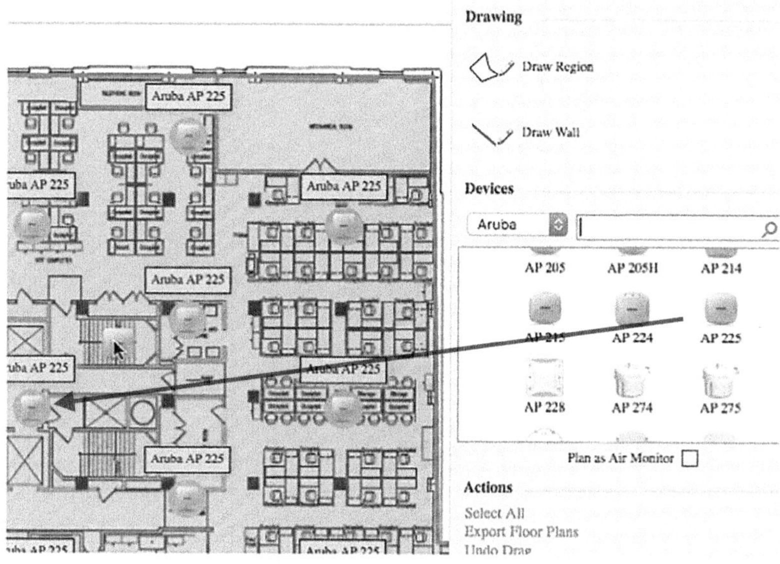

Figure 2-51 Finish planning the Third Floor

5. Remember to add these APs to your count for Building 1, Floor 3.

Activity 2.5: View the Heatmap

Activity 2.5 objectives

You will now view the heatmap to see the coverage.

Activity 2.5 steps

1. On the top right click **View** (see Figure 2-52).

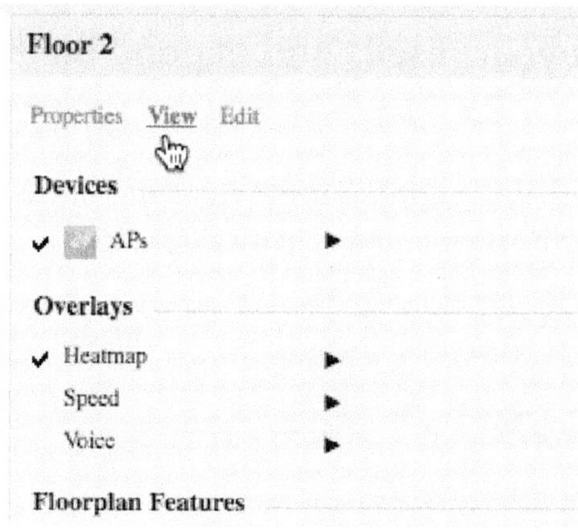

Figure 2-52 Right-click view

2. Then select **Heatmap** (see Figure 2-53).

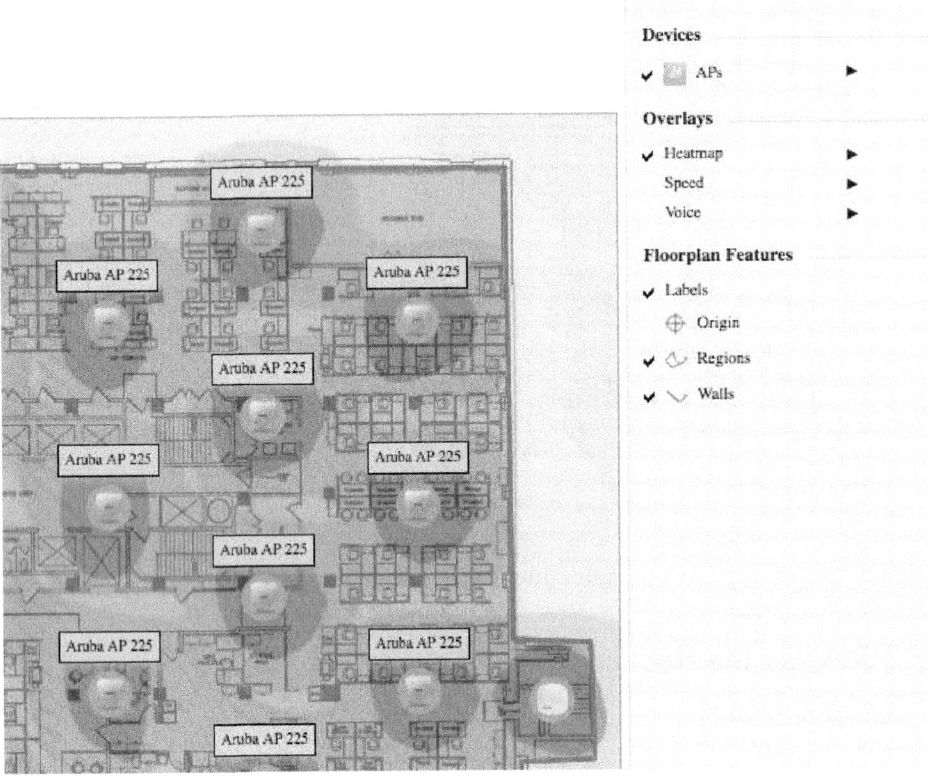

Figure 2-53 Heatmap

3. Here you can get an idea of the coverage. Note this display will not show any abnormalities in the building that may block Wi-Fi signals.

Activity 2.6: Document your results and expand the BOM

Activity 2.6 objectives

You will now tally your results and multiply out the AP count in Iris or the HPE Network Online Configurator. Remember that the steps in this activity are for Iris.

Activity 2.6 steps

1. How many APs on Building 1, Floor 1: ____________
2. How many APs on Building 1, Floor 2: ____________

3. How many APs on Building 1, Floor 3: ___________
4. How many APs on Building 2, Floor 1: ___________
5. How many APs on Building 2, Floor 2: ___________
6. How many APs on Building 2, Floor 3: ___________
7. What is your total count of APs: ___________
8. Document the number of APs in Iris.
   a. Return to Iris.
   b. Select your AP in the workspace.
   c. In the Properties window, select the Attribute tab and enter the number of total AP count in the Quantity Multiplier.
   d. Select True for Create Synced Set. This setting makes it easy for you to quickly draw in connections from your APs to the switches, which you will do in another activity (see Figure 2-54).

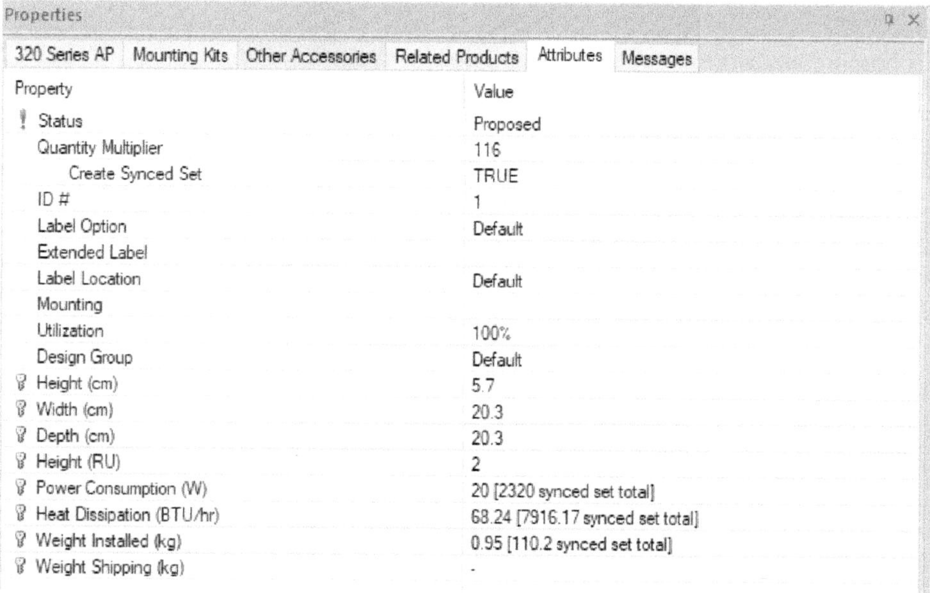

Properties
320 Series AP
Mounting Kits
Other Accessories
Related Products
Attributes
Messages
Property
Value
Status
Proposed
Quantity Multiplier
116
Create Synced Set
TRUE
ID #
1
Label Option
Default
Extended Label
Label Location
Default
Mounting
Utilization
100%
Design Group
Default
Height (cm)
5.7
Width (cm)
20.3
Depth (cm)
20.3
Height (RU)
2
Power Consumption (W)
20 [2320 synced set total]
Heat Dissipation (BTU/hr)
68.24 [7916.17 synced set total]
Weight Installed (kg)
0.95 [110.2 synced set total]
Weight Shipping (kg)
-

Figure 2-54 Select new floorplan

9. Save your project.
10. Select the dollar sign button or select File > Quotation to see your BOM.

Answers to the questions included in activities are provided in "Appendix: Activities and Learning Check Answers."

Summary

You should now understand RF fundamentals including radio and antenna basics and the changes to the 802.11ac standard. You should now be able to use Visual RF and be able to effectively mount APs to provide the best coverage for your customer. Finally, you should now be able to provide a clear RF channel design for your customer.

Learning check

1. You see these specs for an Aruba AP-335: 4x4:4:4:3. How many spatial streams does the AP support and how many MU-MIMO spatial streams?
   a. 3 spatial streams and 4 MU-MIMO spatial streams
   b. 4 spatial streams and 3 MU-MIMO spatial streams
   c. 3 spatial streams and 3 MU-MIMO spatial streams
   d. 4 spatial streams and 4 MU-MIMO spatial streams
2. For which deployment would you suggest an active survey?
   a. A hospital that wants to upgrade its wireless network
   b. A branch site with 20 users
   c. A hospital that does not wants to upgrade its wireless network
   d. A company that has less complex environment

Answers to the learning checks in this study guide are provided in "Appendix: Activities and Learning Check Answers."

3 Aruba Campus Design

EXAM OBJECTIVES

- ✓ Evaluate a customer's needs for a single-site campus, identify gaps and recommend components.
- ✓ Translate a customer's needs into technical requirements.
- ✓ Select the appropriate products based on a customer's needs for a single-site campus.
- ✓ Given a customer's requirements for a single-site campus, design the high-level architecture.
- ✓ Given a customer scenario for a single-site campus, choose the appropriate components that should be included in the BOM.
- ✓ Given the customer's requirements for a single-site environment, determine and document the logic and physical networks.
- ✓ Given the customer's requirements, explain and justify recommended solution.

Assumed knowledge

- Aruba product portfolio
- Aruba 8.x architecture

Introduction

Aruba offers products to meet various deployment needs. This chaper first provides an overview of the Aruba product line. You will then be ready for more details on each of the Mobility Controller (MC) series in the portfolio. You will also learn different deployment scenarios, licensing types,and applications you can use for your network designs.

Product line and portfolio

You will first review the Aruba product line and portfolio.

Product line

The Aruba product line includes a comprehensive offering of APs, controllers, and switches, combined with a variety of platform, application, and cloud-monitoring and management options, enabling a wide variety of enterprise-class networking solutions (see Figure 3-1).

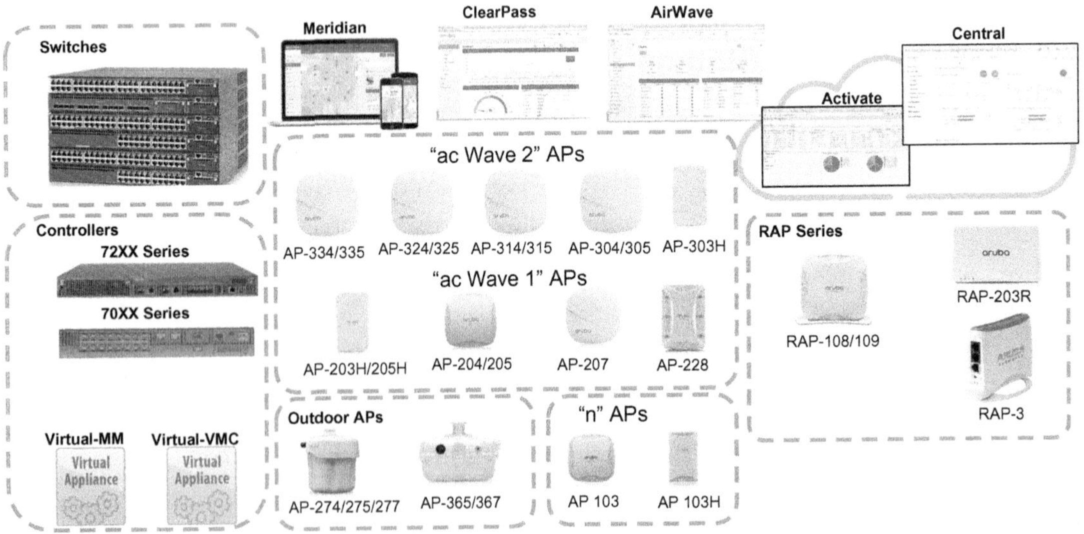

Figure 3-1 Aruba product portfolio

Aruba APs

The 100 series are 802.11n APs.

The 200 series are 802.11ac APs, and the 300 series APs are 802.11ac Wave 2 APs. The "H" APs are hospitality and branch office APs that can be placed on the wall or have stands. They also have wired ports with Power over Ethernet (PoE) capabilities. The 203H is single radio, and the 205H is dual radio. The 303H is 802.11ac Wave 2. The AP228 is an 802.11ac AP for rugged indoor environments.

The 200 series APs come with various antennas specification. All of these APs have the radio omnidirectional down-tilt dual-band antennas. The 305 has three of these antennas, whereas the 310 has four, the 320 has eight, and the 330 has 12 antennas.

The Remote AP (RAP) series boot up initially as Instant APs (IAPs) but can automatically or manually be converted to a RAP. RAP is a great solution as a single Wi-Fi device that can give corporate access to home or a small branch office. The RAPs come with extra Ethernet ports, which can provide wired access to endpoints.

Outdoor APs are built to survive the extremes of an outdoor environment.

You can purchase most APs as either Campus APs (CAPs) or as Instant APs (IAPs). IAPs can be converted to CAPs when desired and returned to IAPs if required. As of AP-340 Series, APs are no longer purchased as CAPs or IAPs but as unified APs (UAPs).

Aruba Controllers

Aruba offers an assortment of controllers that can handle from 16 APs to 2048 APs. Controllers can be a hardware appliance, or you can also get controllers as a virtual mobility controller (VMCs). The Mobility Master (MM) is a new master for AOS 8.X deployments.

Aruba switches

Aruba offers an assortment of switches that support networking protocols such as spanning tree and routing protocols. These switches also support tunneled node.

Aruba management solutions

AirWave is a management platform for monitoring and managing the wireless and wired networks. AirWave also has other capabilities such as Reporting, Visual RF, and rogue detection. AirWave can monitor Aruba networks and products from other vendors as well.

Aruba Central, on the other hand, is a cloud-based management system.

Activate is a cloud-based system that is used for the zero touch deployments.

Aruba ClearPass

Aruba ClearPass is used for network control, access security and advanced features such as Captive Portal, guest login, self-registration as well as onboarding employee owned devices.

Aruba Meridian

The Meridian System is used for location awareness and advertising features. Integrated Bluetooth beacons and radios allow you to pull more value from your Aruba wireless network. This means that BLE-enabled Aruba APs can be a beacon for mobile engagement or a reader for asset tracking. The adoption of smartphones gives venues an opportunity to leverage their existing Wi-Fi network to extract actionable location data. Use built-in or third-party integration to view shopper dwell times, traffic flow, zone analytics, or campaign effectiveness to help drive business decisions.

Review Aruba OS 8.X architecture

In AOS 8.x, you can set up the entire configuration for both the Mobility Master (MM) and Mobility Controllers (MCs) from a centralized point, the MM (see Figure 3-2).

The MM is responsible for all configuration of MCs and multiple high-end features.

APs cannot terminate on the MM, but they can terminate on MCs. MCs can be in the internal network. MCs can also use a virtual private network (VPN) to join the corporate network and MM.

You can deploy the MM on both physical and virtual machine (VM) appliances. The MM can push full configuration, including VLAN, interface configuration, IP routing, and so on, to all the managed MCs. The MM will also validate the configuration.

Once any MC is in communication with the MM, all configuration is done on the MM. You cannot do any configuration directly on the MC.

When connectivity between the managed device and MM is disrupted for a period of time, administrators can enable **Disaster Recovery** mode on the managed device they are using. The MC can then be configured locally.

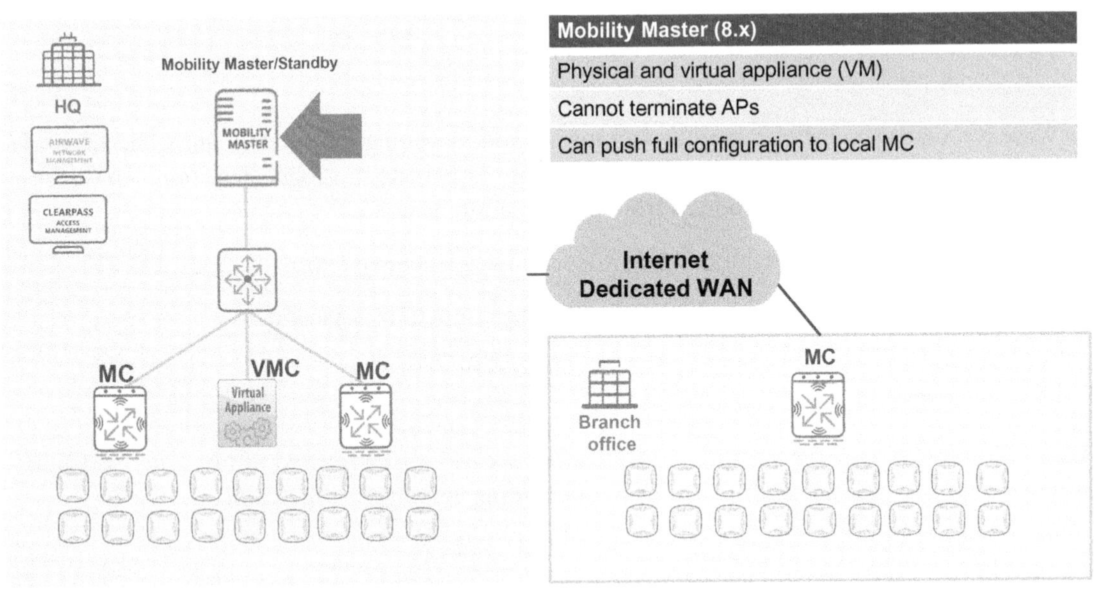

Figure 3-2 Review Aruba OS 8.X Architecture

Review APs deployment

As Figure 3-3 shows, you can configure an AP in many ways to take on specific roles in the WLAN architecture. For example, a CAP (also referred to as an AP or regular AP) is the typical AP that will connect to, and get all of its configuration from, a controller.

Mesh APs are CAPs that use a radio interface as a link to another AP. The Mesh Portal has a physical connection to the corporate network and connects to Mesh Points. The Mesh Point uses its radio to get an uplink to the corporate network.

Air Monitors (AMs) constantly scan the radio environment to gather Intrusion Detection System (IDS) and RF information. Spectrum Analyzers (SAs) are spectrum APs that have been set up, temporarily or permanently, to capture radio signals for analysis.

RAPs act similarly to CAPs but must go through the Internet to reach a controller. This requires the RAP to set up a Virtual Private Network (VPN) tunnel to the controller. A RAP can also be set up as a Remote Mesh Portal. This is basically a RAP with Mesh Portal capabilities; it enables customers to easily extend a wireless network at a branch office.

IAPs do not need a controller. All IAPs in the same subnet will form a cluster and elect a Virtual Controller (VC) so they can operate independently of a hardware controller.

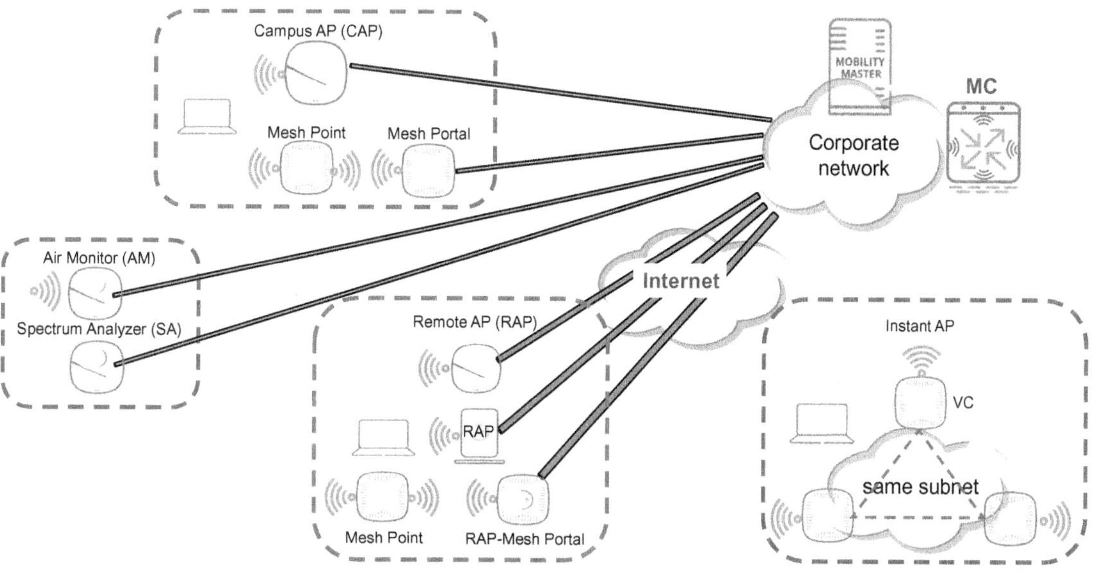

Figure 3-3 Review APs deployment

The user traffic each AP receives typically arrives encrypted. The APs send this traffic, via the GRE tunnel, directly to the controller.

The controller decrypts the users' packets, firewalls the packets, and then switches or routes the packets onto the wired network. Prior to this, the user must authenticate to the network.

Often, there is a network firewall between the AP and the Mobility Controller. If this is the case, the firewall must be configured to allow protocols so the AP and MC can communicate.

An AP issues a Dynamic Host Configuration Protocol (DHCP) request to obtain an IP address and then looks for the MC, in many cases, using Domain Name System (DNS). Normally, these protocols are not blocked in the network.

The AP attempts to communicate with the MC, using the PAPI protocol (UDP8211). This traffic may be encrypted using Aruba's CPSec feature, which uses an Internet Protocol Security (IPsec) tunnel. The AP may need to be upgraded or possibly downgraded; the MC manages its software. For this, you will use TFTP or FTP. Client traffic is encapsulated into a GRE (Protocol 47) tunnel and sent to the MC.

The AP also uses other protocols such as Network Time Protocol (NTP) and syslog, but these are not critical for AP functionality.

In tunnel mode, the APs create GRE tunnels to the mobility controller—one per WLAN. All user traffic the AP receives normally arrives encrypted. The AP sends the encrypted traffic, via the GRE tunnel, directly to the controller.

The controller will de-capsulate the packet from the GRE headers, decrypt the packets, firewall the packets, then switches or routes the packets onto the network, if the firewall permits.

In decrypt tunnel node, the APs will create the GRE tunnels to the mobility controller. The AP will decrypt the packet prior to sending it down the GRE tunnel. The controller will de-capsulate the packet from the GRE headers, firewall the packets, then switches or routes the packets onto the network (if the firewall permits).

For a bridged WLAN, the packet is decrypted, then placed on the APs wired port as a standard 802.3 frame. Aruba does not recommend this deployment because the stateful firewall is not enforced.

Each AP creates one GRE per WLAN, per radio and one extra GRE for keepalives.

In tunnel mode the AP handles all 802.11 association requests and responses, but sends all 802.11 data packets, action frames, and Extensible Authentication Protocol (EAP) over LAN (EAPOL) frames over the GRE tunnel to the mobility controller for processing. The controller removes or adds the GRE headers, decrypts or encrypts the 802.11 frames, and applies firewall rules to the user traffic as usual. You can configure both remote and campus APs in tunnel mode.

In Decrypt-Tunnel mode, the AP decrypts and decapsulates all 802.11 frames from a client and sends the 802.3 frames through the GRE tunnel to the controller, which then applies the firewall policies to the user traffic.

When the controller sends traffic to a client, the controller sends 802.3 traffic through the GRE tunnel to the AP, which converts it to encrypted 802.11 and forwards to the client. This forwarding mode allows a network to utilize the encryption/decryption capacity of the AP while reducing the demand for processing resources on the controller. APs in decrypt-tunnel forwarding mode also manage all 802.11 association requests and responses, and process all 802.11e and 802.11k action frames.

In bridge mode, the 802.11 frames are bridged into the local wired LAN. The AP handles all 802.11 association requests and responses, encryption/decryption processes. The AP also processes the 802.11e and 802.11k action frames and sends out responses as needed. An AP in bridge mode does not support captive portal authentication.

In decrypt-tunnel mode and bridge mode, each r0 key generates up to four r1 keys and the controller pushes each r1 keys to the corresponding AP. Therefore, you must enable the control plane security to be able to send the keys to the AP securely.

Mobility Controllers

You will now review Aruba Mobility Controllers.

Aruba Controller 7000 portfolio

Aruba 7000 series Mobility Controllers optimize cloud services and secure enterprise applications for hybrid WAN at branch offices, while reducing the cost and complexity of deploying and managing the network.

As Figure 3-4 shows, the portfolio combines wireless, wired and hybrid WAN services, support from 16 up to 64 APs, and wired ports with PoE+ switching and routing capabilities. Integrated features include WAN compression, health checks, Zero-Touch Provisioning (ZTP), configuration, and policy-based routing.

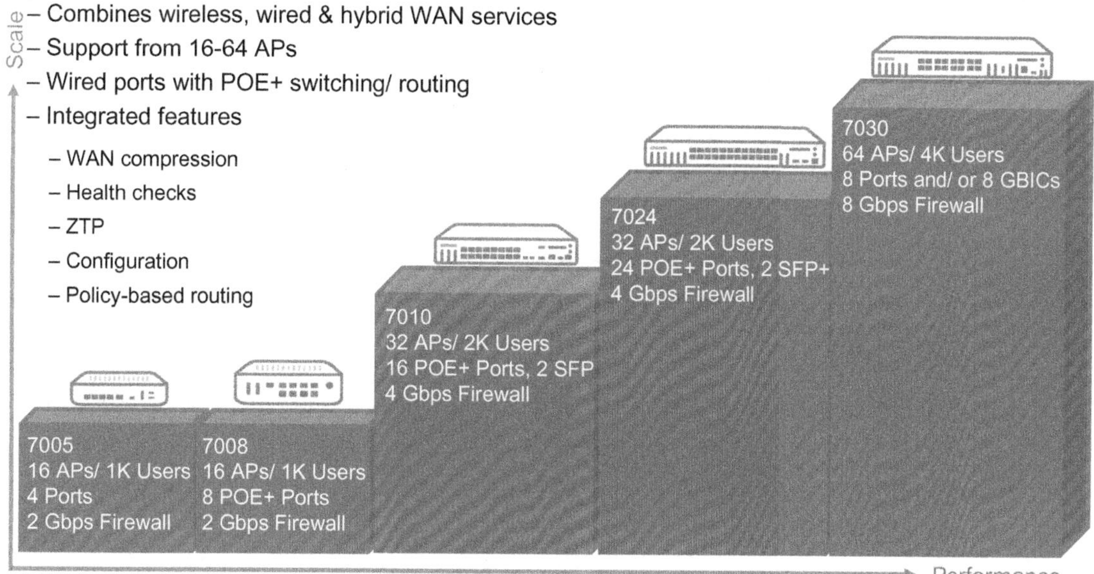

Figure 3-4 Aruba Controller 7000 Portfolio

All controllers support the same features. The difference is the capacity of APs and devices supported.

The smallest controller in the 7000 series portfolio can handle 16 APs, and the largest branch office controller can handle 64 APs. (Although these controllers are targeted for branch offices, they can also meet the needs for smaller main offices.)

The 7005 Mobility Controller is a compact, fanless entry-level branch platform, with four 10/100/1000 base-T ports, and can be powered by a PoE switch.

The 7008 Mobility Controller is a compact, fanless entry-level branch platform with eight integrated PoE and PoE+ switch ports for small branch locations with some PoE requirements. Its capabilities are very similar to those of the 7005, but it provides more integrated PoE+ ports.

The 7010 Mobility Controller is designed for midsized branch deployments that require 16 PoE and PoE+ switch ports.

The 7024 Mobility Controller is designed for mid-sized branch offices that require up to 24 switch ports for unified wired and wireless access.

The 7030 Mobility Controller is ideal for mid- to large-sized branch deployments that require unprecedented scale and performance.

Performance capacity 7000 series

With the 7000 series of controllers, as the series number of the controller augments, so does the capacity of the controller. The 7005 series is the smallest controller and handles 16 APs and 1024 users. On the other side of the 7000 scale, the 7030 handles 64 APs and 4096 users. These are smaller controllers that are good for small and branch offices or even a small network.

Table 3-1 shows many other attributes of these controllers. If you reach any limitation then it may be time to upgrade to a higher series controller.

Table 3-1 Performance and capacity 7000 series

Features	7005	7008	7010	7024	7030
Maximum campus AP licenses	16	16	32	32	64
Maximum remote AP licenses	16	16	32	32	64
Maximum concurrent users/devices	1,024	1,024	2,048	2,048	4,096
Maximum VLANs	4,094	4,094	4,094	4,094	4,094
Active firewall sessions	16,384	16,384	32,768	32,768	65,536
Concurrent GRE tunnels	256	256	512	512	1,024
Concurrent IPsec sessions	512	512	1,024	1,024	2,048
Mobility Access Switch tunneled-node ports	512	512	1,024	1,024	2,048
Firewall throughput	2 Gbps	2 Gbps	4 Gbps	4 Gbps	8 Gbps
Encrypted throughput (3DES, AES-CBC)	1.2 Gbps	1.2 Gbps	2.4 Gbps	2.4 Gbps	2.4 Gbps
Encrypted throughput (AES-CCM)	1.6 Gbps	1.6 Gbps	3.4 Gbps	3.4 Gbps	4.0 Gbps

The maximum campus AP license is the absolute maximum number of APs that the controller can support. The maximum number of APs can be Campus APs, remote APs or a combination of the maximum specified.

The maximum concurrent users feature is the total number of devices associated with APs that terminate on a controller.

Each controller supports up to 4094 VLANs. The number of IP addresses depends on the controller model.

The "Show IP DHCP stats" will show you the limit on DHCP leases. The 7010 and 7024 have 2048 DHCP IP address leases.

Each user may have several sessions running. On average users may have 20 sessions, but these are highly transient.

Each MC has the number of concurrent firewall sessions allowed. APs will create one GRE per WLAN, per radio and one extra GRE for keepalives. With 16 APs this would signify that each AP could support 16 GRE or 8 WLAN on both radios. Normally, Aruba recommends three to four WLANs per AP to not overload the radio with management messages.

The concurrent limit in an IPsec session would be a combination of IPsec tunnels to the MM and RAP and VIA sessions.

Each MC has a limit to the number of ports that switches can support in tunneled node mode.

Firewall throughput is a total measure of firewalled packets that traverse the mobility controller.

Encryption throughput is a measure of the encryption decryption capabilities of the mobility controller.

Aruba Controller 7200 portfolio

Aruba 7200 series Mobility Controllers are powered with a new central processor that employs up to eight cores with four threads each. It is like having a total of 32 virtual CPUs. As a result, the 7200 series controllers support up to 32,000 user devices and perform stateful firewall policy enforcement at 40 Gbps (see Figure 3-5).

The 7205 supports 256 APs and 8000 users. At the other end of the scale, the 7240XM supports 2048 APs and 32,000 users. The 7240XM has more DRAM than its predecessor 7240, and the 7280 has more ports.

The 7200 series Mobility Controllers are typically campus controllers but can also be deployed at the branch office if required. These controllers include a number of features that make them ideal for deployment in customer locations that require maximum availability:

- The 7205 supports two 10 GbE ports as well as four Base-X or Base T ports.
- The 7210/7220/7240/7240XM/7280 support dual, field-replaceable redundant power supplies to maintain uninterrupted network operations. They also include field-replaceable fan trays, providing sufficient cooling and rapid time to repair. These models have four 10GBASE-X (SFP+) ports connections for high availability.
- The 7240XM has an increased DRAM to 16 GB, whereas the 7240 has 8 GB.
- The 7280 support two 40 GbE ports and eight 10 GbE ports for high performance and availability.

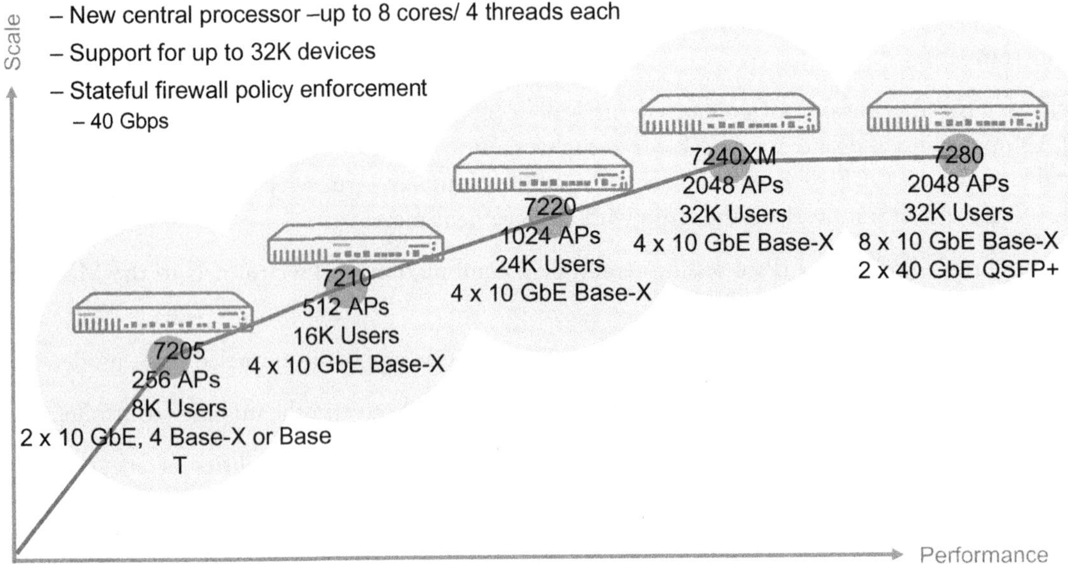

Figure 3-5 Aruba Controller 7200 portfolio

Performance capacity 7200 series

Aruba 7200 series Mobility Controllers are optimized for 802.11ac and mobile app delivery. With a new central processor employing eight CPU cores and four virtual cores, 7200 supports more than 32,000 wireless devices, 2048 access points, two million firewall sessions, and Wi-Fi encryption at speeds up to 40 Gbps.

The 7240XM has an increased DRAM to 16 GB, whereas the 7240 has 8 GB. The 7280 supports two 40 GbE ports and eight 10 GbE ports for high performance and availability.

In the 7000 series the number of APs is limited to a maximum of 64 APs. As Table 3-2 shows, the 7200 series has the capacity for much more traffic and sessions. The 7030 has a limit of 64 APs and 65,000 active firewall sessions. The 7205 supports up to 256 APs and one million active firewall session supporting 8192 concurrent devices.

For new deployments, the 7240XM is generally recommended; its greater memory capacity helps to ensure good performance as future software updates add features.

Table 3-2 Performance and capacity 7200 series

Features	7205	7210	7220	7240/7240XM	7280
Maximum APs (licenses)	256	512	1,024	2,048	2,048
Maximum RAPs	256	512	1,024	2,048	2,048
Maximum concurrent devices	8,192	16,384	24,576	32,768	32,768
VLANs	2,048	4,094	4,094	4,094	4,094
Concurrent GRE Tunnels (System BSSIDs)	8,192	8,192	16,384	32,768	32,768
Concurrent Tunneled Ports	4,096	8,192	12,288	16,384	16,384
Concurrent IPsec sessions	4,096	16,384	24,576	32,768	32,768
Concurrent SSL fallback sessions	4,096	8,192	8,192	8,192	8,192
Active Firewall Sessions (Concurrent sessions)	1,000,000	2,015,291	2,015,291	2,015,291	2,015,291
Wired Throughput (large packets)	12 Gbps	20 Gbps	40 Gbps	40 Gbps	100 Gbps

Virtual Mobility Controller portfolio

The Virtual Mobility Controller (VMC) runs on ArubaOS 8.x, providing a flexible deployment alternative to the mobility controller hardware appliance. As Figure 3-6 shows, there are three types of VMCs.

- The VMC 50 licenses for 50 AP and supports up to 4000 users.
- The VMC 250 can support 250 APs and 8000 users.
- The VMC 1000 can support 1000 APS and 4000 users.

Each VMC has VM requirements to support the number of users and APs. The AP count is also set by the license.

Customers who already have a virtualized environment can benefit from operational cost savings by deploying MC-VA, since it can reside with other VMs sharing the same existing virtualization infrastructure. For high availability, customers can use centralized licensing to install multiple VMs in a cluster without the need to purchase separate controllers.

MC-VA also supports all Aruba's enterprise-critical capabilities such as AppRF, AirGroup, advanced cryptography, IPv4 and IPv6 services, Adaptive Radio Management, ClientMatch, and RFProtect module and wireless intrusion protection.

When an MM manages the VMCs, the VMC licenses are pooled. For example, a customer could have an MM and three VMCs. The customer could install 500 VMC licenses on the MM, and the VMCs could support different numbers of APs up to 500 total. When a VMC is deployed as a standalone device, though, the licenses belong to it and cannot be pooled.

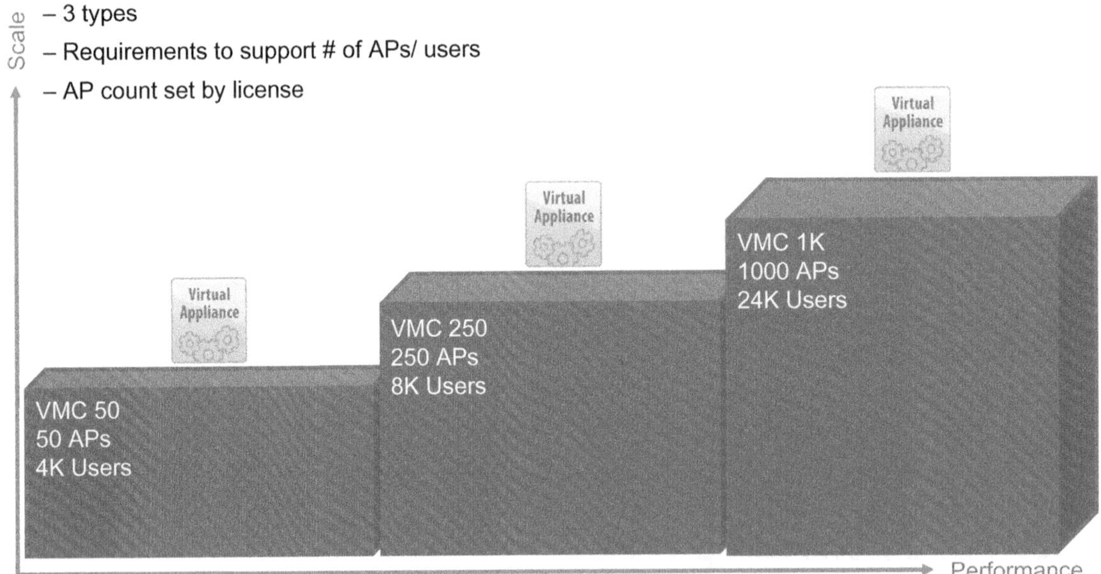

Figure 3-6 Virtual Mobility Controller portfolio

Mobility Master portfolio

The Mobility Master (MM) is the next generation of master controllers that run ArubaOS 8.x and is deployed as a Virtual Machine (VM) or hardened server appliance with large amounts of memory and compute power (see Figure 3-7).

With the Mobility Master, major AOS 8 features can be updated via software without requiring any planned network outages. Each MM has specific VM requirements to support features and clients. The number of devices is measured by the number of Mobility Controllers (MCs) and APs.

You can calculate the amount of clients and MCs an MM can handle with the times by 10 and divide by 10 rule. For example, MM 50 can handle 50 X 10 = 500 clients. The number of MCs is 50/10 = 5 MCs. The actual client count depends on the VM resources allocated to the MM. These ratios are general guidelines. The MM must have sufficient licenses to support the total number of MCs and APs; however, the number of MCs does not have to be exactly one-tenth the number of MM licenses. In addition, no licenses limit the number of clients supported. However, you should follow the general guidelines to ensure that the MM meets performance expectations.

For networks with higher encryption needs, the Mobility Controller appliances scale up to 40 Gbps with integrated network processors to accelerate cryptography and packet forwarding performance supporting all the AOS 8 features.

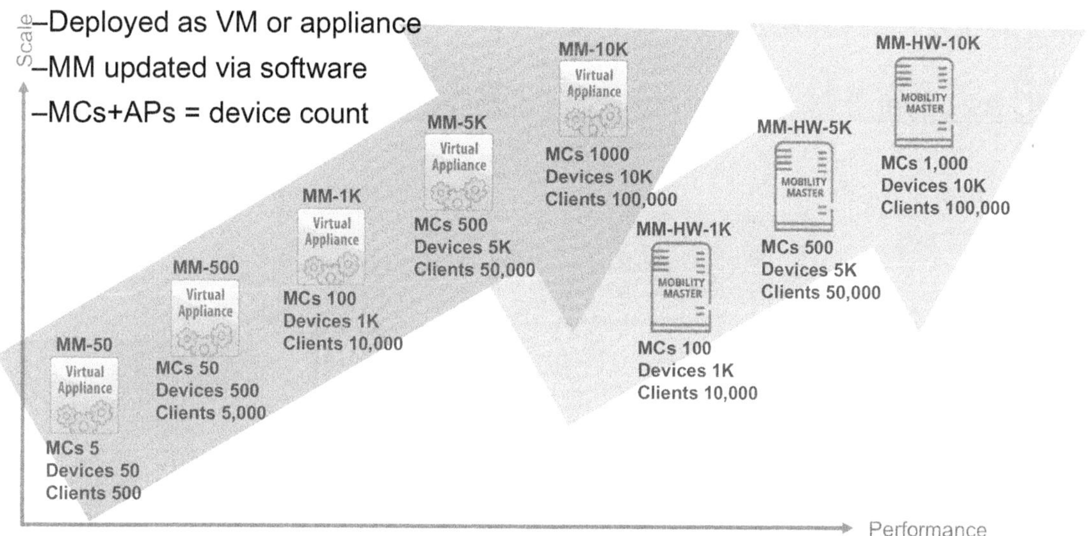

Figure 3-7 Mobility Master Portfolio

MM and MC implementation

You will now review guidelines for implementing MMs and MCs.

MC as Layer 2 switch or Layer 3 router

As Figure 3-8 shows, you can use the MC as a switch with simple VLANs or as a router that provides default gateway services.

If the default gateway for a VLAN is the uplink router or routing switch, the MC will operate at Layer 2. It will switch the user traffic to the Layer 3 switch. If an MC's IP addresses on a VLAN is the VLAN's default gateway IP address, then the MC will act at Layer 3. It will route all user traffic to the uplink routing switch. Layer 3 functionality on the MC is rarely used. However, the MC is capable of acting as a router for VLANs on which it has IP addresses without extra configuration. If the MC is routing, then implementers must configure it with static routes or Open Shortest Path First (OSPF).

What determines whether the MC operates at Layer 2 or Layer 3 is the default gateway IP address assigned to wireless clients. If the upstream routing switch has the default gateway IP address for a VLAN on the MC, then the MC is switching on that VLAN. You should define a trunk between the

MC and the upstream switch. (An ArubaOS switch does not use the "trunk" term for a link that carries tagged VLANs; the link is simply set up to support the appropriate tagged VLANs.) If the default gateway address for a VLAN is defined on the MC, the MC is routing.

You could set up one VLAN to switch and one VLAN to route. Routing or switching all depends on where the default gateway resides.

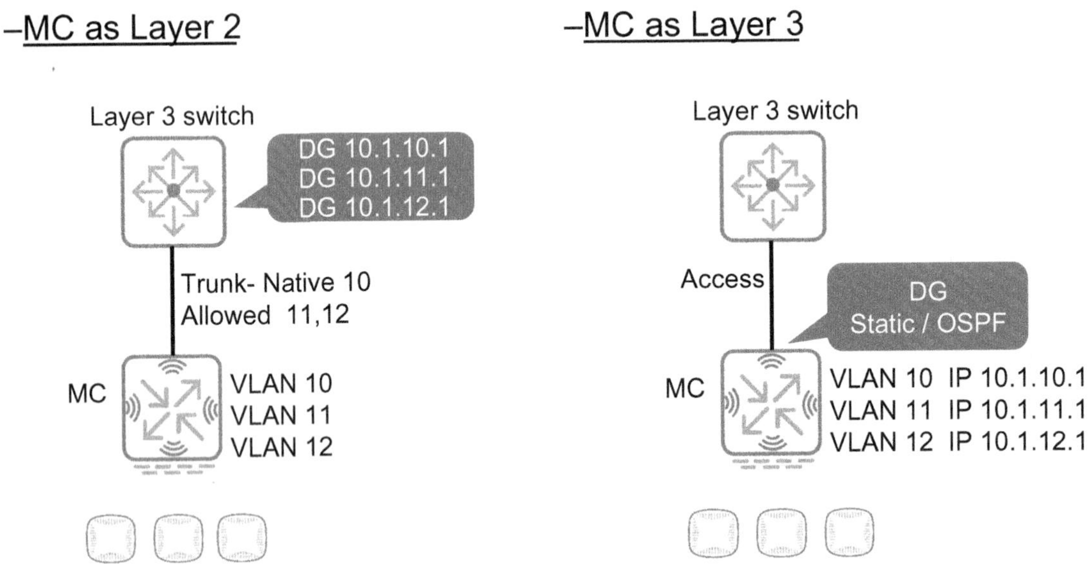

Figure 3-8 MC as Layer 2 switch or Layer 3 router

MC data termination point

You should determine where to deploy the MCs based on where most of the traffic is destined. When a major percentage of traffic is destined for the data center, then the data center or campus core is an appropriate location to install the MCs. All traffic from the APs will be sent via GRE and tunnels back to the MCs. The MCs will then decrypt, firewall, and switch or route the traffic.

Figure 3-9 shows a common deployment for customers that have a centralized system.

Campus networks

Campus deployments are networks that require more than a single active controller to cover a contiguous space. Examples of campus deployments are corporate campuses, large hospitals, and higher-education campuses. In these deployments, the WLAN is often the primary access method for the network, and it is typically used by multiple classes of users and devices.

When you deploy the MC centrally, you will deploy APs in a campus mode on all floors of all buildings for the entire campus. The MM is a VM installed on a server or an appliance that is a pre-configured server. The most likely location of the MM will be the datacenter.

The management layer consists of AirWave or Central to provide a single point of management for the WLAN, including reporting, heatmaps, centralized configuration, and troubleshooting.

Aruba's ClearPass Policy Manager provides role and device-based network access control for employees, contractors, and guests across any multivendor wired, wireless, and VPN infrastructure.

With a built-in context-based policy engine, RADIUS, TACACS+ protocol support, device profiling and comprehensive posture assessment, onboarding, and guest access options, ClearPass is unrivaled as a foundation for network security in any organization.

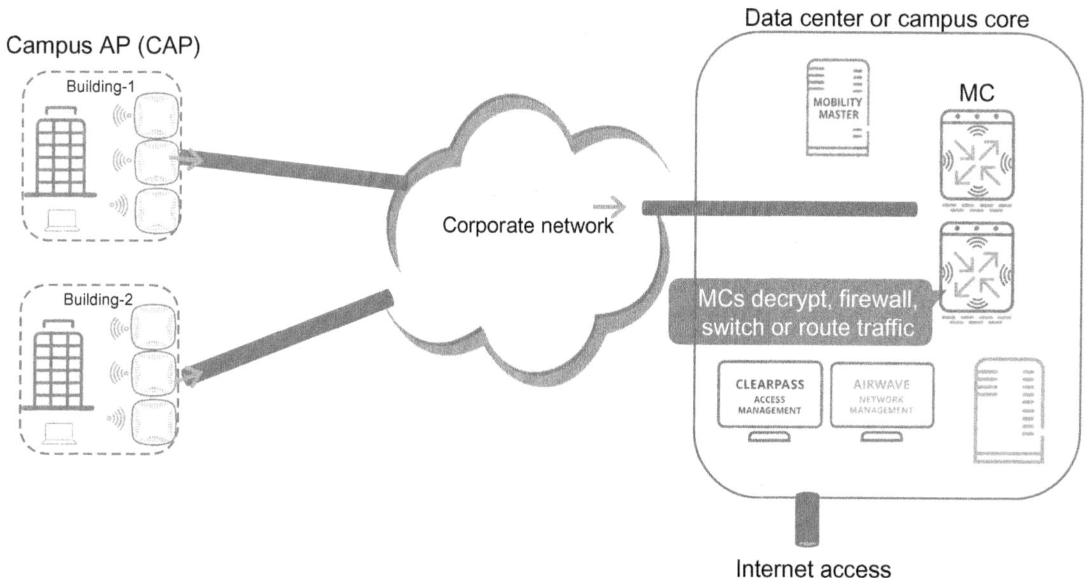

Figure 3-9 MC Data termination point

MCs in a cluster

If you collocate all MCs, then you can easily implement a cluster. Deploying MCs as managed devices in a cluster configuration has many advantages, including live cluster upgrades, client state sync, stateful failover, AP active load balancing, client load balancing, and seamless roaming of clients (see Figure 3-10).

All APs and users are load balanced on the cluster. The cluster MC then decrypts, firewalls, and switches or routes the packet. Roaming is simplified because all user traffic is sent to the MC where the user information lies, regardless of which AP the user roams to.

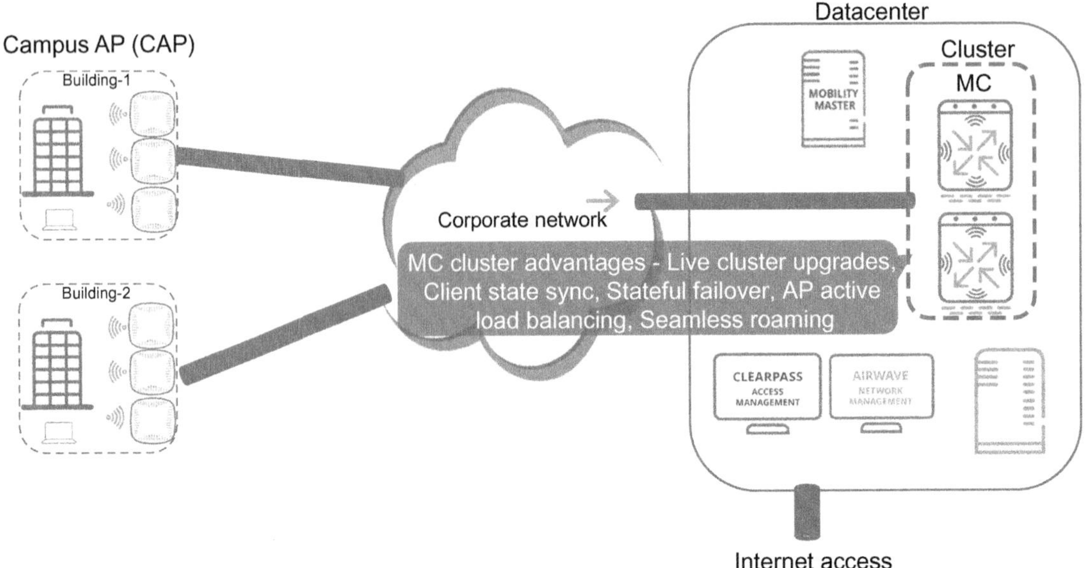

Figure 3-10 MCs in a cluster

With seamless campus roaming, clients remain anchored to a single controller in the cluster throughout associated time even while roaming on campus no matter which AP they connect to. This makes their roaming experience seamless because their Layer 2/Layer 3 information and sessions remain on the same MC.

Thanks to full redundancy within the cluster, in the event of a cluster member failure, the connected clients are failed over to a redundant cluster member without disruption to their wireless connectivity, or their existing high-value sessions.

As mentioned earlier, the clients are automatically load balanced within the cluster. When clients are moved among cluster members, the move is done in a stateful manner.

AP load balancing is a new feature in AOS8.1. The APs are automatically load balanced across cluster members.

The clustering feature is only available in deployments with a Mobility Master (MM).

The cluster members can only be Mobility Controllers (MCs). In other words, at this time Mobility Masters are not supported in a cluster.

There is no license required to turn on the clustering feature.

Cluster members can terminate both Campus APs (CAPs) and Remote APs (RAPs).

MC local

With the MC local model, APs will be deployed in campus mode on all floors of all buildings for the entire campus, as Figure 3-11 shows. The MC, or MCs, will be in the same building. This keeps the traffic local and also improves site survivability. If for some reason the building loses access to the corporate network, the MC can still provide local access.

A branch office can be installed using Zero Touch Provisioning (ZTP). If a branch office has Internet access, then it communicates with Activate and learns the IP address of the MM. The MM will then download a configuration to the MC.

ZTP automates deployment of the managed devices, giving customers a plug-n-play deployment. The managed device now learns the required information from the network and provisions itself automatically.

The MM communicates with the Activate server to obtain a whitelist of managed devices, configuration nodes mapping to the devices, the controller model, and (optional) VPN concentrator information. You can also enter this information manually as part of the configuration device command that is used to add devices to a configuration hierarchy. The MM validates the end devices with the whitelist and pushes the configuration, based on the device-configuration node mapping.

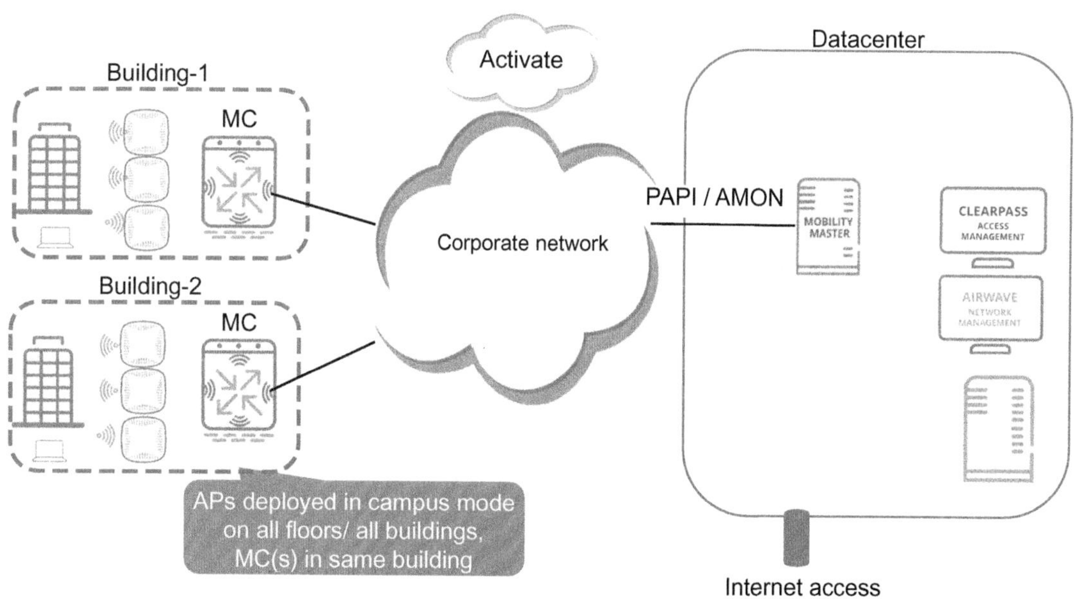

Figure 3-11 MC local

Licenses

You will now focus on the licensing requirements for the solutions you recommend.

Licensing in AOS 8.x

Licensing requirements depend on the controller, AP counts, as well as other features. In AOS 8.X the MM has the central pool of licenses, but you can also allocate dedicated pools to devices in the hierarchy.

Different license capacities are required for the MM. You should review each license type to determine if the features and functionality meet the goals of the organization. With that information it is possible to determine the required feature licensing levels.

MMs must manage the functionality for all other platforms. Licensing unlocks configuration capabilities on the system. However, MMs will not terminate APs or devices.

Types of licenses

Each license can be either an evaluation or permanent license. As the name suggests, a permanent license permanently enables the desired software module on a specific Aruba controller. You obtain permanent licenses through the sales order process only. Permanent software license keys are sent to you via email.

An evaluation license allows you to evaluate the unrestricted functionality of a software module on a specific controller for 90 days, in three 30-day increments. Subscription licenses are a feature-specific license and should be installed for each additional feature as per your requirement. The Web Content and Classification (WebCC) license is a subscription license that enables WebCC features only for the duration of the subscription (1, 3, 5, 7, or 10 years). The subscription time period starts from the time the license key is generated from the licensing website.

The MM maintains the centralized licensing pool. An expired evaluation license will be nonfunctional but remain in the license database until the controller is reset using the command **write erase all** which removes all license keys.

To determine the remaining time on an evaluation license, select the Alert flag ( ) in the WebUI titlebar. The WebUI displays information about evaluation license status. When an evaluation period expires, the controller automatically backs up the startup configuration and reboots itself at midnight (according to the system clock). All permanent licenses are unaffected. The expired evaluation licensed feature is no longer available and will display as Expired in the WebUI.

Thirty days before the license period expires, an alert appears in the banner in the MM WebUI, warning the user that the license is ready to expire. After the license expiration date is passed, the license continues to operate as an active license for an extended grace period of 120 days. After this final grace period elapses, the license permanently expires.

License SKUs

There are several types of licenses within the SKU that you will install on different controllers or the MM.

You install the MM license on the MM and the VMC license on the VMC or the MM (see Table 3-3). You install the AP, PEF RFProtect licenses on either the MM, standalone VMC, or standalone master. Where you install licenses depends on your deployment.

WebCC is a subscription license that works with Brightcloud.

The LIC-ENT SKU is a combination of the AP, PEF, RFP, and AirWave flex licenses. This will generate four separate licenses.

Table 3-3 License SKUs

SKU	Consumption	Installed on:
LIC-MM-VA-xx	MM Platform	VMM
LIC-MC-VA-xx	VMC Platform	VMC or MM if VMC is an MD
AP/PEF/RFP	Flex License	MM or Master or Standalone VMC/Master
VIA (8.2)	Per user	MM or Master or Standalone VMC/Master
PEFV (8.0-8.1)	Per controller	MC/VPN
WebCC	Subscription-based LIC Brightcloud Malicious IP/URL	MM

The MM license is based on the cumulative count of the total number of MCs, VMCs, and all APs in the network. If there are VMCs in the network, then you would install a VMC license, based on the count of each AP on the VMCs, on the MM. If the VMC is a standalone, then the VMC license is the total number of APs, the license you would install the license on the VMC.

The AP, PEF, and RFP licenses are based on the count for each AP in the network. If you have an MM, then you would install these licenses on the MM. For a standalone VMC or controller, you can install licenses directly on the standalone controller.

The PEFV license, pre 8.2, enables VPN/VIA users. This license allows the system to create user roles that are applied to the VPN tunnel.

The VIA license, starting at 8.2, is based on the number of VIA users and is the license you will install on the MM.

WebCC is a subscription-based license that works with Brightcloud. This is a threat Intelligence Service that tracks IP reputation, IP Geolocations, WEB URL Classification, and WEB URL Reputations. You can install the WebCC license on the MM.

Note the AP, PEF, and RFP are flex licenses you can order in any amount. The MM and VMC licenses are platform licenses you must order in increments.

License SKUs—VMM

The MM uses these main types of licenses: the MM, VMC, AP, PEF, and RFProtect licenses. You can install all of these licenses on the MM (see Figure 3-12).

If you have an MM, then the AP, PEF, and RFP licenses are based on the count of all APs on the VMCs.

If there are VMCs in the network, then you can install a VMC license on the MM, based on the count of all the APs on the VMCs.

The MM license is based on the cumulative count of all the number of MCs, all the number of VMCs and all the number of APs in the network.

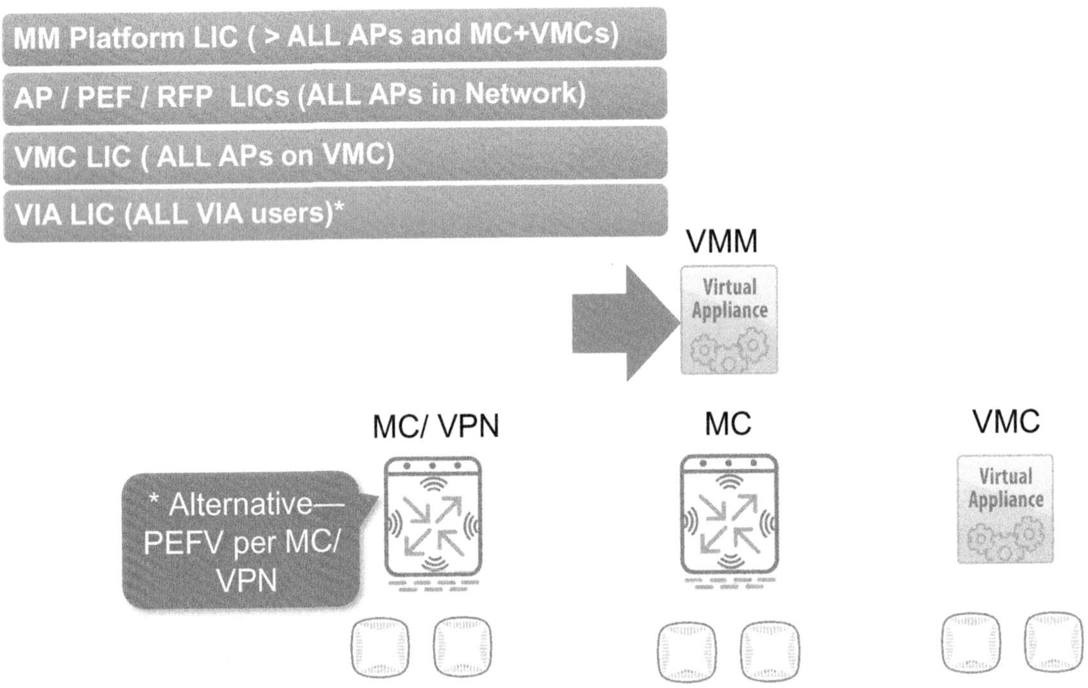

Figure 3-12 License SKUs—VMM

If the customer has remote workers who will use VIA, install one VIA license per VIA user. Each VIA license enables an MC managed by the MM to terminate a VPN connection for one user. The VIA licenses can be transferred from one MC to another as necessary. Alternatively, you can install a PEFV license on each MC that you want to terminate VPN connections for VIA users. In this case, the license is tied to the controller and its capacity for supporting users; the PEFV license cannot be transferred to another MC.

Remember that the AP, PEF, and RFP are FLEX licenses. If you have 123 APs, you can get a license for 123 APs.

The MM and VMC licenses are platform licenses.

The MM license LIC-MM-VA-xx is selected based on the count of all APs all MDs in the network. The MM licenses are purchased in blocks of 50, 500, 1000, 5000, or 10000.

The virtual MCs have licenses in blocks of 50, 250, and 1000.

All these licenses are cumulative. As an example, you could combine two MM licenses of 50 to give you a count of 100.

You can also install the VMC license LIC-MM-VA-xx on the MM for all APs on the VMCs.

License SKUs—Hardware Mobility Master (HMM)

If you are running an MM in the hardware appliance, then the MM license is preinstalled. You still need to add the AP, PEF, and RFP licenses based on the total count of APs (see Figure 3-13).

If the network includes VMCs, then you must add a VMC license on the MM, based the count for each AP on the VMCs.

The MM license LIC-MM-VA-xx will be built-in. There is no need to install it.

The AP/PEF/RFP licenses match the AP count in network. This includes all APs on 7000 or 7200 Series controllers and VMCs.

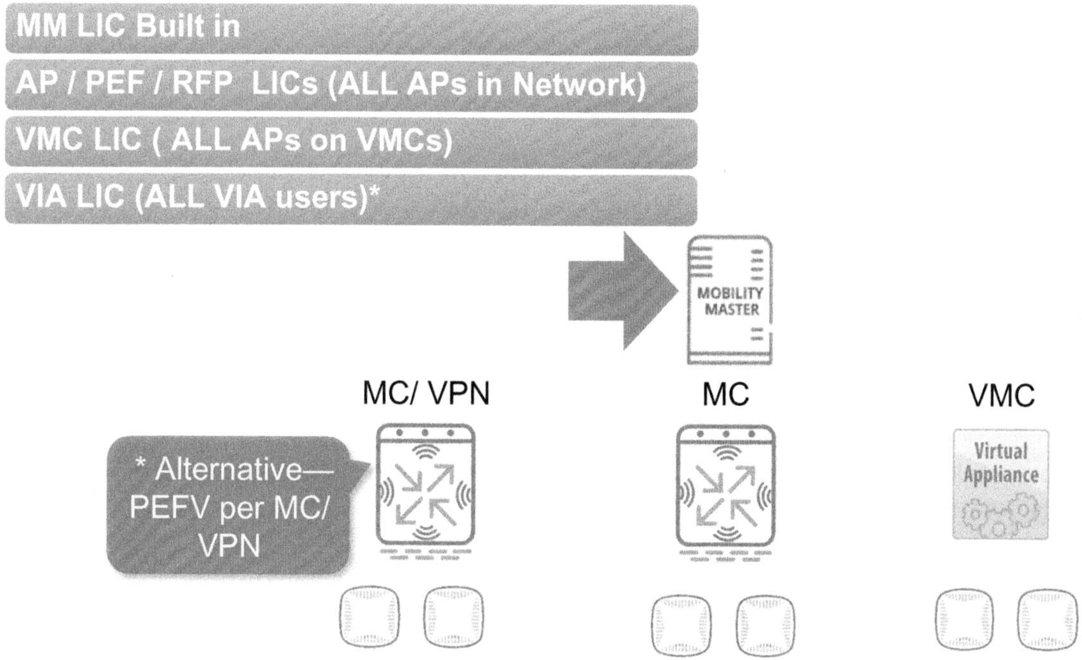

Figure 3-13 License SKUs—Hardware Mobility Master (HMM)

License SKUs for VMC

If you have a standalone VMC, then the VMC license is counted for all APs on the VMC. This is a platform license, so it must be greater than all APs reporting to the VMC (see Figure 3-14).

AP, PEF, and RFP are flex licenses and are based on the total count of APs.

Running an 8.X network with no MM but retaining the master local structure would require AP, PEF, and RFP licenses. In this case, the master is referred to as Mobility Controller Master. The Mobility Controller Master or standalone controller needs the AP, PEF, and RFProtect licenses to match the AP count.

The Master Local setup is not supported as of OS 8.2

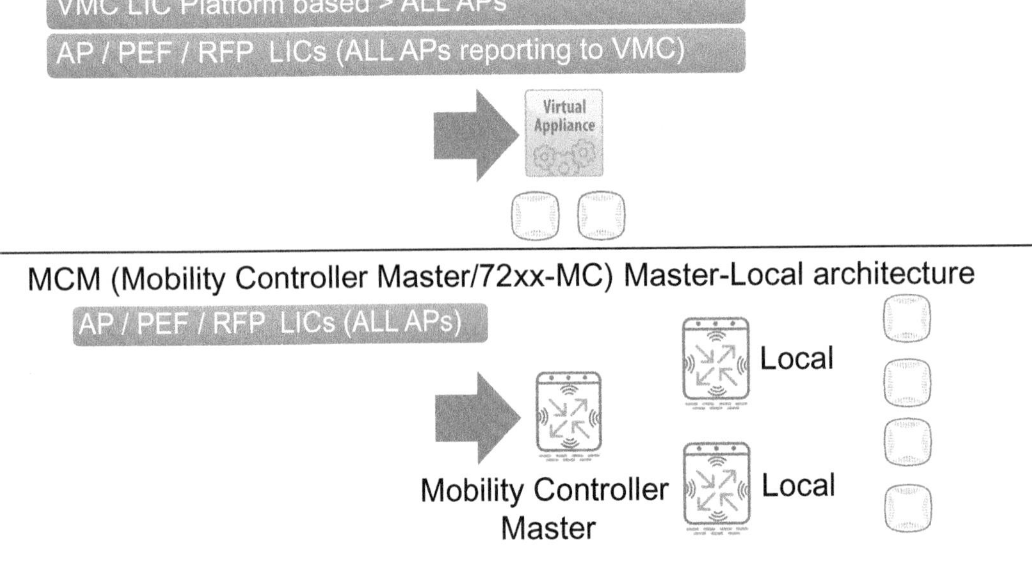

Figure 3-14 License SKUs for VMC

Calculating licensing requirements

Calculating license requirements means counting the anticipated required number of APs and controllers and estimating the need for each license type. For example, suppose you are planning to deploy 20 Mobility Controllers (MCs) along with 200 APs and 100 APs terminating on 10 VMCs. In total you have 30 controllers and 300 APs. The license count for the MM would be 330 MM licenses, including 300 APs and 30 controllers (see Figure 3-15).

You would also need AP, PEF, and RFProtect licenses equaling 300.

Because there are 10 VMCs with a total of 100 APs you must also install a VMC license for 100.

In conclusion, you would install on the MM 1 MM license of 330, 1 VMC license of 100 and AP, PEF and RFProtect licenses of 300.

Note

You cannot purchase 330 MM licenses. MM licenses are purchased in increments of 50, or you can purchase a license of 500, which would prove more economical.

The WebCC and PEFV licenses are optional. While the PEF and RFP licenses are also optional, Aruba highly recommends them. Instead of separate AP, PEF, and RFP licenses, you can propose Enterprise licenses, which include those three licenses, as well as an LIC-AW for AirWave.

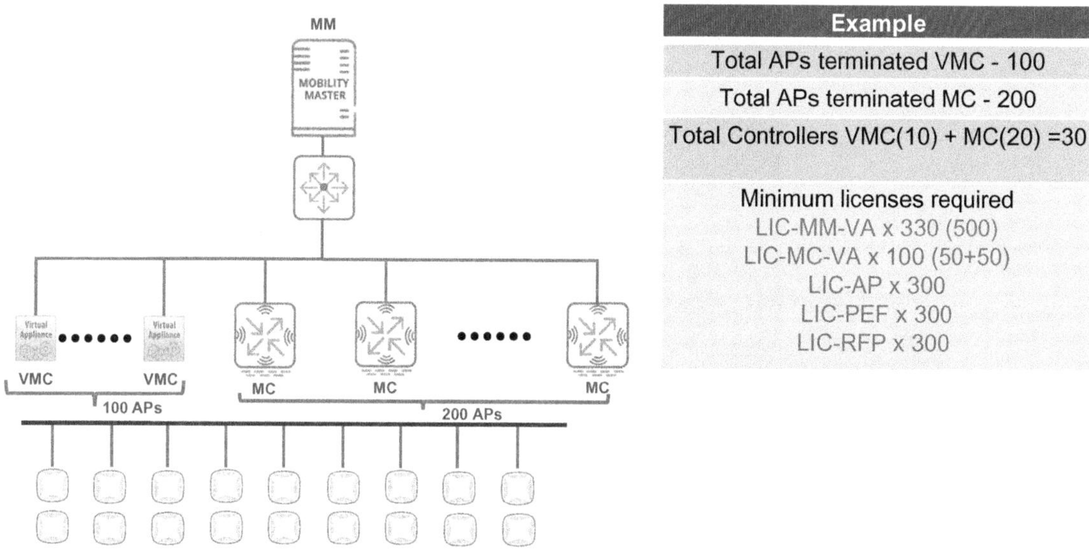

Figure 3-15 Calculating licensing requirements

Dedicated license pool

As Figure 3-16 shows, the MM uses the global pool of licenses to distribute licenses to a large number of managed devices across geographic locations. By default, all MCs associated to an MM share a single global pool of all the sharable licenses added to that MM.

A dedicated pool allows you to create additional licensing pools at a configuration node, allowing a group of managed devices at or below that configuration level to share licenses among themselves, but not with other groups. This pool has a limit and any devices in that pool cannot pass this limit even if more license are available in the global pool. Dedicated pools provide a good option for customers with departments or business units (BUs) that are responsible for paying for licenses out of their individual budgets.

The ArubaOS base operating system contains many features and types of functionality that are needed for an enterprise WLAN network. Aruba uses a licensing mechanism to enable additional features and to enable AP capacity on controllers. By licensing functionality, organizations can deploy the network with the functionality to meet their specific requirements in a flexible and cost-effective manner.

The MM is the centralized license server that is responsible for the global pool of licenses.

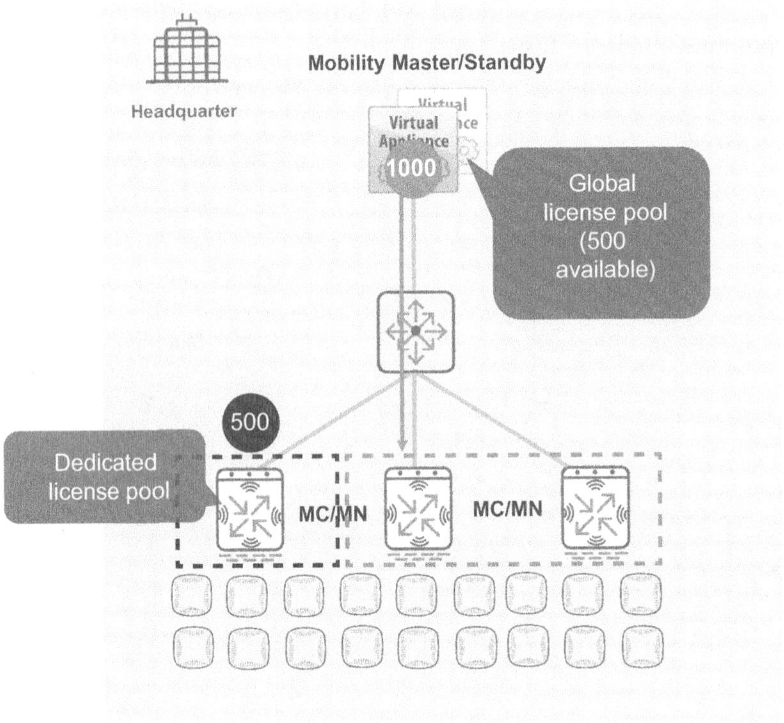

Figure 3-16 Dedicated license pool

Instant AP

You will now review Instant APs.

IAP clusters

IAPs are a set of APs that do not need a controller. All IAPs in a subnet form a cluster and elect one of the IAPs to be the Virtual Controller (see Figure 3-17). The maximum number of IAPs Aruba recommends in a cluster is 128.

IAPs have many of the same capabilities as a controller. An IAP has ARM, client match, spectrum analysis as well as a stateful firewall with roles and policies. IAPs have access to cloud-based content filtering and can categorize user devices and display applications used in the cluster. IAPs also have IDS/IPS for network protection.

You can configure multiple WLANs with 802.1X Auth or MAC Auth or with an internal captive portal. The IAP cluster has a GUI, but managing several clusters becomes difficult. Therefore, Aruba recommends using a management system such as AirWave or Central. Activate is a cloud-based system that can automatically direct the IAPs to AirWave or Central. A new IAP cluster directed to AirWave or Central can receive all of its configuration.

You can convert an IAP into a CAP that controllers manage. If your customer's network grows and requires need controllers, they will not lose their AP investment.

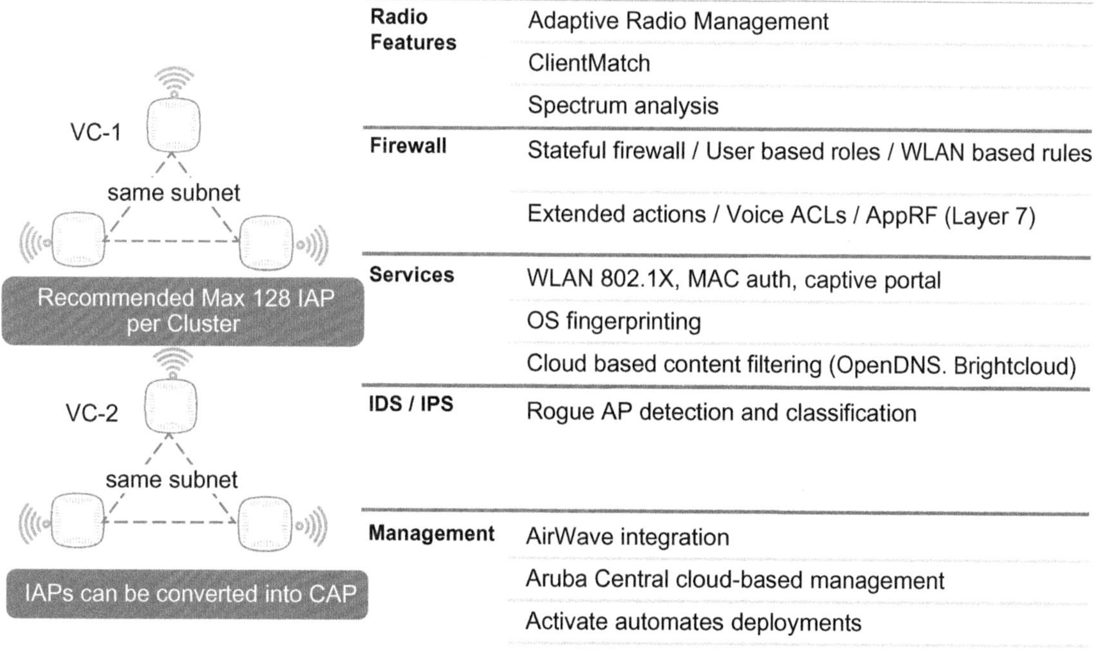

VC-1
same subnet
Recommended Max 128 IAP per Cluster
VC-2
same subnet
IAPs can be converted into CAP
Radio Features
Adaptive Radio Management
ClientMatch
Spectrum analysis
Firewall
Stateful firewall / User based roles / WLAN based rules
Extended actions / Voice ACLs / AppRF (Layer 7)
Services
WLAN 802.1X, MAC auth, captive portal
OS fingerprinting
Cloud based content filtering (OpenDNS. Brightcloud)
IDS / IPS
Rogue AP detection and classification
Management
AirWave integration
Aruba Central cloud-based management
Activate automates deployments

Figure 3-17 IAP clusters

Ease of deployment

One of the key challenges in deployments that include multiple interconnected buildings is the placement of APs. Aruba addresses this challenge with Aruba Activate, which is a cloud-based, zero-touch provisioning system. Aruba Activate provides plug-and-play capability to an Aruba Instant cluster, which allows rapid deployment of Aruba Instant clusters with minimal or no IT expertise.

An IAP or UAP will communicate with the cloud-based Activate server. If the Activate server has a setup configuration for this AP, it sends a message back to the AP. This message will include its setup instructions.

As Figure 3-18 shows, the AP may be directed to a Central cloud-based management system to receive its configuration and after that be managed. Or, the AP may be directed to the AirWave management system. From here, it will receive its configuration and be managed. The AP can also be converted into a Campus AP (CAP) and directed to an MC.

If you need remote access, you can convert the AP to a RAP. It will then be directed to an MC with VPN services.

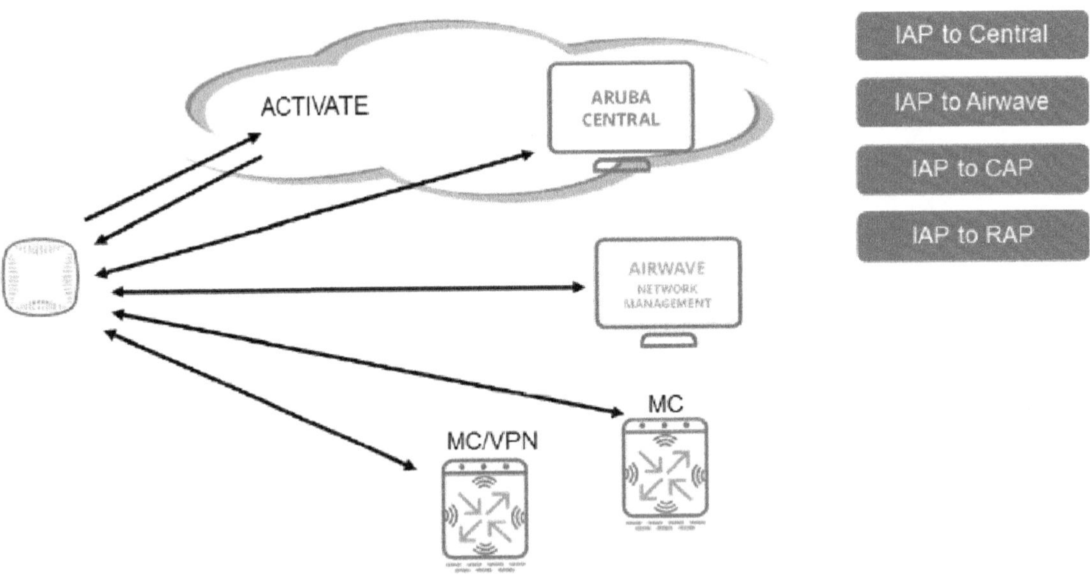

Figure 3-18 Ease of deployment

Setup procedure for UAPs

UAPs have a new setup procedure. First, the UAP searches for a controller with the five standard methods, using the static configuration, or option 43 in a DHCP response. If there isn't static information or an option 43, then it'll attempt a broadcast multicast and DNS query.

If the AP cannot discover a controller, it will listen for IAP master beacons. If it receives a master beacon, the AP will join that Instant cluster.

If it cannot locate an Instant cluster, the AP tries to connect to an AirWave server (if one was provided via DHCP/DNS options). The AP receives its image and configuration from AirWave and joins the appropriate Instant cluster.

If an AirWave server is not found, the AP connects to Activate and looks for a provisioning rule on Activate. If the AP finds a provisioning rule to connect to an AirWave server, it will attempt connection to AirWave and receive its image and configuration accordingly. If the AP finds a provisioning rule to connect to Central, it will communicate with Central and receive its image and configuration. If the AP finds a provisioning rule to connect to a controller as a CAP (or RAP), it will attempt to communicate with the controller and receive its image and configuration.

Even if no provisioning rules are configured on Activate, the UAP will still upgrade its image to the latest available full Instant image. After a reboot, if nothing was found during the discovery process, the UAP becomes the VC and forms its own Instant cluster. It broadcasts an SSID called SetMeUp-xx:xx:xx (where xx:xx:xx are the last hex octets of the UAP MAC address).

After it broadcasts the SSID, if there is no keyboard input or an active WebUI session for 15 minutes, the UAP will reboot and start the discovery process again.

Upon connecting to the SetMeUp- xx:xx:xx SSID, you are presented with the Instant web UI where you can perform the usual Instant-related tasks. Other UAPs that are subsequently brought up will discover this Instant cluster and join it. This ends the discovery process.

There might be scenarios where the UAP does not have Internet connectivity and cannot reach Activate to upgrade its manufacturing image to a full Instant image. In this case, when no controller-based or controller-less networks are discovered, the UAP broadcasts the SSID SetMeUp-xx:xx:xx. However, connecting to this SSID will give you a provisioning web UI and not the Instant web UI. Using this UI, you can upgrade the UAP image to AOS or InstantOS, or perform maintenance and debugging tasks.

As mentioned earlier, if there is no keyboard input or an active WebUI session for 15 minutes after the SSID is broadcast, the UAP will reboot and start the discovery process again.

IAP or controller

In a MC and CAP environment, the client encrypts all traffic and the AP sends it to the MC in a GRE tunnel. The MC receives the frame and decrypts, firewalls and then switches or routes the frame. This is useful in a centralized environment where all traffic is destined for a server farm because all the traffic is encrypted from the client to the MC.

The IAPs receive the client's traffic encrypted and then decrypts and firewalls the frame. The frame is then be placed on the IAP wired port as a standard 802.3 frame. IAPs bridge all traffic locally.

The network admin will need to configure each switch port where an IAP is connected to have the right VLANs (see Figure 3-19).

One advantage of a controller-based system is MC clustering. MC clustering has client state sync, which allows for easy client failover and client roaming. The MC cluster also has the capabilities of seamless upgrades.

The IAPs have roaming capabilities, within an IAP cluster and between IAP clusters. AirMatch calibrates the MC network and has many RF customizable parameters. This is helpful in an RF challenged environment. The IAPs each use ARM to calibrate themselves with their RF environment.

Both the MC and the IAP have mesh capabilities. The MC also includes several customizable parameters if needed.

An MC is required if you are terminating VIA or Branch controllers with site to site. You can configure the IAP cluster with a VPN to a MC VPN giving corporate access.

The cost of an IAP network is less than an MC network. All you are purchasing are APs and no MCs and no licenses.

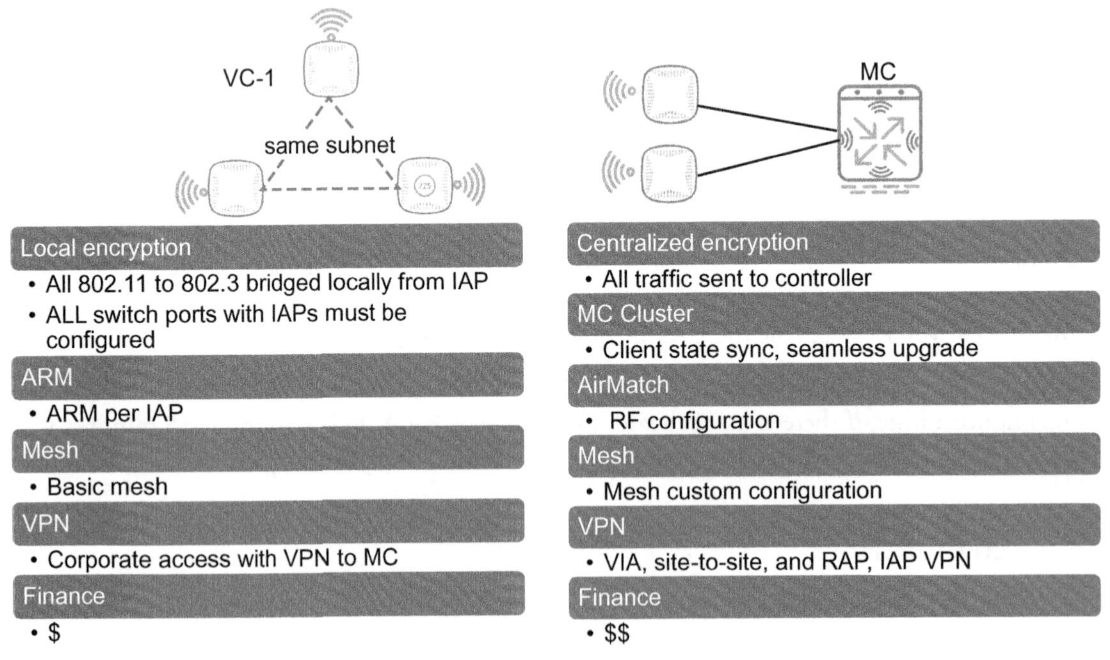
VC-1
same subnet
MC
Local encryption
• All 802.11 to 802.3 bridged locally from IAP
• ALL switch ports with IAPs must be configured
ARM
• ARM per IAP
Mesh
• Basic mesh
VPN
• Corporate access with VPN to MC
Finance
• $
Centralized encryption
• All traffic sent to controller
MC Cluster
• Client state sync, seamless upgrade
AirMatch
• RF configuration
Mesh
• Mesh custom configuration
VPN
• VIA, site-to-site, and RAP, IAP VPN
Finance
• $$

Figure 3-19 IAP or controller

Activity 3: Campus design

In this activity, you will continue to design the network for the customer scenario that was introduced in Activity 1.

In this activity, you will decide how many controllers you need for the customer scenario and where they should be located in the campus. You will also recommend what licenses the customer will need.

To complete this activity, use the information about the scenario provided in Activities 1 and 2 plus the additional information provided below.

Scenario campus design

- There are two closets on every floor. Every closet has extra rack space.
- The datacenter also has extra rack space.
- The old Wi-Fi structure had one small controller on every floor.
- The two buildings are approximately 50 feet (15.2 m) apart with underground fiber linking both buildings together. These links have never failed.
- There are approximately 900 employees on the main campus and another 350 in the various warehouses.
- Most of the corporate traffic is destined for the data center.
- The data center also has a firewall gateway for Internet access as well as a small DMZ.

Corp1 requirements

Corp 1 is replacing all its old APs for the new Aruba installation.

The security department is very interested in the Aruba firewall possibilities and RF protection.

Activity 3.1: Controller design

Activity 3.1 objectives

You will recommend the Mobility Controller(s) you need for the AP count you decided upon in the previous activity. Like Activity 2, this activity was written to be used with Iris. If you do not have access to Iris, however, you can use the HPE Networking Online Configurator. (Visit http://hpe.com/networking/configurator to access this tool.)

Activity 3.1 steps

1. What was your AP count?
   a. Total number of APs: ___________
   b. Total number of APs in Building 1: ___________
   c. Total number of APs in Building 2: ___________
2. What is your total number of wireless client devices? ___________
3. How many controllers are you recommending? ___________
4. Which models of controller(s) are you recommending, and why?
5. What is your recommended location of each controller, and why?
6. What is your recommended setup for the controllers?
   a. Controllers in L2 mode (Y/N): ___________
   b. Controllers in L3 setup (Y/N): ___________
   c. Controllers in Cluster setup (Y/N): ___________
   d. Justify your recommendation.
7. Will you recommend an MM? Explain.
8. You will now add the MC or MCs to your BOM in IRIS or the HPE Networking Online Configurator. (You used one of these tools to complete Activity 2.)
   a. Return to the IRIS application. If you closed the application, make sure to open the site that you already started.
   b. In the Catalog, expand **HPE Networking > Aruba > Mobility Controllers > Campus**.
   c. Select the MC series. Double-click to add it to your topology (see Figure 3-20).

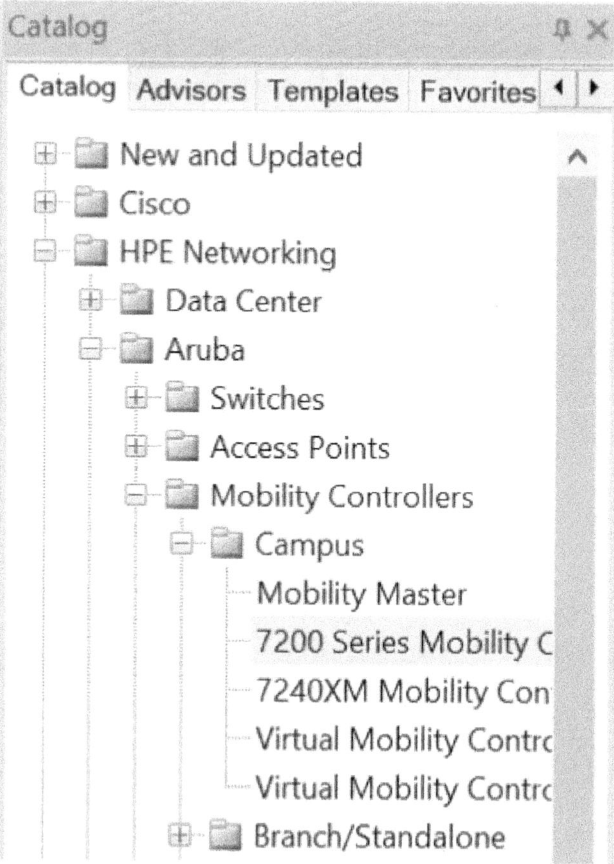

Figure 3-20 Mobility Controller Series in Iris

d. In the first tab of the Properties window, choose your region under Filter by Region.

e. Then select the model that you chose earlier. (Figure 3-21 shows an example, but select the model that you chose.)

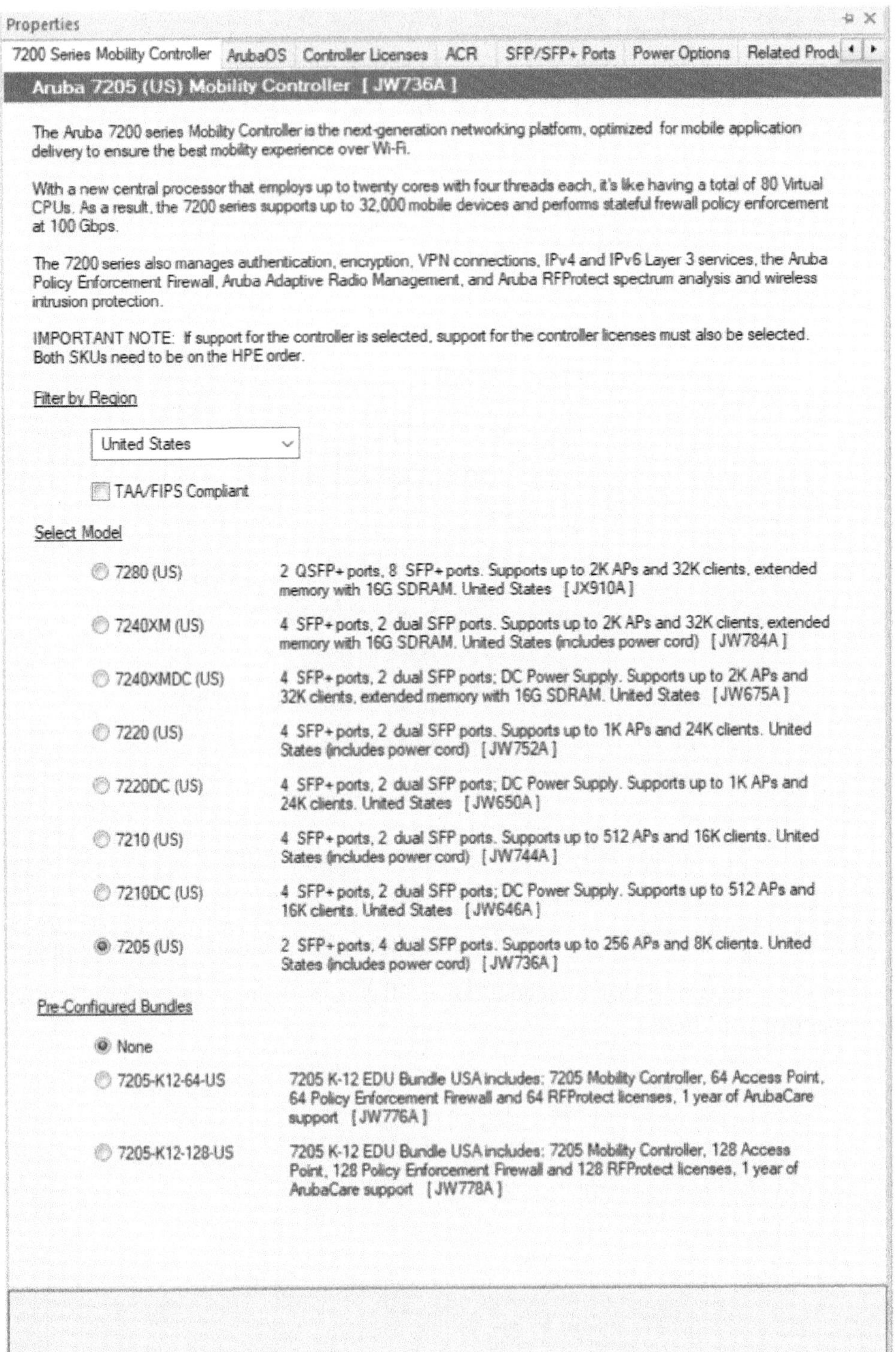

Properties
7200 Series Mobility Controller
ArubaOS
Controller Licenses
ACR
SFP/SFP+ Ports
Power Options
Aruba 7205 (US) Mobility Controller [JW736A]
The Aruba 7200 series Mobility Controller is the next-generation networking platform, optimized for mobile application delivery to ensure the best mobility experience over Wi-Fi.
With a new central processor that employs up to twenty cores with four threads each, it's like having a total of 80 Virtual CPUs. As a result, the 7200 series supports up to 32,000 mobile devices and performs stateful firewall policy enforcement at 100 Gbps.
The 7200 series also manages authentication, encryption, VPN connections, IPv4 and IPv6 Layer 3 services, the Aruba Policy Enforcement Firewall, Aruba Adaptive Radio Management, and Aruba RFProtect spectrum analysis and wireless intrusion protection.
IMPORTANT NOTE: If support for the controller is selected, support for the controller licenses must also be selected. Both SKUs need to be on the HPE order.
Filter by Region
United States
TAA/FIPS Compliant
Select Model
7280 (US)
2 QSFP+ ports, 8 SFP+ ports. Supports up to 2K APs and 32K clients, extended memory with 16G SDRAM. United States [JX910A]
7240XM (US)
4 SFP+ ports, 2 dual SFP ports. Supports up to 2K APs and 32K clients, extended memory with 16G SDRAM. United States (includes power cord) [JW784A]
7240XMDC (US)
4 SFP+ ports, 2 dual SFP ports; DC Power Supply. Supports up to 2K APs and 32K clients, extended memory with 16G SDRAM. United States [JW675A]
7220 (US)
4 SFP+ ports, 2 dual SFP ports. Supports up to 1K APs and 24K clients. United States (includes power cord) [JW752A]
7220DC (US)
4 SFP+ ports, 2 dual SFP ports; DC Power Supply. Supports up to 1K APs and 24K clients. United States [JW650A]
7210 (US)
4 SFP+ ports, 2 dual SFP ports. Supports up to 512 APs and 16K clients. United States (includes power cord) [JW744A]
7210DC (US)
4 SFP+ ports, 2 dual SFP ports; DC Power Supply. Supports up to 512 APs and 16K clients. United States [JW646A]
7205 (US)
2 SFP+ ports, 4 dual SFP ports. Supports up to 256 APs and 8K clients. United States (includes power cord) [JW736A]
Pre-Configured Bundles
None
7205-K12-64-US
7205 K-12 EDU Bundle USA includes: 7205 Mobility Controller, 64 Access Point, 64 Policy Enforcement Firewall and 64 RFProtect licenses, 1 year of ArubaCare support [JW776A]
7205-K12-128-US
7205 K-12 EDU Bundle USA includes: 7205 Mobility Controller, 128 Access Point, 128 Policy Enforcement Firewall and 128 RFProtect licenses, 1 year of ArubaCare support [JW778A]

Figure 3-21 Mobility controller model in Iris

f. If you are planning more than one MC, specify the number. (Select the Attributes tab in the Properties window. Enter the number in the Quantity Multiplier.)

9. If you are planning to deploy an MM, add it now.

   a. Double-click Mobility Master in the catalog (see Figure 3-22).

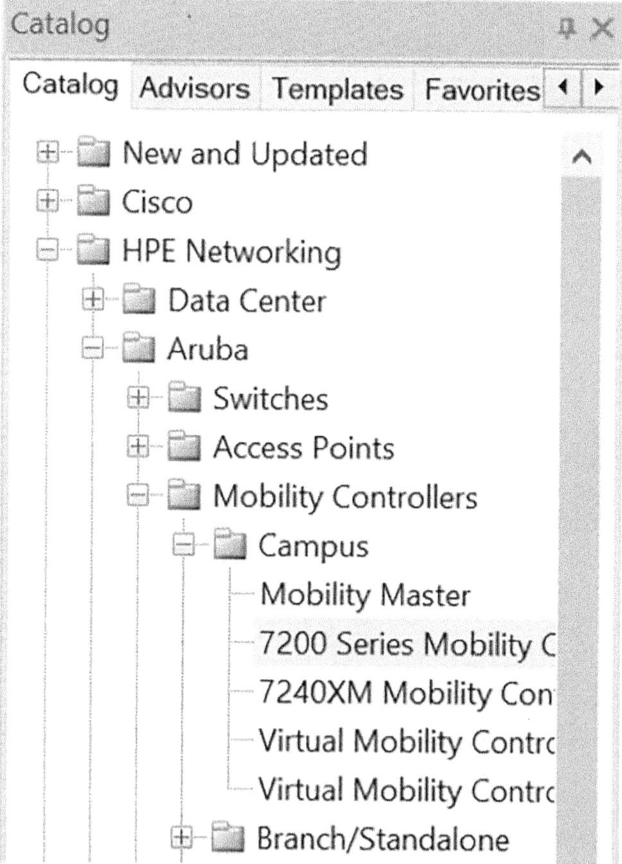

Figure 3-22 Mobility Master in Iris

   b. In the Properties window, select the MM virtual appliance or the hardware appliance of your choice.

Activity 3.2: Licenses

Activity 3.2 objective

You will recommend the licenses you need for this project.

Activity 3.2 steps

1. Based on your AP count and controller count, list the licenses you will need for this project.
2. Explain why you need these licenses and what they are used for:

3. Add the licenses in IRIS.

   a. Select the MM and look at its Properties window. You should be at the Aruba Mobility Master Appliance table.

   b. In the Total number of devices field, enter the number of controllers and APs. IRIS will automatically fill in the current number of MM-VA license bundles (see Figure 3-23).

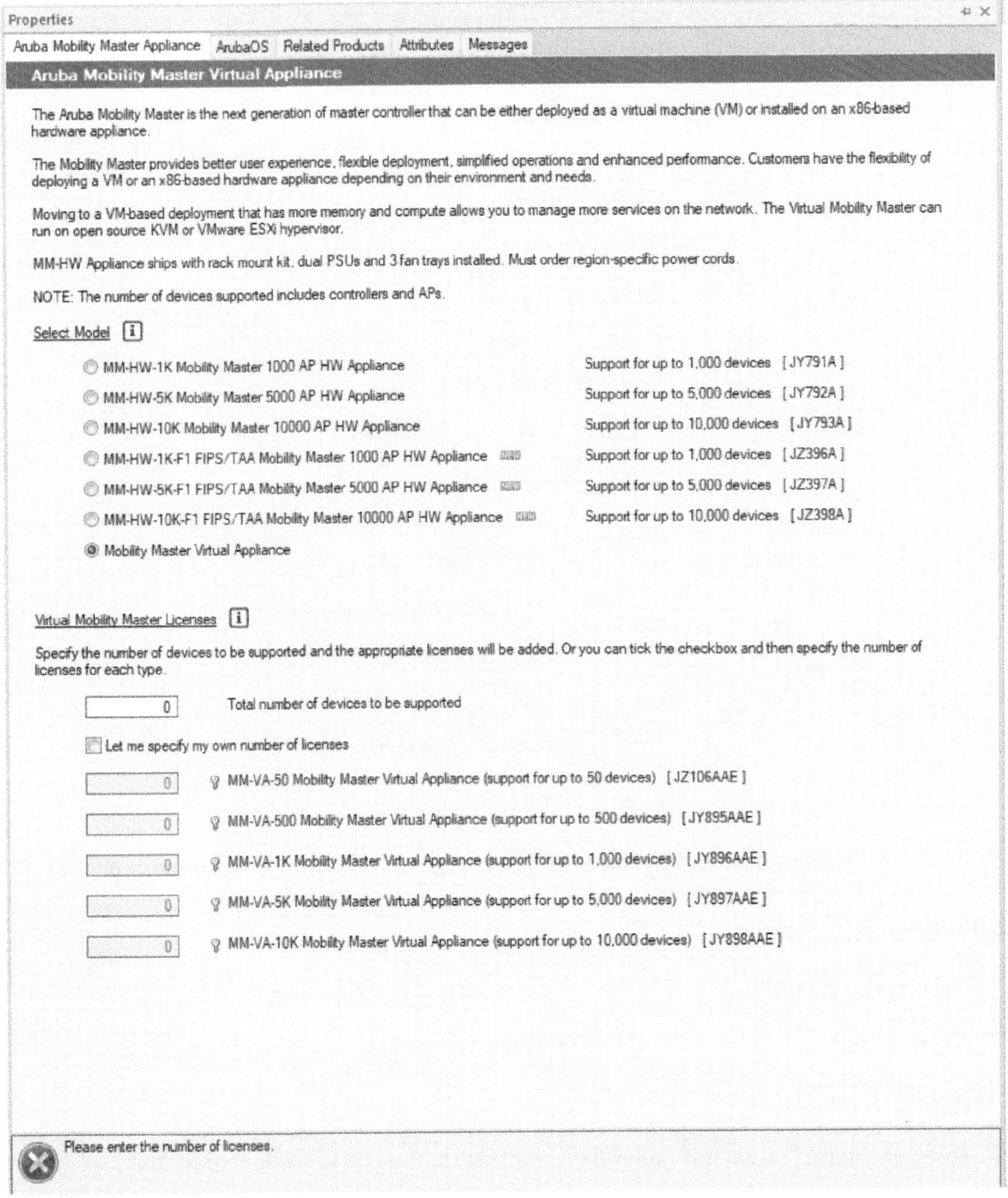

Properties

Aruba Mobility Master Appliance | ArubaOS | Related Products | Attributes | Messages

Aruba Mobility Master Virtual Appliance

The Aruba Mobility Master is the next generation of master controller that can be either deployed as a virtual machine (VM) or installed on an x86-based hardware appliance.

The Mobility Master provides better user experience, flexible deployment, simplified operations and enhanced performance. Customers have the flexibility of deploying a VM or an x86-based hardware appliance depending on their environment and needs.

Moving to a VM-based deployment that has more memory and compute allows you to manage more services on the network. The Virtual Mobility Master can run on open source KVM or VMware ESXi hypervisor.

MM-HW Appliance ships with rack mount kit, dual PSUs and 3 fan trays installed. Must order region-specific power cords.

NOTE: The number of devices supported includes controllers and APs.

Select Model

- MM-HW-1K Mobility Master 1000 AP HW Appliance — Support for up to 1,000 devices [JY791A]
- MM-HW-5K Mobility Master 5000 AP HW Appliance — Support for up to 5,000 devices [JY792A]
- MM-HW-10K Mobility Master 10000 AP HW Appliance — Support for up to 10,000 devices [JY793A]
- MM-HW-1K-F1 FIPS/TAA Mobility Master 1000 AP HW Appliance NEW — Support for up to 1,000 devices [JZ396A]
- MM-HW-5K-F1 FIPS/TAA Mobility Master 5000 AP HW Appliance NEW — Support for up to 5,000 devices [JZ397A]
- MM-HW-10K-F1 FIPS/TAA Mobility Master 10000 AP HW Appliance NEW — Support for up to 10,000 devices [JZ398A]
- Mobility Master Virtual Appliance

Virtual Mobility Master Licenses

Specify the number of devices to be supported and the appropriate licenses will be added. Or you can tick the checkbox and then specify the number of licenses for each type.

0 Total number of devices to be supported

Let me specify my own number of licenses

0 MM-VA-50 Mobility Master Virtual Appliance (support for up to 50 devices) [JZ106AAE]

0 MM-VA-500 Mobility Master Virtual Appliance (support for up to 500 devices) [JY895AAE]

0 MM-VA-1K Mobility Master Virtual Appliance (support for up to 1,000 devices) [JY896AAE]

0 MM-VA-5K Mobility Master Virtual Appliance (support for up to 5,000 devices) [JY897AAE]

0 MM-VA-10K Mobility Master Virtual Appliance (support for up to 10,000 devices) [JY898AAE]

Please enter the number of licenses.

Figure 3-23 Licenses in Iris

4. Refer back to your plan for licenses.
5. If you are using an MM, you will install most licenses on the MM. If you are not using an MM, move to the next step.
    a. Select the MM in the workspace.
    b. Select the ArubaOS tab in the Properties window.
    c. You can either bundle licenses together or specify the number of AP, PEFNG, and RFP licenses separately.
6. If you are planning to use WebCC, you must add licenses on MCs in IRIS even though the licenses can be pooled. Follow the steps below. Otherwise, you are done with the activity.
    a. Select the MC in the workspace.
    b. Select the **Controller Licenses** tab in the Properties window.
    c. Choose the subscription term and specify the number of APs per controller (see Figure 3-24).

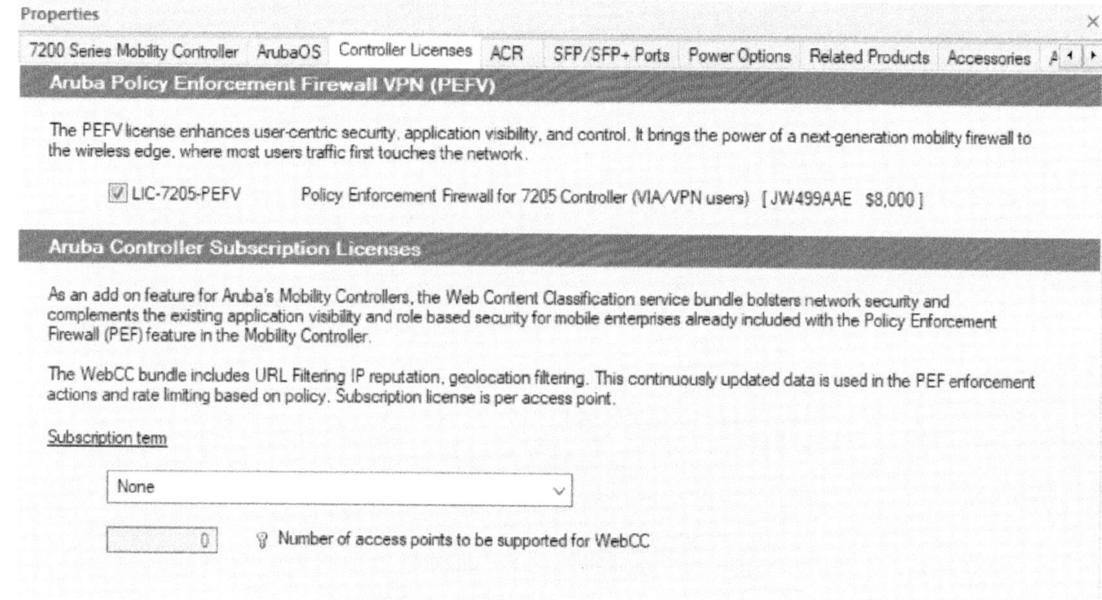

Figure 3-24 Subscription licenses in Iris

7. If you are deploying MCs without an MM, follow these steps.
    a. Select the MC in the workspace.
    b. In Properties window's first tab, you can choose a bundle of the hardware and licenses, which may be a cost-effective way to meet the customers' needs.
    c. Alternatively, you can specify licenses in the ArubaOS tab. In this tab, you can either bundle licenses together or specify the number of AP, PEFNG, and RFP licenses separately.

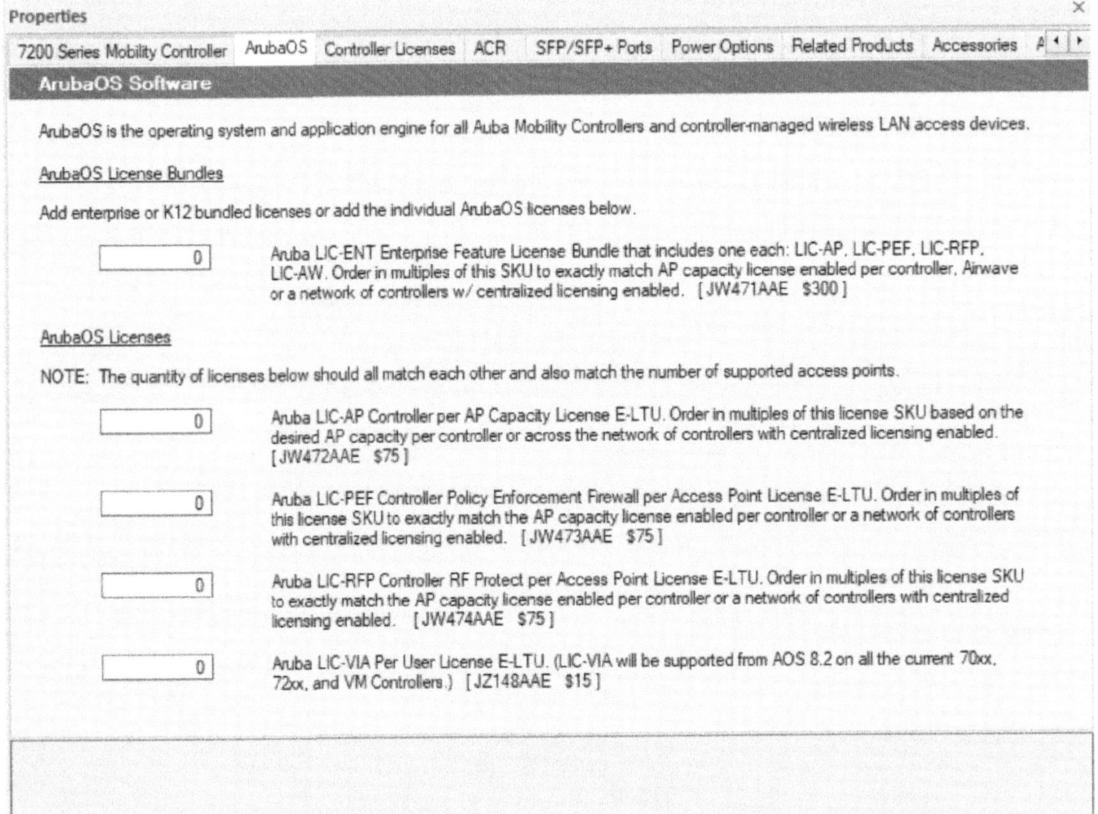

Figure 3-25 Licenses if your plan does not include a mobility master in Iris

    d. If you are planning WebCC subscription licenses, select the **Controller Licenses** tab in the Properties window. Choose the subscription term and specify the number of APs per controller.

8. Save your project.
9. Select the dollar sign button or select File > Quotation to see your BOM.

Answers to the questions included in activities are provided in "Appendix: Activities and Learning Check Answers."

Summary

You should now have a working knowledge of the Aruba product line and the various controller models Aruba offers. You also have various deployment models that you can use for your network design. Finally, you should be able to add licenses to controllers and understand the different features for IAPs and IAP clusters.

Learning check

1. What APs can be converted to a a spectrum analyzer?
    a. Any AP
    b. Only 300 series APs
    c. RAP only
    d. Only 200 series APs
    e. Only dual radio APs
2. You have calculated that you need 500 APs in a building. What are two options for a controller can handle this many APs?
    a. 7210
    b. 7205
    c. VMC 250 + 250
    d. MM 500
    e. 7030
3. Which statements are applicable to an MM?
    a. It cannot be a centralized licensing server.
    b. It can terminate APs.
    c. It can do configuration validation.
    d. It can push a full configuration to managed node (MD/MC).
    e. It can only be deployed on a physical server.

4. You have calculated that you need an MM license of 550. What license(s) do you purchase?
    a. MM 1K
    b. MM 10K
    c. MM 500 + 50
    d. Only AP licenses are needed
5. What features do IAPs support?
    a. ARM
    b. Mesh
    c. Spectrum analyzer
    d. IDS/IPS
    e. Stateful firewall

Answers to the learning checks in this study guide are provided in "Appendix: Activities and Learning Check Answers."

4 Wired Network Design

EXAM OBJECTIVES

- ✓ Given a scenario, evaluate the customer requirements for a single-site campus environment (less than 1000 employees), or for subsystems of an enterprise-wide network, to identify gaps per a gap analysis, and select components based on the analysis results.
- ✓ Given a scenario, translate the business needs of a single-site campus environment (less than 1000 employees) or subsystems of an enterprise-wide environment into technical customer requirements.
- ✓ Given a scenario, select the appropriate products based on the customer's technical requirements for a single-site campus environment (less than 1000 employees) or subsystems of an enterprise-wide environment.
- ✓ Given the customer requirements for a single-site campus environment (less than 1000 employees) or subsystems within an enterprise-wide environment, design the high-level architecture.
- ✓ Given a customer scenario, explain how a specific technology or solution would meet the customer's requirements.
- ✓ Given a customer scenario for a single-site campus environment (less than 1000 employees), choose the appropriate components that should be included on the BOM.
- ✓ Given the customer requirements for a single-site campus environment (less than 1000 employees), determine the component details and document the high-level design.
- ✓ Given the customer's requirements, explain and justify the recommended solution.

Assumed knowledge

- Switching and routing technologies such as:
  - VLANs
  - Link Aggregation Control Protocol (LACP)
  - Rapid Spanning Tree Protocol (RSTP) and Multiple Spanning Tree Protocol (MSTP)
  - IP routing, including static IP routing and Open Shortest Path First

  - Link Layer Detection Protocol (LLDP)
  - Power over Ethernet (PoE) and PoE+
- Aruba switch virtualization technologies, including Virtual Switching Framework (VSF) and backplane stacking
- Basics of managing Aruba OS (AOS)-Switches

Introduction

Even as users untether themselves from the desktop and move toward ubiquitous wireless access, wired network design remains crucial. In addition to traffic from remaining wired devices, all wireless traffic passes through the wired network at the backend. The hard work that you put into a wireless design can be compromised by a wired network that fails to follow best practices.

By designing a simple, but resilient architecture based on the latest AOS technologies, such as backplane stacking and Virtual Switching Fabric (VSF), you can ensure that the campus network meets your customer's expectations.

To get started, you will examine two-tier and three-tier wired architectures and reasons for deploying on one versus the other.

Next, you will review VSF and backplane stacking and how to use these technologies as a fundamental part of your wired network architecture.

Then you will take a closer look at planning each layer of the network, beginning with the access layer, learning how to select the switch series that meets your needs, making sure it meets requirements for edge ports. You will also learn how to plan appropriate oversubscription for uplinks and VSF links as well as a PoE budget.

Lastly, you will explore planning the core and optional aggregation layer.

Wired architectures

This section presents wired architectures at a high level.

Two-tier versus three-tier topology

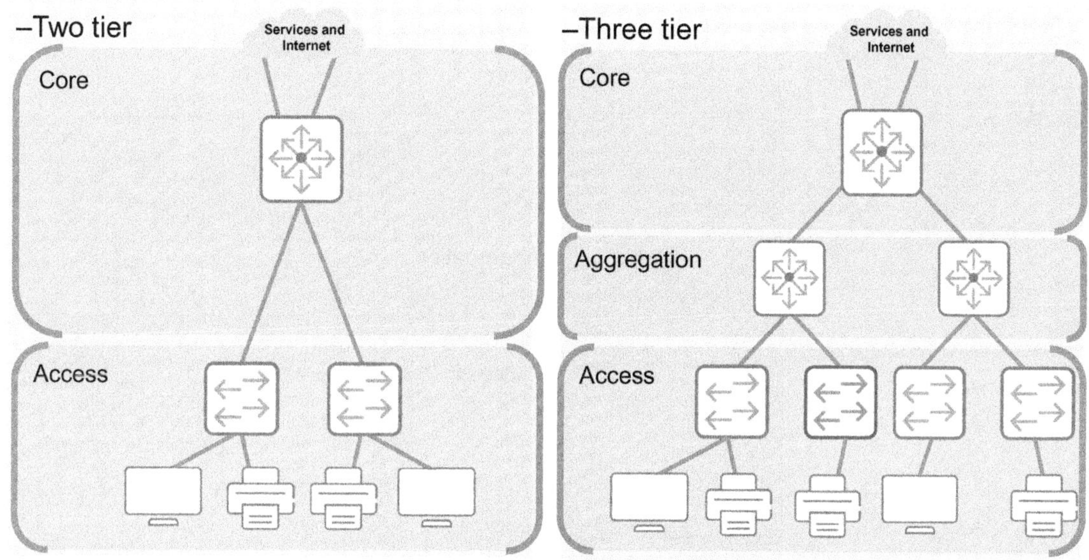

Figure 4-1 Two-tier versus three-tier topology

When you plan a wired network, you need to select products and design how they will fit together. You can classify the architecture based on the number of tiers or layers.

For example, Figure 4-1 shows a two-tier topology. This topology includes an access layer to which endpoints connect and a core which interconnects the access layer switches. The core might connect the access layer to external resources or provide connections for centralized servers as well.

A three-tier topology, also shown in Figure 4-1, has an access layer divided into groups or zones. Each group of access layer switches connects to an aggregation layer switch, which aggregates links and often applies various policies. The aggregation switches then connect to the core.

Often the aggregation layer switch accomplishes this aggregation while still maintaining good performance by supporting ports of different speeds. In other words, the aggregation switch might connect to access layer switches on 1 GbE or 10 GbE links and to the core on 10 GbE or 40 GbE links.

Also, note that the services and Internet cloud could represent servers that connect directly to the core and a direct Internet connection. Or companies with more servers might place the servers in a dedicated data center, which would have a two- or three-tier topology of its own. If the campus core does connect to the Internet, a third-party firewall would typically lie between it and the data center.

Redundant two-tier and three-tier architectures

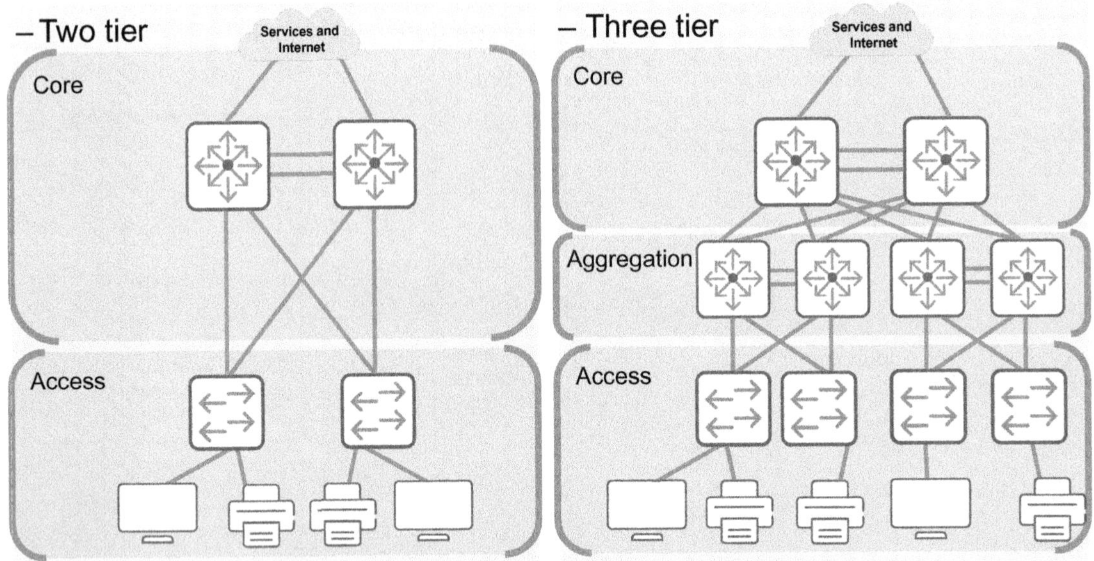

Figure 4-2 Redundant two-tier and three-tier architectures

Many customers require redundancy at the core and aggregation layer. The example in Figure 4-2 shows the physical links for establishing a fully redundant topology.

You must implement the correct technologies on top of these links to load balance traffic over redundant links, eliminate Layer 2 loops and handle failover. Without VSF and backplane stacking, you must use traditional technologies such as VRRP and MSTP. However, as you will explore in more detail, backplane stacking and VSF let you simplify and eliminate these technologies.

These examples show the core connecting to a cloud. The campus core might connect to a few servers and then to a firewall and the Internet. Or the company might have a complete data center. In that case, the campus core should connect to the data center core switches. Not shown in these examples are the mobility controllers (MCs), which could connect at the aggregation layer or core, based on the customer requirements.

Benefits of a two-tier topology

Aruba recommends a two-tier topology for most small and medium campuses because this topology provides the simplicity and best performance.

A three-tier topology typically requires routing between the aggregation layer and the core. However, a two-tier topology can permit the extension of a Layer 2 network across the campus, permitting greater flexibility in VLAN design.

In a two-tier topology, traffic flows directly from the edge to the core, reducing latency. The customer also avoids buying several aggregation layer switches and possibly purchases fewer transceivers.

A flat topology that permits the extension of VLANs can be particularly important for virtualized data centers; however, even in the campus, the flexibility to extend VLANs can simplify the design. For example, the customer may want to extend a security camera VLAN across the campus.

The lower latency of a two-tier topology might also offer benefits for a campus in which users run a lot of UC applications.

When three-tier architectures may be required

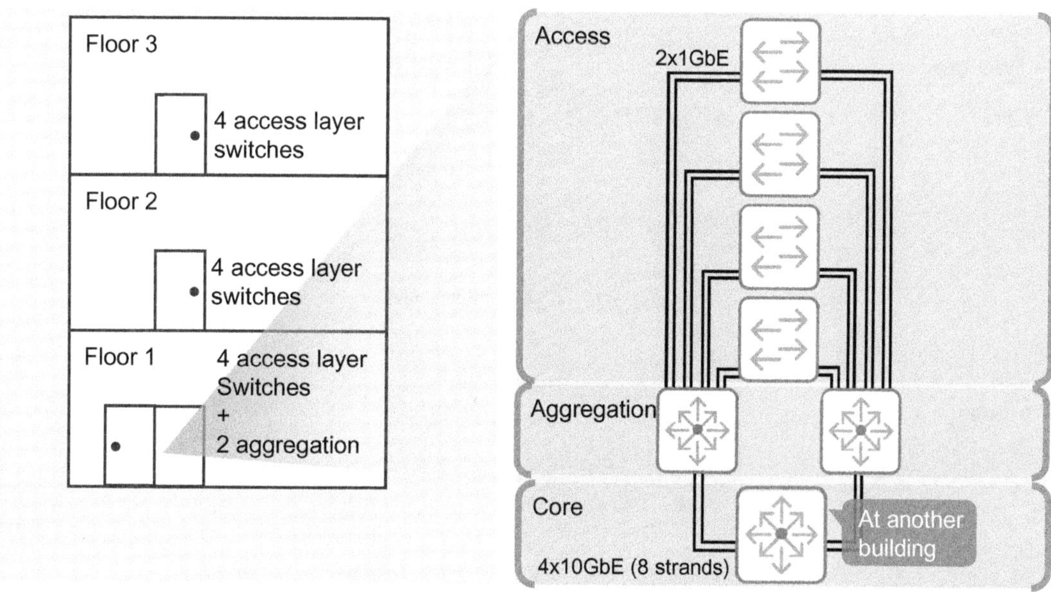

Figure 4-3 When three-tier architectures may be required

A customer's physical site and cabling restraints are the primary factors that determine if you can use a two-tier topology, which requires the access layer to connect directly to the core.

Access layer switches typically reside in distributed closets, while core switches reside in a central location. You will need to examine the distances between the switches and types of cabling available for each connection.

Single-building campuses can often use a two-tier topology because it is simpler to pull sufficient cable to connect access layer switches to the core. However, multibuilding campuses may pose a challenge. Large sites may not have enough expensive building-to-building fiber links to connect each access layer switch to the core, particularly when each switch needs a redundant link.

Using an aggregation layer and three-tier topology has traditionally addressed this issue. In each building, access layer switches connect to one or two aggregation layer switches. The aggregation layer switches then connect the entire building to the core.

In the example shown in Figure 4-3, a building has 12 access layer switches, but limited fiber to connect the access switches to the core of another building. You can use an aggregation layer to aggregate the many 1 GbE connections from the access layer into just four 10 GbE connections, which fit on the limited cabling.

Backplane stacking and VSF can also help solve cabling limitation issues in some cases.

Options for implementing routing

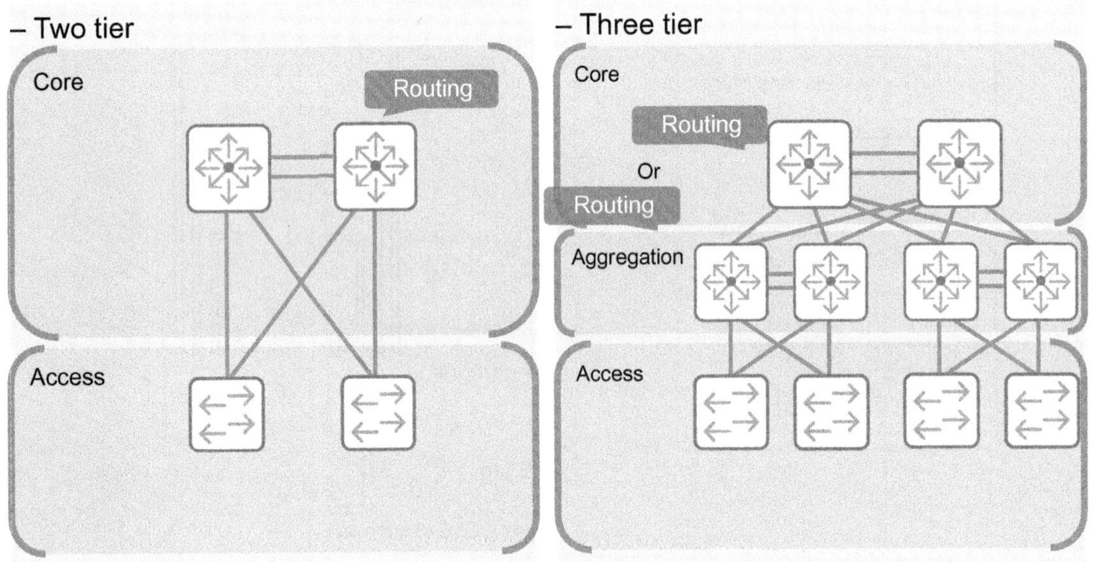

Figure 4-4 Options for implementing routing

Once you select the architecture you want to use, you also need to choose where to implement routing. Figure 4-4 shows some options based on the type of architecture.

Traditionally, you implemented routing between the aggregation and core layer in a three-tier topology to prevent issues from broadcasting domains that extend too far. However, because you can now use VSF and backplane stacking to eliminate spanning tree, you do not need to worry as much about larger Layer 2 domains. Therefore, you can alternatively implement routing at the core.

In a two-tier topology, you normally only route at the core. However, many Aruba access layer switches support routing, which allows you to implement routing there as well.

In the past, some architects preferred routing traffic at the access layer to avoid implementing a spanning tree. When properly designed, routed links load balance traffic better and failover faster than links using a spanning tree.

VSF and backplane stacking can provide many of the same benefits without requiring routing at the access layer.

VSF and backplane stacking

The previous section laid out some of the important choices for you to make as you design a wireless network. Virtual Switching Framework (VSF) and backplane stacking can affect your architectural decisions and play an important part in simplifying the architecture while enhancing its resiliency.

VSF and backplane stacking review

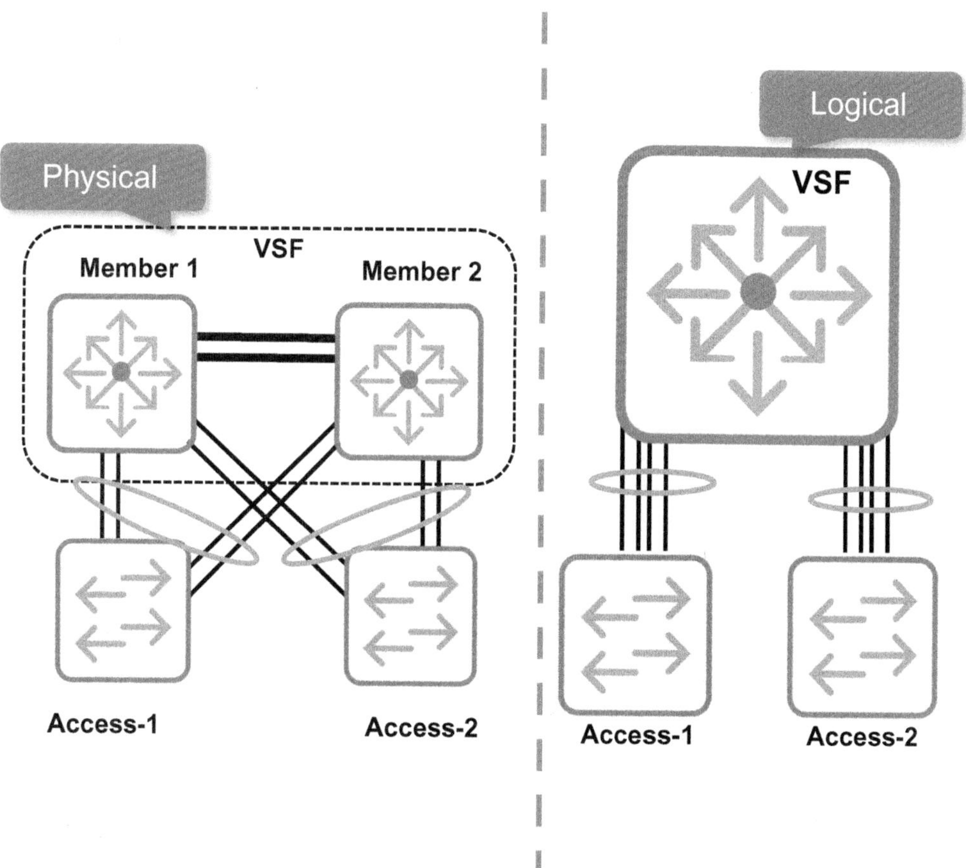

Figure 4-5 VSF and backplane stacking review

VSF and backplane stacking are Aruba technologies that allow you to combine multiple switches into a single virtual switch. You can then manage the combined switches as a single entity. You can also treat switches as a single entity from the viewpoint of designing Layer 2 and 3 protocols, as shown in Figure 4-5. This provides significant advantages as you design a wired network.

Aruba 5400R zl2 and 2930F Series switches support VSF, and Aruba 3810M and 2930M switches support backplane stacking.

A VSF fabric must consist of switches that are the same series, such as two Aruba 5406R switches or two Aruba 2930F switches. For the 2930F Series, you can combine different models in the same fabric. For example, you can combine a 2930F-48G PoE+ switch with a 2930F-48G switch.

You cannot combine 5406R and 5412R switches in the same VSF fabric. The two switches can support different interface modules, although it is recommended that you install the same modules in both switches for consistency.

Additional considerations

For 5400R zl2 switches, the VSF fabric must use only v3 modules—both for VSF links and other links. The switch must operate in v3 only mode. In fact, when you enable VSF on the switch, it automatically changes to v3 only mode, disabling any v2 modules installed in the switch.

When part of a VSF fabric, a 5400R switch can only use one management module; enabling VSF deactivates a second management module. We recommend you install only one management module in the switch. If the management module of the commander fails, the standby member takes over.

You can combine up to four 2930F switches in a fabric, and two 5400R Series switches in a fabric. For latest features and specifications please refer to references available through product support and community resources outlined here, Next Steps.

As for backplane stacking, up to ten 3810M and 3800 switches can be combined in a backplane stack. However, if you want to implement a mesh topology—in which every switch is connected to every other switch in the backplane stack—the backplane stack is limited to five switches. Only 3810M and 3800 switches support the mesh topology. You can combine up to ten 2930M Series switches and up to four 2920 Series switches in a backplane stack.

VSF versus backplane stacking—Similarities

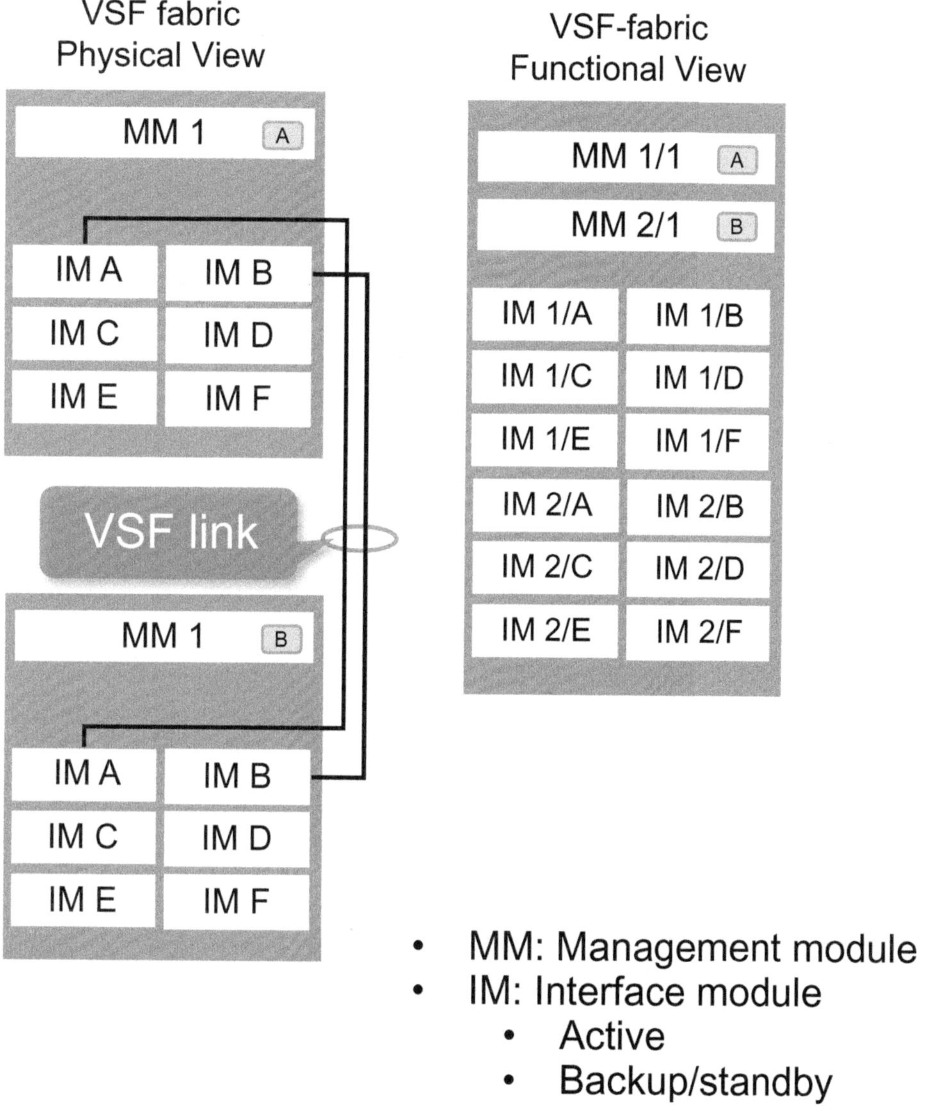

Figure 4-6 VSF versus backplane stacking—Similarities

Within the VSF fabric or backplane stack, one member functions as the commander. This switch's management plane and control plane are the active planes for the entire stack (see Figure 4-6). The commander runs routing protocols, builds forwarding tables, manages the configuration, and so forth. The commander proxies the control plane to the standby member.

Each member in the VSF fabric or backplane stack has its own forwarding plane, which remains active, allowing it to process and forward packets.

This architecture enables the switches in the fabric or stack to act as one logical unit, which can establish link aggregations that spread across multiple members. VSF fabrics and backplane stacks can establish aggregated links that use ports on multiple members. This establishes a highly redundant topology.

The 5400R zl2 Switch series supports redundant management modules. If a 5406R zl2 switch includes two management modules and is configured in a VSF fabric, only one of the management modules will be active; the other management module will be disabled. If the active management module fails, a reboot is required to activate the redundant management module.

VSF versus backplane stacking—Differences

VSF and backplane stacking are implemented in different ways.

For one thing, VSF is a front-plane stacking technology, meaning that you can connect the switches in the VSF fabric using regular wired switch ports.

Backplane stacking, in contrast, is not implemented through wired ports; you must purchase a stacking module and stacking cable to implement this technology.

With VSF fabrics, then, you will need to plan the VSF links to ensure there is adequate bandwidth to handle the traffic that members exchange.

For backplane stacking, all the members are connected through dedicated, high-bandwidth stacking links (40 GbE).

Because VSF members are connected via wired ports, VSF fabrics support long distances between members. However, the distance between backplane stacking members is limited to a maximum of 9.8 feet (3 m) (you can order .5 m, 1 m, or 3 m cable). For this reason, backplane stacking is intended for deployment between members that are relatively close together such as in the same closet.

These two technologies also differ in how they handle traffic forwarding. Within a VSF fabric, deep-level queuing and forwarding is implemented at a member level. With backplane stacking, queuing and forwarding are implemented at a stack level. In other words, members can forward traffic on their own, but all traffic has a stacking module for processing, enabling the stack to operate even more deeply as a single entity.

Keep a few more things in mind as you consider the differences between VSF and backplane stacking technologies.

For VSF, all ports assigned to the same VSF link must be the same speed (such as all 10 GbE or all 40 GbE). On 5400R zl2 Series switches, you connect the switches using 10 GbE or 40 GbE links. On the 2930F Series switches, you can connect the switches using 1 GbE or 10 GbE links; however, 10 GbE links are recommended in most cases. For backplane stacking, each stacking port provides 42 Gbps of bandwidth in both directions. VSF has the advantage of supporting longer lengths for connections.

For the greatest redundancy, you should ensure that the members are connected via a ring topology or mesh topology (if you are using 3810M or 3800 Series switches).

Also, consider how VSF and backplane stacking handle traffic differently in more depth.

With VSF fabric, members always forward unicasts on a local link in the link aggregation, if available. If a local link is not available, the member can send the traffic over VSF links to another member. If the member needs to forward a multicast, it forwards it over the VSF link and lets other members decide whether they need to forward the multicast on any local ports.

Backplane stacking, on the other hand, uses the same load-balancing algorithm as aggregated links on a single switch to select a link. That is, it hashes information about the conversation and then uses that hash to choose one of the link in the link aggregation. All links are equally likely to be chosen, regardless of whether the link is on this member or another member. Therefore, each member might end up forwarding a relatively large amount of traffic on backplane stacks to other members. The backplane stacking links offer a great deal of bandwidth to handle this traffic. For the multicast example, the member that receives the multicast calculates which ports on all members need to forward the multicast and sends the multicast to members that need it.

Benefits of VSF and backplane stacking

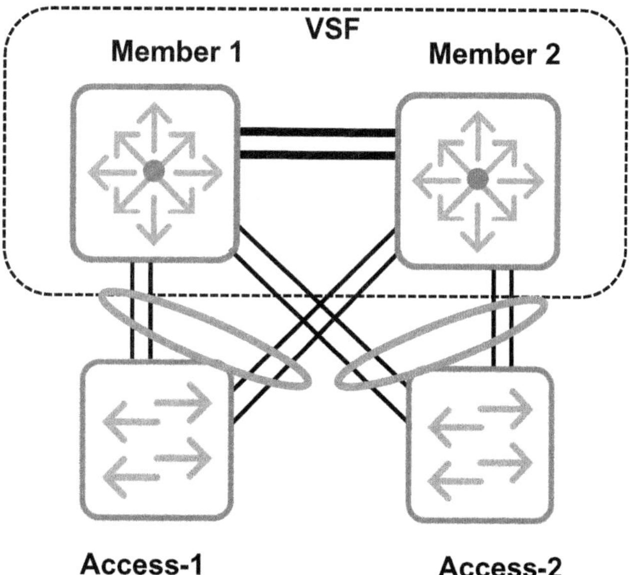

Figure 4-7 Benefits of VSF and backplane stacking

VSF and backplane stacking provide many benefits.

For example, they simplify management. You no longer need to connect, configure, and manage each switch individually. Instead, you can access one of the switches (typically the commander) in the VSF fabric or backplane stack to make configuration changes. The configuration changes are then distributed to all members automatically, considerably simplifying network setup, operation, and maintenance.

VSF and backplane stacking also increase resiliency for the network design and at the same time simplify it. For example, you can eliminate spanning tree and connect different layers of the network together with link aggregations, as shown in Figure 4-7. The next sections describe how VSF and backplane stacking increase resiliency and simplify the design in more detail.

Simpler design—VSF and backplane stacking at the core and aggregation layer

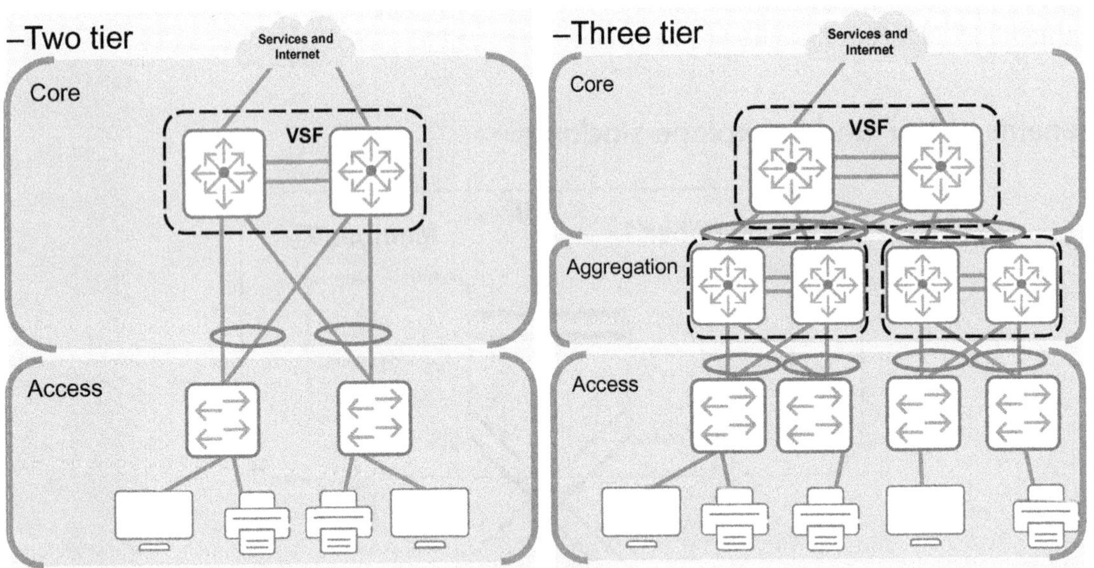

Figure 4-8 Simpler design—VSF and backplane stacking at the core and aggregation layer

VSF and backplane stacking simplify the design at the core and in a three-tier topology at the aggregation layer and core, as shown in Figure 4-8.

Redundancy for the vital core and aggregation routing switches is a must in many environments. VSF and backplane stacking provides the most effective way to deliver redundancy by:

- Increased resiliency through the creation of multiple, load-balancing links and the elimination of all single point of failures at the network core
- Simplified topologies based on aggregated links rather than complex MSTP solutions

- Simplified management through a single management IP address for the virtual switch
- Simplified router redundancy using a single IP address rather than complex VRRP solutions

VSF and backplane stacking can achieve these benefits because they have built-in redundancy. Therefore, you do not have to run network protocols that are designed to provide redundancy. For example, you do not need to run Multiple Spanning Tree Protocol (MSTP) or Rapid Spanning Tree Protocol (RSTP) to provide redundant links without causing a network loop or Virtual Router Redundancy Protocol (VRRP) to provide gateway redundancy. Layer 2 and Layer 3 redundancy are built into the VSF fabric and backplane stack. This can improve network performance because network protocols such as RSTP and MSTP have relatively slow convergence times.

Eliminating the need to run these protocols further simplifies management as well. MSTP and RSTP take time to configure properly, especially in a large network. Without VSF and backplane stacking, you must manage the switches individually. And you need to set up spanning-tree instances on each switch in turn, making sure that the parameters for one switch match those of the rest of the region. The real problem emerges in maintenance. Troubleshooting spanning-tree-related issues is no easy task, usually requiring a great deal of time to locate the root cause of the failure.

In addition, VSF and backplane stacking provides better utilization of redundant links. RSTP blocks all but one set of redundant parallel paths. MSTP permits the use of some redundant links but does not load balance very granularly. Half (or more) of the available system bandwidth can be squandered in a backup role, off-limits to data traffic—not a very good use of the network infrastructure investment.

With VSF and backplane stacking, however, you can create link aggregations that span both switches in the fabric. Spanning tree is no longer necessary to eliminate loops on switch-to-switch links because at a logical level no loop exists. In this way, these virtualization technologies provide all the advantages of a completely link-aggregation based Layer 2 design:

- Faster failover
- Load-balancing
- Simple setup and design
- Better stability and less need for troubleshooting

Although, in these examples, the switch-to-switch topology does not have any loops, you should still implement a form of loop protection to prevent broadcast storms if a loop is introduced accidentally. You could choose to enable spanning tree on the access layer switches or loop protection (an AOS-Switch proprietary feature) on the edge ports.

VSF and backplane stacking also simplify routing environments in addition to providing redundancy. For example, they reduce the number of logical routing devices and make the topologies simpler.

Simpler design—VSF and backplane stacking at the access layer

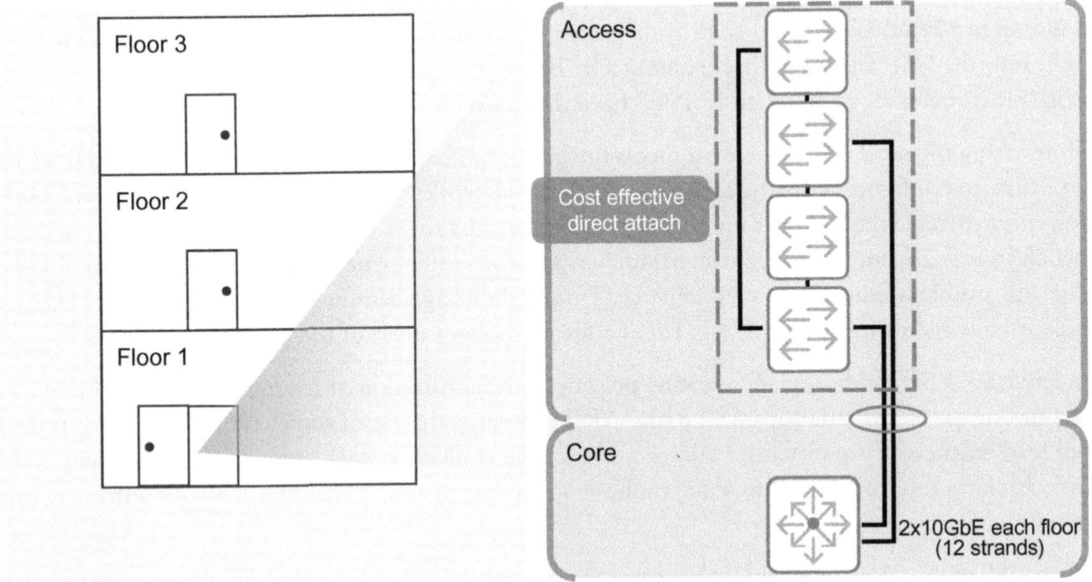

Figure 4-9 Simpler design—VSF and backplane stacking at the access layer

Although cabling constraints can cause you to choose a three-tier topology, you should keep in mind how ArubaOS-Switches may help you create a two-tier design.

For example, as shown in Figure 4-9, rather than connecting four switches to an aggregation layer switch, you connect them to each other in a VSF fabric or backplane stack. Then the entire group shares a couple of high-bandwidth uplinks to the core. As a result, a building with twelve 48-port switches can connect to the core on six 10 GbE connections without an aggregation switch—a more feasible proposition than 24 1 GbE connections.

The direct attach or stacking cables required for such a solution are relatively cost-effective when you compare them to the additional fiber optic transceivers and other fiber cabling required for a three-tier design.

This approach would not work well for less-capable access layer switches without high-throughput connections between them. Traffic would bottleneck on the links between the access layer switches, reducing performance. In addition, a failed link or switch could affect connectivity for the entire building. However, AOS-Switches in a backplane stack are designed to work this way. They have nonblocking stacking links and support resilient ring and mesh topologies. You can also plan VSF fabrics for similar benefits, taking care to plan adequate bandwidth between the members.

Sometimes you can also use modular switches in a similar way. For example, a smaller building might require just three modular switches, which can connect directly to the core, as opposed to 12 fixed-port switches, which would need an aggregation layer to connect.

Choosing between backplane stacking and VSF

Although VSF and backplane stacking provide similar benefits, they are not interchangeable technologies. There are reasons why you might choose one technology over another.

Sometimes your choice of technology will depend on which switch model you select for other reasons. For example, you may select 5400R zl2 because the company wants a modular switch, with the ability to add modules in the switch as the company grows.

Sometimes, you will be choosing between two similar models, one that supports backplane stacking and one that supports VSF. In this case, you need to consider if a front-plane or a backplane stacking solution meets the customer's needs. VSF provides a low-cost stacking solution that does not require a stacking module and VSF fabrics can span multiple wiring closets.

Because backplane stacking uses dedicated modules and cabling, it provides a high-performance virtual switch. And though backplane stacks must be relatively close together, they support more members (depending on the switch model).

Table 4-1 summarizes these considerations.

Table 4-1 VSF and backplane stacking comparison

Initial Selection Factors	VSF	Backplane Stacking
Switch model chosen for other reasons	• 5400R zl2 Series Switch • 2930F Series Switch	• 3810M Series Switch • 2930M Series Switch
Cost versus higher performance	• Low-cost stacking solution	• Dedicated hardware • Higher performance
Stack requirements	Multi-wiring closet stack	Stack size • Up to ten 3810M and 2930M switches

General best practices for VSF fabrics and backplane stacks

You should make sure that your plan for VSF fabrics and backplane stacks follows best practices. For example, you must plan adequate bandwidth for the links between members; guidelines vary based on where you are deploying the fabric or stack.

When you configure a VSF fabric, you should also make sure that it has a unique domain number. This will ensure that switches join the appropriate VSF fabric.

And you need to protect the fabric or stack against a split, where members cannot reach each other over the VSF or stacking links. You can establish protection by building redundancy into the links and using a ring topology (or mesh when available). On modular switches, create a resilient VSF link with multiple physical links that are connected on interfaces on different modules.

As a secondary defense, set up a form of Multi-Active Detection (MAD) for the VSF fabric. A backplane stack uses built-in mechanisms to shut down isolated members if a split occurs. For these functions to work (if the stack splits in equal-sized fragments) make sure to connect the members over a network on their out of band management (OOBM) ports.

Implementing MAD is a best practice even when you create resilient VSF or stacking links due to the negative effects of a split fabric. If a split occurs, the standby member loses contact with the commander. From the standby member's point of view, the commander has failed, so the standby member becomes commander as well. Now multiple devices are active; in other words, the VSF fabric has a split brain. Both devices use the same IP addressing and run the same protocols, which can cause issues with routing. Traffic can also go astray because members cannot forward it between each other.

When MAD is operating, the VSF link can fail without disrupting traffic. The standby still loses contact with the commander on the VSF link. However, the standby member can use MAD to detect that the commander is still up. The standby member then becomes an inactive fragment and shuts down all of its interfaces. The commander remains the up and an active fragment. The link aggregations all fail over gracefully to the links connected to the active fragment.

VSF supports three types of MAD: OOBM, LLDP, and VLAN. You can configure one method on each VSF fabric. The 5400R zl2 switches support OOBM MAD and LLDP-MAD. The 2930F switches support LLDP-MAD and VLAN MAD.

Split stacks can also occur with backplane stacking. A split could occur if, for example, one member fails in a four member chain topology. You can create a more resilient stack by using a ring or a mesh topology. A chain topology creates a greater risk of stack fragments occurring.

As mentioned, because a backplane stack has built-in mechanisms to handle splits, you do not have to configure a protection mechanism. However, it is still important that you enable and configure OOBM on each member of the stack. If a link or a switch failure occurs, separating a stack into two groups, the smaller group, or fragment, becomes inactive by default. All of the user ports on the switches in this stack fragment will be disabled. However, the OOBM and stack ports will be active, allowing you to manage the switches.

The stack fragment that includes more switches will become the active stack and will inherit the MAC address and IP addresses assigned to the stack. This allows more of the user ports to remain operational.

If the two stack fragments are the same size, meaning they have the same number of members, a problem can occur if OOBM is not enabled. In this case, both fragments are active and will attempt to use the global IP addresses configured on the backplane stack. If OOBM is enabled and configured on all members in the stack, the fragment that contains the commander is active. To protect the network from this problem, you should ensure that OOBM is configured on all of the switches.

Plan the access layer

You will now turn your attention to planning the access layer of the wired network.

Aruba access layer switches

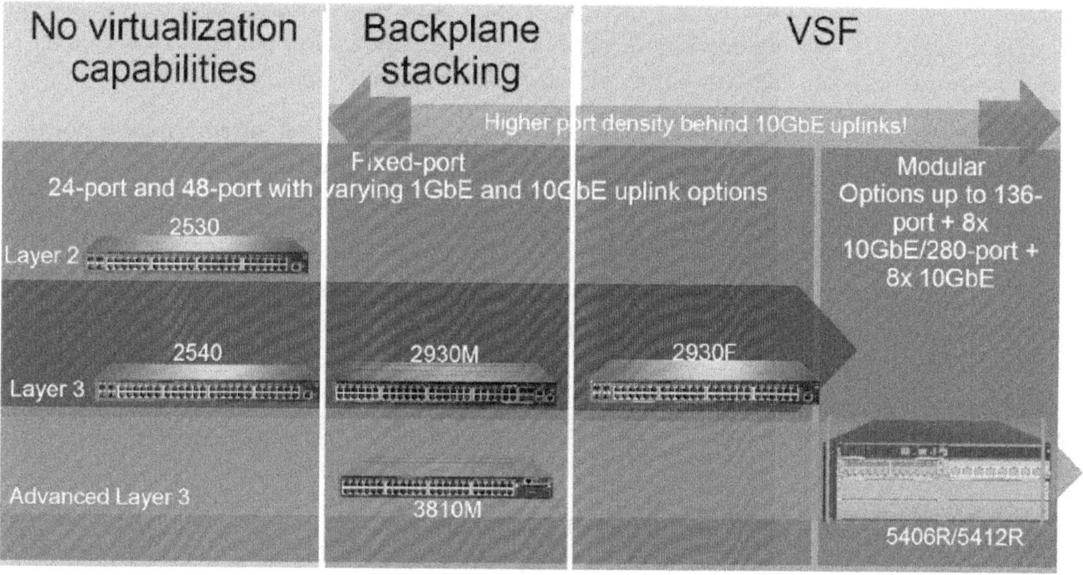

Figure 4-10 Aruba access layer switches

Aruba's latest generation access layer switches are classified into two basic types: fixed-port and modular. You can also classify the switches by the types of virtualization technologies that the switches support. Whether you choose to implement VSF, backplane stacking, or neither technology at the access layer will affect other access-layer choices.

Deploying modular switches, backplane stacks, or VSF fabrics enables the access layer to support a higher density of edge ports behind two 10GbE links. You might choose these options to support a high density of ports in a wiring closet.

The 5400R Series supports VSF. Although these switches might use VSF more at the aggregation or core than they do at the access layer, you can combine modular switches with VSF at the access layer. A single modular switch and a backplane stack can often offer similar scalability and performance. The one you choose will depend on customer preference and the best price point for the particular scenario.

Some customers have switch-type preferences and may choose the same switch type across their organizations. Others will be open to mixing switch types to meet specific needs. You should select the general type of access layer switches to use before you move on to other considerations. However, you might return to your choice and reconsider based on other factors.

In many ways, a modular switch and a backplane stack act as similar building blocks, so choosing between a 5400R or a 3810M backplane stack, for example, can come down to which can most cost-effectively meet the customer's needs. The modular 5400R does have the advantage of flexibility. For example, you can select some PoE+ modules and other non-PoE modules. The 5412R switch can also scale to support a very high density of ports. Figure 4-10 shows an example of the number of ports that these models can support, but many other combinations are possible.

Some of the fixed-port switches, however, also offer some flexibility. They often feature a slot into which customers can install their choice of modules for the uplinks. And you can mix different switch models of the same series in the same fabric or stack.

Turning back to customer-preference considerations, you should know that in addition to modular versus fixed-port preferences, the desire to standardize might drive a customer's choices. For example, a customer might have a main office with high port-density wiring closets but also many branch sites. The customer might choose the same fixed port switches across the line. You would then use VSF or backplane stacking to combine the switches at the main office into larger groups. These virtual switches can work as well as modular switches in wiring closets with a high port density.

Overview of choices for access layer switches

When you have the general type of switches in mind, you can make more precise choices. To choose the correct switch models and accessories, you should consider the media, count, and speed for both edge ports and uplinks. Then look at planning the PoE budget.

You may need to check whether your proposed switch supports all of the features the customer requires. However, the typical features that customers require at the access layer—such as VLANs, QoS, MAC-Auth, and 802.1X—have broadly consistent support across the portfolio.

Also, keep in mind that 5400R switches support modules with different versions, and some features such as VSF are not supported with legacy modules. You should only select v3 modules or later for new deployments.

Consider the physical design

You should always keep wiring restraints in mind when planning connections. A site has two types of cabling: horizontal cabling and vertical cabling (see Figure 4-11.)

Horizontal cabling consists of the cable runs out across a floor to wired jacks. The cable is typically run from the wired jack to a patch panel in a wiring closet. To activate a jack, a patch cable connects the patch panel to a switch port. Horizontal cable can also be run directly to APs.

Each wiring closet can be called an intermediate distribution frame (IDF). Vertical or backbone cabling connects the IDF to a main distribution frame (MDF). Again, typically cable is run from a patch panel on the IDF to a patch panel on the MDF. Patch cable can then connect to switches. In this way, an access layer switch in a closet on Floor 3 can to a core switch in the server room on Floor 1, for example.

Calculate length
Cable run + 16.4 feet (5m) for each patch cable

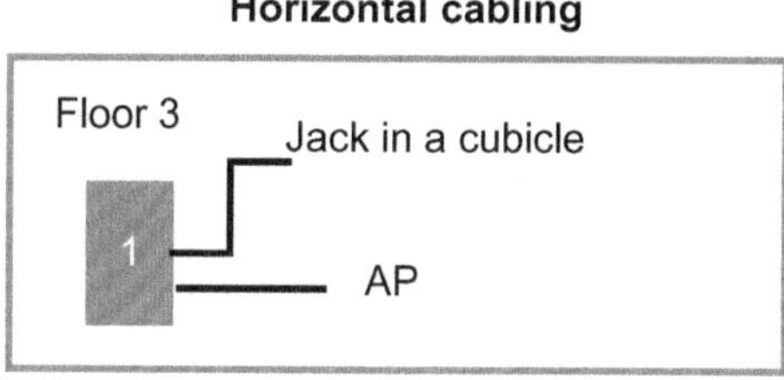

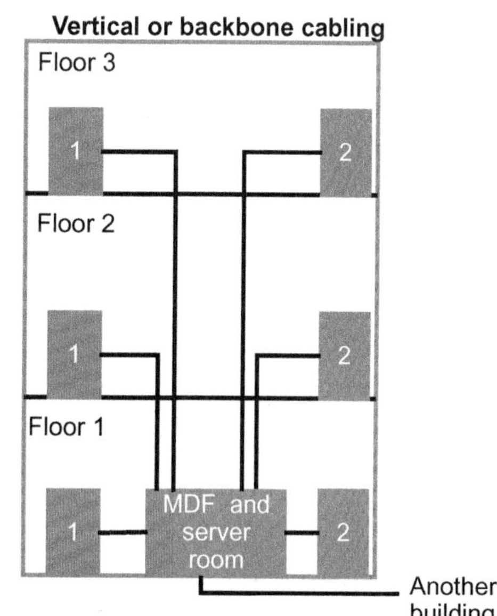

Figure 4-11 Consider the physical design

Vertical cabling can also establish inter-building connections.

To plan connections, it is important to know the cable length so that you can verify that the speed you require is supported on that cable type over that length. You should take into account the length of cable run as well as about 16.4 feet (5 m) to account for each patch cable in the connection.

Usually, this means adding 10 m to account for 5 m at each side of the connection.

While vertical cabling between two closets often connects switches in those closets, a closet can also be matched through the MDF to another IDF. For example, a switch in Floor 3, Closet 1 could connect to a switch in Floor 3, Closet 2 through the MDF as long as there is sufficient cable for the connection and the distance does not exceed the maximum for the cable type. Similarly, a closet might be able to be patched through the MDF to the inter-building cable. However, you must check with the customer carefully to understand how the cabling at their site terminates and which runs can be patched together.

Consider the media

You can now consider the media type.

Copper

Horizontal cabling usually uses BASE-T copper cable. In some smaller building, vertical cabling might also be copper. However, copper only supports high-enough speeds for these links over limited distances.

Also be aware that most BASE-T ports on Aruba switches are 10/100/1000 Mbps ports. Use switches with Smart Rate ports if you need to connect the switches at 10 GbE.

Because access layer uplinks in modern networks often require at least 10 GbE, if your customer site uses copper cable between closets, make sure that it will support 10 GbE speeds at the required distances. CAT5e cable is not rated for 10 GbE. For unshielded twisted pair (UTP) CAT6 cable, Aruba Smart Rate ports support 10 GbE only up to 180.5 feet (55 m). But they support 10 GbE at the full 328 feet (100 m) on CAT6a or CAT7 cable, as well as on shielded twisted pair (STP) CAT6 cable.

Smart Rate ports also support 2.5 GbE operation or 5 GbE operation on CAT5e cable. On unshielded CAT5e, 2.5 GbE is supported up to 328 feet (100 m) and 5 GbE up to 180.5 feet (55 m). On CAT6 cable or better, as well as shielded CAT5e, both 2.5 GbE and 5 GbE are supported up to 328 feet (100 m).

Fiber and direct attach cable

The other main type of cable used in Ethernet network is fiber optic cable. To support fiber optic connections, a switch requires special ports that are designed to hold a transceiver, as well as the transceiver itself. For example, to support 1 GbE fiber, a switch has SFP ports, and it also needs an SFP transceiver.

A fiber cable often has multiple strands such as 6, 12, or 24. Both 1 GbE and 10 GbE transceivers require two fiber strands per-connection.

In rare cases, buildings use fiber optic horizontal cabling, requiring edge ports to provide fiber SFP transceivers. You may also need a switch with a higher density of SFP ports if the network design includes outdoor APs with SFP ports such as the Aruba AP-370 Series.

But fiber cabling is most often used for vertical cabling between closets and between buildings, in which case switches just need a few SFP or SFP+ ports for uplinks. Depending on the fiber type and transceiver, fiber can extend for hundreds of meters or even many kilometers.

One exception exists for the rule that an SFP+ port requires a transceiver. You can use cost-effective Direct Attach cable (DAC) for direct links between SFP+ ports on switches in the same wiring closet—such as member switches of a VSF fabric. SFP+ DACs are a maximum of 22.9 ft. or 7 m.

Like copper cable, fiber cable is often run to fiber patch panels, which switches connect to.

If you do need to support a high density of SFP ports at the edge, you can select an Aruba 5400R switch with 24-port SFP modules.

Plan fiber optic transceivers for 1GbE and 10GbE

You are probably entering an environment with existing fiber cabling, in which case you must choose transceivers that support it. In addition to the fiber type, you must consider the required bandwidth and the distance for links when you select the transceiver. You have different options for 1 GbE and for 10GbE.

For uplinks within a campus LAN, you are typically dealing with distances less than 300 m; longer distances are provided for reference.

Table 4-2 1 GbE transceivers

Fiber Type	Max distance	Transceiver
Multi-mode OM1	275 m	1000BASE-SX
	550 m	1000BASE-LX
Multi-mode OM2, OM3, OM4	550 m	1000BASE-SX
		1000BASE-LX
Single-mode	10 km	1000BASE-LX
		1000BASE-BX10
	70 km	1000BASE-LH70

Table 4-3 10 GbE transceivers

Fiber Type	Max distance	Transceiver
Multi-mode OM1	33 m	10GBASE-SR
	220 m	10GBASE-LRM
Multi-mode OM2	82 m	10GBASE-SR
	220 m	10GBASE-LRM
Multi-mode OM3	220 m	10GBASE-LRM
	300 m	10GBASE-SR
Multi-mode OM4	300 m	10GBASE-SR
Single-mode	10 km	10GBASE-LR
	40 km	10GBASE-ER

1 GbE fiber connections

Table 4-2 indicates the appropriate transceivers for particular types of fiber at various distances. For example, if you have multi-mode OM3 fiber and a link of less than 550 m, you can use either 10000BASE-SX or 1000BASE-LX. However, if you use 1000BASE-LX on multimode fiber over distances longer than 300 m, you might need to add a conditioning patch cord.

You might choose the 1000BASE-LX transceiver when the customer expects to upgrade to single-mode fiber, as this transceiver also supports that type. On single-mode fiber, Aruba 1000BASE-LX transceivers can extend up to 10 km.

The 1000BASE-BX10 transceiver also supports distances up to 10 km over single mode fiber. It uses one fiber to transmit and one to receive. Each fiber uses a different wavelength, so you must select different transceivers on either end of the connection. One transceiver receives at 1310 nm and transmits at 1490 nm, and the other receives at 1490nm and transmits at 1310nm.

For distances up to 70 km, propose 1000BASE-LH70 transceivers.

All the transceivers use two fiber strands.

10 GbE fiber connections

You can establish 10 GbE connections on both multimode and single-mode fiber. The 10 GbE connection requires two fiber strands. Table 4-3 indicates the appropriate transceivers for particular types of fiber and particular distances.

For example, suppose you are working with multi-mode OM2 fiber and need a connection of 100 m. The 10GBASE-long reach multimode (LRM) transceiver is an appropriate transceiver for multimode OM2 fiber over distances up to 220 m. A 10GBASE-SR transceiver only supports up to 82 m on OM2 fiber, though. Therefore, you should choose the LRM transceiver. A 10GBASE-LRM transceiver also supports connections up to 220 m on OM3 and OM4 grade multimode fiber. However, if you have that fiber grade and are selecting new transceivers, you could also use 10GBASE-short range (SR), which supports up to 300 m con the higher-grade fiber.

Also keep in mind, that you might be able to upgrade the fiber. In the example in which you need a 10 GbE connection over 100 m, you might upgrade the multimode OM2 fiber to multimode OM4 fiber or single-mode fiber and select a different transceiver. The choice depends on factors such as the customer environment and how easy it is to make changes.

Also note that a patch conditioning cord is required with LRM transceivers and OM1 or OM2 MM fiber, but not with OM3 fiber.

Note

Make sure the desired switch supports the transceiver type that you want. Not every Aruba switch supports every transceiver type.

Meet edge port bandwidth requirements

Although many users and wired devices typically use less than 100 Mbps, Gigabit ports have become the typical choice for enterprise customers. Deploying Gigabit to the edge ensures that the network can support bandwidth-hungry modern apps such as multimedia apps and accommodate bursts. In addition, APs can require hundreds of Mbps, particularly 802.11n and 802.11ac APs, driving the need for Gigabit to the edge.

Access layer switches that provide enterprise features come with standard 10/100/1000 Mbps ports. You need to make sure that the site has a CAT-5e cable or above, which most sites are at this point. If your customer's site needs updating, discuss pulling new cables for at least the APs. The switch port will automatically negotiate down to 10 Mbps or 100 Mbps on other ports if the cable and connected device cannot handle higher bandwidth.

Identify AP requirements

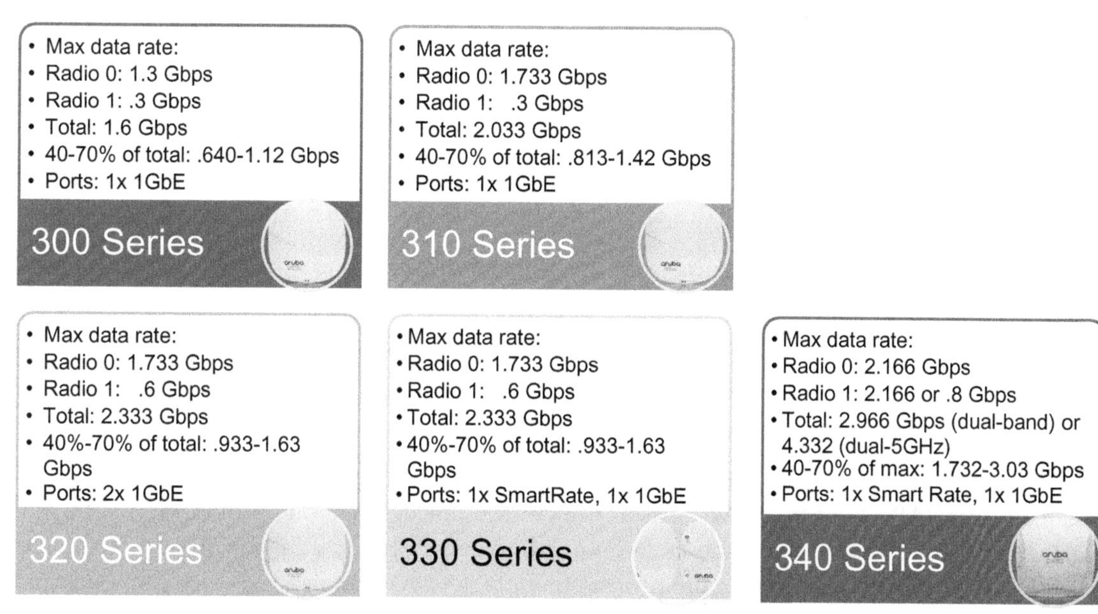

• Max data rate:
• Radio 0: 1.3 Gbps
• Radio 1: .3 Gbps
• Total: 1.6 Gbps
• 40-70% of total: .640-1.12 Gbps
• Ports: 1x 1GbE
300 Series
• Max data rate:
• Radio 0: 1.733 Gbps
• Radio 1: .3 Gbps
• Total: 2.033 Gbps
• 40-70% of total: .813-1.42 Gbps
• Ports: 1x 1GbE
310 Series
• Max data rate:
• Radio 0: 1.733 Gbps
• Radio 1: .6 Gbps
• Total: 2.333 Gbps
• 40%-70% of total: .933-1.63 Gbps
• Ports: 2x 1GbE
320 Series
• Max data rate:
• Radio 0: 1.733 Gbps
• Radio 1: .6 Gbps
• Total: 2.333 Gbps
• 40%-70% of total: .933-1.63 Gbps
• Ports: 1x SmartRate, 1x 1GbE
330 Series
• Max data rate:
• Radio 0: 2.166 Gbps
• Radio 1: 2.166 or .8 Gbps
• Total: 2.966 Gbps (dual-band) or 4.332 (dual-5GHz)
• 40-70% of max: 1.732-3.03 Gbps
• Ports: 1x Smart Rate, 1x 1GbE
340 Series

Figure 4-12 Identify AP requirements

Each Aruba AP model delivers a maximum data rate. An AP radio's actual throughput is typically under 50% of the theoretical data rate. However, some APs could still require over 1 Gbps from their wired connections to ensure that the wired backend does not limit the wireless-side bandwidth in which the customer is investing.

Different models require different approaches to ensure adequate bandwidth on the AP's side. Aruba 220 and 320 series APs have two 1 GbE ports, which you can combine in a single link aggregation.

Aruba 330 and 340 series APs have a single Smart Rate port and one 1 GbE port. The Smart Rate technology enables 2.5 GbE operation on the same CAT5e or above cable traditionally used for 1 GbE. The cable can still extend 100 m and provide PoE+.

On 330 and 340 series APs, you can combine the Smart Rate port and the normal port in a link aggregation. In that case, though, the Smart Rate port operates at 1 Gbps. If the Smart Rate port is operating at 1 Gbps, the connected switch does not have to support Smart Rate. However, it is often a good idea to propose switches that do for future proofing.

Note that the data rates and bandwidth ranges in Figure 4-12 are to give you an idea of the type of bandwidth that an AP might need to burst at a particular moment. They are not meant to indicate the actual throughput that an AP will deliver in a particular environment, which depends on many factors, including client location, density, and capabilities.

Meet the AP port requirements

You have three choices for accommodating APs.

You will typically deploy the APs on a single 1 GbE link, which is the least expensive and complex choice. However, customers should know that the AP will not scale past 1 GbE throughput on both radios, despite its theoretical capabilities.

For customers with special performance or high-availability needs, you could plan to connect both of the AP's 1 GbE ports to the switch. This approach requires twice as much access layer equipment.

For customers with higher-performance needs, a better option is to connect the APs on a single Smart Rate port, which can scale up to 2.5 Gbps. This approach lacks redundancy on the AP link, but most customers can tolerate a single lost AP.

For this approach, make sure that the switch supports Smart Rate ports. Aruba offers some fixed-port models with 8 Smart Rate ports and 40 normal ports, making a more cost-effective option for the customer.

When an Aruba AP connects on two ports, it can create a link aggregation on those ports. The link aggregation sends all traffic from one radio on one link and all traffic from the other radio on the other link. Therefore, load balancing of upstream traffic is not very granular. However, the connected switch can use its typical conversation-based load balancing mechanisms, providing better load balancing for downstream traffic to the AP.

For the 330 and 340 Series, you can take any of the approaches. These APs support two ports, one of which is a Smart Rate port and one of which is not. You should be aware that on Aruba switches, Smart Rate ports that are configured in a link aggregation function as normal ports.

The Aruba 330 and 340 APs support dual hitless PoE failover. If both ports are connected to PoE+ sources, the AP draws power from the first port and minimal power from the second port so that the port is ready to supply power. If the first link fails, the AP can quickly start drawing power from the second port, ensuring that the AP can survive the failure without rebooting. This feature provides the most resiliency when the AP is connected to two different switches in the same backplane stack or VSF fabric. By comparison, on other APs connected on two ports, the AP reboots if the port supplying the PoE power fails.

Uplink requirements and oversubscription

If all of the edge devices were operating at near full capacity, you would need to plan the uplink bandwidth to match the bandwidth provided to the edge ports. However, because most wired devices do not require nearly all of their 1Gbps port, you can oversubscribe the uplinks, as shown in Figure 4-13.

The following formula describes the oversubscription ratio:

Number of edge ports times edge port bandwidth divided by uplink bandwidth equals oversubscription.

Figure 4-13 provides an example. An access layer switch has 48 Gigabit edge ports and 2 Gigabit uplinks. The oversubscription ratio is 48 times 1 Gbps divided by 2 Gbps, or 24. This ratio is usually reported in this format, 24:1.

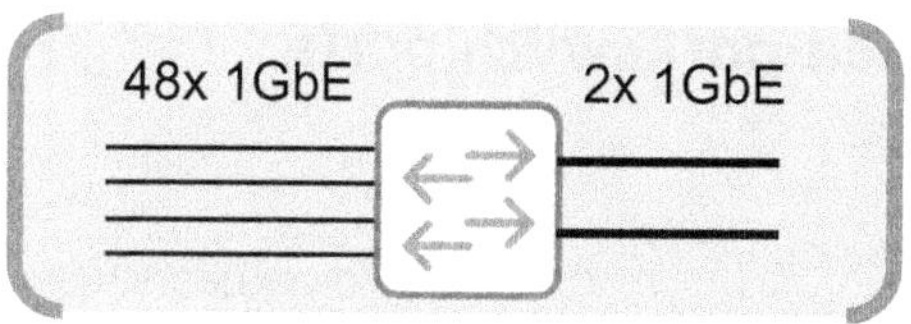

(Number of edgeports*Edge port bandwidth)/Uplink bandwidth= Oversubscription

Example

48	*1Gbps	/2Gbps	=24:1

Figure 4-13 Uplink Requirements and Oversubscription

Note

This ratio assumes that the switch can use most of the bandwidth on the uplinks, which it can if the uplinks are combined in a link aggregation.

Using VSF and backplane stacking at the aggregation layer and core makes it possible to combine all access layer uplinks in a link aggregation. If you were using an MSTP-based architecture, traffic would tend not to be distributed as evenly over the uplinks.

Factors that affect appropriate oversubscription

Several factors affect an appropriate oversubscription ratio, including traffic patterns and connected devices' average and maximum bandwidth requirements. For example, a typical office employee might rarely need more than 40 Mbps. If the switch connects to 48 such endpoints, the uplink could support just under 2 Gbps to deliver all of the traffic.

You should also consider traffic patterns. For example, are the endpoints devices IP security cameras that send a constant stream of data, or are they user devices that send and receive bursts of traffic as users download an attachment or browse the Internet? If you thought that all 48 devices would require 40 Mbps at all times, you would probably want to plan more than 2 Gbps to give yourself a buffer. For this example though, 40 Mbps per device defines the peak requirements, which the uplink can meet.

In more normal circumstances, some devices will have lower demands while others have higher, and the uplink will only operate at 50% utilization or less. This gives you a buffer for future increases in demands.

Customers will have the best experience when you plan to meet peak requirements on all devices and provide a buffer for expansion. Make sure the solution will continue to meet customers' needs for at least 3 years.

Example designs for fixed-port switch ports

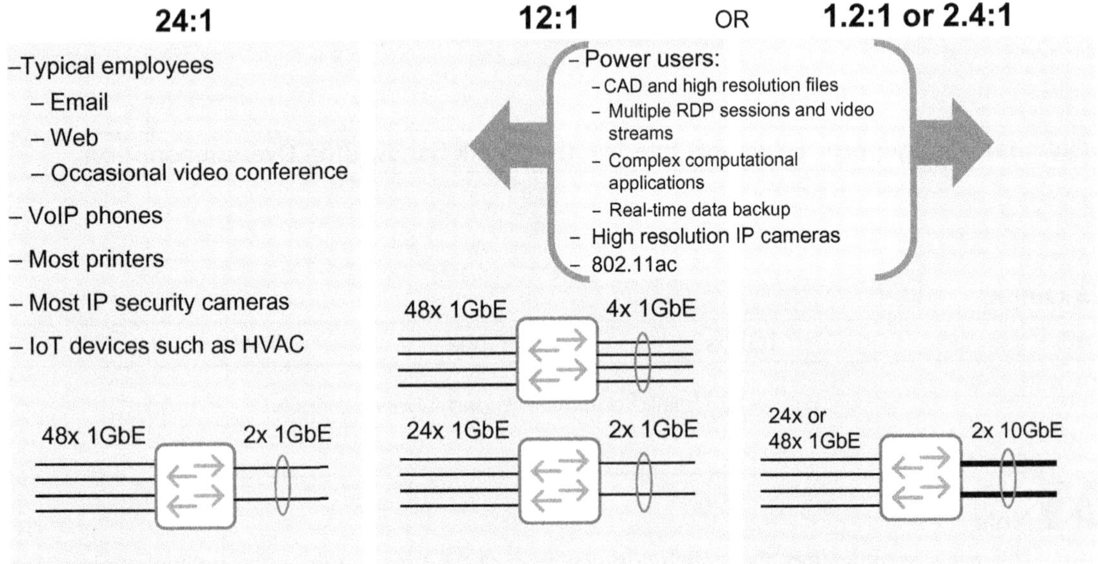
24:1
12:1
OR
1.2:1 or 2.4:1
–Typical employees
– Email
– Web
– Occasional video conference
– VoIP phones
– Most printers
– Most IP security cameras
– IoT devices such as HVAC
48x 1GbE
2x 1GbE
– Power users:
- CAD and high resolution files
- Multiple RDP sessions and video streams
- Complex computational applications
- Real-time data backup
- High resolution IP cameras
- 802.11ac
48x 1GbE
4x 1GbE
24x 1GbE
2x 1GbE
24x or 48x 1GbE
2x 10GbE

Figure 4-14 Example designs for fixed-port switch ports

Generally, the most oversubscription that you should plan is 24:1, which a 48-port switch with two 1 GbE links can provide, as shown in Figure 4-14. This works when most devices have peak requirements under 30 Mbps–40 Mbps and lower average requirements. Examples include desktops used by typical employees—such as users sending email or browsing the Internet. Even users running video conferencing applications fall below these levels as long as they are not running several other intensive applications at the same time. VoIP phones, printers, most IT security cameras, and many IoT devices also fall below these ranges.

Power users might require more bandwidth, as shown in Figure 4-14. Examples of power users include users who work with large CAD files or high-resolution medical graphics. Power users might also run many RDP sessions or multiple HD streaming videos. Very high-resolution cameras also require as much as 70 Mbps or 80 Mbps.

When devices have higher peak requirements, you must decide whether to expand the number of 1GbE links to four for a 12:1 oversubscription ratio or upgrade to 10 GbE links—two of them to maintain redundancy.

Enterprise APs can require hundreds of Mbps of throughput. If you are deploying a switch dedicated to APs, it will require a low oversubscription ratio, which two 10 GbE links can provide.

The leap to 10 GbE can more than satisfy requirements now and in the future with a very generous 2.4:1 oversubscription ratio. In fact, depending on the environment, the customer might even prefer two 10 GbE links to multiple 1 GbE links to reduce the amount of cable required.

Typically, you spread APs across several switches to prevent a single failure from wiping out the wireless coverage in an area. However, even if a switch only connects to eight APs, the uplink bandwidth requirements might push past the ability of the switch to deliver with 1 GbE, so you would decide to use 10 GbE connections.

Example designs for 5400R switches at the access layer

About 20:1 or lower

- –Typical employees
- –VoIP phones
- –Most printers, cameras, and IoT devices

About 6 or 7:1 or lower

- –Power users
- –High resolution IP cameras
- –Mix of 802.11ac APs and typical employees
- –Add more 10GbE links to accommodate more 802.11ac APs

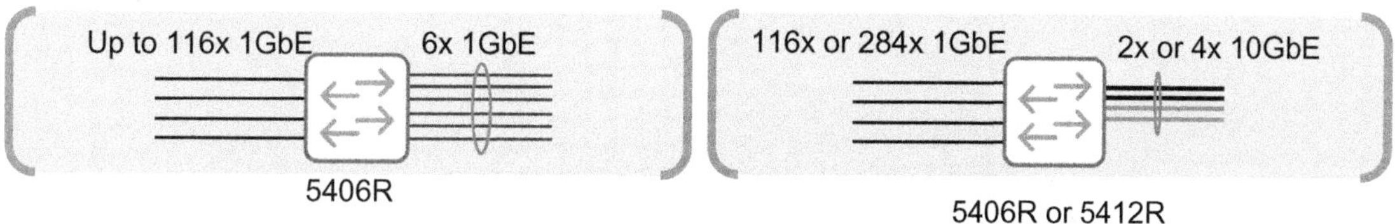

Figure 4-15 Example designs for 5400R switches at the access layer

If you deploy the Aruba 5400R Series at the access layer, a single switch can support a large number of ports with a great deal of flexibility. The two examples in Figure 4-15 show the high end of edge ports that a single 5406R switch might support.

In the first example, a 5406R switch has 8 1 GbE uplinks and up to 136 1 GbE downlinks, delivering an oversubscription ratio of about 20:1 (actually 17:1).

In the second example, the uplinks upgrade to 10 GbE, providing support for more power users and APs. A fixed port switch with, at most, 48 ports will often under use the two 10 GbE links. With the modular 5406R, you can increase the number of edge ports to around 140. Even with 140 edge ports, the switch will provide about a 7:1 oversubscription ratio, which can meet the needs of power users, high-resolution IP cameras, and a mix of typical employees with a few 802.11ac APs. The 5412 could have up to 284 1 GbE ports and four 10 GbE ports for an oversubscription ratio of about 7:1.

You might choose a design like the one in the first example when you want to provide high density of wired connections to devices with relatively low bandwidth needs. You can create this design with two modules that each provide 20 10/100/1000 Mbps ports and four SFP ports, and four other modules, each of which could provide up to 24 10/100/1000 Mbps ports.

The design in the second example could also meet the needs of a mixed environment with several APs that support the bulk of the traffic, a few printers, a few IoT devices, and some desktops. You can create this design with two modules that each provide 20 10/100/1000 ports and four SFP+ ports; and four (or 10) other modules, each of which could provide up to 24 10/100/1000 Mbps ports. Note that this design also provides plenty of extra SFP+ ports for additional uplink bandwidth, as well as redundancy with uplinks on different modules. It provides 136 edge ports on 5406R switches and 280 ports on 5412R switches; if you wanted 140 or 288 edge ports, you would need to change one of the modules with four SFP+ ports to a module with 24 10/100/1000 Mbps ports.

Also, note that the 5400R series supports 40 GbE ports, so you can deploy 40 GbE to the edge if needed.

Uplink plan for backplane stacks and VSF fabrics

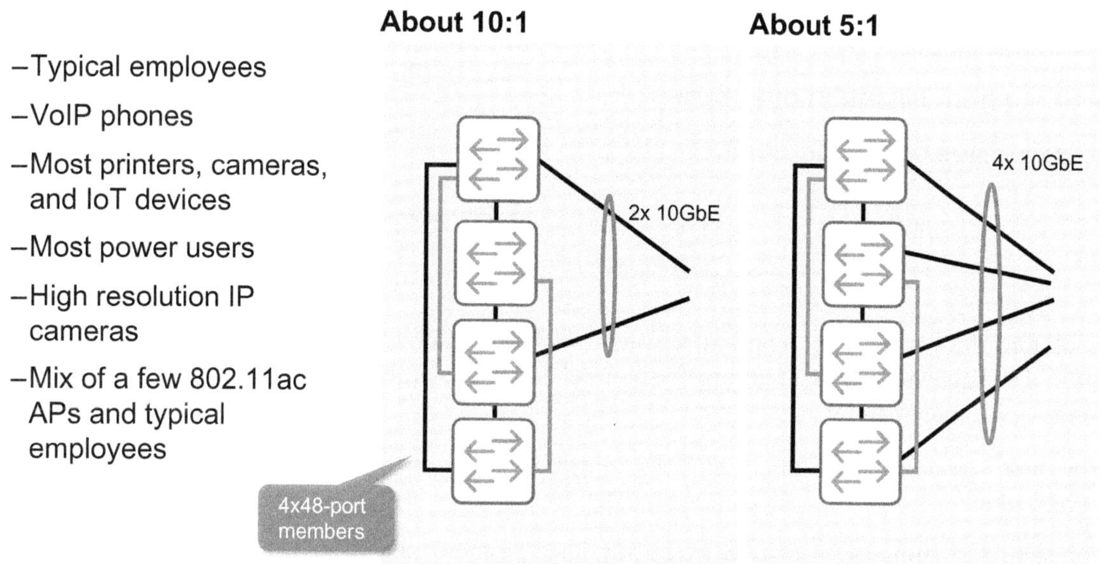
About 10:1
About 5:1
–Typical employees
–VoIP phones
–Most printers, cameras, and IoT devices
–Most power users
–High resolution IP cameras
–Mix of a few 802.11ac APs and typical employees
2x 10GbE
4x 10GbE
4x48-port members

Figure 4-16 Uplink plan for backplane stacks and VSF fabrics

A backplane stack or VSF fabric allows multiple switches to share 10 GbE uplinks and to be managed as a single device. The example plan in Figure 4-16 features a four-member backplane stack or VSF fabric. The stack has 192 downlink ports and two 10 GbE uplinks distributed on two different members that are one hop apart. This plan provides an oversubscription ratio of under 10:1, which is suitable for all of the applications and devices except a very high density of high-bandwidth 802.11ac APs.

You can add 10 GbE links to other members if you need to increase the bandwidth.

You can also add more members to the group and create a higher oversubscription ratio. However, although you can combine some switch models in larger groups, you should cap the members at four or five in most cases. You should also loop the fabric in a ring topology or full mesh (if supported).

Using current traffic information to assess a plan's validity

If your customer has a current solution, you can begin with the customer's current oversubscription ratio as a baseline.

You can then work with network managers to analyze current traffic volumes and patterns to determine how well the current oversubscription is working. You can use a network management solution, such as Aruba AirWave, as well as CLI diagnostic commands, like **show interface**. Figure 4-17 shows an example of output that you can see in AirWave.

Look for congestion:

–Peak utilization over 80%

–Sustained utilization over 60%

–Error rates over 1 percent

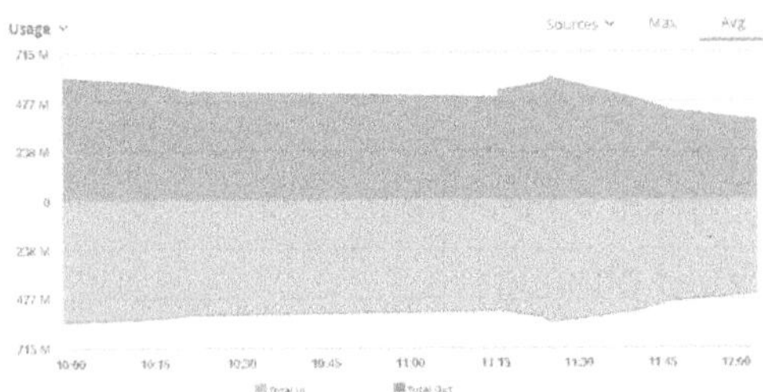

Port utilization monitoring on Aruba AirWave

Figure 4-17 Using current traffic information to assess a plan's validity

Look for information such as peak usage on access layer switch uplinks. A wired link can use about 80% of its bandwidth. If you see usage at this level, the link is probably dropping traffic and you need to provision more bandwidth.

The link should not sustain over 50% or 60% usage for either transmitting or receiving information.

You should examine the error rates over a sustained period of time. High error rates (over one percent) indicate that congestion is occurring.

When you detect congestion, you can make a plan to resolve the issue. For example, the customer has a 48-port Gigabit switch, which has two 1 GbE uplinks each. The current oversubscription is 24:1. You might plan to upgrade to a backplane stack with two 10 GbE uplinks, achieving less than half the current oversubscription, as shown in the example in Figure 4-18.

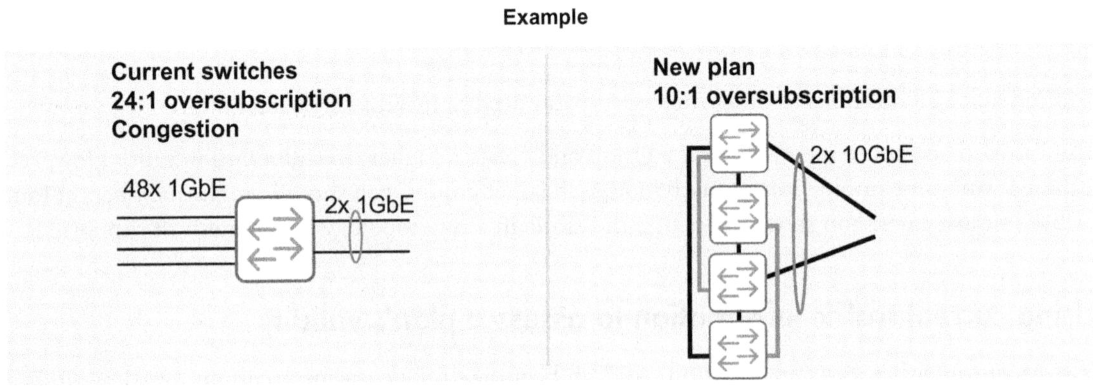

Figure 4-18 Example of creating a new plan

Although you can use CLI diagnostic commands to find out how the customer's oversubscription is working, you might keep in mind that network management solutions consolidate information, show trends, and otherwise help you to analyze the data.

Where peak usage is concerned, be aware that brief peaks do not matter as much as peaks that are sustained over longer periods. Generally, if a link occasionally but rarely rises to 50% or 60% usage, you do not need to worry. However, if the link is often used at over 50%, you might want to provision more uplink bandwidth.

Note that you are looking at both transmit and receive usage on the access layer edge ports. Often the receive usage will be higher because traffic tends to flow from servers toward endpoints.

You might find lower usage on one uplink and unacceptably high usage on another uplink. A spanning tree can introduce this type of inefficient load balancing. If this is the case, rather than add more bandwidth, you might take steps (such as implementing VSF at the aggregation layer or core) for improving load balancing across the links.

You can also look at average usage. However, peak usage usually gives the best information about the traffic that the switch needs to support.

Finally, consider variations over different switches. You should look at the usage on different switches to determine whether traffic is distributed more or less evenly or whether particular locations handle a higher traffic load. You might need to plan different oversubscription ratios for different switches to account for variable usage patterns.

Planning adequate bandwidth for access layer VSF fabric: Upstream

- Traffic sent on local link in upstream aggregation;
  or, if no local, selected by hash
- VSF links must support roughly half of upstream (this example)
- Can reduce by adding uplinks
- But keep uplinks divisible by two

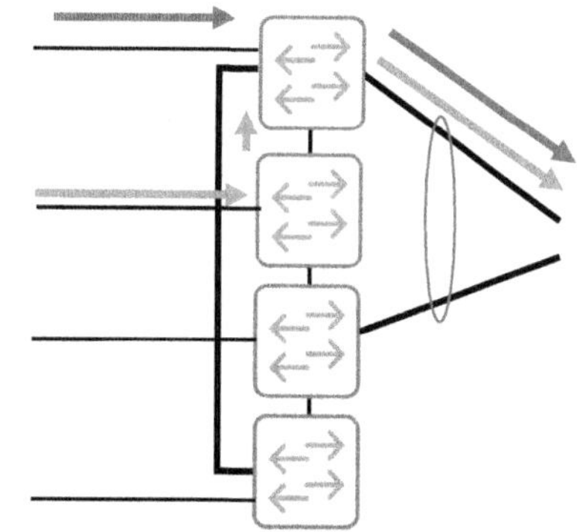

Figure 4-19 Planning adequate bandwidth for access layer VSF fabric: Upstream

Now turn your attention to the links between the switches in the fabric stack. Special stacking cables provide sufficient bandwidth that you usually do not need to worry about the backplane stacking links as long as you create the recommended ring or complete mesh. However, when you use VSF, you need to make sure to provision enough bandwidth for the VSF links.

When a VSF member needs to forward traffic on a link aggregation, it favors the aggregation's local link. This cuts down on the inter-member traffic flowing from endpoints up toward the core.

In a VSF fabric where half of the members have uplinks, roughly half of the upstream traffic can be forwarded locally while the other half of the traffic must cross a VSF link to a member with an uplink one hop away.

In this example, each member has two 10GbE VSF links and the uplinks as a whole only support 20 Gbps. Therefore, the VSF bandwidth should be adequate.

You can reduce upstream traffic on VSF links by adding uplinks to more members. However, you should keep the total number of uplinks even to balance them across the typical two aggregation or core switches. Do not make your plan until you consider downstream traffic.

In the example, the VSF links are oversubscribed in the sense that member 2 could be receiving 48 Gbps of traffic from endpoints and has 20 Gbps of VSF bandwidth to forward the traffic. However, this design assumes that a 10:1 oversubscription ratio amply meets the needs so this level of oversubscription should not negatively affect performance.

Planning adequate bandwidth for access layer VSF fabric: Downstream

–Traffic might arrive on any link
- –Must be forwarded to member connected to destination

–VSF links must support roughly ¾ of downstream traffic (this example)

–Cannot reduce by adding uplinks

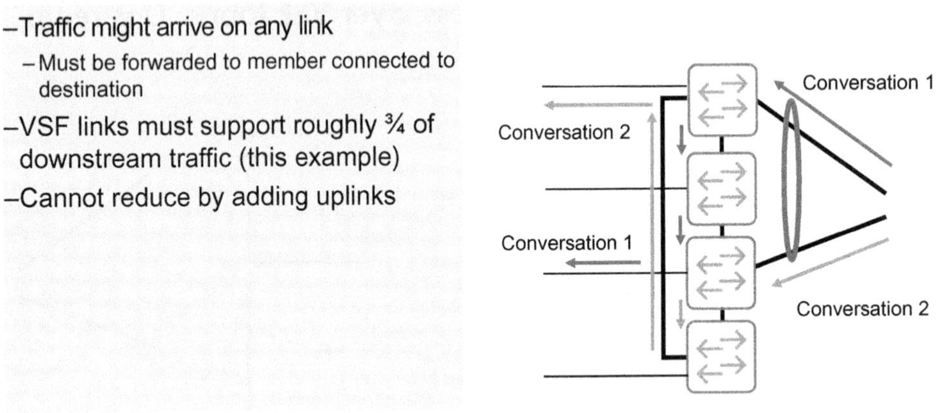

Figure 4-20 Planning adequate bandwidth for access layer VSF fabric: Downstream

Now examine the traffic arriving from the core and destined to the endpoints in Figure 4-20. Downstream traffic, which often forms the bulk of traffic in a campus LAN, might arrive on any of the VSF fabric's uplinks regardless of the destination because the upstream switch selects the link.

The member that receives the traffic must forward it to the member that connects to the destination endpoint. In a four-member group, the VSF links will carry roughly three-fourths of the downstream traffic, and you cannot reduce that by adding uplinks.

In this example, the VSF fabric has two 10 GbE uplinks, and each member has two 10 GbE VSF links for a total of four links between them. As long as the 10 GbE uplinks are not congested, the VSF links should be adequate to the demands. However, if the VSF fabric requires more uplink bandwidth, such as a 10 GbE link on each member, it might require more bandwidth on the VSF links.

The example VSF fabric should be able to handle traffic with one 10GbE link for each logical connection between members—but only as long as each member is able to reach other members within two hops, or preferably one.

A VSF fabric that requires more uplink bandwidth might also require more bandwidth on VSF links. Although placing an uplink on each member reduces the upstream traffic that must flow between the members, it does not reduce the amount of downstream traffic that flows between the members.

Edge port PoE and PoE+ requirements

You will now move on to another important feature for many access layer switches: the customer's need for PoE and PoE+.

First, you need to identify the devices that require PoE or PoE+ so that you can plan which edge ports need to support this feature. The current line of AOS-Switches supports PoE+ and PoE on switches that provide PoE.

Devices that require PoE include VoIP phones, security cameras, some IoT devices, and some enterprise APs. Devices that require PoE+ include dual-radio 802.11n or 802.11ac APs and sometimes higher resolution video phones.

Considerations for the PoE budget

Once you determine whether the customer requires PoE or PoE+, you can plan the power budget for each access layer switch.

In addition to clarifying how many total PoE/PoE+ devices the network needs to support and how these devices are distributed, you need to determine how much power each device will draw.

Is the switch expected to support all PoE devices drawing power at their peak requirements? When you are talking about critical devices such as APs and VoIP phones, the switch should meet peak requirements even if it is unlikely that all devices will draw the maximum power at the same time.

Also consider how the devices are distributed. Will a switch support a mix of PoE/PoE+ devices and non-PoE devices, or will it need to deliver PoE on every port?

PoE and PoE+ Review

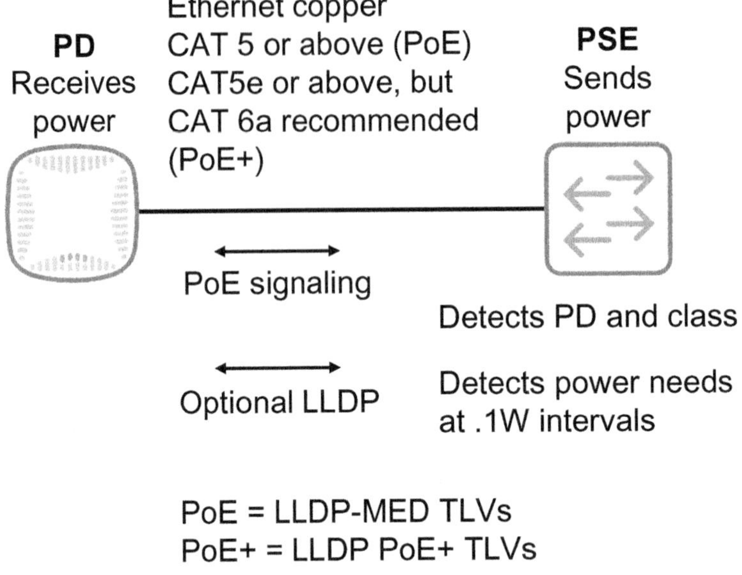

Figure 4-21 PoE and PoE+ Review

You must plan how switches allocate power to ports. To understand your options, you should understand a little about how devices request power and how switches allocate power to them.

Figure 4-21 shows an example Powered Device (PD), which is a device that receives PoE power. Power Sourcing Equipment (PSE) provides the power. When devices connect to PSEs, the PSEs detect whether the devices are PDs.

Classes indicate the power ranges that PDs require. The power available to the PD is less than the power required at the PSE due to cable loss. PDs can indicate their classes when they connect. Only PoE+ devices can use class 4.

PSEs classify PDs that do not indicate a class as 0, which can receive power up to the PoE maximum of 12.94W (perhaps slightly higher if the cable loss is low).

Table 4-4 PoE and PoE+ Classes

Class	Power available to PD (W)	Power required at PSE (PoE switch) (W)
0	0.44-12.94	15.4
1	0.44-3.84	4
2	3.84-6.49	7
3	6.49-12.95	15.4
4 (802.3at or PoE+ only)	12.95-25.5	30

PSEs and PDs can negotiate the power that PDs require more precisely at .1W intervals using LLDP messages. PoE uses LLDP-MED TLVs for this purpose. PoE+ includes power management through LLDP as part of the standard.

The PSE supplies power to the PoE PD over a CAT 5 cable or above and to the PoE+ PD over a CAT 5e cable or above. PSEs use the resistance between the powered pairs of wires in these cables to detect whether the device is a PD, in which case the PSE knows that it should power the device.

It is also by specific signature resistance that a PD indicates its class.

Some PDs might support Cisco Discovery Protocol (CDP) rather than LLDP for negotiating required PoE levels. AOS-Switches also support this protocol.

Note that although CAT5e cable can deliver PoE+, CAT6a cable is recommended for less loss and better delivery.

How AOS-Switches allocate power—No LLDP-MED

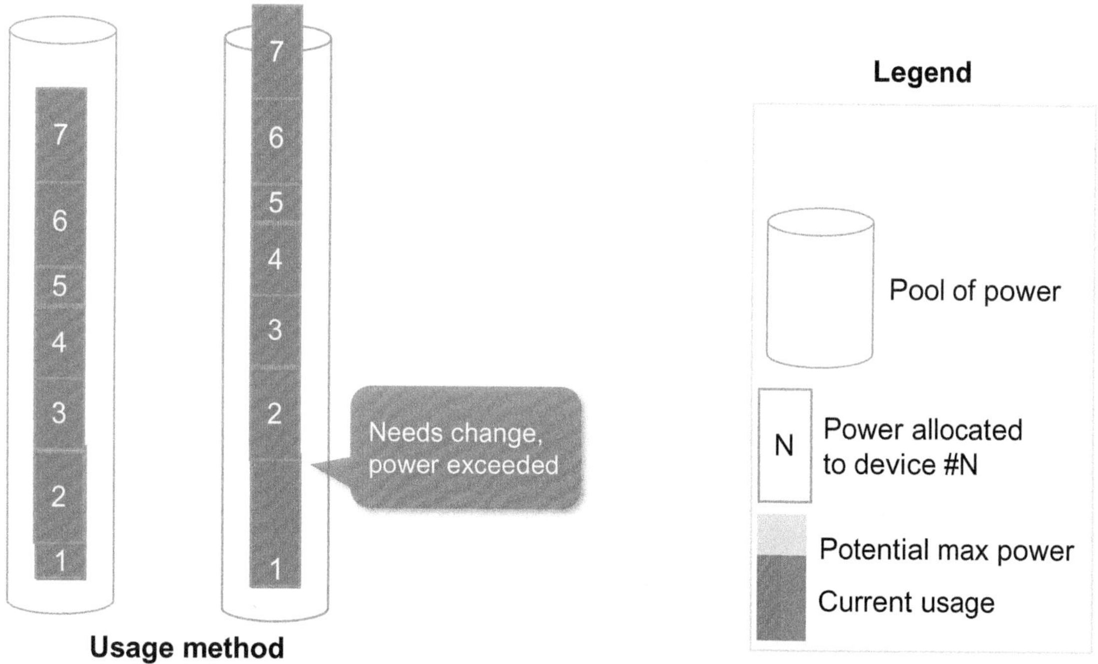

Figure 4-22 How AOS-Switches allocate power – No LLDP-MED and usage method

As you plan the budget, consider how the switch will allocate power to powered devices (PDs). Most AOS-Switches, by default, supply power to each port according to the usage method. With the usage method, it allocates just as much power as the PD currently draws, allowing it to meet the needs of as many PDs as possible.

Make sure that the switch has enough power to meet the maximum requirements of every device that might connect to it. Otherwise, if some PDs start to draw more power, the switch could run out of power, as shown in Figure 4-22. (The examples shown here show fewer devices than an AOS-Switch could actually support for purposes of space.) If the switch runs out of power, it supplies power to the ports according to their priority (low, high, or critical). If multiple ports have the same priority, the switch prioritizes the ports according to port ID.

You can use the class allocation method, illustrated in Figure 4-23, to avoid this issue. With this method, the switch allocates to the port the maximum power associated with the PD's class. In this way, the switch guarantees that it can fully support every PD to which it starts to deliver power. However, the switch could reserve more power for PDs than they need, compromising its ability to support as many PDs.

As an example of how the class method can cause the switch to be able to support fewer devices, consider APs with a maximum power consumption of 26W. These APs are class 4 devices, so the switch would reserve 33W when using the class allocation method. The reserved but unused power would translate to fewer APs supported.

You should also think about how port priority can affect operation on a switch that uses the class allocation method. With this method, the switch cannot use power it has reserved via the class allocation method even if the PD is not drawing power at that level. However, if a new PD connects on a port with a higher priority (or same priority, but better port ID), the switch stops supplying power on the lowest priority port and starts serving the new PD. You might want to set high or critical priority on the critical ports that are using the class option. If a new PD connects to a low priority port, the new PD does not cause another port to power down simply because the port has a lower ID.

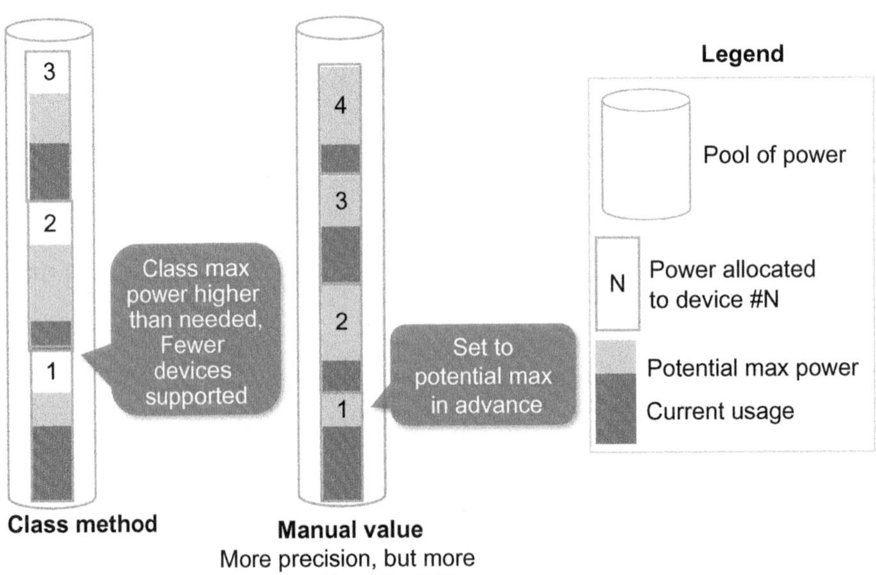

Figure 4-23 How AOS-switches allocate power—No LLDP-MED and class or manual method

You can prevent the switch from over-allocating power on a port by setting a manual power on each port. Setting the power manually also helps when you want to guarantee a specific power to a device that does not specify a PoE class. However, this method takes more planning and configuration effort and offers less flexibility.

Careful power planning with the usage method can also ensure that enough power is available for all PoE devices that might connect to your switch. You must simply make sure that you know which devices are connecting to each switch and the maximum amount of power that they might draw.

How AOS-Switches allocate power—LLDP-MED

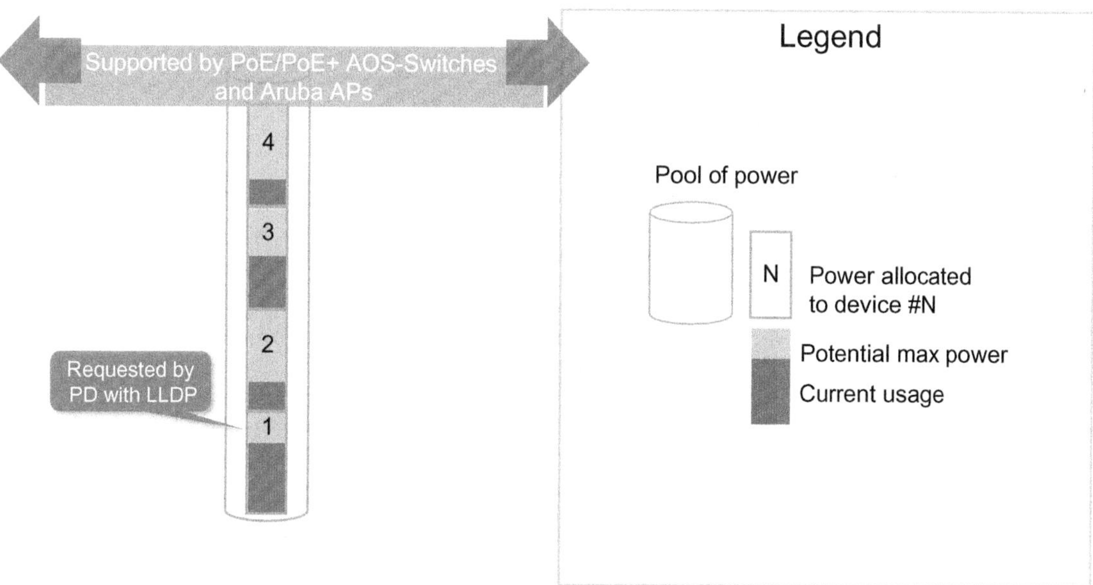

Figure 4-24 How AOS-Switches allocate power—LLDP-MED

LLDP-MED can offer a best of all worlds approach to allocating the power. The PD uses LLDP to indicate its power requirements. As shown in Figure 4-24, the AOS-switch reserves that amount of power for the PD and confirms the amount of power that it will supply with another LLDP message. Because the LLDP request is more fine-tuned than a class indication, the switch only allocates power that the device truly needs. As long as you enable LLDP-MED or PoE+ TLVs on a port, the AOS-Switch uses LLDP to allocate power regardless of the allocation method you set on the port. You should verify whether your PDs support LLDP-MED.

Aruba APs do support this feature.

You should know that PoE and PoE+ LLDP TLVs are different, and AOS-Switch ports only support PoE+ TLV by default (lldp config <int_ID_list> dot3TlvEnable poeplus_config). If the PD is a PoE device that supports LLDP-MED instead, admins can enable the port to accept the LLDP-MED messages with the **interface <int_ID_list> poe-lldp-detect enable** command. (If both types of TLVs are enabled and received, the PoE+ TLVs take precedence.)

When a PD sends a power request using LLDP TLVs and the port supports that type of TLV, the allocation method for the port becomes "lldp" regardless of the PoE allocation method configured on the port. The port reserves the number of watts the PD requests, subtracting that amount from the available power. The switch keeps the requested power in reserve for the port even if the PD isn't drawing all of the requested power at the moment. This behavior can help to prevent the switch from beginning to power too many devices and running out of power when the PD needs more power.

In some cases, you might want to set a manual value rather than accept the LLDP value. For example, you might have less important devices that request more power than they need when they are using the features that you want.

Calculating switch PoE budget

To calculate the power required per-switch, add up the maximum amount of power that the AOS-Switch might allocate to each PD. When you look up the maximum requirements in specs, make sure to look at the maximum power that the PD could draw from the PSE, which is higher than the power that is actually available to the PD due to cable loss. If necessary, add about 10% to account for the loss. Also remember that if the switch is using the class allocation method, you are summing the maximum power for the device's class rather than for the device.

As you sum the requirements, leave out the device with the lowest requirement. Add 17W or 33W instead, because an AOS-Switch needs at least 17W to supply power to a PoE PD and at least 33W to supply power to a PoE+ PD, regardless of how much power the PD actually needs. You cannot get too close to the PoE limits or the switch cannot bring up the PoE devices.

Here is an example of how to determine the minimum power required for 8 APs. The APs use LLDP-MED to request their power, and they need a max of 26 W each. You multiply 26 by 7, which totals 182W. You then add 33W. The minimum power requirement is 215W.

Now review an example of failing to consider the power required to bring up the last PD. A switch might have 380W available, so you believe that it can power 19 APs that draw up to 20W. If 18 APs are connected and drawing max power, only 20W is left. A nineteenth AP only needs 20 W, but the switch cannot start supplying power to the device unless 33W are available.

You should also consider the implications of the power required to bring up a device when multiple devices connect at the same time. For example, a customer connects 20 phones to ports at the same time. Each phone might only require 5 W to 7 W, but it takes a moment for the switch to detect the requirements and allocate the correct amount of power (whether by usage, class, or LLDP level). Therefore, the switch might not be able to start bringing some ports up until it has started allocating less power to other ports, freeing up enough power to start the process on other ports. A switch reboot, when every device is coming up at the same time, is a more extreme example. You can use priorities to ensure that the switch allocates power to the most important ports first.

Meeting the PoE budget requirements

Once you know the maximum power required on the switch, you can check against the maximum power the switch model supplies, which Iris reports in the switch properties. If the switch meets your requirements, all is well.

If the switch cannot meet the requirements, you have a few options. In some cases, you can select a different power supply or add a redundant power supply. For example, the 2930M, 3810M, and 5400R switches support power supplies of varying power, as well as redundant power supplies. As you are choosing power supplies, also consider the redundancy needs. Rather than choose a higher powered supply, you might choose two lower-powered supplies for the extra redundancy. Or if you require the switch to deliver full PoE power in a failover situation, you could choose two higher-powered supplies.

In some cases, you might need to select a different switch model than the one you were considering because the original model does not supply enough power. For example, the power supplied by the 2930F model depends on the model itself.

Or you might be able to stagger the PoE devices across different switches, mixing them with non-PoE devices, and reducing the burden on any one device.

AP and switch power specifications reference

To help you determine how much power your APs need—and which power supplies your switches need to meet the requirements—the sections below provide some power specifications for Aruba devices. Note that you should always check these specifications against the latest datasheets or information in Iris.

Note that you will also need to find power requirements for non-Aruba PoE/PoE+ devices. IP phones often list their power requirements as a PoE class. A class 1 devices requires up to 4W whereas a class 3 device requires a maximum of 15.4W (refer to Table 4-4).

Example Aruba AP power requirements

Depending on the model of the AP, the PoE+ requirements vary. As an example, the AP 340 series require 25.1W of maximum power whereas the AP 303 series requires 13W of maximum power.

Example AOS-switch power specifications

The current AOS-Switch Series offer a variety of non-PoE and PoE+ models. The PoE+ models can deliver both PoE+ and PoE.

The 2930F has an integrated power supply. Some of the models can supply PoE+. The maximum power supplied ranges from 125W to 740W depending on the model.

The 2930M and 3810M PoE+ models have similar power specifications. The PoE+ models in these series support dual modular supplies. Each switch can take either one or two 680W or 1050W power supplies. You cannot mix the type of power supplies on the same switch. The dual power supplies provide N+1 redundancy for the switch power as well as additional power for PoE+.

The 2930M and 3810M PoE+ models can supply from 370W to 1440W, depending on the number and type of power supplies installed. The max PoE+ power supplied also depends on the number of PoE+ ports on the switch. For example, you could install two 1050W power supplies in both a 2930M 48G PoE+ switch and a 2930 24G PoE+ switch; however, the former switch will deliver up to 1440W while the latter will deliver only up to 840W because each port can deliver a maximum of just over 30W. If one of the power supplies fails on the switches, both models could then deliver just 740W—the drop off in power is lower for the 24G PoE+ switch in this scenario. The maximum PoE+ power delivered by the 2930M or 3810M 24 SR switch is 860W with two 1050W power supplies.

The 5406R has dual modular power supplies of 700W, 1100W or 2750W. The 5412R supports the same power supplies, but it has four power supply slots; at least two power supplies are required. For both models, all installed power supplies must be the same type.

The maximum amount of PoE+ power that the 5400R switch can deliver depends both on the power supplies and the interface modules that are installed in the switch. The power supplies power the switch itself, and leftover power forms a PoE+ reserve. Different interface modules have different power requirements, so different interface modules combinations might alter the power remaining to the PoE+ reserve. Table 4-5 indicates the maximum PoE+ power delivered when a switch has no interface modules installed. The best way to determine the actual PoE+ reserve is to use Iris. Populate the switch with the desired interface modules and then check the PoE+ reserve in the Properties window. You can also choose a level of redundancy, and Iris will indicate the PoE+ power for which that level of redundancy can be provided.

Table 4-5 5400R power supplies

Power supply type	Number	Max PoE+ power (no interface modules and no redundancy)
700W	1	275W
	2	508W
	3	762W
	4	1016W
1100W	1	875.5W
	2	1664W
	3	2496W
	4	3328W
2750W	1	2380.3W
	2	4624W
	3	6936W
	4	9248W

Activity

You will now compete a design activity in which you apply what you have learned so far in this chapter. This activity builds on the scenario introduced in earlier chapters.

You will now plan an upgrade to the customer's wired network to meet the new needs. In particular, this upgrade will support the new wireless network, but also to meet expanding performance requirements.

Detailed scenario for the wired design

Use the scenario information from previous chapters plus the additional information in the scenario below to complete the activity.

Wired upgrade goals

Corp1 needs a wired upgrade to support the upgraded Wi-Fi network. In addition, although the company is shifting toward wireless, the customer wants to continue support for the currently active wired jacks. Employees who want a wired connection can continue to use their docking stations. In later phases, the customer plans to repurpose some of the wired ports to support IoT devices such as smart HVAC devices.

In discussions, the customer has clarified that the campus aggregation layer switches are routing the campus traffic, but a pair of FlexFabric switches are acting as a combined core for both the data center and the campus. As part of the campus upgrade, the customer wants to introduce separation between the data center core and a proper campus core. It is important for the customer that end-user wired and wireless traffic be routed by campus switches. Also, they want administrators to manage the complete campus topology using the same software and the same management tools such as AirWave.

The campus currently features chassis-level redundancy at the aggregation layer and core; the customer wants to provide this same level of redundancy in the new campus design.

Logical topology

You have collected information about the campus network. Figure 4-25 shows the current logical topology.

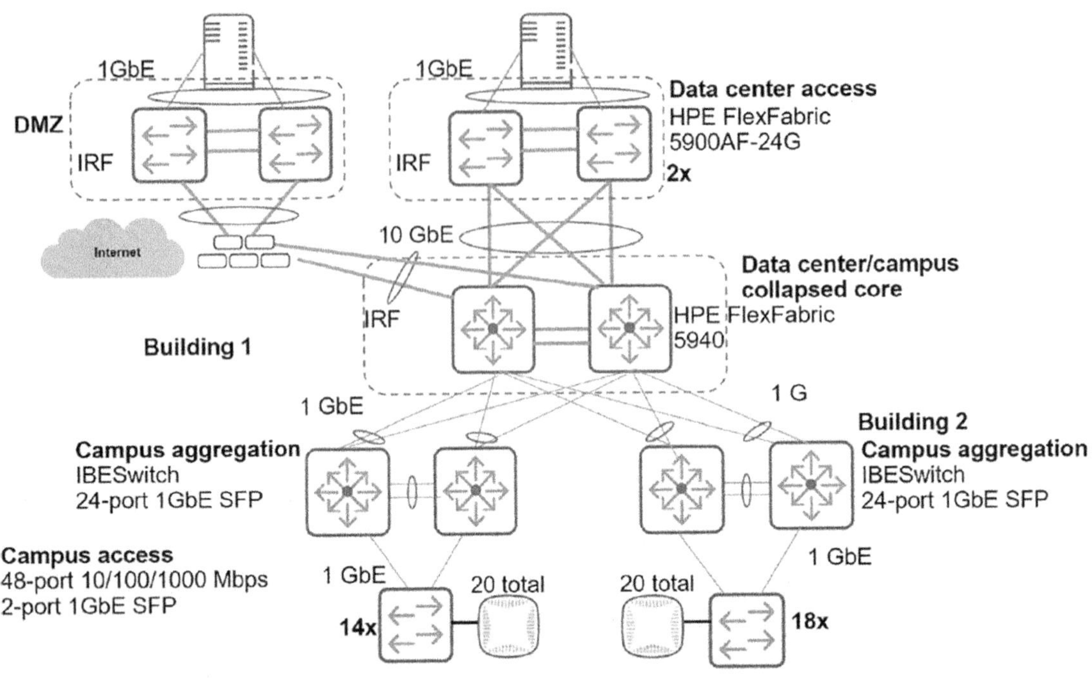

Figure 4-25 Existing logical topology

Campus physical topology

Figure 4-26 shows the physical topology. Each wiring closet in Building 1 connects with 12 strands of OM3 multimode fiber to the main distribution frame (MDF) in the data center, where the existing Building 1 aggregation switches are located. This rack is within 5 m of the MDF.

Each wiring closet in Building 2 connects to an IDF in Building 2 Floor 1 Closet 1 with 12 strands of OM3 multimode fiber. This IDF also connects to the MDF in Building 1 on 12 strands of OM3 multimode fiber.

Figure 4-26 indicates the distances for cable runs between each closet.

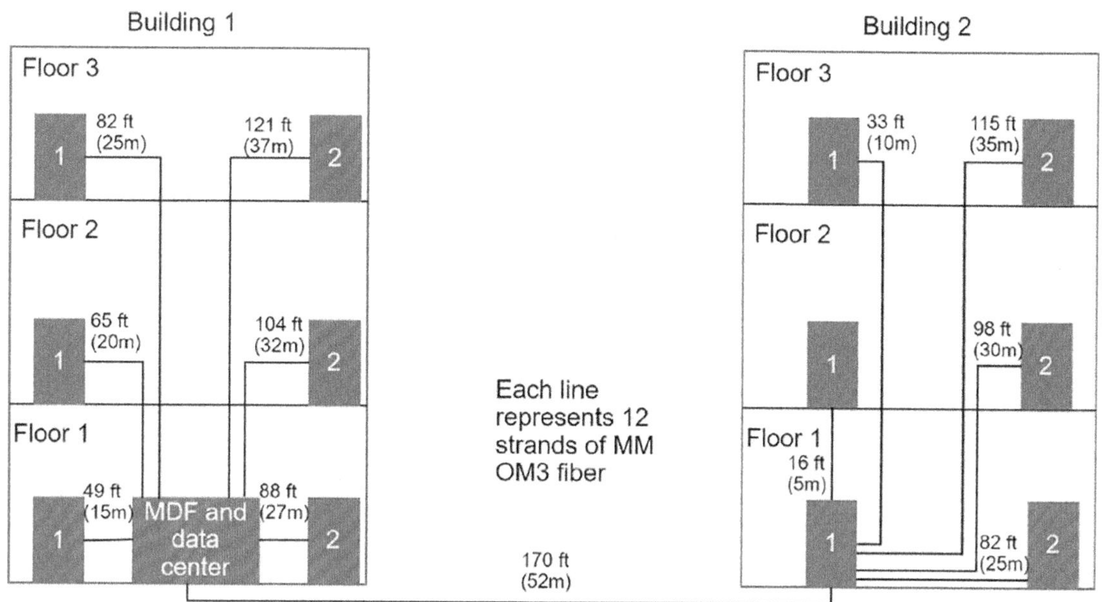

Figure 4-26 Existing physical topology

Wiring closets are located in the same place on each floor. Figure 4-27 gives an example. The cable drops use CAT5e cable. The maximum distance from a far corner of the floor to any wiring closet is 294 feet (89.6 m).

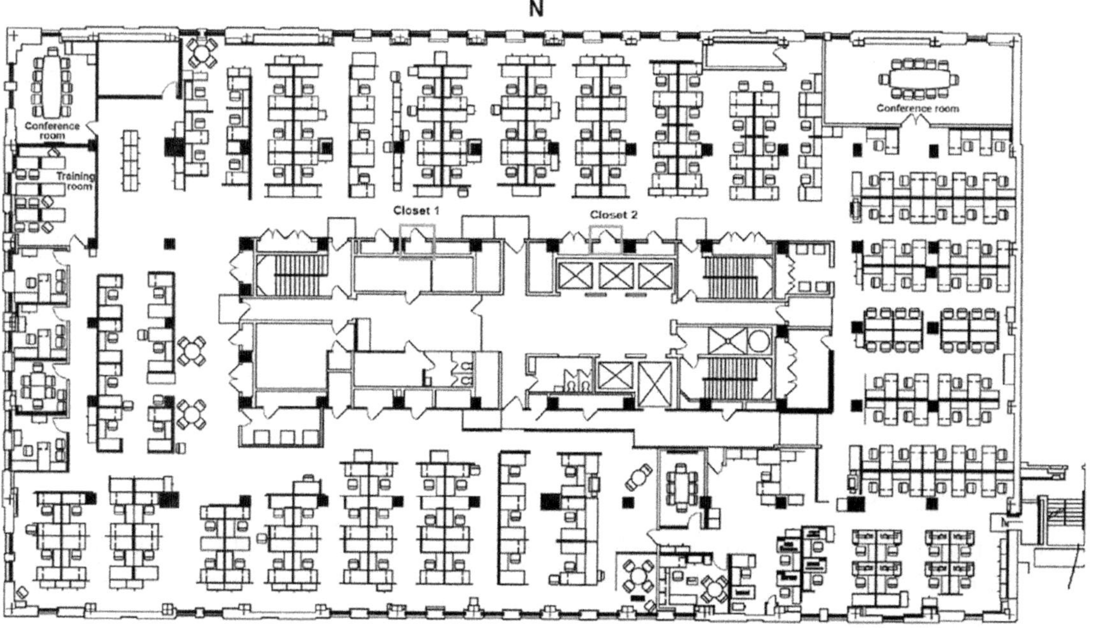

Figure 4-27 Floor with wiring closets marked

Additional campus network information

Links in the campus have been experiencing congestion, and users are noticing performance issues with the new applications. The campus aggregation layer switches only support 1 GbE. The campus switches all need to be refreshed.

This is the information that you have collected about the current network.

a. Wired network to be replaced: Y
b. If yes, which segments: Campus
c. Datacenter network vendor: FlexFabric
d. Campus network vendor: IBESwitch
e. Number of wired closets: 12
f. Rack space in each closet: 20U
g. Number of access layer switches per closet: 3 in 10 and 1 in 2
h. Number and speed of ports on access layer switches: 48 10/100/1000Mbps and 2 1 GbE (SPF)
i. Current access layer switch capabilities such as PoE and PoE+: PoE, 8 ports only; Layer 2
j. Is every desk wired (Y/N): Y
k. Wired network diagram supplied (Y/N): Y
l. Give a brief explanation of the customer's present-wired issues. Also note features that might be required that the current switches do not support:

   Users are experiencing performance issues. The current switches only support up to 1GbE. They do not support PoE+, which the new APs require. They also lack authentication capabilities. The aggregation layer switches also lack support for 10 GbE and 40 GbE and have relatively small MAC and ARP tables.

You determined that the existing wired network does not have enough drops for the upgraded Wi-Fi network, which has a higher AP density that the existing deployment and also places APs in new locations.

Data center network

The customer has a data center with several server racks. The data center network has been expanding was recently upgraded to HPE FlexFabric switches.

The two HPE FlexFabric 5940 switches at the core are deployed with Intelligent Resilient Framework (IRF) technology. Like VSF, IRF enables other devices to connect to both members of the IRF fabric over a link aggregation.

Wired network requirements

You have collected this information about the wired network requirements.

a. Excluding APs, how many drops per closet must the new switches support?

    i. Building 1

        1. Floor 1 Closet 1:35
        2. Floor 1 Closet 2: 35
        3. Floor 2 Closet 1: 125
        4. Floor 2 Closet 2: 125
        5. Floor 3 Closet 1: 125
        6. Floor 3 Closet 2: 125

    ii. Building 2

        1. Floor 1 Closet 1: 125
        2. Floor 1 Closet 2: 125
        3. Floor 2 Closet 1: 125
        4. Floor 2 Closet 2: 125
        5. Floor 3 Closet 1: 125
        6. Floor 3 Closet 2: 125

b. User types: Employees

c. Device types: HP EliteBook Folio G1 laptops in docking stations, HD printers on Floor 1 in both buildings, video conference equipment in conference rooms on Floor 2 of Building 1, board rooms on Floor 3 of Building 1, and conference rooms on all floors in Building 2

d. Types of applications in use: Web, Email, print, access to shared files, access to inventory software, video conferencing on special equipment

Activity 4.1: Plan the access layer

Activity 4.1 Objective

For your first activity, you will select new switches for the wiring closets in Building 1 and Building 2 to support the new customer requirements—in particular, the wireless deployment.

You will also implement your design in Iris to create a BOM for the access layer. If you do not have access to Iris, you can use the HPE Network Configurator instead. You will not be able to complete the steps for connecting devices together, but you can add the devices to your BOM.

Activity 4.1 Steps

1. The customer needs new cable drops to the new APs. What cable type do you recommend?

   ____

2. Calculate the number of edge ports per closet with the exception of Building 1 Floor 1 (B1F1) closets. Assume that you will connect half of the APs per floor to each closet. If you have an odd number of APs, plan for the closet that must support one more AP. Similarly, if you have a slight variation in range for the number of APs per floor, plan for the floor with the highest number of APs.

   125 + ½ APs per floor = ____________________________ number of edge ports per closet

3. Calculate the number of edge ports per closet on B1F1.

   35 + ½ APs per floor = ____________________________ number of edge ports per B1F1 closet

4. Begin to plan the switches for all closets except the ones in B1F1. These questions do not have a single correct answer, but be prepared to justify your answers.

   - Will you place APs on their own switches or on the same switches that support wired ports?
   - Will you use fixed-port (24-port or 48-port) or modular switches?
   - If you choose fixed-port switches, do you want to combine them with backplane stacking, VSF, or neither?

   ____

   ____

   ____

   ____

   ____

   ____

   ____

   ____

   ____

5. Based on your answers to the questions above, as well as the requirements that you have collected, which switch series will you use?

___

6. To further select the model from the series, you should consider PoE needs. Based on the scenario, which devices require PoE or PoE+? What additional questions should you ask your customer to make sure that you plan the correct number of PoE/PoE+ ports?

___
___
___
___
___
___
___
___
___

   You should have identified that the APs require PoE+. You should also pay attention to the fact the customer is planning to gradually repurpose wired connections for IoT devices. These devices might require PoE (802.1af). Assume that you have had further discussions with the customer about this need, and the customer wants to support PoE on all switch ports.

7. Do your APs have any other special requirements for their switch port?

___
___
___
___

   You will now use Iris to fill out your plan.

8. Return to your existing site within Iris.

9. Select your switch series from the Catalog (find the series by expanding the folders shown in Figure 4-28). Double-click to add the switch to your topology workspace.

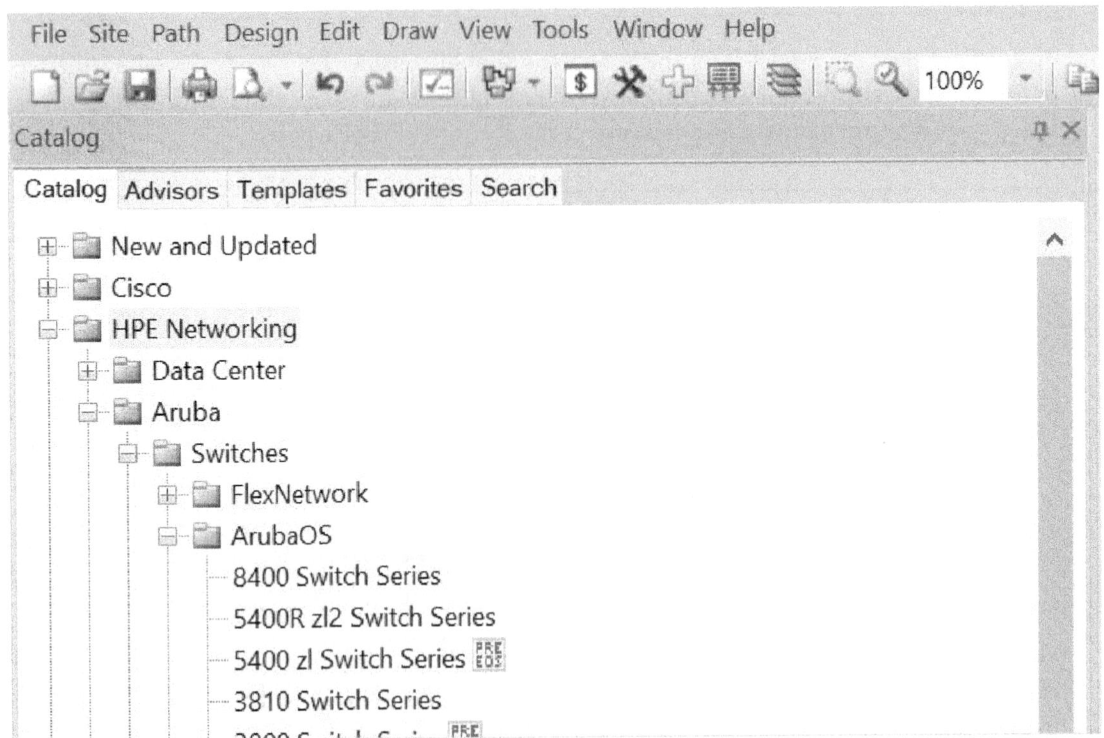

Figure 4-28 Activity 4.1: Plan the access layer

10. In the Properties window, scroll down and select a model that meets your needs for PoE+ ports and Smart Rate ports, if any. (Figure 4-29 gives one example of what the window might look like. You might have selected a different series.)

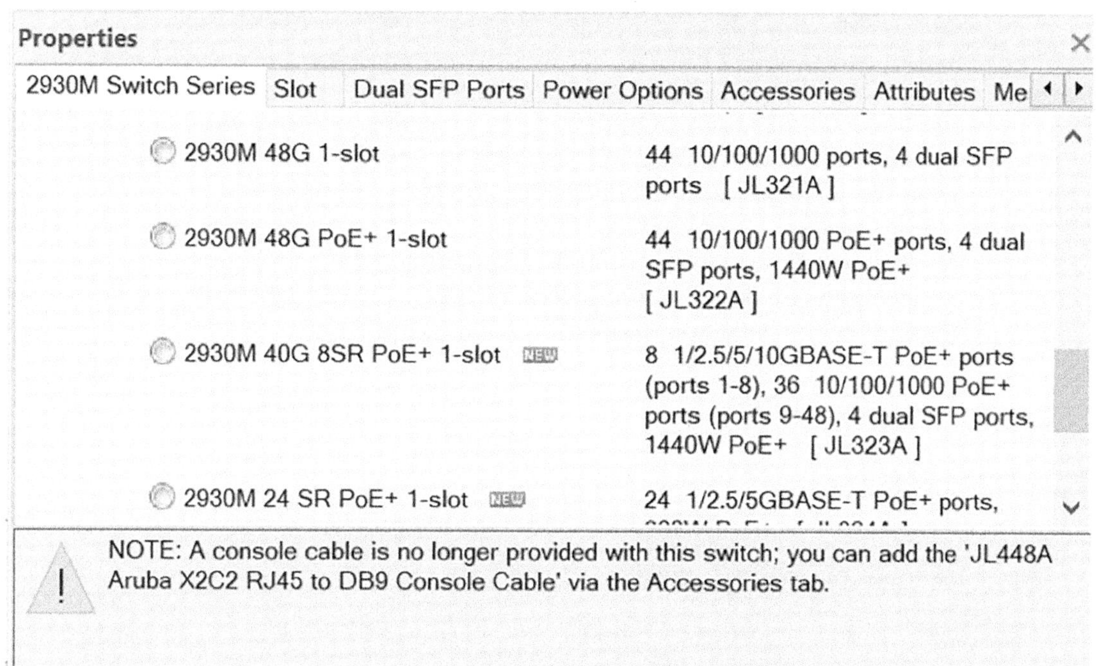

Figure 4-29 Activity 4.1: Plan the Access Layer

11. If you want to use backplane stacking, make sure that you also select the check box for the stacking module.

12. You only need to complete this step if you are planning a 5400R switch. Add a power supply.

    Select the Power Options tab and choose a Power Supply Type (see Figure 4-30).

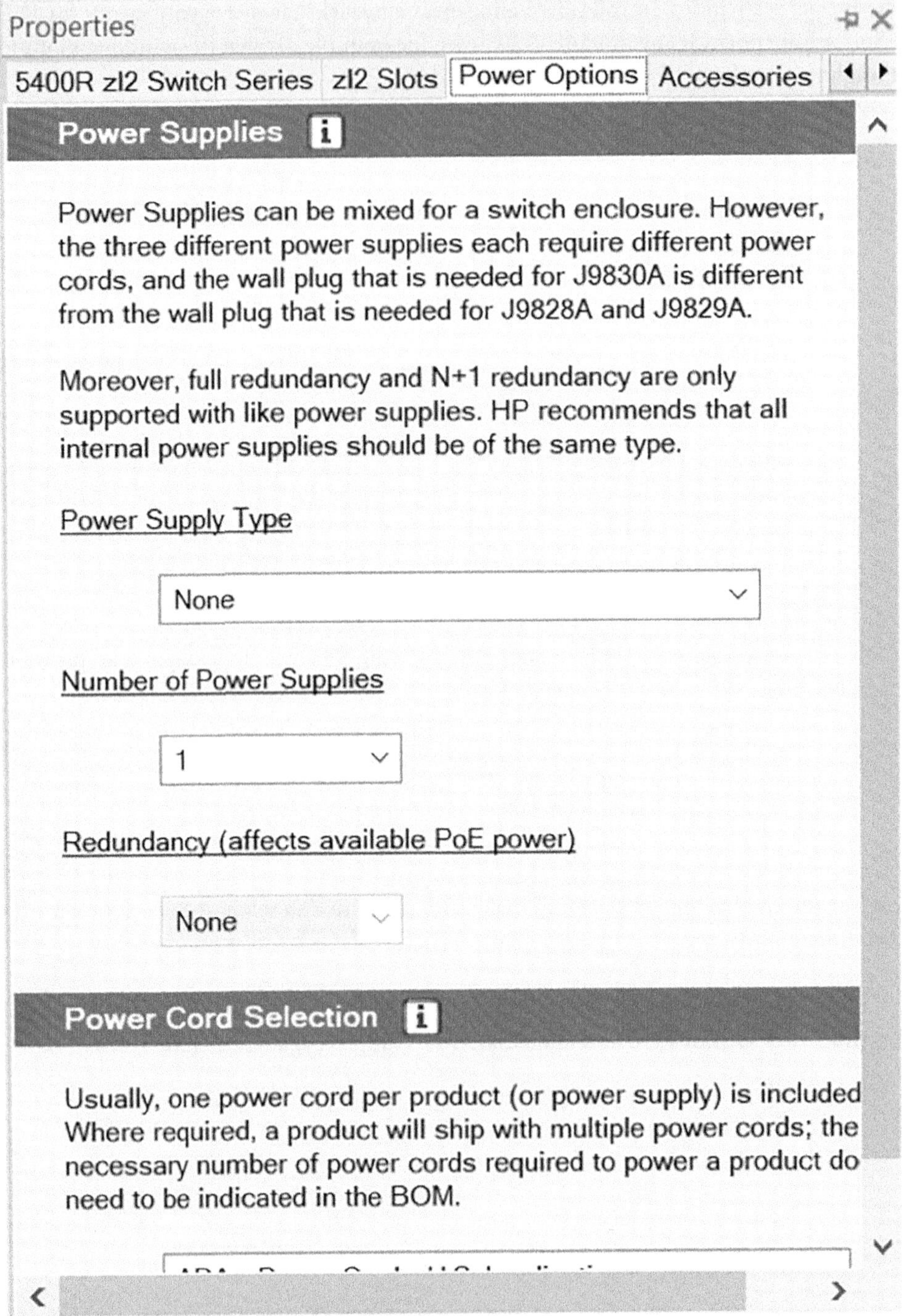
Properties
5400R zl2 Switch Series
zl2 Slots
Power Options
Accessories
Power Supplies
Power Supplies can be mixed for a switch enclosure. However, the three different power supplies each require different power cords, and the wall plug that is needed for J9830A is different from the wall plug that is needed for J9828A and J9829A.
Moreover, full redundancy and N+1 redundancy are only supported with like power supplies. HP recommends that all internal power supplies should be of the same type.
Power Supply Type
None
Number of Power Supplies
1
Redundancy (affects available PoE power)
None
Power Cord Selection
Usually, one power cord per product (or power supply) is included
Where required, a product will ship with multiple power cords; the
necessary number of power cords required to power a product do
need to be indicated in the BOM.

Figure 4-30 Activity 4.1: Plan the access layer

13. Next fill out the requirements for the edge ports:

    a. If you are using a fixed-port switch, add additional switches until you have met the edge port requirements for the closet (any except B1F1). As you do, think about how you will distribute APs across the switches and the requirements for the AP ports.

       You can right-click a switch in the workspace and select copy. Then right-click the workspace background and paste. You can then change the model if you want a slightly different model for one of your switches.

    b. If you are using a 5406R switch, select Slots and select the modules that you want for each slot.

14. Take notes on your plan below, including the exact switch models and modules you selected.

    ____

    ____

    ____

    ____

    ____

    ____

15. For how many closets can you use this plan (hint: all closets except the two on Building 1 Floor 1).

    ____

16. Multiply out the number of switches by setting the quantity multiplier for each switch that you have planned for this closet. (But you are keeping the switches in the closet separate, rather than using a quantity multiplier for them, so that you can plan VSF/stacking links and different uplinks, if you want.)

    a. Select all of the switches (hold down **[Ctrl]** as you select them).

    b. Enter the number of closets for Quantity Multiplier, as shown in Figure 4-31.

    c. Select True for Create Synced Set.

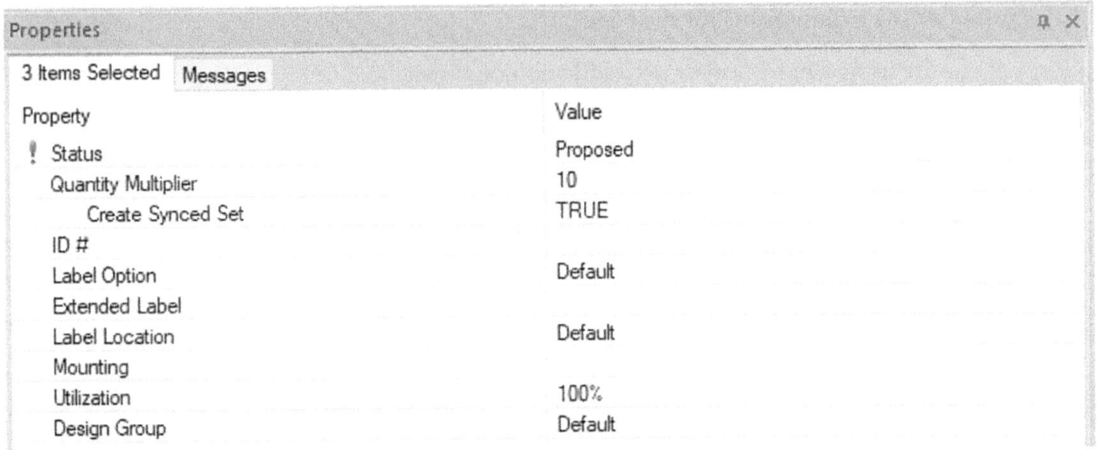

Figure 4-31 Activity 4.1: Plan the access layer

You should see "(set of 10)" next to your switches, as shown in Figure 4-32.

Figure 4-32 Activity 4.1: Plan the Access Layer

17. How will you adjust your plan for the B1F1 closets? Again, think about how you will distribute APs across the switches and the requirements for the AP ports, as well as other wired devices.

    ____

    ____

    ____

18. Add the switch or switches for *one* of the B1F1 closets. Then set the Quantity Multiplier to 2 to create a synced set for both closets.

19. Drag the B1F1 switches to another area of the workspace to keep them separate. You can label your closets as follows:

    a. Select Draw > Text box.

    b. Double-click the text box and enter the label in the Content tab. Select OK.

    c. Drag the text box where you want it to go.

    Figure 4-33 shows an example.

Figure 4-33 Activity 4.1: Plan the access layer

Activity 4.2: Access layer uplinks and VSF/stacking links

Activity 4.2 objective

You will now plan the uplinks for switches in each closet, taking care to plan adequate bandwidth based on the customer requirements. If you are planning to use VSF or backplane stacking, you will also plan those links. Remember to refer to the scenario as you answer questions.

Activity 4.2 steps

1. For Building 1 Floor 3 Closet 2, how many uplinks will you plan and of what speed?

   ____

2. Based on the scenario, what media do these links need to use?

   ____

3. What level of oversubscription does your plan give per-switch, VSF fabric, or backplane stack?

   ____

4. Based on the scenario, is this an appropriate level of oversubscription?

5. Will the same plan be appropriate all other areas of the campus? Adjust your plan if necessary.

6. How many fibers does your plan take per closet? Make sure that your plan fits with the current cabling indicated in the scenario.

7. If you are using a VSF fabric or backplane stack, plan how you will distribute the uplinks across the members.

___

___

___

___

You will now implement your plan in Iris.

8. Return to Iris and select one of the switches under "All closets except B1F1."
9. Check whether the switch supports the uplink speeds that you require. For a 2930F switch, you might need to change the switch model in the Properties window. For other switches, you may need to add a module with the ports.
10. If the switch has the required ports without the addition of a module, follow these steps. Otherwise, go to step 11.
    a. If necessary, change the switch model.
    b. Select the tab with the name of the ports. One example in Figure 4-34.
    c. In the drop down menu for one of the ports, choose the correct type of transceiver based on what you learned in the module and the distance between B1F3C2 and the data center where core switches are installed (see the wiring information in the scenario).
    d. Repeat the steps to add transceivers for other uplinks. If you plan to install the other uplink or uplinks on another switch in the closet, make sure to select that switch in the workspace first.
    e. Go to step 12.

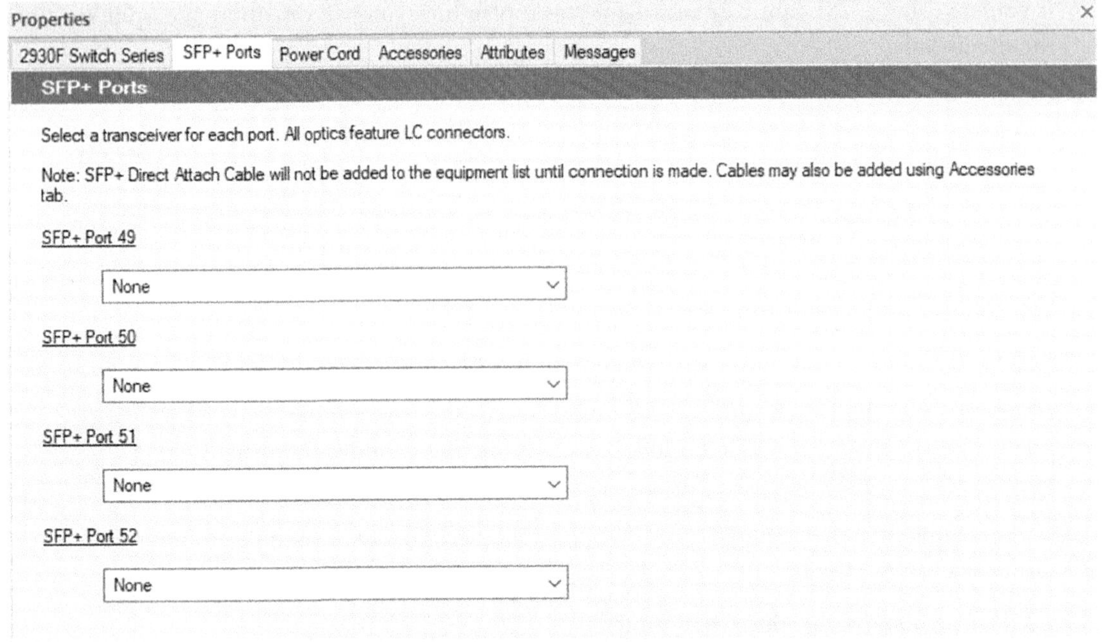

Figure 4-34 Activity 4.2: Access layer uplinks and VSF/Stacking links

11. If you need to add a module, follow these steps:

    a. Choose the module in the Slots tab. One example is shown in Figure 4-35.

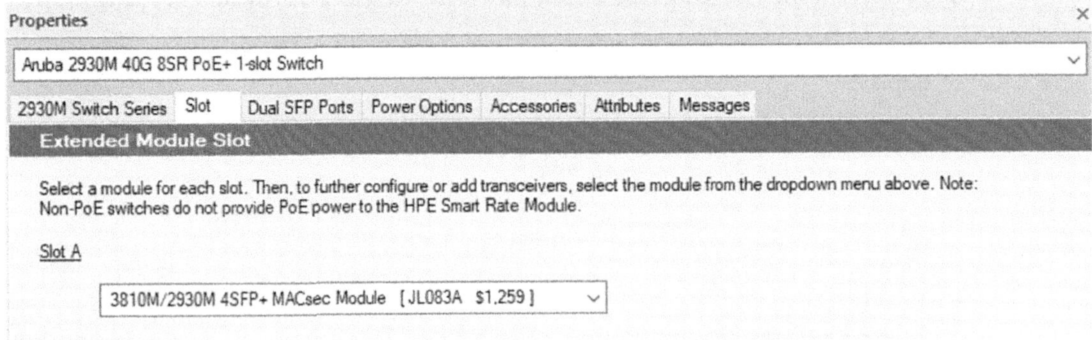

Figure 4-35 Activity 4.2: Access layer uplinks and VSF/Stacking links

b. Select the module that you added from the drop-down menu at the top of the Properties window (see Figure 4-36).

Figure 4-36 Activity 4.2: Access layer uplinks and VSF/Stacking links

c. Choose the correct type of transceiver based on what you learned in the module and the distance between B1F3C2 and the rack in the data center where campus aggregation or core switches will be installed (see the wiring information in the scenario). Fill in the desired number of transceivers in the field for the correct transceiver types.

d. Repeat the steps to add transceivers for other uplinks. If you plan to install the other uplink or uplinks on another switch in the closet, make sure to select that switch in the workspace first.

12. Which type of transceiver did you select?

____

13. The transceivers that you added are automatically multiplied across all 10 switches in the synced sets.

14. Follow step 10 or 11 again to add the uplinks to the switches in the B1F1 synced set.

15. Total the number of uplinks in each building. If all closets are using the same uplink plan, you can multiply the number of uplinks in B1F3C2 by six to find the number required in each building. Adjust as necessary if some closets are using a different plan.

    This information will be useful later when you plan the core and possible aggregation layers. You might want to tear out this page and keep it on hand.

    Building 1: ______________________________

    Building 2: ______________________________

    Continue to the next page.

    Complete the following steps if you are planning to use VSF or backplane stacking. Otherwise, move on to Activity 4.3.

16. What topology will you plan for your fabric or stack?

____

17. If you are using VSF, you must choose ports to dedicate to the VSF links. How many ports of what speed will you use? Do you already have ports available on the module that you chose for the uplinks, or do you need to add a module?

___

___

You will now implement your VSF or backplane stacking plan in Iris.

18. If necessary, add a module to your switches.
19. If you are planning a VSF fabric between switches in different closets, you might be using the same icon to represent the switches and cannot draw the VSF links. However, make sure to add necessary transceivers for the links. (For example, adding two transceivers to a switch multiplies out the number across the set.)
20. If you are planning a backplane stack or VSF fabric *within* a closet, draw the connections to add the DACs or stacking cables.
    a. Select the connect icon, which is shown in Figure 4-37. Select one of the switches and drag to another switch to establish a connection.

Figure 4-37 Activity 4.2: Access layer uplinks and VSF/Stacking links

    b. Choose the stacking module or the type of port for the VSF link.
    c. Fill in .5 m for the connection length.
    d. Select 10 for Quantity to indicate that you are connecting switches in all 10 closets.
    e. Select Connect (see Figure 4-38).
    f. The stacking cables or DACs are automatically added to the BOM.

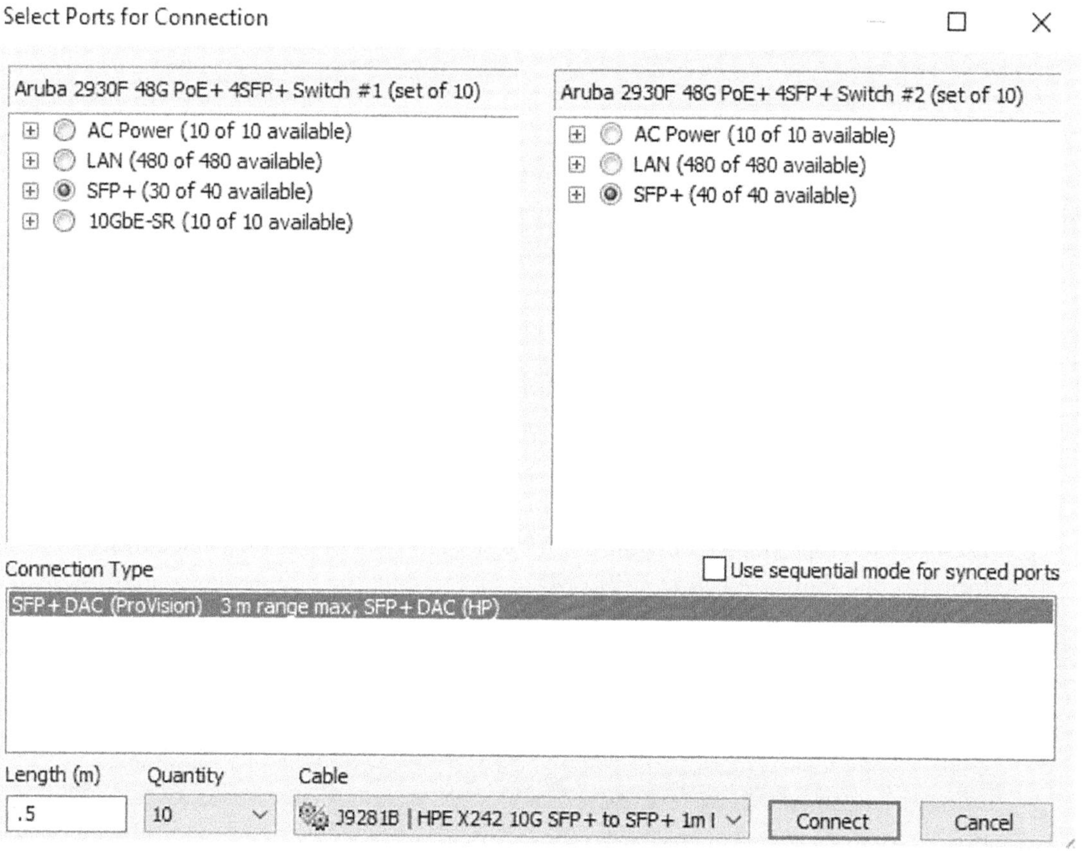

Select Ports for Connection

Aruba 2930F 48G PoE+ 4SFP+ Switch #1 (set of 10)

- AC Power (10 of 10 available)
- LAN (480 of 480 available)
- SFP+ (30 of 40 available)
- 10GbE-SR (10 of 10 available)

Aruba 2930F 48G PoE+ 4SFP+ Switch #2 (set of 10)

- AC Power (10 of 10 available)
- LAN (480 of 480 available)
- SFP+ (40 of 40 available)

Connection Type

Use sequential mode for synced ports

SFP+ DAC (ProVision) 3 m range max, SFP+ DAC (HP)

Length (m) .5 | Quantity 10 | Cable J9281B | HPE X242 10G SFP+ to SFP+ 1m I | Connect | Cancel

Figure 4-38 Activity 4.2: Access layer uplinks and VSF/Stacking links

g. Follow the same steps to finish the connections for the fabric or stack.

You now have a BOM for the access layer switches and APs in both buildings.

Activity 4.3: PoE Budget

Activity 4.3 objective

One of the reasons that you are proposing an access layer upgrade to the customer is that the customer's current switches do not support PoE+. In this task, you will plan the PoE budget for your APs. You will also make sure that the switches can provide enough power to meet future requirements.

Activity 4.3 steps

1. What are the maximum PoE requirements for the APs that you are proposing?

   You can use Iris to find out.

   a. Select the AP in the workspace.

   b. Look in the Properties window on the AP series tab (you might be using a different series from the one shown in Figure 4-39.) Make sure to note the 802.3at PoE+ maximum consumption, not the PoE consumption, which does not allow the AP to operate at full capacity.

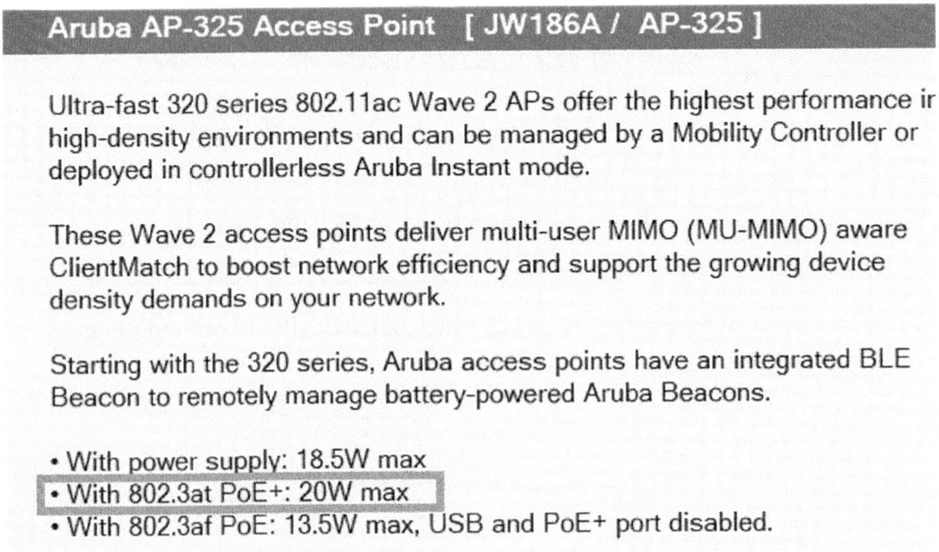

Aruba AP-325 Access Point [JW186A / AP-325]

Ultra-fast 320 series 802.11ac Wave 2 APs offer the highest performance ir high-density environments and can be managed by a Mobility Controller or deployed in controllerless Aruba Instant mode.

These Wave 2 access points deliver multi-user MIMO (MU-MIMO) aware ClientMatch to boost network efficiency and support the growing device density demands on your network.

Starting with the 320 series, Aruba access points have an integrated BLE Beacon to remotely manage battery-powered Aruba Beacons.

• With power supply: 18.5W max
• With 802.3at PoE+: 20W max
• With 802.3af PoE: 13.5W max, USB and PoE+ port disabled.

Figure 4-39 Activity 4.2: Access layer uplinks and VSF/Stacking links

If you looked at the data sheet, you would see that the AP can actually draw more power if it uses a USB, but you do not need a USB for this scenario.

2. For all closets except those on floor 1, how many APs are you planning to connect to per closet? (Remember to divide the number per floor by two for the number per closet.)

   And how do you plan to distribute the APs across the switches?

   ____

   ____

   ____

   ____

3. The Building 1 Floor 1 closets probably have a slightly different design. How many APs are you planning to connect to each switch in these closets?

   ____

   ____

   ____

4. Use the answers above to calculate how much power each representative switch in your topology requires. For example, if you are planning to connect four APs per switch on most switches but nine APs on a couple switches, calculate the requirements for four APs and for nine APs.

   ____

   ____

   ____

5. Select a switch and look at the bottom of its properties window for its current PoE reserve, as illustrated in Figure 4-40.

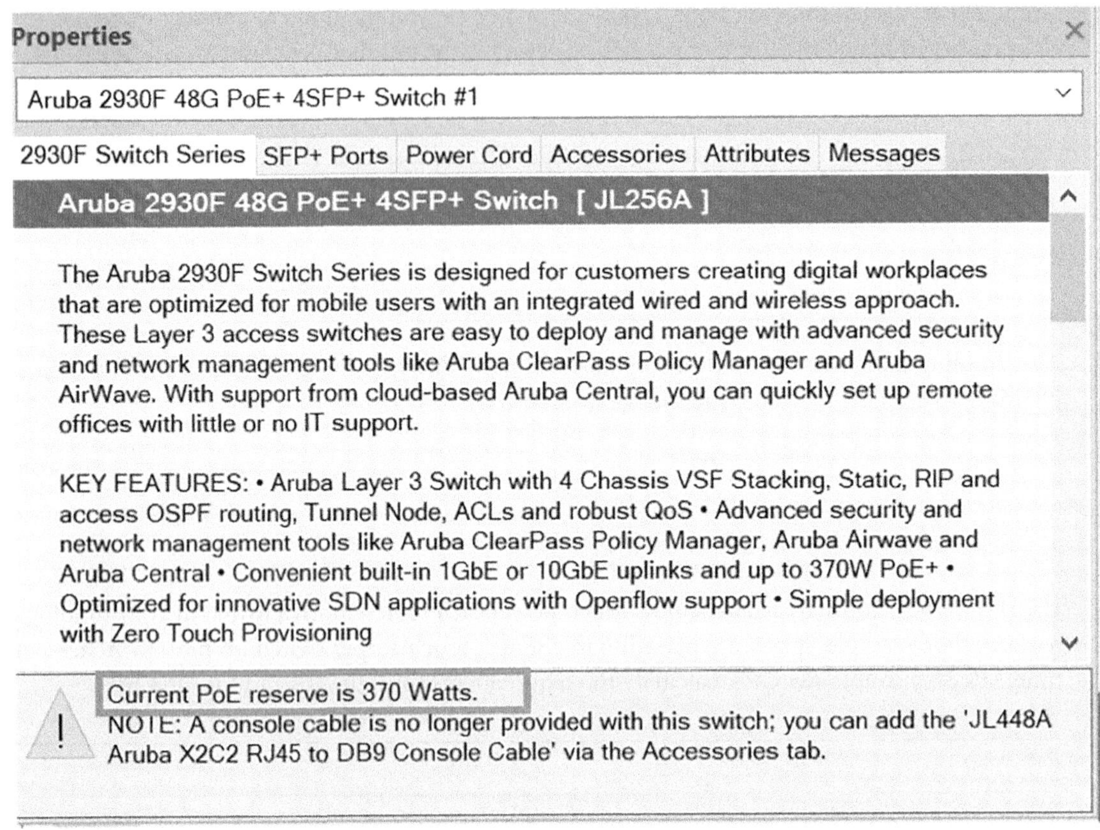

Properties
Aruba 2930F 48G PoE+ 4SFP+ Switch #1
2930F Switch Series
SFP+ Ports
Power Cord
Accessories
Attributes
Messages
Aruba 2930F 48G PoE+ 4SFP+ Switch [JL256A]
The Aruba 2930F Switch Series is designed for customers creating digital workplaces that are optimized for mobile users with an integrated wired and wireless approach. These Layer 3 access switches are easy to deploy and manage with advanced security and network management tools like Aruba ClearPass Policy Manager and Aruba AirWave. With support from cloud-based Aruba Central, you can quickly set up remote offices with little or no IT support.
KEY FEATURES: • Aruba Layer 3 Switch with 4 Chassis VSF Stacking, Static, RIP and access OSPF routing, Tunnel Node, ACLs and robust QoS • Advanced security and network management tools like Aruba ClearPass Policy Manager, Aruba Airwave and Aruba Central • Convenient built-in 1GbE or 10GbE uplinks and up to 370W PoE+ • Optimized for innovative SDN applications with Openflow support • Simple deployment with Zero Touch Provisioning
Current PoE reserve is 370 Watts.
NOTE: A console cable is no longer provided with this switch; you can add the 'JL448A Aruba X2C2 RJ45 to DB9 Console Cable' via the Accessories tab.

Figure 4-40 Activity 4.2: Access layer uplinks and VSF/Stacking links

6. Do your switches meet the PoE budget needs for the APs?

____

____

7. If you need more power, you can select the Power Options tab (see Figure 4-41) and change power supply type or set the number to 2. What supplies have you selected?

____

____

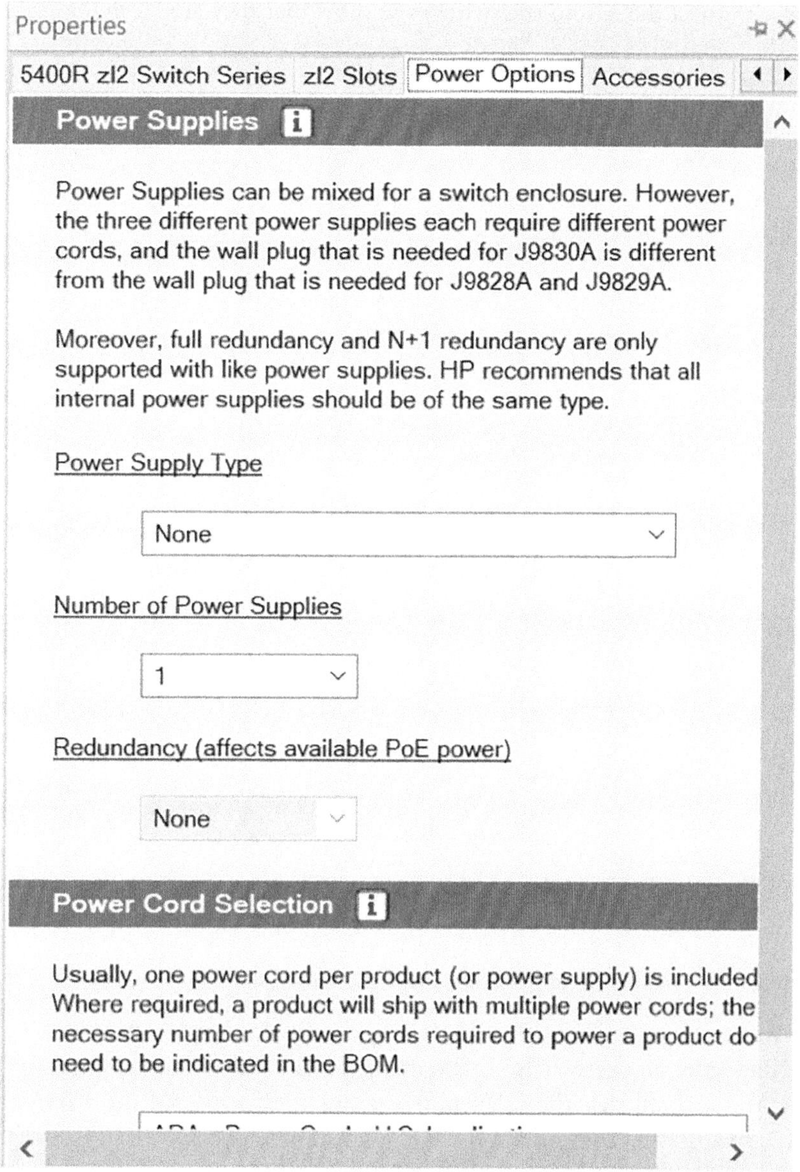

Properties

5400R zl2 Switch Series | zl2 Slots | Power Options | Accessories

Power Supplies

Power Supplies can be mixed for a switch enclosure. However, the three different power supplies each require different power cords, and the wall plug that is needed for J9830A is different from the wall plug that is needed for J9828A and J9829A.

Moreover, full redundancy and N+1 redundancy are only supported with like power supplies. HP recommends that all internal power supplies should be of the same type.

Power Supply Type

None

Number of Power Supplies

1

Redundancy (affects available PoE power)

None

Power Cord Selection

Usually, one power cord per product (or power supply) is included Where required, a product will ship with multiple power cords; the necessary number of power cords required to power a product do need to be indicated in the BOM.

Figure 4-41 Activity 4.2: Access layer uplinks and VSF/stacking links

8. You will now connect the APs to the switches to show that they are receiving power.

   a. Select the connect icon, as shown in Figure 4-42.

Figure 4-42 Activity 4.2: Access layer uplinks and VSF/Stacking links

   b. Select the AP icon and drag to a switch under "B1F1 closets" (see Figure 4-43).

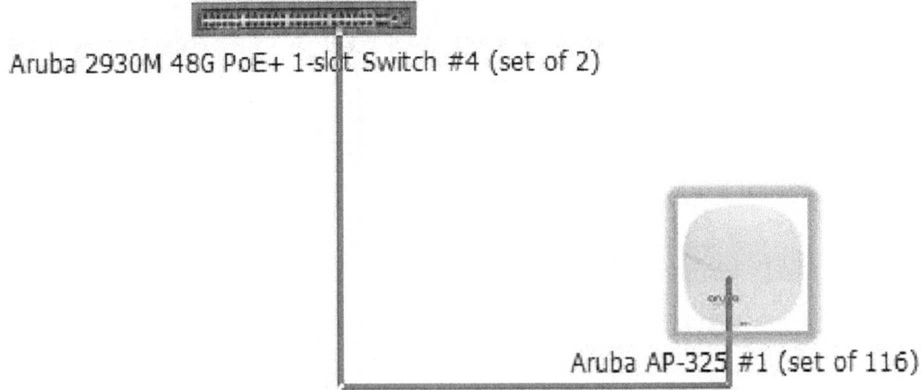

Figure 4-43 Activity 4.2: Access layer uplinks and VSF/Stacking links

   c. Choose a LAN PoE+ port on the AP and the switch (see Figure 4-44).

      If you want to connect Smart Rate ports together, make sure to select an AP's SR port and a Smart Rate port on the switch. This port might be labeled "Smart Rate" or "LAN" on the switch. You can tell that the LAN port is a Smart Rate port if the Connection Type box shows 2.5BASE-T as an option.

   d. Fill in 95 m for the connection length as the maximum length.

   e. For Quantity, specify the number of APs on Building 1 Floor 1.

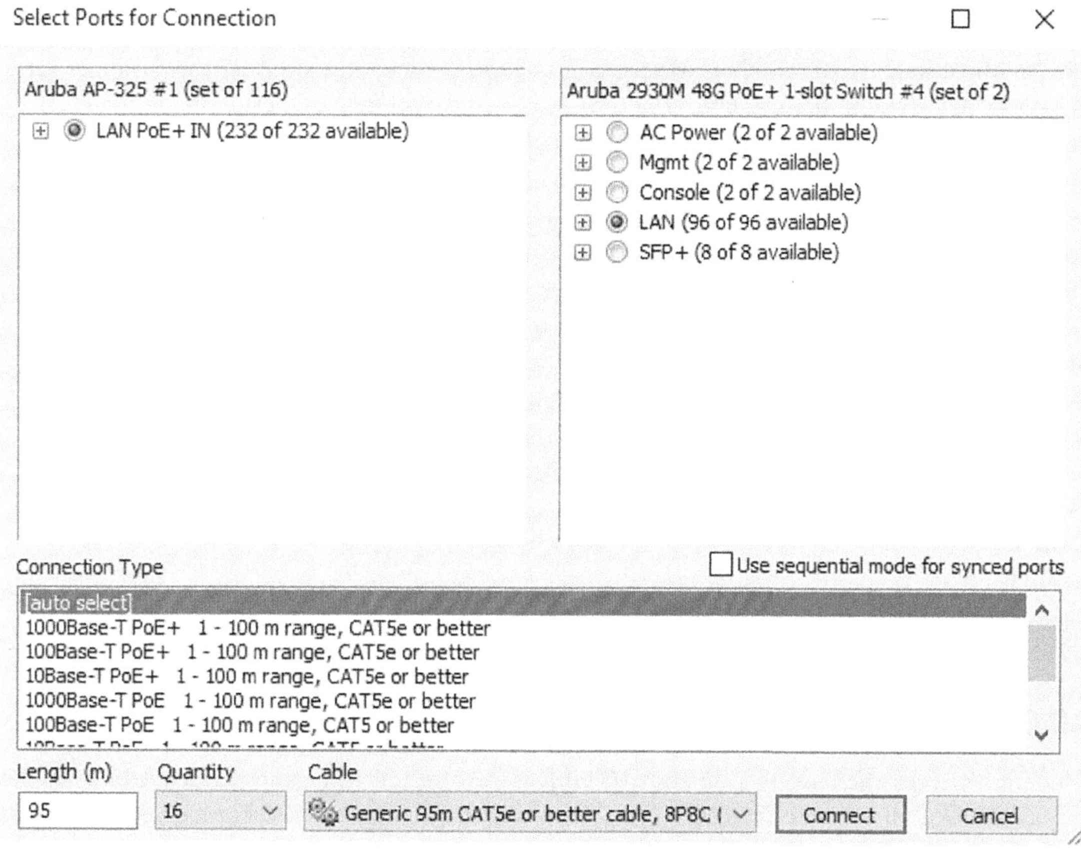

Figure 4-44 Activity 4.2: Access layer uplinks and VSF/Stacking links

f. Iris automatically distributes the connections across each device in the synched set.

   For example, if you selected 16 connections, have two switches in the synched set, and 116 APs in that synched set, Iris connects one AP to the first switch, the next AP to the second switch, the next AP to the first switch again, and so on.

g. Follow the same steps to connect the rest of your APs to the switches in the remaining closets. (Do not clone the connection because the number will not properly multiply out.)

   – If these closets have one switch each, you can simply draw the connection to the single icon under "All closets except B1F1" and specify the remaining number of APs for Quantity.

   – If these closets have three switches each, draw three connections. For each connection, specify the Quantity as about one-third of the remaining APs.

   To create a completely accurate topology, you would need to create separate icons in Iris for each closet and then connect the precise number of APs to each switch. However, for the purpose of creating a BOM, this simpler approach suffices.

9. You should now see that the AP icon is no longer red because all APs in the set are powered.

10. As a final step, make sure that switches will continue to meet the needs as the customer introduces IoT devices that use PoE. The devices that the customer is considering are PoE class 1 devices. How many class 1 devices could each of your representative switches support in addition to its APs?

____

____

____

____

____

Plan the Aggregation/Core Layers

You have completed the activity. You will now receive guidelines in planning the core and optional aggregation layers.

40 GbE and 100 GbE Media Considerations

Modern networks often use 40 GbE and 100 GbE at the core and aggregation layers. Aruba switches can use 40 GbE direct attach cables to connect switches in the same closet. For longer connections, they use fiber.

For multimode fiber, the 40GBASE-SR4 standard uses MPO connectors that require 12-fiber strands, eight of which are used (four for transmitting and four for receiving). You can achieve the longest distances with eSR4.

Alternatively, you can use LC BiDi transceivers. These transceivers can support connections as far as SR4 over only two fiber strands.

Running 100 GbE over multimode fiber requires 12 strands per connection. Single-mode fiber lets you use two strands.

Refer to Table 4-6 and 4-7 for a summary of the distances that various types of transceivers support over various types of cable.

Table 4-6 40 GbE media considerations

Fiber Type	Max distance	Transceiver	Strands
Multi-mode OM3	100 m	QSFP+ MPO SR4	12
		QSFP+ LC BiDi	2
	300 m	QSFP+ MPO eSR4	12
Multi-mode OM4	150 m	QSFP+ MPO SR4	12
		QSFP+ LC BiDi	2
	400 m	QSFP+ MPO eSR4	12
Single-mode	10 km	QSFP+ LC LR4	2

Table 4-7 100 GbE Media Considerations

Fiber Type	Distance	Transceiver	Strands
Multi-mode OM3	70 m	QSFP28 MPO SR4	12
Multi-mode OM4	100 m	QSFP28 MPO SR4	12
Single-mode	10 km	QSFP28 LC LR4	2

It is important to keep track of the fiber strand requirements. Because MPO connectors require 12 fiber strands, you do not achieve the same fiber consolidation when you move from 10 GbE to 40 GbE that you do when you move from 1 GbE to 40 GbE unless you use LC BiDi transceivers. In other words, when you move from 1 GbE to 10 GbE, you achieve 10 times the speed on the same amount of fiber as before. But when you move from 10 GbE to 40 GbE over MPO, you achieve four times the speed on six times the fiber. To achieve higher speeds on less fiber, use LC BiDi.

Long distance 40 GbE and 100 GbE links require single mode fiber and LC LR4 transceivers. The LC LR4 transceivers use wavelength division multiplexing (WDM) to multiplex multiple transmit or receive lanes on one fiber strand, permitting them to use only two fiber strands as well.

Design the core for a two-tier topology

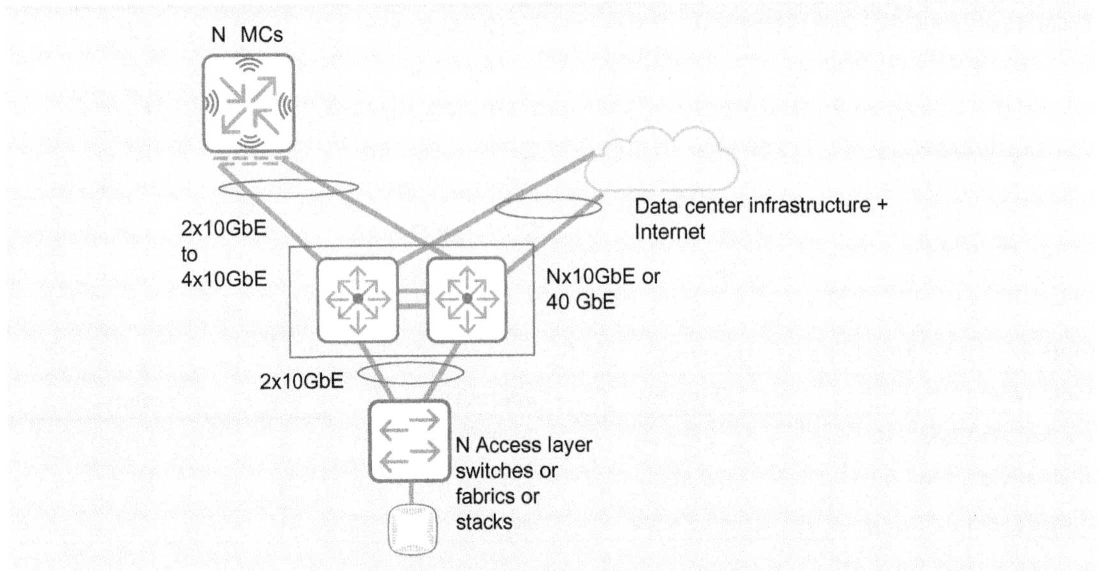

Figure 4-45 Design the core for a two-tier topology

To support enterprise resiliency requirements, you should generally deploy two core switches combined in a VSF fabric or backplane stack, as shown in Figure 4-45. In a two-tier topology, the network core takes responsibility for connecting every component, including the access layer and Aruba Mobility Controllers (MCs), to the data center and the Internet. The access layer itself could consist of individual switches, VSF fabrics, or backplane stacks.

The typical bandwidth between the access layer and core can vary, depending on whether the access layer switches operate individually or combine together in stacks or VSF fabrics. Usage requirements also affect this bandwidth. Links between the core and the MCs tends to be about two to four 10 GbE links. These links could scale from two 1 GbE to eight 10 GbE links or even 40GbE links, depending on MC capabilities and environment bandwidth requirements. However, the ranges in the example are typical.

The example shows how the campus core connects to the data center. For larger networks and data centers, the campus core connects to a data center core, which supports the data center's two-tier or three-tier topology. Or the campus and data center cores might be collapsed with the same pair of core switches supporting the campus access layer and a couple of server access switches. The data center also typically provides the connection to the Internet.

Consider the wired traffic flow in a two-tier topology

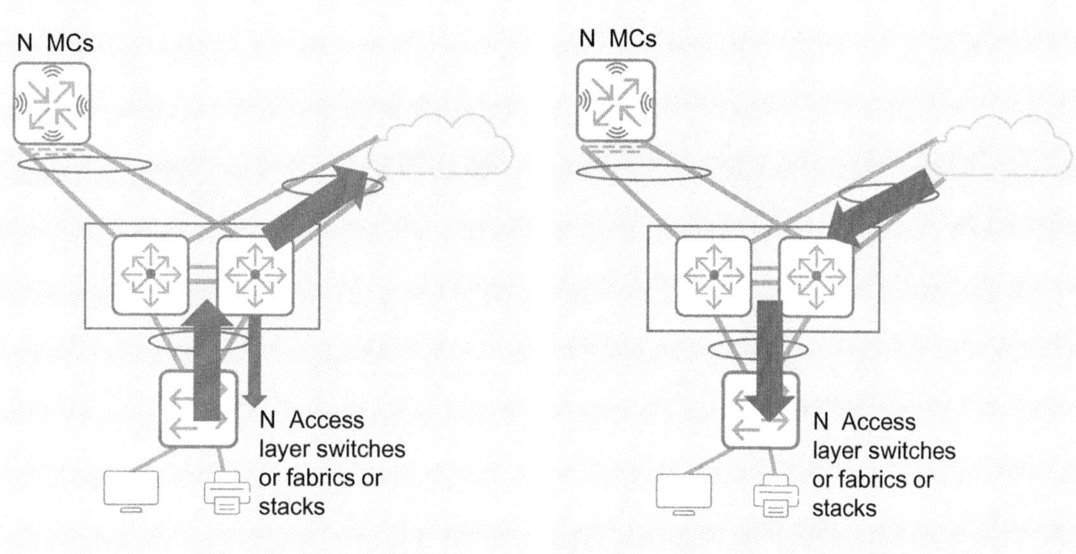

Figure 4-46 Consider the wired traffic flow in a two-tier topology

To understand the bandwidth requirements, you should understand how most traffic in your network will flow. The wired traffic flow is simple to understand. Most traffic from wired endpoints will flow to the core, and then to the data center and Internet, as shown on the left of Figure 4-46.

In some cases, some traffic will flow from the access layer to the core and then to another access layer switch. For example, one softphone might contact another softphone.

The traffic from the data center and Internet to wired endpoints flows through the core as well, as shown on the right of Figure 4-46.

Consider the wireless traffic flow in a two-tier topology

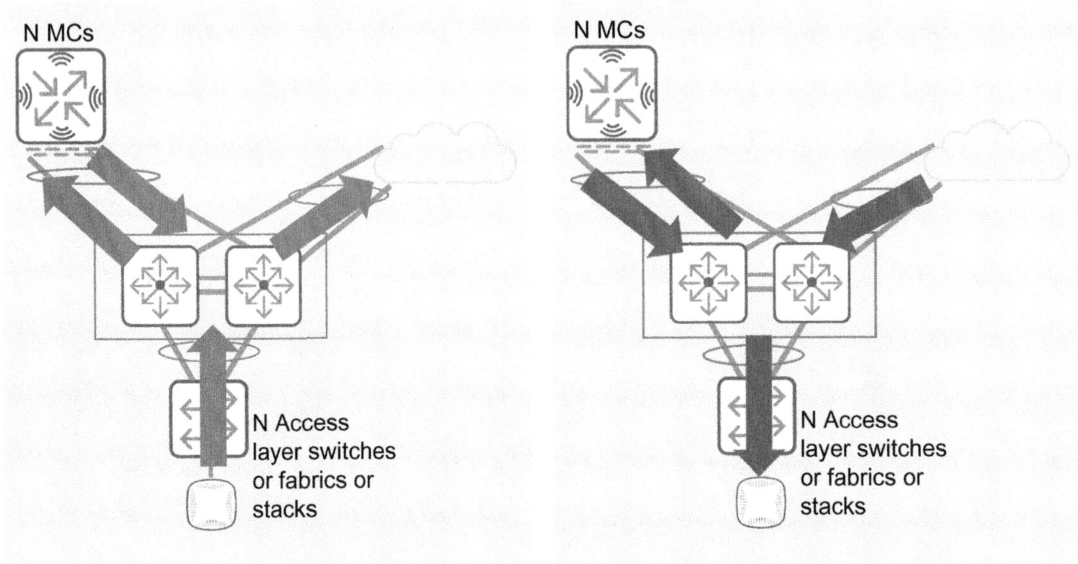

Figure 4-47 Consider the wireless traffic flow in a two-tier topology

It is important to understand that all wireless traffic flows through the MCs. Traffic from wireless endpoints flows through the access layer to the core and MCs then back to the core and out to the data center or Internet, as shown on the left of Figure 4-47.

Traffic outbound to wireless clients from the data center or Internet also flows both inbound and outbound on the MCs' links (see the right of Figure 4-47).

Plan appropriate oversubscription

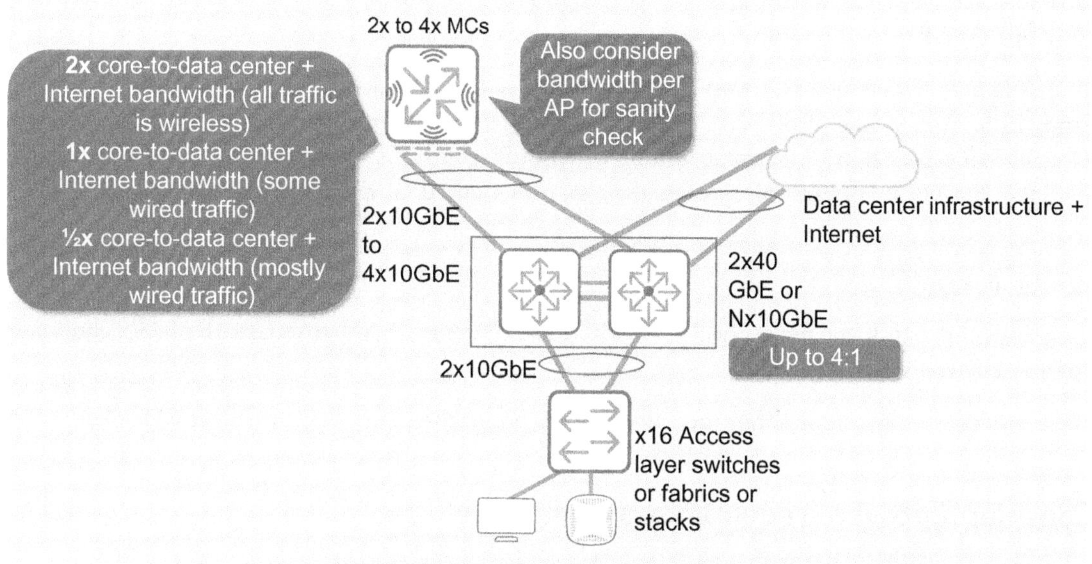

Figure 4-48 Plan appropriate oversubscription

Figure 4-48 helps you to put together what you have learned.

Traffic to and from the access layer typically passes over the links between the core and the data center. Because the core is collapsed in a two-tier topology, the network can tolerate some oversubscription on links to the data center.

For example, you might have 16 access layer switches (or fabrics or stacks) with two 10 GbE links each, for a total bandwidth of 320 Gbps. However, you planned generous bandwidth at access layer uplinks, so you know that the links to the data center can tolerate some oversubscription. You might plan two 40 GbE links on the core VSF fabric, which yields 4:1 oversubscription. This is the maximum amount of oversubscription you should plan here. This design works when you expect the access layer uplinks to operate well below their capacity most of the time.

The MCs' bandwidth depends on wireless traffic. For example, if all traffic is wireless, plan about twice as much bandwidth to the MCs as to the data center. This is because all traffic to and from wireless endpoints must travel over the links in both directions. Here, you are counting the total bandwidth to all of the MCs, not to individual MCs.

You can also consider the bandwidth per AP for a sanity check. If you are planning to scale the MCs' bandwidth to the maximum possible, but this maximum bandwidth does not seem to be enough for your environment, you might need to reconsider your MC plan. You could plan to deploy more MCs with fewer APs per-MC.

For your reference, this is the bandwidth per AP if you deploy the maximum number of APs on an MC and also give the MCs its maximum bandwidth. The 7205 provides 48 Mbps per AP. The 7210 and 7220 provide 40Mbps per AP. The 7240 provides 20Mbps per AP. And the 7280 provides 50 Mbps per AP.

In an example with four MCs, you might plan four 10 GbE links for each, yielding 160 Gbps of bandwidth. (You would also select MCs with 40 Gbps throughput for the firewall.) On the other hand, if you expect that some traffic will be wired only, you can scale down the bandwidth requirements on the MCs. For example, you might plan only two 10 GbE links on each of your four MCs.

If you expect light wireless traffic, you might plan to deploy just two MCs, and they can have two 10 GbE links each.

If, on the other hand, you expect heavy wireless traffic, you might add an MC with 10 GbE links and distribute fewer APs to each MC.

These are just some examples, but they give you a sense of the issues that can affect your plan.

Consider hardware requirements at the core

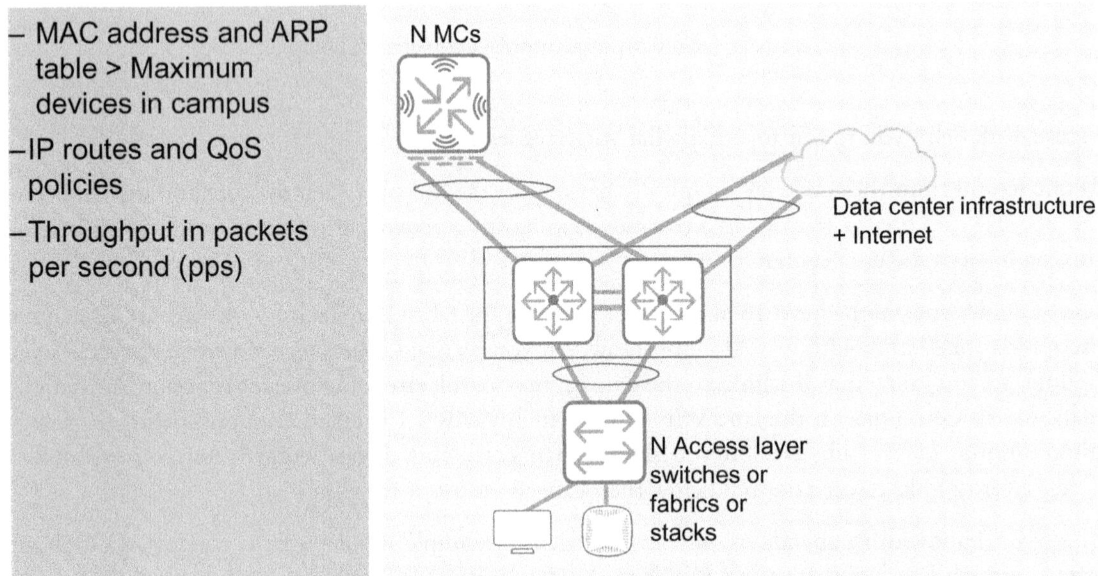

Figure 4-49 Consider hardware requirements at the core

In a two-tier campus topology, such as the one illustrated in Figure 4-49, the core switches (or VSF fabric) typically route all traffic in the campus network. Therefore, the core switches must support all of the VLANs, and their MAC forwarding tables and ARP tables must be large enough to store the MAC address of every device on the campus.

You should also consider factors such as the number of IP routes and QoS policies the switch can support. However, these capabilities are less likely to act as bottlenecks in most environments.

You should verify that your selected switches can support the throughput you expect the switches to handle. Aruba switch backplanes support the maximum bandwidth provided on the ports and far beyond. If the customer requires high performance, though, you might look beyond switching capacity to throughput in packets per second for small packets, which can give you a better sense of how the switch will perform.

Recommended two-tier core switches

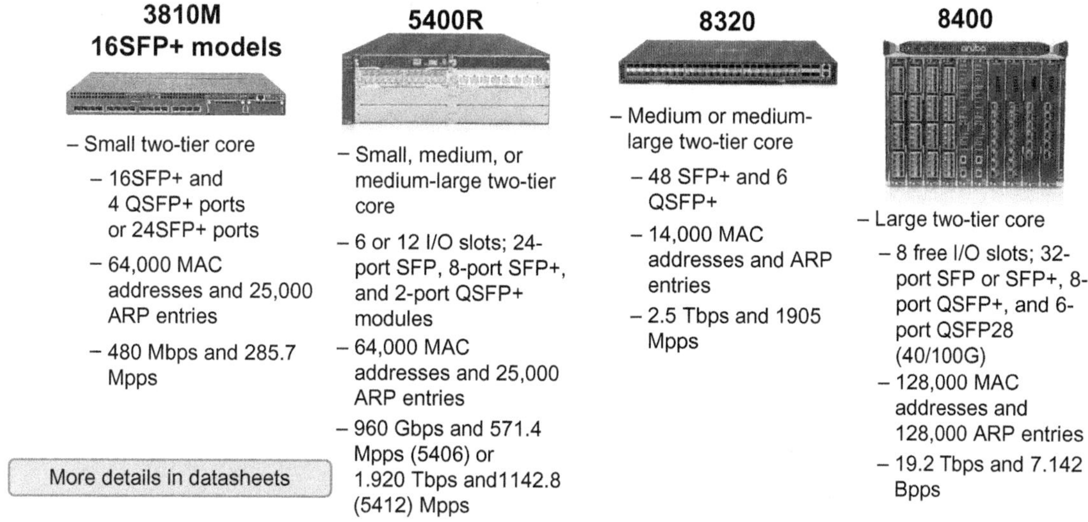

3810M
16SFP+ models
– Small two-tier core
– 16SFP+ and 4 QSFP+ ports or 24SFP+ ports
– 64,000 MAC addresses and 25,000 ARP entries
– 480 Mbps and 285.7 Mpps
More details in datasheets
5400R
– Small, medium, or medium-large two-tier core
– 6 or 12 I/O slots; 24-port SFP, 8-port SFP+, and 2-port QSFP+ modules
– 64,000 MAC addresses and 25,000 ARP entries
– 960 Gbps and 571.4 Mpps (5406) or 1.920 Tbps and1142.8 (5412) Mpps
8320
– Medium or medium-large two-tier core
– 48 SFP+ and 6 QSFP+
– 14,000 MAC addresses and ARP entries
– 2.5 Tbps and 1905 Mpps
8400
– Large two-tier core
– 8 free I/O slots; 32-port SFP or SFP+, 8-port QSFP+, and 6-port QSFP28 (40/100G)
– 128,000 MAC addresses and 128,000 ARP entries
– 19.2 Tbps and 7.142 Bpps

Figure 4-50 Recommended two-tier core switches

Aruba offers several switches for the core of a two-tier network. The main options are outlined in Figure 4-50.

The Aruba 3810M-16SFP+ model can form the core of a small two-tier network. The Aruba 5400R Series provides the most flexibility. You can deploy these switches at the core of two-tier networks from the small to the medium-large. Aruba offers another option for medium or medium-large two-tier cores: the 8320. For the largest networks, Aruba offers the 8400 series.

The 8320 uses a shared table for IPv4 ARP and IPv6 ND entries. Each ND entry takes the space for two ARP entries, so it can support up to 7000 IPv6 ND entries.

For a list of additional hardware capabilities for all switches, refer to the switches' datasheets and details in Iris.

Additional Considerations for Choosing the Core Switches

You can choose between the 5400R, 8320, and 8400 Series for larger networks. Each of these series has its advantages.

The 5400R series supports VSF and uses the same AOS software that other Aruba switches. It can be a good choice for medium network or larger networks with fewer performance demands.

The 8320 delivers a bit higher performance than the 5400R Series. While it does not support quite as many ports as the 5412R switches, it delivers higher throughput in packets per second, although it does not support as many devices. You should choose the 8400 when you require the highest scalability and performance. For example, you can use the 8400 when the core or aggregation layer must support more than 64,000 MAC addresses or when you need to scale a high density of 40G and 100G.

The 8320 and 8400 do not support VSF as of the publication of this book, but they do have options for many redundancy features. These switches support multichassis link aggregations (M-LAGs), which enable a pair of switches to connect to another device using a single link aggregation. Therefore, you can deploy these switches with a simpler topology much like that enabled by VSF. The switches also come with redundant fans and power supplies. Finally, you can add a redundant management module to an 8400 switch. (You can also add redundant management modules to 5400R Series switches if you do not want to deploy them with VSF.)

Also keep in mind that the 8320 and 8400 switches run ArubaOS-CX, an open, Linux-based software platform that is programmable through RESTful APIs. Companies can use these APIs to support cloud management and automation. The software's Network Analysis Engine enables customers to load python scripts that monitor current traffic conditions and even take steps to deal with issues.

Design a three-tier network

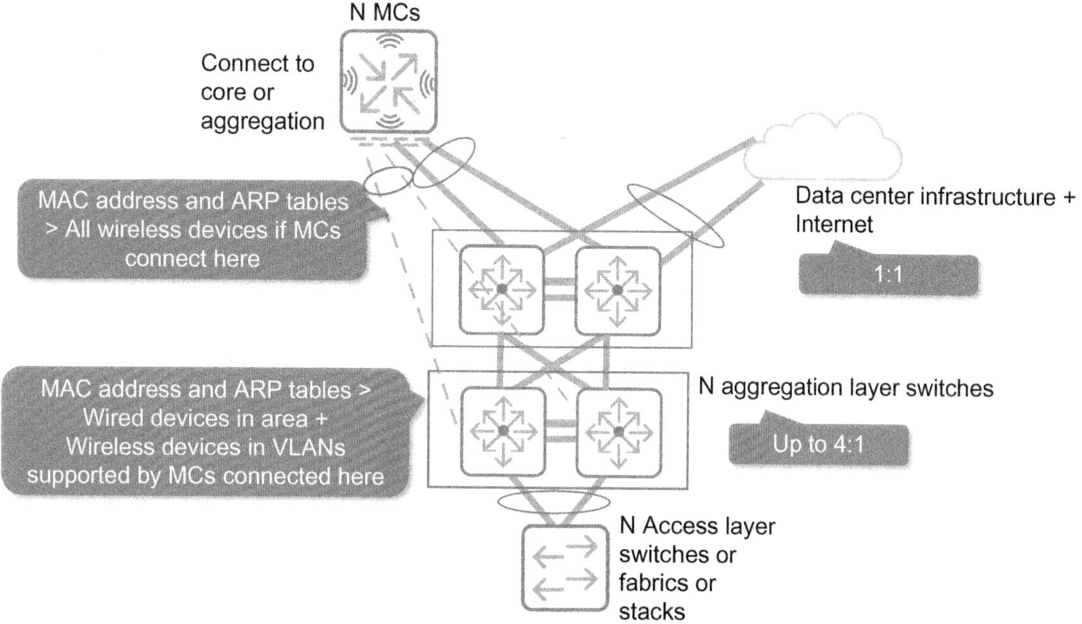

Figure 4-51 Design a three-tier network

Many of the same considerations that apply to the core of two-tier networks also apply to the core and aggregation layer of three-tier networks, shown in Figure 4-51.

When a network has a separate aggregation layer, you should introduce the oversubscription at that layer. The core should not introduce any more oversubscription.

Aruba MCs might connect at the aggregation layer or the core, depending on the traffic flow, roaming needs, and the customer's desire for local resiliency. Keep in mind that where you connect the MCs will affect the number of MAC addresses that the aggregation and core switches must support. Typically, you should implement routing between the aggregation layer and the core, which means that VLANs for wired devices only extend to the aggregation layer. This limits the size of the MAC address tables for those VLANs.

When you plan the MC bandwidth, you must keep in mind what percentage of the traffic you expect to be from wireless devices and plan enough bandwidth for this traffic to enter and leave the MC in each direction.

Because APs tunnel traffic from wireless devices to MCs, if the MCs connect to the core, all wireless VLANs exist in the core. The core must have a large enough MAC address table to support all of the devices. On the other hand, if the MCs connect to the aggregation layer, the wireless VLANs can be segmented like the wired VLANs. However, this might affect the roaming or clustering plan. You could also extend a wireless VLAN across the aggregation layer and core rather than routing between them. In this case, every core and aggregation layer switch would need to have a large enough MAC address to support all of the devices in the wireless VLANs.

Recommended three-tier core and aggregation switches

The same switches that work at the core of a two-tier network can provide the aggregation layer or core of a three-tier network.

The 3810M-16SFP+ modules can provide a small to medium aggregation layer.

5400R switches can provide the core of a small to medium-large three-tier network. They can operate at the aggregation layer of networks of all sizes from small to large.

For medium and large three-tier networks, the 8320 switches provide an ideal aggregation layer. However, they can also provide the core of a small or medium network.

The 8400 switches can form both the aggregation layer and the core of large three-tier networks.

Example medium-large three-tier design

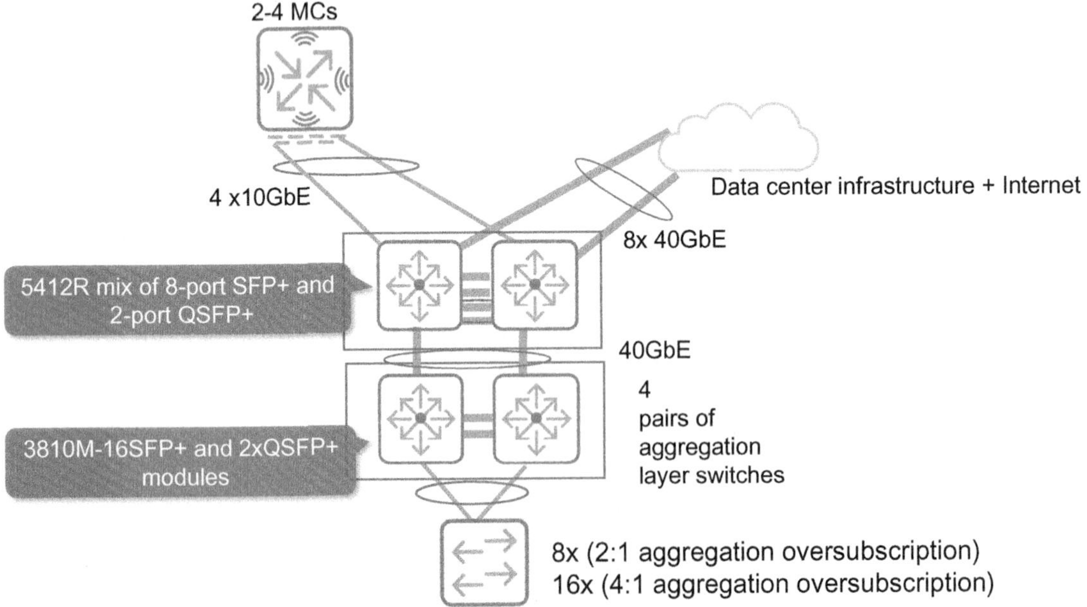

Figure 4-52 Example medium-large three-tier design

In the example shown in Figure 4-52, the aggregation layer consists of pairs of 3810M-16SFP switches, each with two QSFP+ modules, and deployed as a backplane stack. The aggregation layer fabric connects to the access layer, which could be a single switch, a backplane stack, or VSF fabric on two 10 GbE links. The aggregation layer backplane stack can support between 8 and 16 access layer units, depending on the level of oversubscription you want.

At the core is a pair of 5412R switches deployed in a VSF fabric. It supports up to eight 40 GbE links in its link aggregation to the data center core, matching the amount that it supports for the links to the aggregation layer switches. This allows the core to support up to four pairs of aggregation layer switches with two 40 GbE ports each.

When the access layer connects on two generous 10 GbE links, 4:1 oversubscription often works well.

The access layer switch model is not shown, but could be switches such as 2930Fs or 2930Ms.

You could add another two 40 GbE links between the core and the aggregation layer. In that case, the core could only support two pairs of aggregation layer switches to maintain nonblocking forwarding. Because the aggregation layer switches would then support twice as much bandwidth to the core, they could support less oversubscription for the links to the access layer.

Example large three-tier design

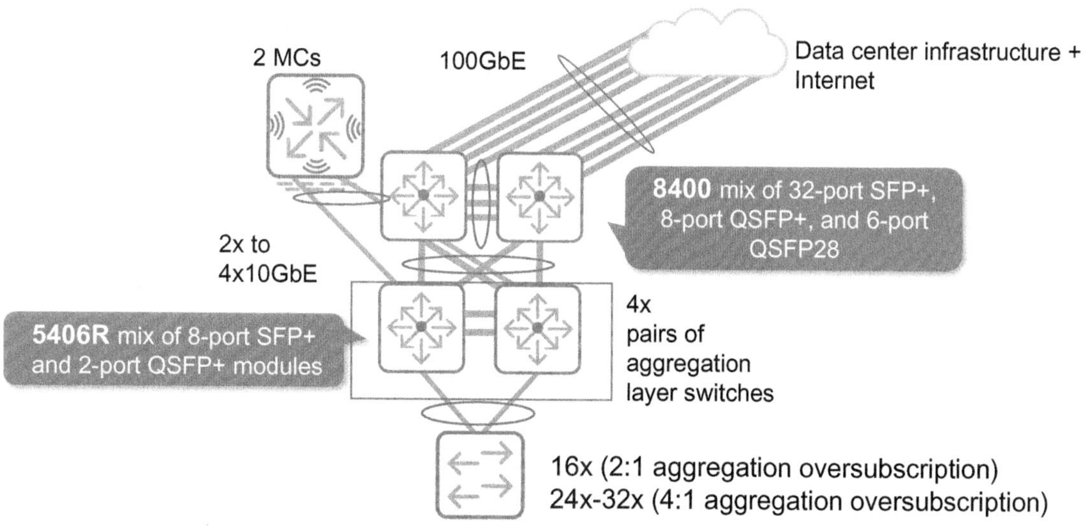

Figure 4-53 Example large three-tier design

In the design shown in Figure 4-53, 5406R switches operate in VSF fabrics at the aggregation layer. They have 160 Gbps uplink bandwidth (and could scale higher). This VSF fabric can support up to 32 access layer switches (or fabrics or stacks), each with two 10 GbE links, at a 4:1 oversubscription rate. However, in this example, the Aruba MCs connect at the aggregation layer, taking up some of the available access layer connections and leaving the VSF fabric capable of supporting 24 or 28 access layer switches.

At the core are two 8400 switches, which can connect to the data center core on high-speed 100 GbE link aggregations. The 8400 switches can use M-LAGs to connect to the aggregation layer VSF fabrics as well as other devices.

If you chose to connect the MCs to the core instead of the aggregation layer, you could also connect them to both core switches on M-LAGs.

Again, you could vary this basic pattern by adding or removing links between the aggregation layer and the core to support the desired number of switches at the access layer and to maintain the required level of oversubscription.

Plan VSF links at the aggregation layer and core

- Unicast traffic forwarded locally
- Multicast and broadcast traffic forwarded on VSF links
- All traffic forwarded on VSF link if uplink fails

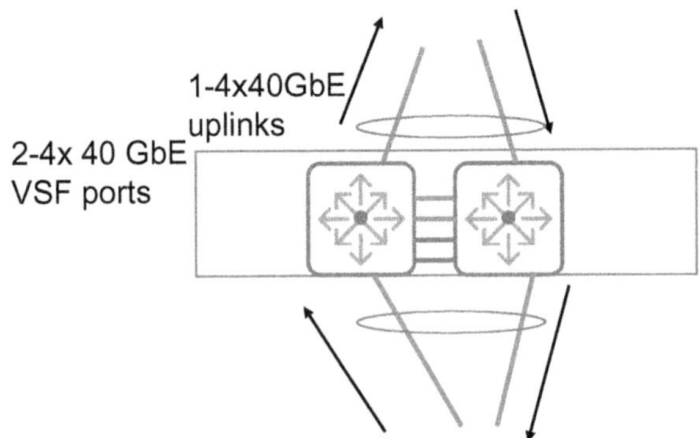

Figure 4-54 Plan VSF Links at the aggregation layer and core

For all of these designs, you need to plan adequate bandwidth for VSF and stacking links. Make sure to design the network such that links in each link aggregation are balanced over both members, as shown in Figure 4-54. This helps you to minimize traffic on the VSF links.

Because each member of the VSF fabric has a local link in every link aggregation to forward traffic, no unicast traffic crosses the VSF link during normal operation.

The VSF link *does* need to handle multicast and broadcast traffic. In this example, one member receives a multicast and sends it over the VSF link. The other member happens to be the designated multicast forwarder for the outgoing link aggregation, so it forwards the traffic. (A link aggregation on a VSF fabric has a single member that acts as the designated multicast forwarder. The fabric automatically picks which member will play this role for each link aggregation.)

Remember also that if all of a member's local links in an aggregation fail, the switch must forward traffic destined out that aggregation over the VSF link.

As a general rule, plan two 40 GbE links for the VSF link for 5400R switches at the distribution layer or the core of small-to-medium networks. Plan up to four 40 GbE links for VSF links for 5400R switches at the core of larger networks.

Plan adequate bandwidth for aggregation layer and core backplane stacks

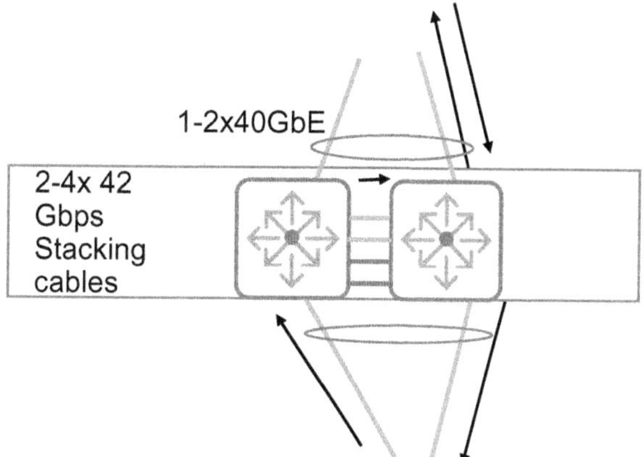

Figure 4-55 Plan adequate bandwidth for aggregation layer and core backplane stacks

A backplane stack chooses the link for unicast traffic based on a hash without preference for a local link, as shown in Figure 4-55.

You should keep in mind that the pair of switches might exchange a significant amount of traffic. You will receive the best performance if the stacking links provide more bandwidth to the core than the uplinks. Plan two stacking cables for most deployments and up to four for higher bandwidth networks.

Activity

You will now practice planning the core and possible aggregation layer of a customer's network. Use the same scenario that you used earlier in this chapter for this activity.

Activity 4.4: Need for aggregation layer

Activity 4.4 objective

The customer wants to replace the current aggregation layer switches and add a campus core so that user traffic can be routed within the campus and the staff can manage the complete campus network using a consistent OS and tools. You will now plan whether to deploy aggregation layer switches or to connect the access layer switches directly into the core and eliminate the aggregation layer. You may decide to take a hybrid approach in the two different buildings.

You can use Figure 4-56 to help you plan as you complete the sections below.

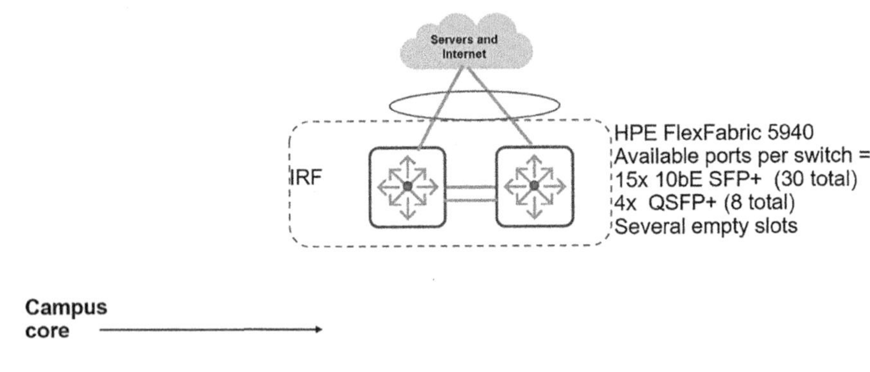

Figure 4-56 Activity 4.4: Need for aggregation layer

Activity 4.4 steps

1. Refer back to the number of access layer uplinks that you planned for Building 1 in Activity 4.2. Also refer to the cabling information in the scenario. Will you plan an aggregation layer for this building? Why or why not?

2. Refer back to the number of access layer uplinks that you planned for Building 2. Also refer to the cabling information in the scenario. Will you plan an aggregation layer for this building? Why or why not?

Activity 4.5: Aggregation layer design

Activity 4.5 objective

You will select appropriate switch models for the aggregation layer, plan the required uplink bandwidth, and choose transceivers. Only complete this activity if you chose to create an aggregation layer for one or both buildings. Otherwise, move on to Activity 4.6.

Activity 4.5 Steps

1. Refer back to the number of downlinks planned in the building for which you are planning the aggregation layer. How many uplinks to the core will you plan and of what speed? Explain the reasoning for your plan.

____

____

____

____

____

____

2. If you are planning to connect an MC or MCs at the aggregation layer, add these ports to your plan as well. If you are not planning to connect MCs here, move on to step 3.

   If you do not know the ports supported on your MC, you can use Iris to check the ports.

   a. Select the MC in the workspace.

   b. Check the Properties window and the first tab. You can see the number and type of ports supported by the model. Figure 4-57 shows an example.

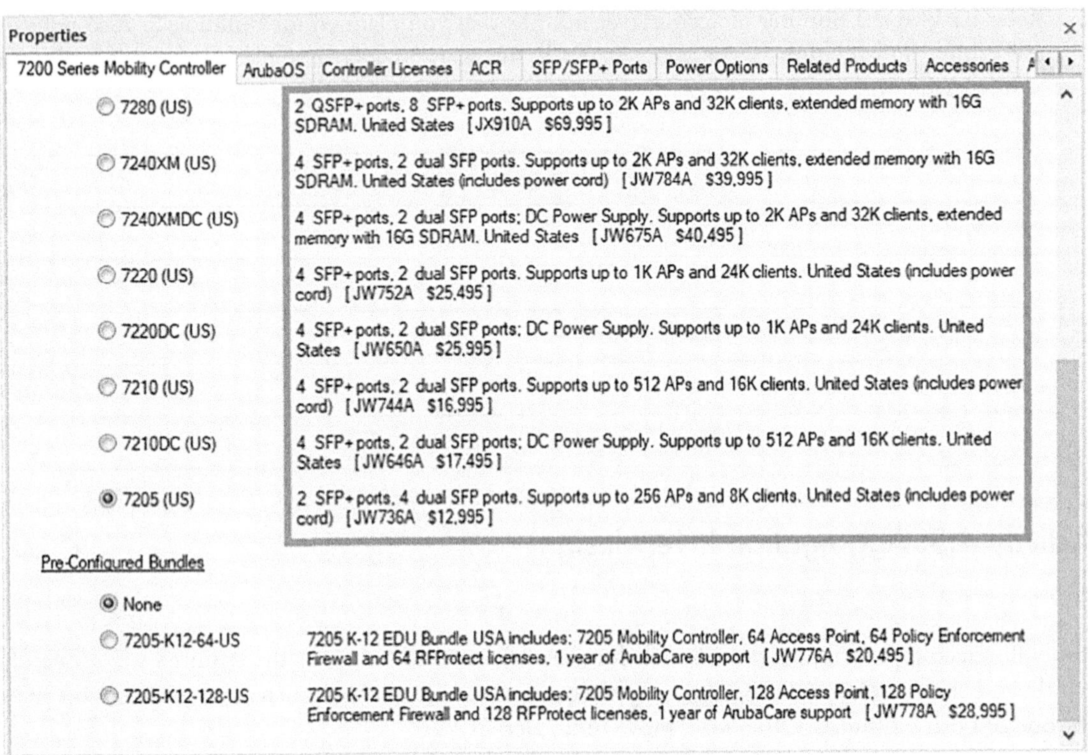

Figure 4-57 Activity 4.5: Aggregation layer design

c. How many ports of which type do you plan to use? Note that you will need to use transceivers because you cannot use DACs between the AOS-Switches and the controllers.

____

____

d. Select the **SFP/SFP+ Ports** tab and select the planned transceivers. Figure 4-58 below shows one example.

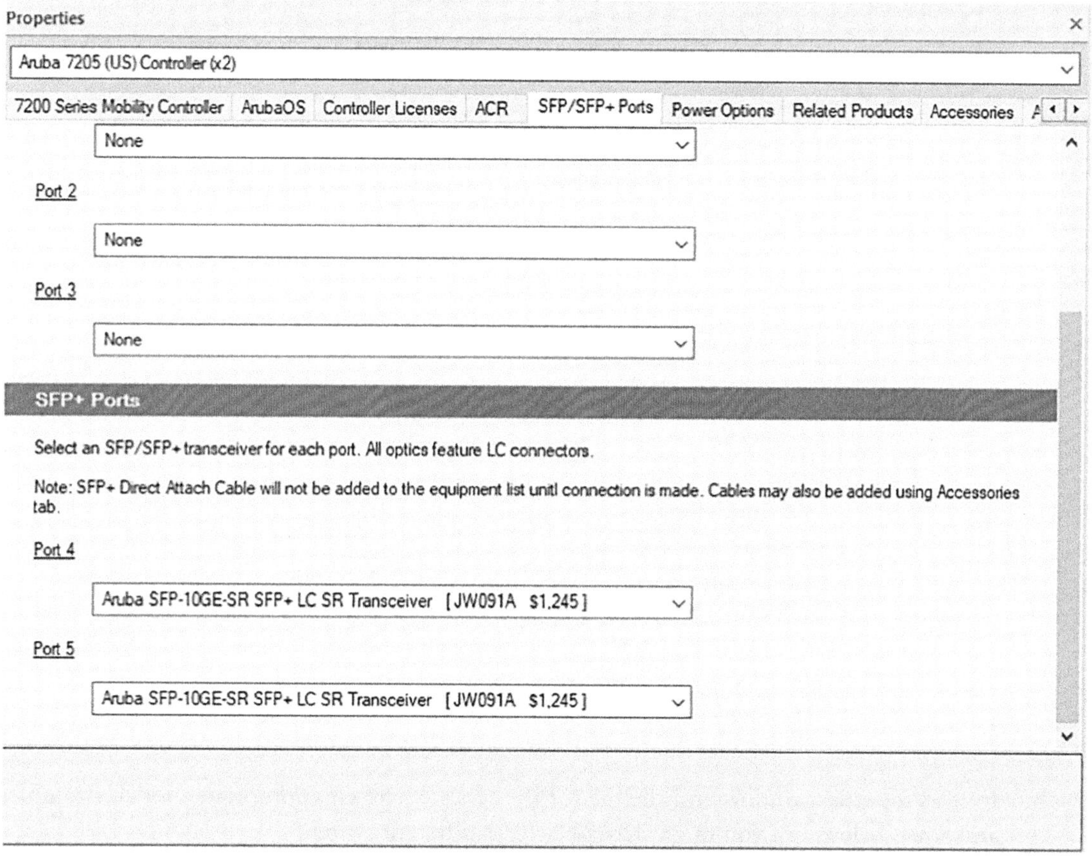

Properties
Aruba 7205 (US) Controller (x2)
7200 Series Mobility Controller | ArubaOS | Controller Licenses | ACR | SFP/SFP+ Ports | Power Options | Related Products | Accessories
None
Port 2
None
Port 3
None
SFP+ Ports
Select an SFP/SFP+ transceiver for each port. All optics feature LC connectors.
Note: SFP+ Direct Attach Cable will not be added to the equipment list unitl connection is made. Cables may also be added using Accessories tab.
Port 4
Aruba SFP-10GE-SR SFP+ LC SR Transceiver [JW091A $1,245]
Port 5
Aruba SFP-10GE-SR SFP+ LC SR Transceiver [JW091A $1,245]

Figure 4-58 Activity 4.5: Aggregation Layer Design

3. Determine whether to deploy one aggregation layer switch or a pair based on the customer scenario.

____

____

4. Select a switch series.

____

You will now add your aggregation layer in Iris.

5. Add a switch of the correct model.

   a. In Iris, expand the Catalog and find your selected switch series (see Figure 4-59). Add it to the workspace.

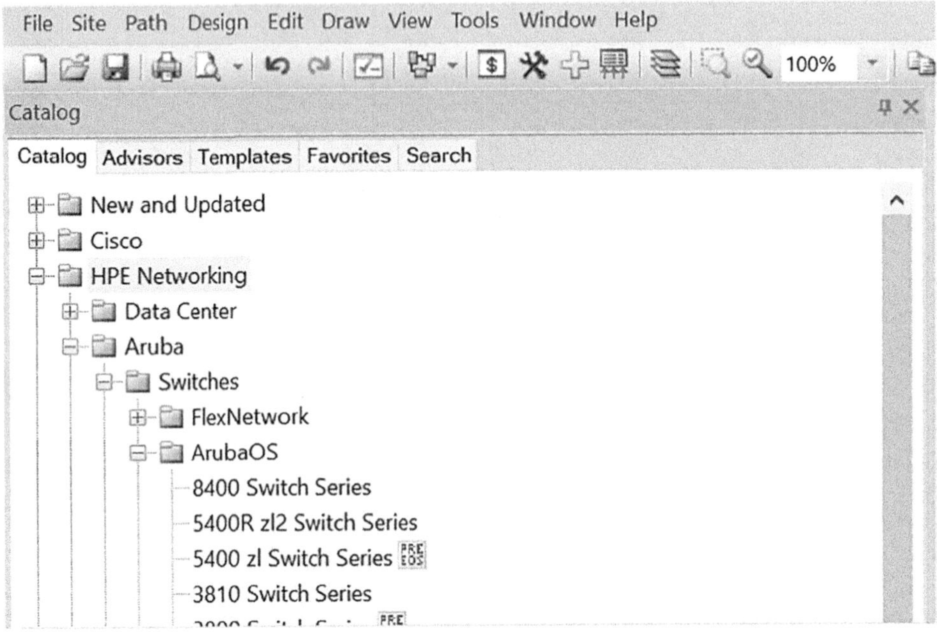

Figure 4-59 Activity 4.5: Aggregation layer design

   b. In the Properties window and the tab for the series, select an appropriate model. Examples are shown below, but you might have selected a different switch.

   c. If you selected a backplane stacking switch, remember to select the check box for the stacking module, as shown in Figure 4-60.

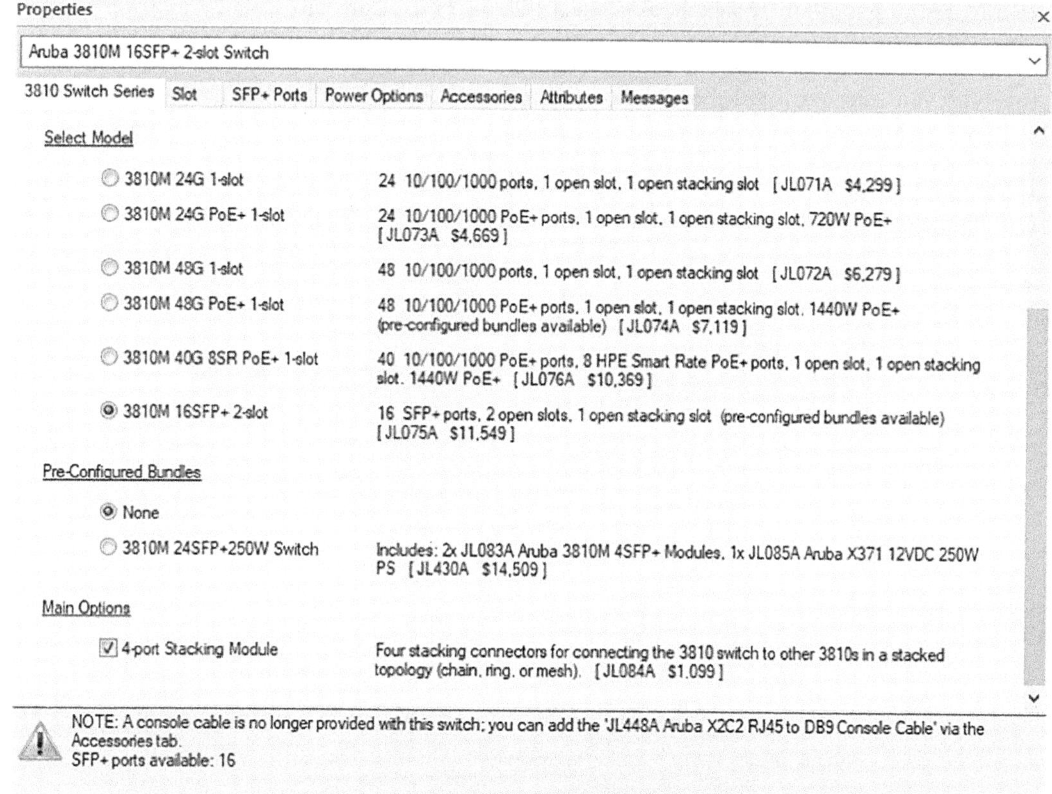

Figure 4-60 Activity 4.5: Aggregation layer design

6. If you chose a modular switch or your plan requires you to add an uplink module to the switch, select the Slots or zl2 Slots tab in the Properties window (see Figure 4-61). Choose the modules required for your plan.

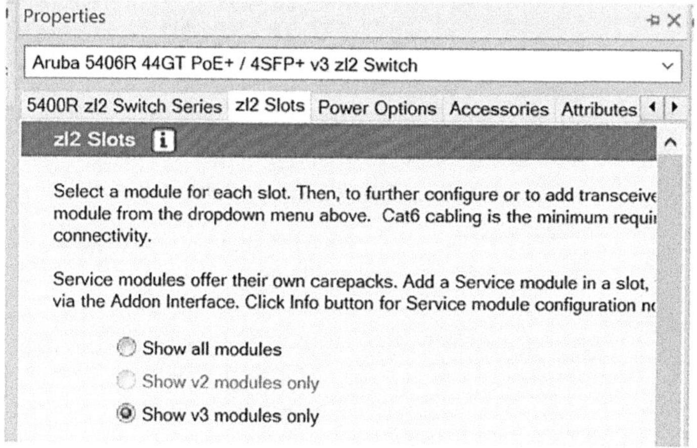

Figure 4-61 Activity 4.5: Aggregation layer design

7. If you are using a modular switch, also select the Power Options tab and add a power supply.
8. Add the transceivers required for connecting to the access layer. Remember to match the type to the type you selected for the access layer switches. If you are planning two aggregation layer switch switches, add just half of the transceivers.

   If the ports for these transceivers are on a module, select the module from the drop-down menu (see Figure 4-62). Otherwise, select the tab for the port type (see Figure 4-63).

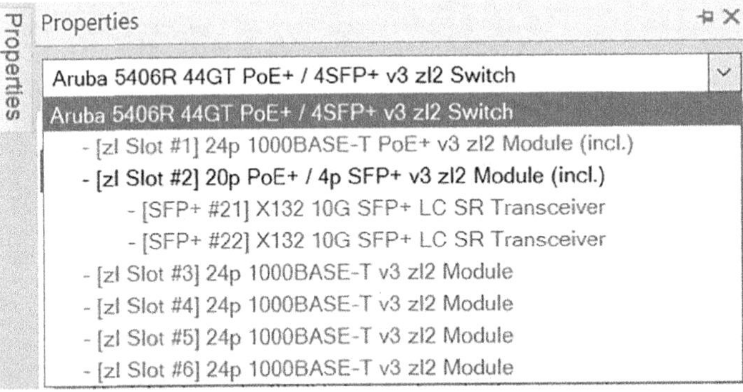

Figure 4-62 Activity 4.5: Aggregation layer design

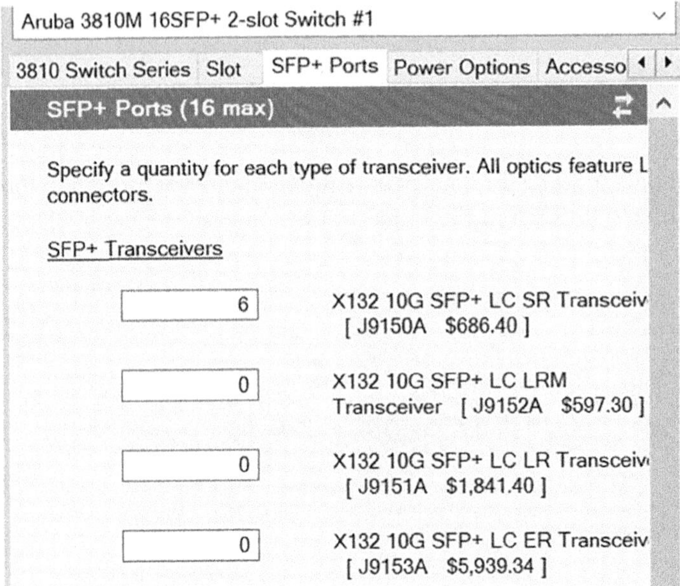

Figure 4-63 Activity 4.5: Aggregation layer design

9. Add the transceivers required for connecting to the core. Choose the correct type based on the number of fiber strands available, the desired speed, and the distance.

   If the ports for these transceivers are on a module, select the module from the drop-down menu. Otherwise, select the tab for the port type.

10. Record the type of transceivers that you chose.

    ____

    ____

11. If you are planning a pair of switches, copy and paste the switch. Then draw a VSF or stacking connection between the switches, as you learned how to do before.

12. If you decided to make an aggregation layer for both buildings, set the Quantity Multiplier to 2 in the Attributes.

Activity 4.6: Core design

Activity 4.6 objectives

You will now design the campus core, as well as plan how to connect all components of the campus network together.

Activity 4.6 steps

1. Record the total number of ports required on the core switches to support the aggregation layer, if any. Also note the transceiver type.

   ____

   ____

2. Record the total number of ports required on the core switches to support directly connected access layer switches, if any. Also note the transceiver type.

   ____

   ____

   ____

3. Add the answers to 1 and 2 to obtain the total number of downstream ports that your core switches must support.

   ____

   ____

4. How many links of what bandwidth will you plan to connect the campus core to the data center core? As a reminder, ports available on the data center core are:

    - 15 10 GbE SFP+ ports per-switch (30 total)
    - 4 40 GbE QSFP+ ports (8 total)

    The data center core switches also have slots open for new interface modules.

    ___

    ___

    ___

5. If you are planning to connect an MC or MCs at the campus core, add these ports to your plan, as well.

    If you do not know the ports supported on your MC, you can use Iris to check the ports.

    a. Select the MC in the workspace.

    b. Check the Properties window and the first tab. You can see the number and type of ports supported by the model. Figure 4-64 shows one example.

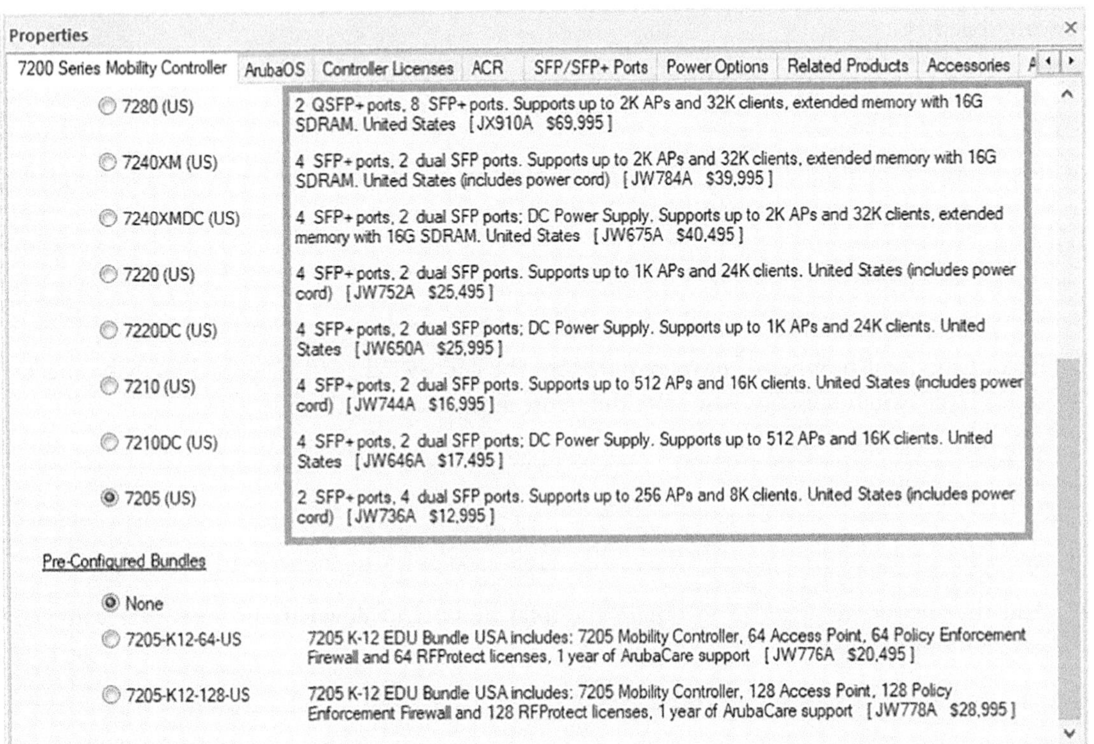

Figure 4-64 Activity 4.6: Core design

c. How many ports of what type do you plan to use? Note that you will need to use transceivers because you cannot use DACs between the AOS-Switches and the controllers.

____

____

d. Select the **SFP/SFP+ Ports** tab and select the planned transceivers. Figure 4-65 provides an example.

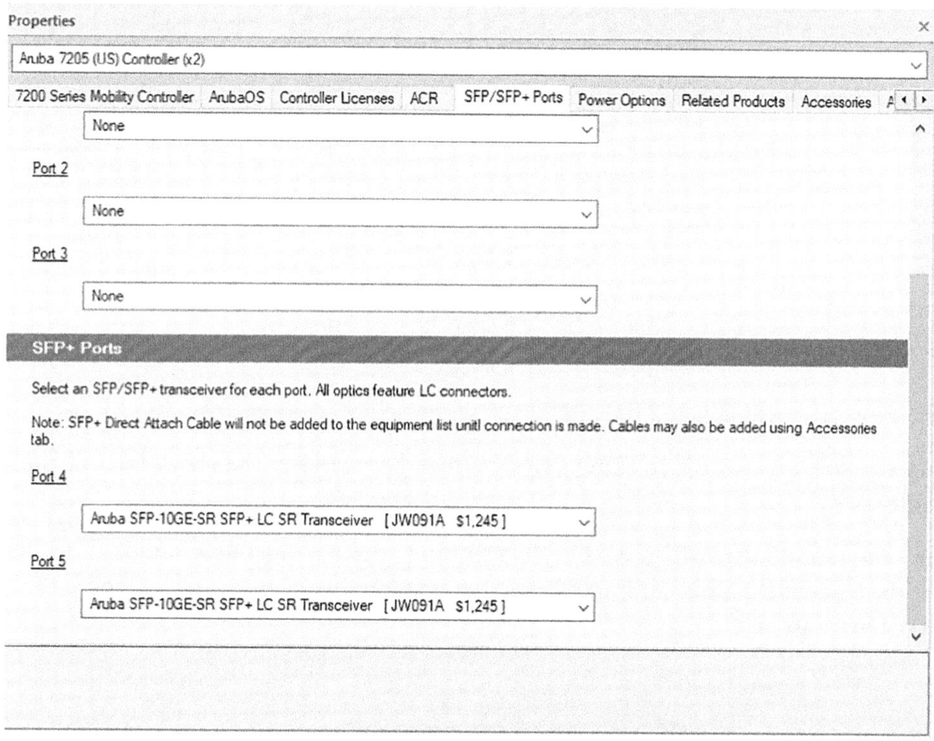

Figure 4-65 Activity 4.6: Core design

6. Determine whether to deploy one core switch or a pair based on the customer scenario.

____

____

____

7. Choose a switch series for the campus core.

____

____

8. In Iris, expand the Catalog and find your selected switch series (see Figure 4-66). Add it to the workspace.

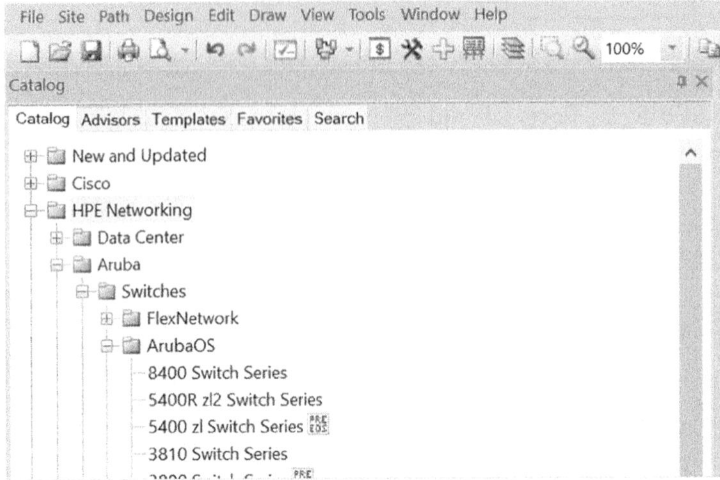

Figure 4-66 Activity 4.6: Core design

9. In the Properties window and the tab for the series, select an appropriate model.

10. If you selected a backplane stacking switch, remember to select the check box for the stacking module.

11. If you chose a modular switch, or your plan requires you to add an uplink module to the switch, select the Slots or zl2 Slots tab in the Properties window (see Figure 4-67). Choose the modules required for your plan.

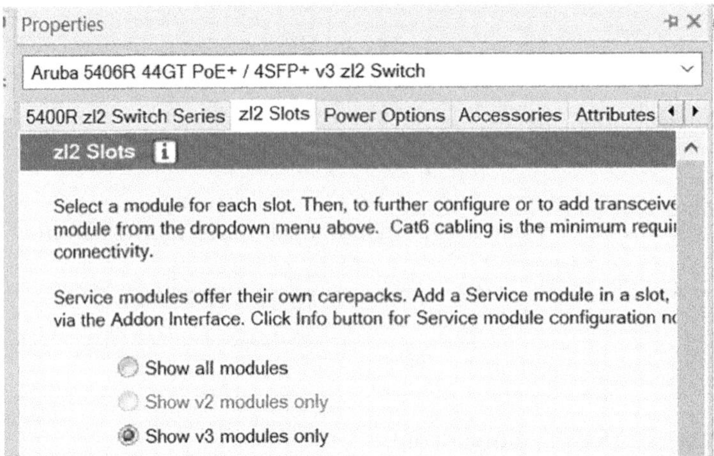

Figure 4-67 Activity 4.6: Core design

12. If you are using a modular switch, also select the Power Options tab and add a power supply, as shown in Figure 4-68.

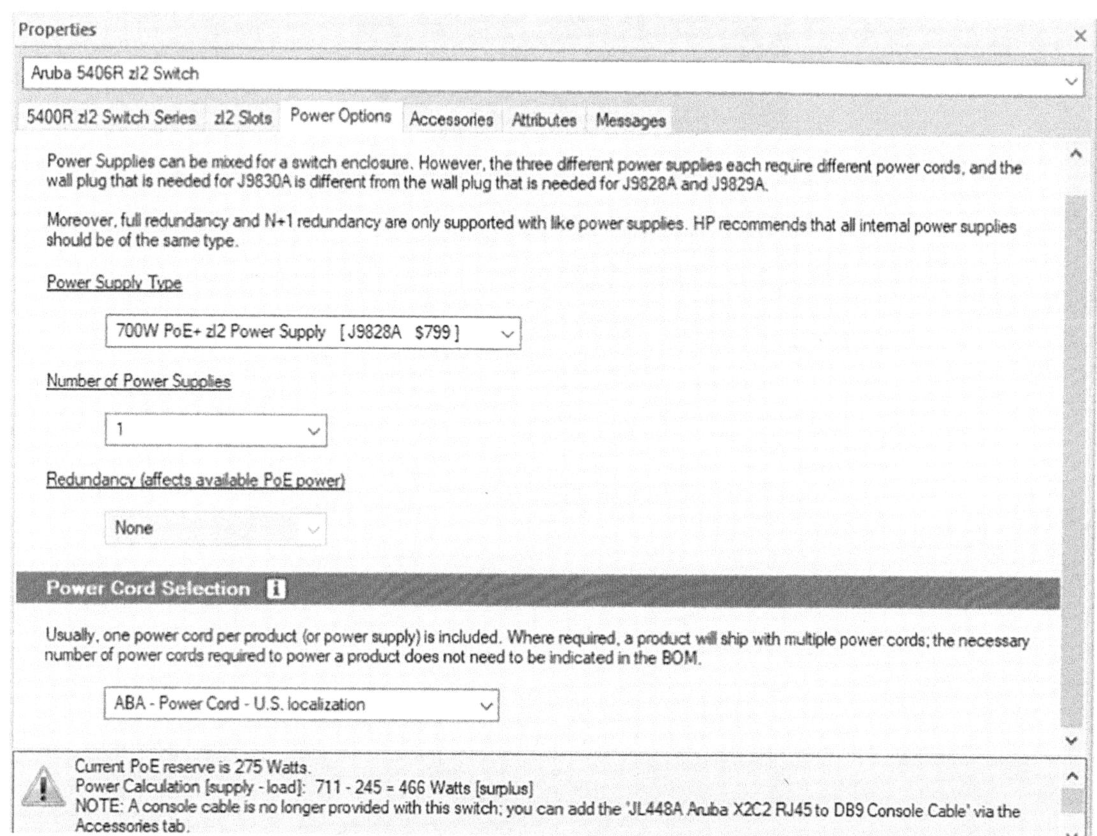
Properties
Aruba 5406R zl2 Switch
5400R zl2 Switch Series
zl2 Slots
Power Options
Accessories
Attributes
Messages
Power Supplies can be mixed for a switch enclosure. However, the three different power supplies each require different power cords, and the wall plug that is needed for J9830A is different from the wall plug that is needed for J9828A and J9829A.
Moreover, full redundancy and N+1 redundancy are only supported with like power supplies. HP recommends that all internal power supplies should be of the same type.
Power Supply Type
700W PoE+ zl2 Power Supply [J9828A $799]
Number of Power Supplies
1
Redundancy (affects available PoE power)
None
Power Cord Selection
Usually, one power cord per product (or power supply) is included. Where required, a product will ship with multiple power cords; the necessary number of power cords required to power a product does not need to be indicated in the BOM.
ABA - Power Cord - U.S. localization
Current PoE reserve is 275 Watts.
Power Calculation [supply - load]: 711 - 245 = 466 Watts [surplus]
NOTE: A console cable is no longer provided with this switch; you can add the 'JL448A Aruba X2C2 RJ45 to DB9 Console Cable' via the Accessories tab.

Figure 4-68 Activity 4.6: Core design

13. Add the transceivers that you planned in questions 3 through 5. If you are planning two core switches, add just half of the transceivers.

    If the ports for these transceivers are on a module, select the module from the drop-down menu (see Figure 4-69). Otherwise, select the tab for the port type (see Figure 4-70).

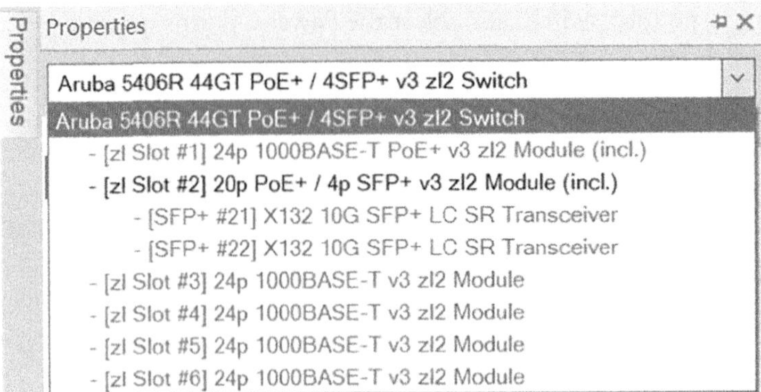

Figure 4-69 Activity 4.6: Core design

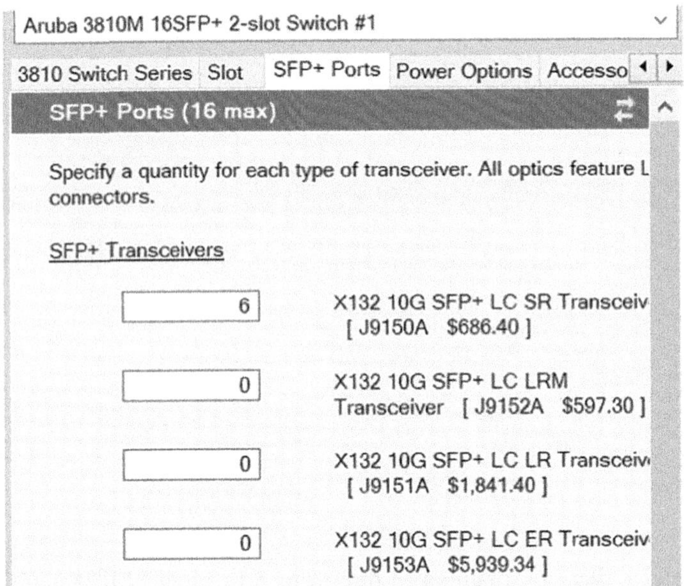

Figure 4-70 Activity 4.6: Core design

14. If you are planning a pair of switches, copy and paste the switch. Then draw a VSF or stacking connection between the switches.

15. Iris helps you to verify that you made the correct choice for transceivers when you draw connections between switches. For example, follow these steps to check your choice of transceiver for the link between an access layer switch and the upstream switch. For this example, plan the connection between B1F3C2 and its upstream switch (either at the aggregation layer or core, depending on your plan).

a. Select the Connect icon, shown in Figure 4-71.

Figure 4-71 Activity 4.6: Core design

b. Connect an access layer switch to the upstream switch; select the first switch, and drag to the second switch.

c. In the window that is displayed, which should resemble Figure 4-72, select the option for the transceivers that you selected for this link.

d. Enter the distance, which is 47 m (37 m plus 5 m at each end for patch cable).

e. You should see cable that matches the cable at your site shown as a valid Connection type: multi-mode, 50 um, 2000M Hz*km. If you do not, then you have not selected the correct transceiver.

f. Select none for the Cable because the site already has the cabling.

g. Select **Connect**.

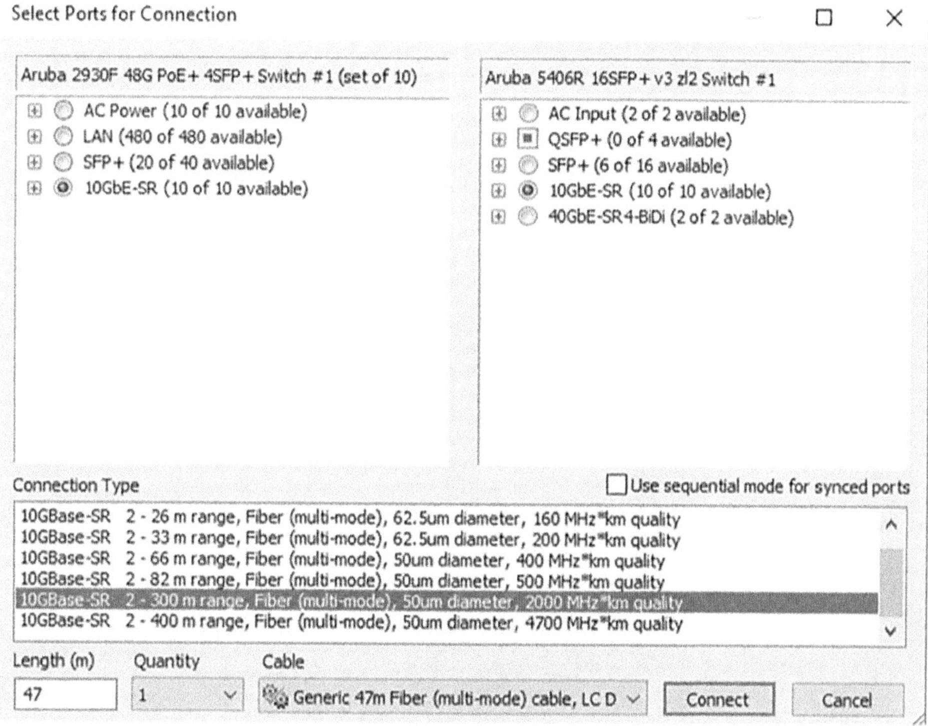

Figure 4-72 Activity 4.6: Core design

You could use Iris to draw all the connections between your switches, but this is not required for the purposes of this lab.

16. Draw your topology in Figure 4-73.

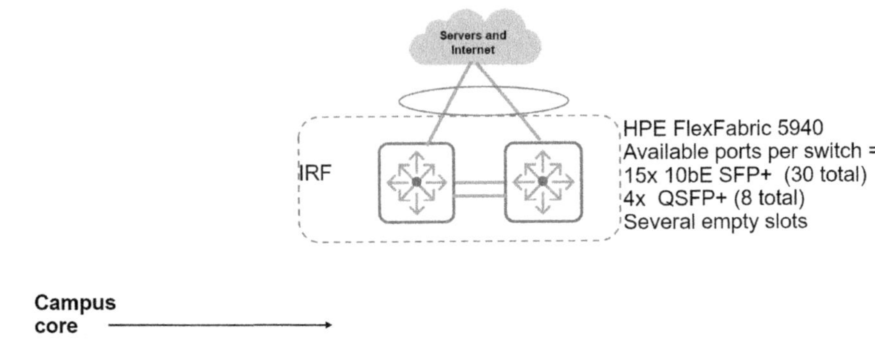

Figure 4-73 Activity 4.6: Core design

Save your project.

17. Select the dollar sign button or select File > Quotation to see your BOM. Figure 4-74 shows an example.

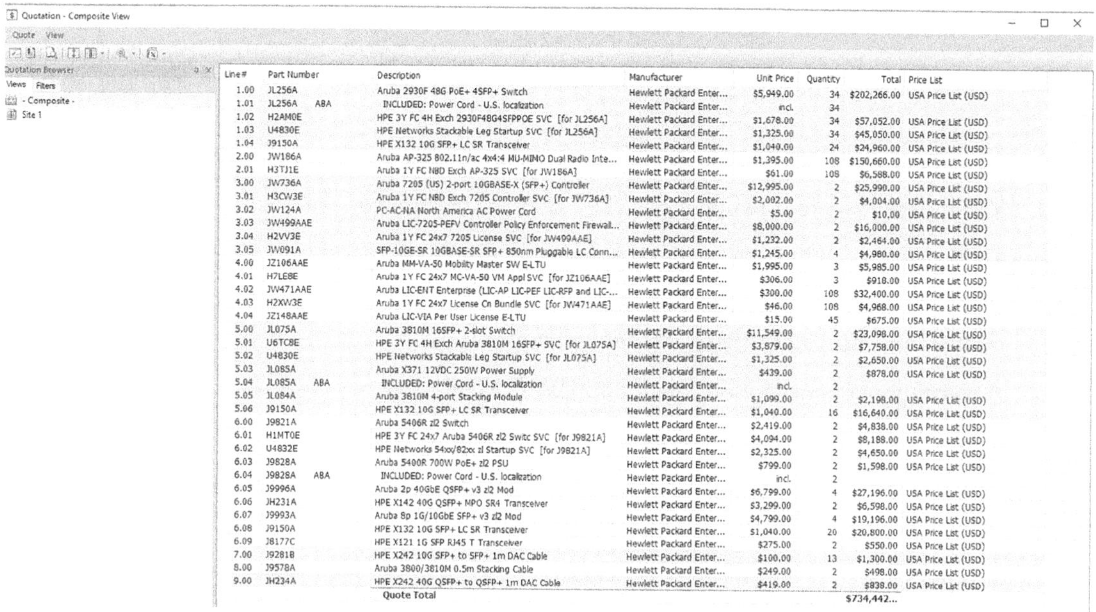

Quotation - Composite View

Quote View

Quotation Browser

Views Filters

- Composite -

Site 1

Line#	Part Number	Description	Manufacturer	Unit Price	Quantity	Total	Price List
1.00	JL256A	Aruba 2930F 48G PoE+ 4SFP+ Switch	Hewlett Packard Enter...	$5,949.00	34	$202,266.00	USA Price List (USD)
1.01	JL256A ABA	INCLUDED: Power Cord - U.S. localization	Hewlett Packard Enter...	incl.	34		
1.02	H2AM0E	HPE 3Y FC 4H Exch 2930F48G4SFPPOE SVC [for JL256A]	Hewlett Packard Enter...	$1,678.00	34	$57,052.00	USA Price List (USD)
1.03	U4830E	HPE Networks Stackable Leg Startup SVC [for JL256A]	Hewlett Packard Enter...	$1,325.00	34	$45,050.00	USA Price List (USD)
1.04	J9150A	HPE X132 10G SFP+ LC SR Transceiver	Hewlett Packard Enter...	$1,040.00	24	$24,960.00	USA Price List (USD)
2.00	JW186A	Aruba AP-325 802.11n/ac 4x4:4 MU-MIMO Dual Radio Inte...	Hewlett Packard Enter...	$1,395.00	108	$150,660.00	USA Price List (USD)
2.01	H3TJ1E	Aruba 1Y FC NBD Exch AP-325 SVC [for JW186A]	Hewlett Packard Enter...	$61.00	108	$6,588.00	USA Price List (USD)
3.00	JW736A	Aruba 7205 (US) 2-port 10GBASE-X (SFP+) Controller	Hewlett Packard Enter...	$12,995.00	2	$25,990.00	USA Price List (USD)
3.01	H3CW3E	Aruba 1Y FC NBD Exch 7205 Controller SVC [for JW736A]	Hewlett Packard Enter...	$2,002.00	2	$4,004.00	USA Price List (USD)
3.02	JW124A	PC-AC-NA North America AC Power Cord	Hewlett Packard Enter...	$5.00	2	$10.00	USA Price List (USD)
3.03	JW499AAE	Aruba LIC-7205-PEFV Controller Policy Enforcement Firewall...	Hewlett Packard Enter...	$8,000.00	2	$16,000.00	USA Price List (USD)
3.04	H2VV3E	Aruba 1Y FC 24x7 7205 License SVC [for JW499AAE]	Hewlett Packard Enter...	$1,232.00	2	$2,464.00	USA Price List (USD)
3.05	JW091A	SFP-10GE-SR 10GBASE-SR SFP+ 850nm Pluggable LC Conn...	Hewlett Packard Enter...	$1,245.00	4	$4,980.00	USA Price List (USD)
4.00	JZ106AAE	Aruba MM-VA-50 Mobility Master SW E-LTU	Hewlett Packard Enter...	$1,995.00	3	$5,985.00	USA Price List (USD)
4.01	H7LE8E	Aruba 1Y FC 24x7 MC-VA-50 VM Appl SVC [for JZ106AAE]	Hewlett Packard Enter...	$306.00	3	$918.00	USA Price List (USD)
4.02	JW471AAE	Aruba LIC-ENT Enterprise (LIC-AP LIC-PEF LIC-RFP and LIC-...	Hewlett Packard Enter...	$300.00	108	$32,400.00	USA Price List (USD)
4.03	H2XW3E	Aruba 1Y FC 24x7 License Cn Bundle SVC [for JW471AAE]	Hewlett Packard Enter...	$46.00	108	$4,968.00	USA Price List (USD)
4.04	JZ148AAE	Aruba LIC-VIA Per User License E-LTU	Hewlett Packard Enter...	$15.00	45	$675.00	USA Price List (USD)
5.00	JL075A	Aruba 3810M 16SFP+ 2-slot Switch	Hewlett Packard Enter...	$11,549.00	2	$23,098.00	USA Price List (USD)
5.01	U6TC8E	HPE 3Y FC 4H Exch Aruba 3810M 16SFP+ SVC [for JL075A]	Hewlett Packard Enter...	$3,879.00	2	$7,758.00	USA Price List (USD)
5.02	U4830E	HPE Networks Stackable Leg Startup SVC [for JL075A]	Hewlett Packard Enter...	$1,325.00	2	$2,650.00	USA Price List (USD)
5.03	JL085A	Aruba X371 12VDC 250W Power Supply	Hewlett Packard Enter...	$439.00	2	$878.00	USA Price List (USD)
5.04	JL085A ABA	INCLUDED: Power Cord - U.S. localization	Hewlett Packard Enter...	incl.	2		
5.05	JL084A	Aruba 3810M 4-port Stacking Module	Hewlett Packard Enter...	$1,099.00	2	$2,198.00	USA Price List (USD)
5.06	J9150A	HPE X132 10G SFP+ LC SR Transceiver	Hewlett Packard Enter...	$1,040.00	16	$16,640.00	USA Price List (USD)
6.00	J9821A	Aruba 5406R zl2 Switch	Hewlett Packard Enter...	$2,419.00	2	$4,838.00	USA Price List (USD)
6.01	H1MT0E	HPE 3Y FC 24x7 Aruba 5406R zl2 Switc SVC [for J9821A]	Hewlett Packard Enter...	$4,094.00	2	$8,188.00	USA Price List (USD)
6.02	U4832E	HPE Networks 54xx/82xx zl Startup SVC [for J9821A]	Hewlett Packard Enter...	$2,325.00	2	$4,650.00	USA Price List (USD)
6.03	J9828A	Aruba 5400R 700W PoE+ zl2 PSU	Hewlett Packard Enter...	$799.00	2	$1,598.00	USA Price List (USD)
6.04	J9828A ABA	INCLUDED: Power Cord - U.S. localization	Hewlett Packard Enter...	incl.	2		
6.05	J9996A	Aruba 2p 40GbE QSFP+ v3 zl2 Mod	Hewlett Packard Enter...	$6,799.00	4	$27,196.00	USA Price List (USD)
6.06	JH231A	HPE X142 40G QSFP+ MPO SR4 Transceiver	Hewlett Packard Enter...	$3,299.00	2	$6,598.00	USA Price List (USD)
6.07	J9993A	Aruba 8p 1G/10GbE SFP+ v3 zl2 Mod	Hewlett Packard Enter...	$4,799.00	4	$19,196.00	USA Price List (USD)
6.08	J9150A	HPE X132 10G SFP+ LC SR Transceiver	Hewlett Packard Enter...	$1,040.00	20	$20,800.00	USA Price List (USD)
6.09	J8177C	HPE X121 1G SFP RJ45 T Transceiver	Hewlett Packard Enter...	$275.00	2	$550.00	USA Price List (USD)
7.00	J9281B	HPE X242 10G SFP+ to SFP+ 1m DAC Cable	Hewlett Packard Enter...	$100.00	13	$1,300.00	USA Price List (USD)
8.00	J9578A	Aruba 3800/3810M 0.5m Stacking Cable	Hewlett Packard Enter...	$249.00	2	$498.00	USA Price List (USD)
9.00	JH234A	HPE X242 40G QSFP+ to QSFP+ 1m DAC Cable	Hewlett Packard Enter...	$419.00	2	$838.00	USA Price List (USD)
		Quote Total				$734,442...	

Figure 4-74 Activity 4.6: Core design

Summary

Congratulations! Now you have the essential knowledge and skills to begin designing wired network solutions.

You learned how to plan two-tier and three-tier wired architectures and the appropriate situations for deploying each. You learned how to use VSF and backplane stacking as a fundamental part of your wired network architecture. You can now use technologies to simplify your architecture while also enhancing its redundancy and resiliency.

You also gained insight into planning each layer of the network. Beginning with the access layer, you learned how to select the switch series that meets the customer needs. You can plan to meet edge port requirements, as well as design appropriate oversubscription for uplinks and VSF links. You can also ensure that access layer switches meet the customer's PoE requirements.

Next, you explored planning the core and optional aggregation layer. You learned best practices for selecting core and aggregation switches, and you also examined example designs.

Learning check

1. What is one advantage of deploying VSF on redundant core switches?
    a. VSF makes it easier to configure and deploy technologies such as VRRP.
    b. The topology can use link aggregations rather than spanning tree to handle redundant links.
    c. VSF uses specialized backplane stacking modules and high bandwidth stacking cables.
    d. VSF makes it easier to implement VRRP and MSTP together.
2. Which switch is most likely to require the highest bandwidth on its uplinks?
    a. A switch that connects to employees who browse the Internet and access Word documents and spreadsheets on centralized servers
    b. A switch that connects to printers and HVAC IoT devices
    c. A switch that connects to a high density of 802.11ac APs
    d. A switch that connects to employees who occasionally run video conferencing softwareWhat are advantages of designing an OSPF AS to use multiple areas?
3. What is the most appropriate oversubscription ratio between the aggregation layer and core?
    a. 4:1
    b. 12:1
    c. 20:1
    d. 24:1

5 Access Control and Security

EXAM OBJECTIVES

- ✓ Given the customer's requirements for a single-site environment, determine and document a detailed network security solution.
- ✓ Given the customer scenario, determine and document licensing requirements.
- ✓ Select the appropriate products based on a customer's needs for a single-site campus.
- ✓ Translate a customer's needs into technical requirements.
- ✓ Evaluate a customer's needs for a single-site campus, identify gaps and recommend components.
- ✓ Given a scenario for a single-site campus, choose the appropriate components to be included in a Bill of Materials (BOM).

Assumed knowledge

- Authenticatin, Authorization, and Accounting (AAA)
- 802.1X
- MAC-Auth
- Extensible Authentication Protocol (EAP)
- Digital certificates
- Intrusion Detection System (IDS) and Intrusion Prevention System (IPS)
- Virtual Private Network (VPN)
- Wireless security, including WPA/WPA2 and WEP
- Familiarity with Aruba ClearPass and Aruba IDS and Wireless Intrusion Prevention System (WIPS) features

Introduction

Customers know that the risks of security breaches are too real for them to ignore. By protecting network access and providing a first line of defense against denial of service (DoS) attacks and malware, Aruba solutions can help customers avoid downtime, lost intellectual property, and regulatory fines.

In this chapter, you will learn how Aruba ClearPass helps to secure customer environments. You will look at the many scenarios for which you can recommend ClearPass, including access control for both AAA-ready and non-AAA-ready environments, guest access scenarios, and Bring Your Own Device (BYOD) and endpoint integrity.

You will also learn how to size ClearPass, recommend appliances, and determine the appropriate licenses.

You will learn about designing controls for employee wireless access to ensure that each employee receives access to the correct resources. You will learn how Aruba ClearPass plays a vital part in your plans, allowing you take a role-based approach to assigning rights to users.

You will also learn how to incorporate wired users and devices into the access control plan using authentication, authorization, and optional tunneled-node features.

Finally, you will review the Aruba Intrusion Detection System (IDS) and Wireless Intrusion Prevention System (WIPS) features and learn how to plan the proper Access Point (AP) modes for supporting these functions.

Aruba ClearPass features

You will first focus on understanding ClearPass features.

Understanding connectivity options

You should understand the many scenarios for which you can propose Aruba ClearPass. Almost all customers need to manage which devices connect to the network. Many customers think of wireless first when they hear "access control." However, as many as 50% of Internet of Things (IoT) devices might be wired. Because these devices and their connections can be vulnerable to unauthorized access, customers can benefit from a solution that can authenticate devices, profile their type, and apply appropriate access control policies. Some devices support 802.1X, and some do not, but ClearPass can provide authentication for any device.

Aruba ClearPass OnConnect for wired non-RADIUS enforcement

ClearPass is a natural addition to a proposal for Aruba wireless and wired solutions. However, because ClearPass supports any customer infrastructure, you can also propose it for customers who are not ready to upgrade part of the infrastructure (see Figure 5-1).

For example, a customer might have an Aruba wireless network but is not yet ready to upgrade the wired network. Aruba ClearPass OnConnect enables access control for non-AAA ready infrastructures and is easy to configure on legacy switches from many vendors. ClearPass uses Simple Network Management Protocol (SNMP) to integrate with the legacy infrastructure and learn when a new device connects. ClearPass then does the heavy lifting to determine the device type and even the connected user. It then determines whether to let the device connect, using SNMP to enforce that decision on the switch by leaving the port up or taking it down.

The legacy infrastructure uses SNMP to inform ClearPass of the connected device's MAC address. ClearPass can then integrate with a Windows directory at the backend to determine which user is logged in on a particular device. In this way, it can add user-based information to its decisions.

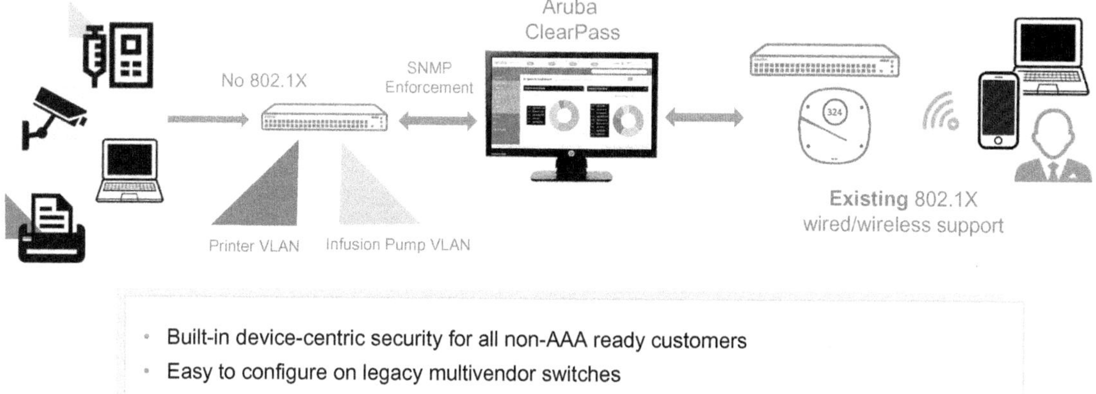
Aruba
ClearPass
No 802.1X
SNMP
Enforcement
Printer VLAN
Infusion Pump VLAN
Existing 802.1X
wired/wireless support
Built-in device-centric security for all non-AAA ready customers
Easy to configure on legacy multivendor switches
Leverages ClearPass profiling for wired/wireless - IoT, laptops, mobile phones.

Figure 5-1 Aruba ClearPass OnConnect for wired Non-RADIUS enforcement

Secure connections—authorization before access

You can also recommend ClearPass for any 802.1X-ready customer environment (see Figure 5-2). Ideally, you will be able to propose a unified Aruba wireless and wired solution for the customer. These solutions, with their built-in AAA capabilities, form the ideal foundation for ClearPass.

In addition to supporting secure, encrypted wireless access, ClearPass can extend access control to wired connections. ClearPass enables authentication for smartphones, tablets, PCs, printers, security cameras, and more. It extends to IoT devices (such as connected sensors) across a multitude of vertical specific devices. Easy-to-create policies give customers the power to control their users and devices exactly as their business requirements dictate.

ClearPass allows network administrators to enforce different network access policies based on multiple factors gathered before and during the access and authentication process.

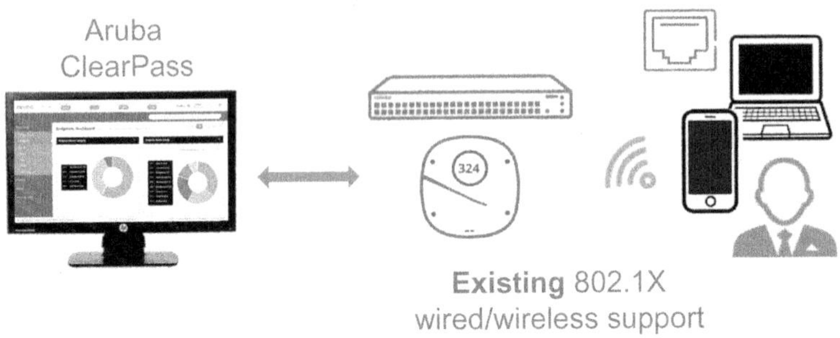

- Multivendor support for all 802.1X ready wired and wireless customers
- Secure encrypted wireless access
- Built-in ClearPass profiling - IoT, laptops, mobile phones
- Easy to use policy creation templates

Figure 5-2 Secure connections

Multi-factor authentication (DUO workflow)

If your customer is one of the many companies that already use Duo for two-factor authentication, you can use ClearPass's ability to integrate with DUO as a selling feature.

Figure 5-3 shows a user with Duo, accessing the Captive portal and selecting Duo Push. An approval request is then pushed to a known device such as the user's cell phone. The user can then approve the login request, proving the user's identity. After this, the user can return to the portal and enter their password and successfully log in to the network.

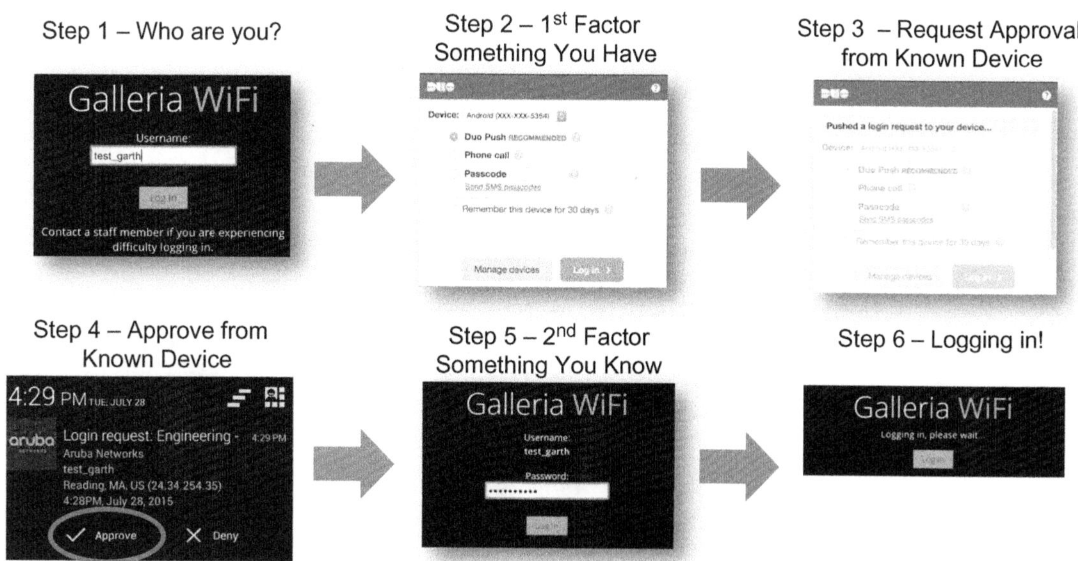

Figure 5-3 Multi-factor authentication (DUO Workflow)

Web Authentication with ClearPass guest

ClearPass Guest provides a Web Authentication solution, typically used for guests.

Web Authentication is performed at Layer 3 and presents a user with a web-browser-based login page, sometimes called a Captive Portal, in order to gain network access. A back-end customized RADIUS server processes the web authentication and allows you to tailor the web pages to the organizations' style sheet and graphics. Most Web authentication systems use a user self-registration page so users can create and manage their own user accounts.

NAS devices, such as wireless APs and wired switches on the edge of the network, use the RADIUS protocol to ask the Web authentication server to authenticate the username and password a guest provides when logging in to the network. If authentication is successful, the guest is authorized to access the network.

Authorized access uses the concept of roles. Each visitor is assigned a role, which consists of a group of RADIUS attributes. These attributes are used to control every aspect of the guest's network session, effectively defining a security policy that controls what the guest is permitted to do on the network. You can also use vendor-specific attributes to configure the finer details of the NAS security policy.

Self-registration with sponsor example

Figure 5-4 shows an example of how ClearPass provides a good workflow for companies. You can have back-up sponsors and send the user an SMS message or email that contains login credentials for use later in the day.

In retail, this is a good way to provide different access for employees and contractors that are not in an active directly. ClearPass can send shoppers to a portal without sponsor approval that provides faster access.

Self-registration allows a visitor or guest user to create their own account in the RADIUS user account database. This simplifies administration and enables the guest user to manage their own information and credentials.

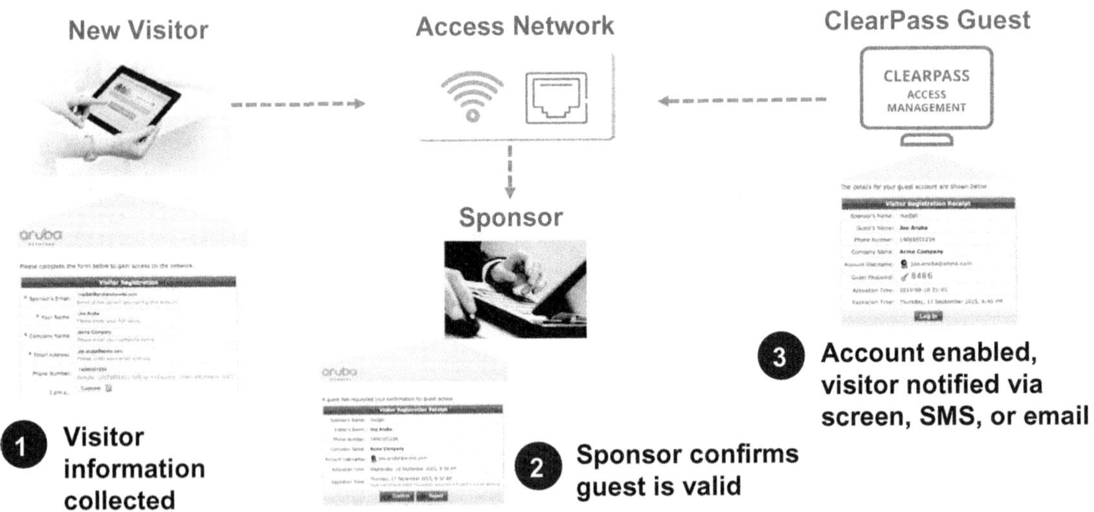

Figure 5-4 Self-registration with sponsor example

Customizable portal features

In ClearPass Guest, you can customize portals to include the organization's logo and style sheet.

You can talk about the SKIN service and how you will deliver a portal design that works on laptops and smart devices. You can also use the built-in advertising module to promote a mobile app. Other features include:

- The ability to show a different portal per location or department
- Integration with third-party systems for hotels and hospitals
- Customizable fields (user name, use of social login, and acceptance of use)
- Backend enforcement rules

Why ClearPass Guest?

In summary, you should propose ClearPass Guest as the best solution for a broad range of customer environments.

ClearPass offers self-service with per user credentials, sponsor approval, social login, and a way for a sponsor to pre-create logins if needed. The portal branding fits laptops, mobile devices, and other user devices. The backend policy manager integration allows for differentiated access based on guest type, Quality of Service (QoS) options, expiry of credentials, and much more.

Lastly, you can use PEAP-Public and Black Hat with ClearPass. These use an encrypted authentication for guest access and you can use them with portal access as well as per user tracking.

ClearPass Onboard and OnGuard

You have focused on how ClearPass can meet customers' needs for authentication and access control of employees, guests, IoT devices, and other specialized devices. ClearPass also provides two more important capabilities.

For customers who need a Bring Your Own Device (BYOD) solution, ClearPass Onboard makes it simple to get the devices connected security without IT involved. ClearPass provides a self-guided portal through which users can easily provision their devices and obtain unique certificates for them, making it easier for customers to enforce strong authentication on devices of all types. Recommend ClearPass OnGuard for customers who need to protect their trusted network against threats that might be lurking on unmanaged devices, particularly ones that move on and off premises and sometimes connect to insecure networks.

ClearPass OnGuard leverages persistent and dissolvable agents to perform advanced endpoint posture assessments over wireless, wired, and VPN connections. OnGuard's health-check capabilities ensure compliance and network safeguards before devices connect. OnGuard supports a broad array of devices, including Windows, macOS, and Linux devices.

Note that Onboard works for Windows, macOS, and Linux devices, as well as iOS, Android, and ChromeOS.

ClearPass server design

In this section, you will learn how to size ClearPass, recommend appliances, and determine the appropriate licenses.

ClearPass appliance options

You can sell ClearPass as an appliance or a virtual machine (VM). The choice of the appliance depends primarily on the performance and load required. As Table 5-1 shows, the C1000 supports a maximum of 1000 concurrent sessions, the C2000 supports a maximum of 10,000, and the C3000 supports a maximum of 50,000 concurrent sessions.

If your customers prefer to use or build their own server, you can propose a VM. The VM should have similar specifications to the C1000, C3000, or C5000 appliance to support a similar load.

You can find these specifications in the *Aruba ClearPass Scaling & Ordering Guide.*

You might also add an appliance to support dedicated features such as reporting for a large deployment.

In addition to detailed hardware specifications, this guide provides more detailed information about maximum guest logins, RADIUS performance, TACACS performance, and performance for other ClearPass features.

The VM works with several different hypervisors, including VMware ESXi and Microsoft Hyper-V. Refer to the current *Aruba ClearPass Scaling & Ordering Guide* for exact versions.

Table 5-1 Maximum concurrent sessions for ClearPass appliances

ClearPass appliances	Maximum concurrent sessions
Aruba ClearPass C1000 S-1200 R4 HW Appl	1000
Aruba ClearPass C2000 DL20 Gen9 HW Appl	10,000
Aruba ClearPass C3000 DL360 Gen9 HW Appl	50,000
Aruba ClearPass Cx000V VM Appl E-LTU	Depends on specifications

ClearPass 6.7 licensing

With ClearPass 6.7, licenses and appliances are uncoupled. You can propose appliances based on the performance requirements and then choose the licenses that the customer needs more granularly. Each hardware or virtual appliance does require a platform activation key (PAK), which is perpetual and does not expire. (The PAK for a hardware appliance is registered with no change licenses and is generated based on serial number.)

As Figure 5-5 shows, you add other licenses to the ClearPass appliance to support the desired functions.

Access licenses support ClearPass's Policy Manager and Guest functions. ClearPass requires one license for each endpoint that is concurrently authenticated or authorized, no matter how the device authenticates.

If the customer is using OnGuard, you should propose one license per endpoint to which ClearPass deploys an agent. Onboard licenses are calculated per-user. A user can onboard several devices, but still only requires one license.

Each of these licenses is sold in increments ranging from the granular 100 up to 10,000. You can also offer them as perpetual licenses or one-year, three-year, or five-year subscriptions. One exception applies: OnGuard licenses do not support a five-year subscription.

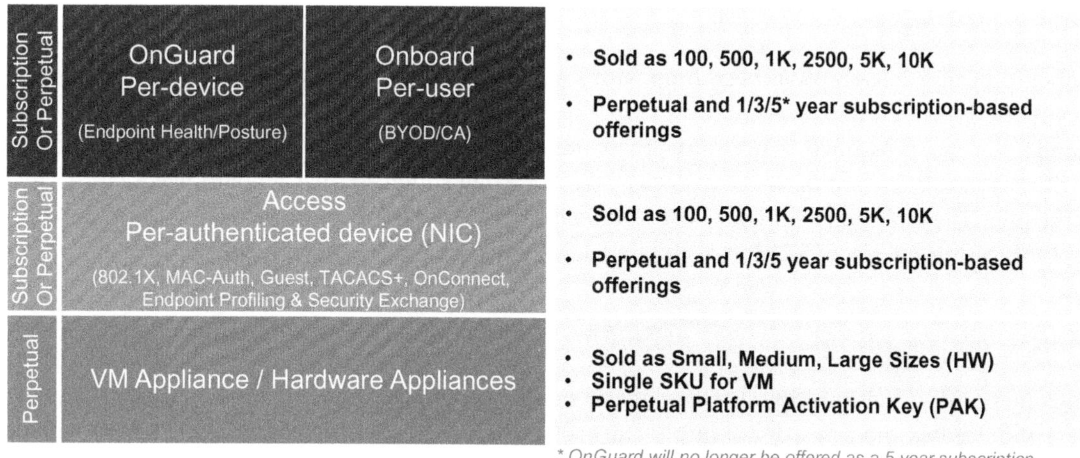

Figure 5-5 ClearPass 6.7 licensing

The sections below provide more details about how licenses are consumed.

Access licenses

Every time a user connects a device to the network and is authenticated and authorized by ClearPass, the device uses one Access license. An employee laptop could authenticate with 802.1X, an IoT device could authenticate with MAC-Auth, and a guest could login through the captive portal. Each endpoint would consume one Access license. TACACS+ authorization of managers also uses Access licenses. If a user is authenticated on more than one endpoint at the same time—such as a laptop and a smartphone—each endpoint uses one Access license. In fact, the license actually applies to the NIC on the endpoint (identified by MAC address). If a user is connected to the network on both a wired connection and wireless connection at the same time, the device uses two Access licenses. However, more typically, a device might be connected to the Ethernet network, disconnect, and then connect to the wireless network. In this case, the device uses only one Access license.

Security Exchange and Endpoint Profiling are enabled when any Access license is installed but not restricted to any licensed capacity limits.

OnGuard licenses

OnGuard license consumption is based upon a per-endpoint model and applies to both persistent and dissolvable agents. For example, within a 24-hour period, ClearPass installs persistent agents on two endpoints and uses dissolvable agents on three endpoints; five OnGuard licenses are required.

Plan one license for each endpoint to which the customer wants to apply posture assessments.

Onboard licenses

Onboard license consumption beginning with ClearPass 6.7 is based upon an active certificate per-user model. For example, if a given user has four devices with an active certificate each, only one Onboard license is required. If over time, three out of the four devices are retired, and their associated certificates revoked, the fourth active device certificate will still keep the Onboard license associated to the user. The intentional onboarding of large numbers of devices by a single user to avoid purchasing Onboard licenses is a violation of the End-User Software License Agreement.

License aggregation and subscriptions

Customers can install multiple licenses of the same type on a standalone appliance or cluster for increased licensed capacity—for example 100+100+100 = 300. Subscription licenses (which include support) are tracked both on licensed capacity and term. If a customer installs two subscription licenses six months apart, the total term will be one and a half years; however, in the last 6 months, the licensed capacity will drop to the remaining valid subscription.

What is concurrency?

Customers require sufficient Access licenses for all of the concurrently authenticated or authorized devices. Under the concurrency model, a user or device that is authenticated and authorized on the network consumes an Access license during an active session. After the active session ends, the license returns to the pool. For most forms of authentication, ClearPass determines when the active session starts and stops using RADIUS accounting start and stop messages (Table 5-2).

Session checks are performed every 15 minutes.

Table 5-2 Concurrency for license usage

Method	Session Begins	Session Ends
802.1X	RADIUS Accounting START	RADIUS Accounting STOP
MAC-Auth	RADIUS Accounting START	RADIUS Accounting STOP
Guest (anonymous, self-reg, socal, and so on)	RADIUS Accounting START	RADIUS Accounting STOP
VPN	RADIUS Accounting START	RADIUS Accounting STOP
OnConnect	MAC Learned (mac-notify or switch link-up)	MAC Removed/Aged (mac-notify or switch link-down)
Note	License usage calculated every 15 min	

If the session end cannot be identified (for example, the NAS does not use accounting), the license will be removed from the pool 24 hours after the time it was consumed.

NAS devices such as MCs and switches can also send interim accounting messages. However, ClearPass does not require these in order to determine when a session starts and stops.

For OnConnect, which does not use RADIUS, ClearPass uses SNMP messages to determine when it learns about the authorized MAC address and when the MAC address ages out.

Whether the customer requires one Access license for every device that could possibly connect or fewer depends on the type of customer and users' connection patterns. For example, for many corporate offices, it is safest to plan one license for every device. A user could easily be connected through both a smartphone and a laptop at the same time. But in other environments, such as airports and hotels, device count must be estimated from network data such as the maximum DHCP pool size and lease length and maximum firewall session usage.

Sample BoM #1—University

You will now look at some examples to ensure that you understand how the licenses work.

In this first example, a university has a maximum 30,000 concurrently connected devices; 100 of these devices are guest devices. The university has 8000 users who need to onboard several devices each.

As Figure 5-6 shows, a solution using virtual appliances could use two appropriately sized Cx00Vs to provide AAA and one Cx00V as a dedicated Insight reporting node due to the size of this deployment. The customer requires 30,000 Access licenses and 8000 Onboard licenses.

It does not matter that 100 of the concurrently connected devices are used by guests; guest logins consume one Access license just other authentications. And it does not matter how many devices each user onboards; the Onboard license is per-user.

As you see, the appliance number is also calculated based on the feature needs, not the licensing requirements.

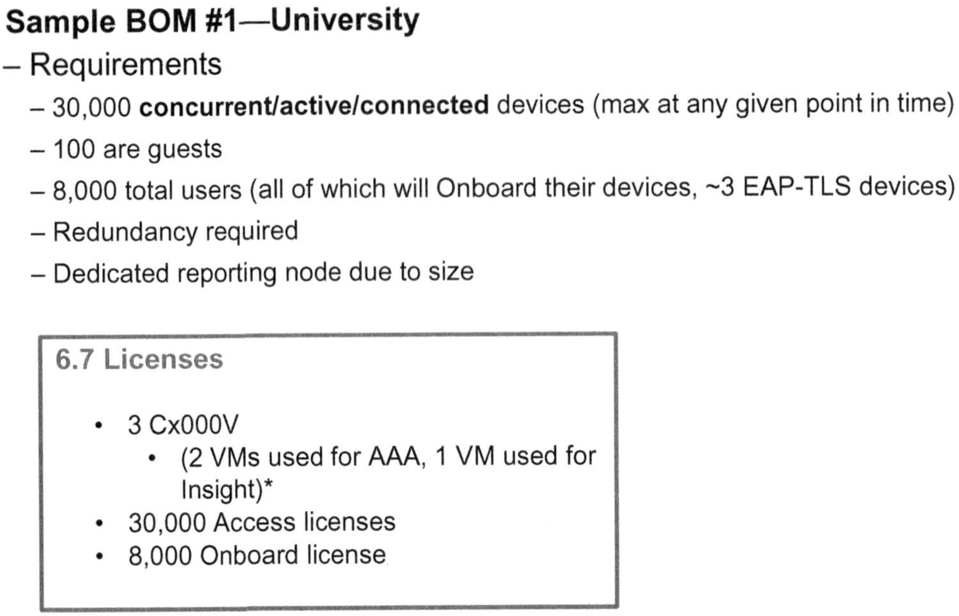

Sample BOM #1—University

– Requirements

– 30,000 **concurrent/active/connected** devices (max at any given point in time)

– 100 are guests

– 8,000 total users (all of which will Onboard their devices, ~3 EAP-TLS devices)

– Redundancy required

– Dedicated reporting node due to size

6.7 Licenses

- 3 Cx000V
  - (2 VMs used for AAA, 1 VM used for Insight)*
- 30,000 Access licenses
- 8,000 Onboard license

Figure 5-6 Sample BoM #1—University

Sample BoM #2—Corporate

In this example, the customer has up to 10,000 concurrently connected devices, 100 of which are guests. Again, the Access license number matches the concurrent device number. Two thousand users will onboard, and these users will have on average two devices each. Despite the number of devices, the Onboard license number matches the user number. The customer also has 100 contractors and wants to run posture assessments on their laptops. For this, the customer requires 100 OnGuard licenses.

Figure 5-7 summarizes the licenses for this example. Because the customer wants redundancy, you should propose two appliances for AAA services. Finally, the customer wants to use ClearPass to analyze inbound events from other solutions, so add an appliance for the Ingress Event Engine (IEE).

You could also propose hardware appliances rather than the virtual ones.

Sample BOM #2—Corporate

- Requirements
  - 10,000 **concurrent/active/connected** devices (max at any given point in time)
  - 100 are guests
  - 2,000 users will Onboard (~2 devices each)
  - 100 active contractors who require posture assessment on their laptops
  - Inbound events from other solutions
  - Redundancy required (2 VMs)

6.7 Licenses

- 3 Cx000V
  - (2 VMs used for AAA, 1 VM used for IEE)*
- 10,000 Access licenses
- 2,000 Onboard licenses
- 100 OnGuard licenses

Figure 5-7 Sample BoM #2—Corporate

Conversion

When a customer updates an existing ClearPass appliance to version 6.7, its Policy Manager license (whether 500, 5K, or 25K) is converted to a pre-activated PAK. The update also auto-installs 1000 Access, 100 Onboard, and 100 OnGuard evaluation licenses in the system with an expiration of six months. These licenses are intended to ensure that the system continues to operate even if the customer has some initial issues converting licenses. The auto-installed evaluation licenses are set to expire in 180 days.

Customers often need more licenses than these on their upgraded appliance. However, the user interface will not lock out and services will not be impacted until 180 days expire. With the six-month licensing buffer, customers do not need to worry about converting their existing licenses the day of upgrade. However, customers do need to convert the licenses before those six months elapse.

MNP will be the first method to convert licenses, but functionality will also be present in ASP at launch. And TAC can help customers and partners to convert Enterprise and Subscription-based licenses.

HOW IT WORKS

Existing customers will get a 1:1 license exchange

- Legacy ClearPass 25K (e.g. CP-VA-25K) = 25,000 Access Licenses
  - Also includes one set of 25 licenses for each feature (Access, Onboard, OnGuard)
- Legacy ClearPass Guest 500 = 500 Access Licenses
- Legacy ClearPass Onboard 10K = 10K Onboard Licenses (new key)
- Legacy ClearPass OnGuard 5K = 5K OnGuard Licenses (new key)
- Legacy ClearPass Enterprise 100 = New xAccess/yOnboard/zOnGuard Licenses in multiples of 25
  - For example, 25 Access + 50 Onboard + 25 OnGuard = 100
  - Enterprise license conversion is a one-time, one-time way process per license key

Existing customers will continue to pay support on the original product purchased.

Figure 5-8 Conversion

Within 180 days after the upgrade to 6.7, customers must convert their existing licenses to the new licensing model as shown in Figure 5-8. The conversion process is a one-time process per key and cannot be reversed.

Figure 5-8 shows some examples of how existing licenses convert to new ones. A Legacy CP-VA-25K was a ClearPass VM with 25,000 Policy Manager licenses. These convert to 25,000 Access licenses. In addition, 25 licenses for Access, Onboard, and OnGuard are provided.

Legacy Guest licenses convert to Access licenses one to one. Legacy Onboard and OnGuard licenses convert to the new versions of these licenses one to one, but they have new keys that need to be installed.

Customers can choose the type of license to which their legacy Enterprise licenses convert, allocating them to Access, Onboard, and OnGuard as desired in multiples of 25. For example, a customer with 100 legacy Enterprise licenses could obtain keys for 25 Access, 50 Onboard, and 25 OnGuard licenses.

Also note that customers will continue to pay for and receive the support for the original product that they purchased.

Wireless employee access control

You will now learn best practices for enforcing authentication and access control for wireless employees, including how to integrate the wireless solution with ClearPass.

Five phases to join a WPA/WPA2 WLAN

As Figure 5-9 shows, a user must go through five phases to access a WPA/WPA2 security-based WLAN. Using WPA or WPA2, the client will first perform 802.11 negotiation and then user authentication using WPA or WPA2. WPA/WPA2 can use the 802.1X framework. This form of authentication, which is performed at Layer 2, can use ClearPass as the backend authentication server. After the user authenticates to the network, the user is assigned an IP address. Then the security policy is applied, and the user is granted network access, and any firewall policies are applied to the user.

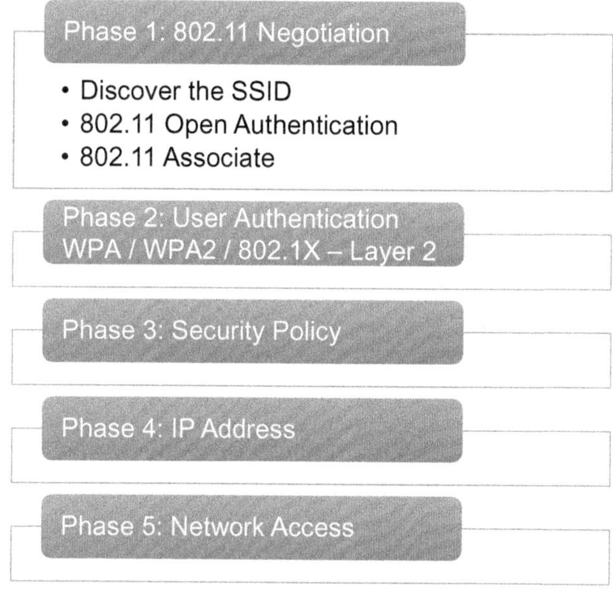

Phase 1: 802.11 Negotiation
• Discover the SSID
• 802.11 Open Authentication
• 802.11 Associate
Phase 2: User Authentication
WPA / WPA2 / 802.1X – Layer 2
Phase 3: Security Policy
Phase 4: IP Address
Phase 5: Network Access

Figure 5-9 Five phases to join a WPA/WPA2 WLAN

Confidentiality through encryption

Encryption Suites ensure privacy of data as it is transmitted through the air. You can also use specialized encryption schemes to ensure integrity of the data. In a WLAN environment, encryption is performed at Layer 2 using AES, TKIP, or WEP and at Layer 3 using a VPN.

There are two broad methods of encryption—Symmetric or Asymmetric Key. Symmetric Key encryption uses the same key to encrypt and decrypt traffic, while Asymmetric key encryption uses a public/private key pair for encryption where one key is used to encrypt and a different key is used to decrypt traffic.

User authentication

Authentication is the process of identifying and determining the identity of a user or device that attempts to connect to the network. In most cases, you want to ensure only authorized users with proper network credentials are allowed to use the network. This is similar to checking a driver's license credential to verify someone's identity. Most forms of authentication rely on a RADIUS server, which is preconfigured with the proper credentials for each user. When users connect their devices to the network, they submit their credentials and are only allowed access if their credentials match those stored on the server. Some forms of authentication are bi-directional; the user also authenticates the server to ensure that it is legitimate.

You may use a MAC address, passphrase, password, or certificate to authenticate a user. The credentials are shown in Figure 5-10 in general order of security. MAC addresses are typically the most insecure credential because they are easily spoofed, while X.509 certificates are typically the most secure credential.

Figure 5-10 also shows the forms of WLAN authentication associated with each credential type.

MAC-Auth authenticates devices based on their MAC address. Although a MAC address is theoretically unique per device, hackers can easily sniff the wireless traffic, detect an authorized MAC address, and spoof that address on their device.

WPA2 Personal uses a passphrase that is shared by all users on the WLAN. This form of authentication does not use a RADIUS server. Instead, the passphrase is configured on the WLAN. This form of authentication is prone to leaks in which authorized users expose the passphrase to unauthorized users. In addition, if a user becomes unauthorized—for example, an employee leaves the company—it is difficult to change the passphrase, so the network remains open to that user. WPA2-Personal is also vulnerable to exploits that help hackers crack the passphrase.

WPA2 Enterprise uses 802.1X with various forms of Extensible Authentication Protocol (EAP). Based on the particular EAP method, WPA2 Enterprise can use usernames and passwords for authentication or X.509 certificates. Usernames and passwords are more secure than a simple passphrase because the password is unique per user and can be updated periodically or revoked. However, users can still share their passwords with unauthorized users.

Certificates are the most secure option. A certificate is uniquely associated with a user or a device. In either case, the certificate is securely stored on the machine with encryption and is not easily accessible by the user. A user might also have to enter a password to use the certificate, which adds further protection in case a device is stolen.

Various forms of authentication use encryption to secure the credential in transit, which is particularly important for authentication on WLANs.

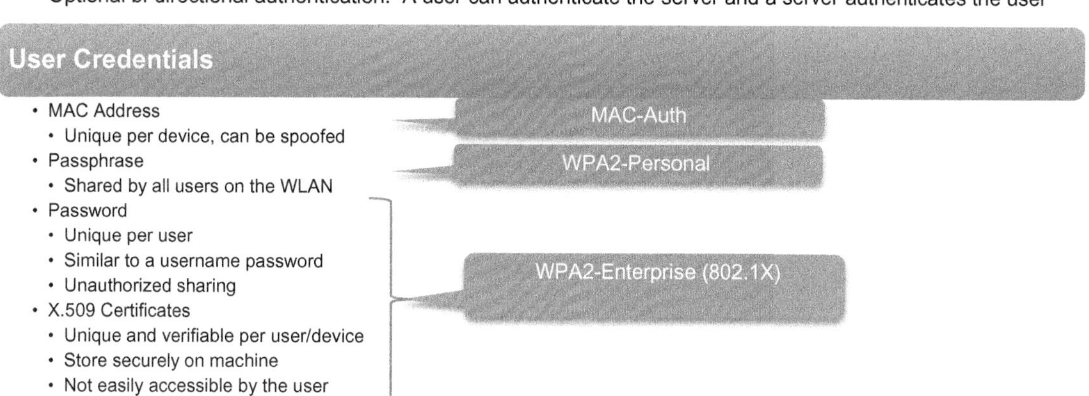

Authentication – Who are you?
• User or device based
• Ensures that only authorized users with proper credentials are allowed to use the network
• Optional bi-directional authentication. A user can authenticate the server and a server authenticates the user
User Credentials
• MAC Address
• Unique per device, can be spoofed
• Passphrase
• Shared by all users on the WLAN
• Password
• Unique per user
• Similar to a username password
• Unauthorized sharing
• X.509 Certificates
• Unique and verifiable per user/device
• Store securely on machine
• Not easily accessible by the user
MAC-Auth
WPA2-Personal
WPA2-Enterprise (802.1X)

Figure 5-10 User authentication

More details on WPA2-Enterprise authentication

You will now examine WPA2-Enterprise security in more detail. It uses 802.1X to authenticate users. Because most client operating systems today support 802.1X, WPA2-Enterprise is recommended for almost all enterprise deployments (see Figure 5-11).

As you saw, WPA2-Enterprise can use different types of credentials. The particular credential depends on the Extensible Authentication Protocol (EAP) method. Protected EAP (PEAP)-MSCHAPv2 requires usernames and passwords to authenticate users while EAP-TLS uses certificates to authenticate users.

Because 802.1X acts at Layer 2, it is the best form of authentication for wireless and wired networks. When a device connects to the network, it cannot send any traffic except EAP messages, which the network infrastructure forwards to the RADIUS server.

PEAP-MSCHAPv2 and EAP-TLS are bidirectional. Both EAP methods use certificates to authenticate the RADIUS server.

Another EAP method is EAP-TTLS, which uses usernames and passwords to authenticate users, much like PEAP.

The EAP method used depends on the methods configured on the RADIUS server and the client. The two negotiate what they will use for the authentication session.

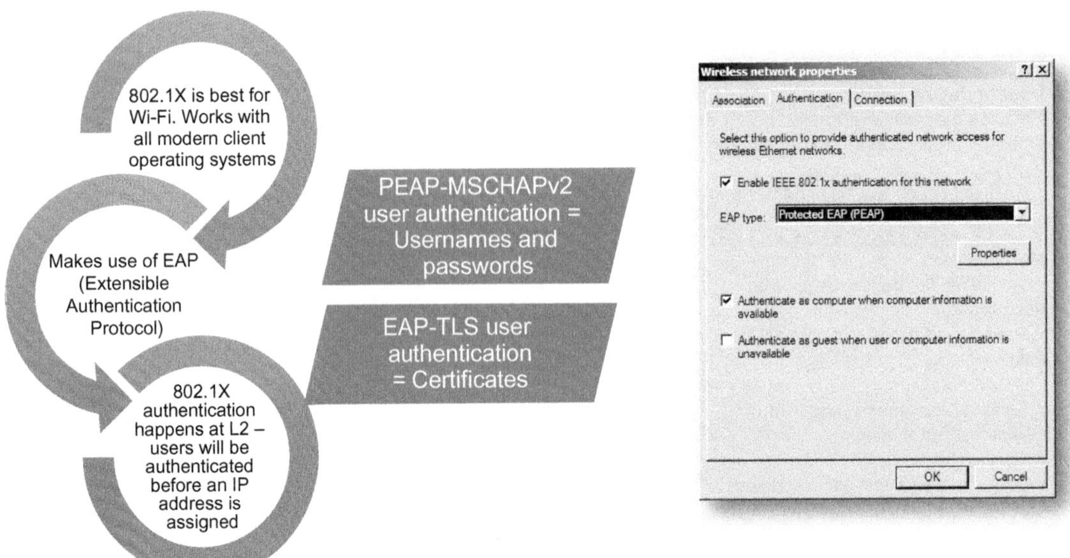

Figure 5-11 More details on WPA2-Enterprise authentication

Encryption and authentication best practices

In summary, a wireless network should use the best practices outlined in Figure 5-12.

For encryption, use WPA2 and AES.

In an enterprise WLAN environment, you should implement 802.1X/EAP for all internal client devices that employees use. You can also implement ClearPass to apply access control security policies and apply roles to WLAN users. Windows clients support machine authentication if they are part of the Microsoft Active Directory. Machine authentication can be important because it lets a device gain the network access required for a user to log into the device with domain credentials. Machine authentication has a limitation of not identifying the user logged onto the system, so Aruba recommends using both machine and user authentication, which can occur after machine authentication.

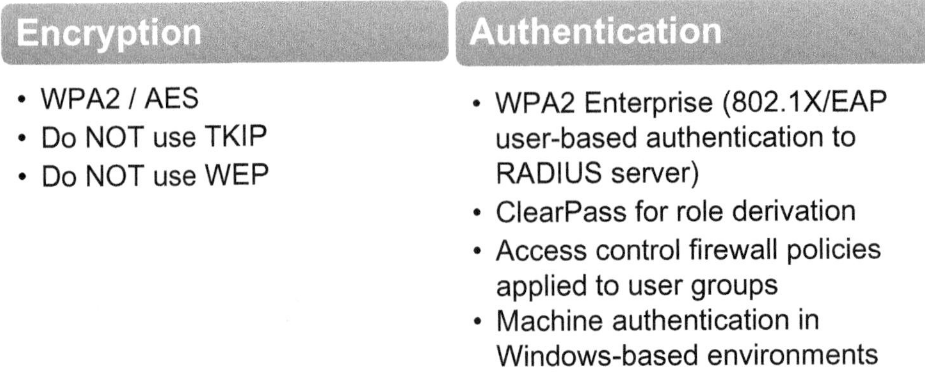

Encryption
• WPA2 / AES
• Do NOT use TKIP
• Do NOT use WEP
Authentication
• WPA2 Enterprise (802.1X/EAP user-based authentication to RADIUS server)
• ClearPass for role derivation
• Access control firewall policies applied to user groups
• Machine authentication in Windows-based environments

Figure 5-12 Encryption and authentication best practices

WPA2-Enterprise best practices

You should also follow the best practices summarized in Figure 5-13 with WPA2-Enterprise.

Whether the clients are using PEAP or EAP-TLS, make sure that they are configured to validate the RADIUS server certificate. Best practice is to have the certificates publicly signed. In addition, clients should check the server common name (CN) included in that certificate. The clients will then authenticate the RADIUS server and verify that it is legitimate before they send their credentials and become authenticated. It is also best practice to configure clients not to allow users to add new CAs or trust new servers.

Following these guidelines can help to prevent clients from exposing information to rogue APs with rogue RADIUS servers.

When using multiple RADIUS servers, you can use a common CN for all of them, which eliminates the need to obtain a different certificate on each server and configure clients to trust each of those certificates.

In a Windows domain, group policies can enforce these best practices.

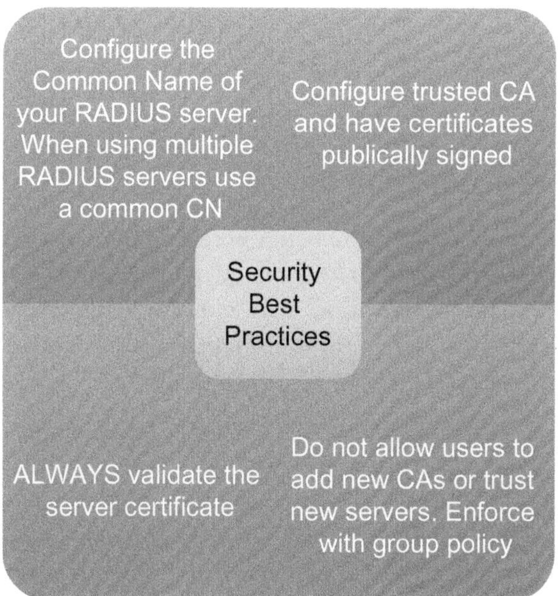

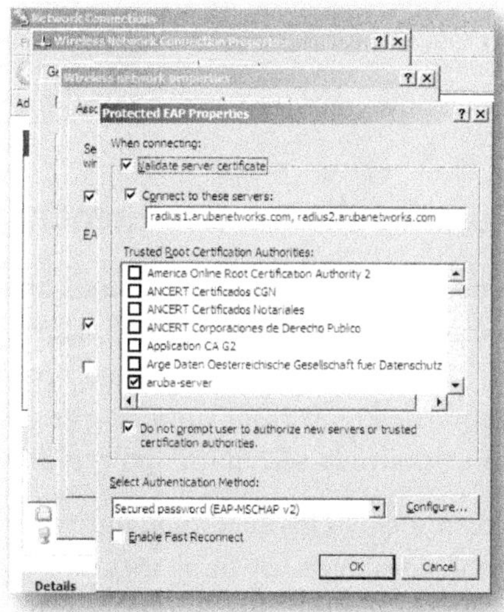

Figure 5-13 WPA2-Enterprise best practices

Role-based access controls

Administrators have traditionally used VLANs to apply restrictions to network traffic. For example, they can place employees in the IT department in one VLAN, users in the financial department in another VLAN, and contractors in a third VLAN. User devices in different VLANs have different IP addresses. The infrastructure might then have access control lists (ACLs) applied to control traffic based on those VLANs and IP addresses. However, VLAN-based access controls have drawbacks. If the VLAN is applied manually, it is difficult to ensure that a user is assigned to the correct VLAN. ClearPass and other RADIUS servers can also apply VLANs to user sessions dynamically. However, this approach still requires correlating IP addresses with access controls, which tends to make those controls less flexible. In addition, from an infrastructure design perspective, it can often be simpler to place all employees' wireless devices in a single VLAN.

Therefore, it is recommended that you use role-based access control. Roles are much more flexible than VLANs alone. ClearPass returns the role to the MC, and the MC then uses the role to apply a variety of controls to the user traffic. In addition to VLAN, these controls include firewall policies, bandwidth controls, QoS, and traffic shaping policies. When you use roles, wireless users can be assigned to the same VLAN and subnet, but controlled in very different ways.

Wired user access control

You will now consider access control for wired users.

Determine the need to authenticate wired users and devices

You need to determine if wired users and devices need to authenticate before accessing the network. You can recommend this level of security to customers who need to keep confidential data and network resources from unauthorized access or damage. Imposing this security can also help prevent unauthorized users from launching attacks. Preventing them, for example, from performing network reconnaissance to search for and exploit vulnerable systems.

If a customer has invested in identity control and role-based security in the wireless network, emphasize the advantages of implementing consistent wired and wireless security and a consistent user experience. You can then apply consistent security policies across the entire network, so users have a reliable and consistent experience, whether they access the network through a wired or wireless connection. Authenticating wired users also allows you to apply granular user-based access controls and deliver identity information to security solutions. Most companies today must demonstrate regulatory compliance, which often includes the ability to audit network access and activity.

User-based access control helps to ensure users access only to the data and network resources they need. This reduces the exposure of highly confidential data by limiting access to only selected users. Furthermore, identity-based information can be provided to security solutions that can implement advanced protection measures such as monitoring users' privileges and removing any excess privileges.

Options for port-based authentication

You have several options for enforcing port-based authentication, whether with local forwarding or with per-user tunneled-node. For choice of authentication method, you will have similar considerations to those for wireless networks.

Because 802.1X provides the tightest security measures to prevent unauthorized users from accessing the network, Aruba recommends it for most enterprises whenever possible. For devices that do not have a user interface or support for 802.1X, you may need to use MAC-Auth. If you are implementing a captive portal solution with ClearPass Guest, you must implement RADIUS-based MAC-Auth on any switch port to which users might connect their devices.

You can also combine 802.1X and MAC-Auth on a single port to provide flexibility in the type of device that can connect to the port.

When you combine 802.1X and MAC-Auth, keep in mind that the security on the port falls to the level the MAC-Auth provides. Because it is trivial for a hacker to discover and spoof MAC addresses, authentication policies should typically provide the lowest level of access possible to devices that authenticate with MAC-Auth. If users pass 802.1X, ClearPass can then authorize them for higher levels of access.

 **Note**

The switch submits both MAC-Auth and 802.1X requests for all devices that connect when you enable both forms of authentication. A device receives access if it passes one form of authentication. If it passes both, the 802.1X authorization settings take precedence.

You can also combine 802.1X and MAC-Auth when the switch integrates with a captive portal solution on ClearPass Guest. In this case, ClearPass Onboard can deploy the correct configuration settings and certificates for supporting 802.1X to devices connected to the captive portal.

Note also that authentication is not generally associated with encryption in the wired world.

Options for authorization on AOS-switches

You can control authenticated users by applying a number of dynamic settings to the user session such as VLAN assignment, ACLs, rate limit, and QoS priority. With AOS-Switches, you can take one of two approaches (see Figure 5-14). With the default approach, the RADIUS server such as ClearPass has settings such as VLAN assignments and ACLs configured on it as RADIUS standard attributes or vendor-specific VSAs. When a user successfully authenticates, ClearPass sends these attributes in the Access-Accept message to the switch, and the switch then applies them. Alternatively, the RADIUS server can simply send the switch the name of the user's role in the Access-Accept message. The role name matches a role configured on the switch, and this role defines settings such as VLAN assignment, ACL, rate limit, and QoS priority, which the switch then applies to the user session. This second approach is called role-based authorization and must be enabled on the switch.

It is important for you to understand that you must choose just one of these approaches for all users. When you enable role-based authorization on the switch, it rejects RADIUS Access-Accepts that have any VSA besides the user role.

Implementers can look up the RADIUS attributes supported by AOS-Switches in the switch Security and Access Control guides.

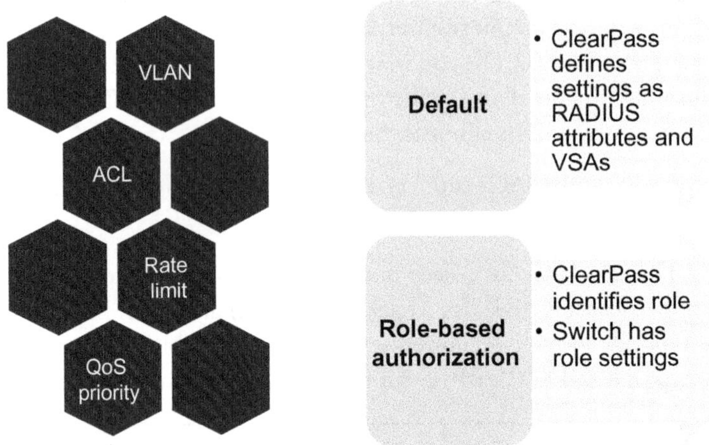

Figure 5-14 Options for authorization on AOS-switches

Determine the need for controller-based security features

You next need to decide whether the security provided by authentication and the authorization settings applied locally on the switch is sufficient.

For some networks, wired devices that successfully authenticate to the network are considered trusted and secure enough for the switch to forward their traffic locally, as conventional. The network might also already include security solutions for wired traffic such as ACLs at the edge, stateful firewalls, Intrusion Detection System/Intrusion Protection System (IDS/IPS), and content filtering.

In addition, some applications running on the wired network require low latency and may not be able to tolerate the small overhead tunnel-node imposes.

In other cases, though, you may want to implement additional security features available from the MC.

The tunneled-node feature allows you to apply the same traffic control capabilities that an Aruba controller-based solution traditionally provides for wireless traffic to wired traffic as well. The tunneled-node feature helps unify access control for users so that the same policies apply, regardless of whether users have a wired or wireless connection. By having the switch act like an AP and tunnel all the client's traffic to the MC, you can take advantage of the controller's security features to protect users, devices, and the network.

These features include a role-based stateful firewall, deep packet inspection (DPI), and application filtering, as well as web content filtering.

For example, you might use tunneled-node if a company has vulnerable devices that require extra filtering for their traffic. Surveillance cameras, payment card readers, and some IoT devices can be easily hacked or blocked by DoS attacks because they do not have an internal firewall. You may use local forwarding for IP phone traffic because it requires low latency. On the other hand, some customers prefer to tunnel IP phone traffic through MCs so that they can collect call quality statistics. The choice depends on the customer preferences.

When a customer has invested IT budget and expertise in deploying identity-based access control and rule-based firewall policies, the customer will often see the benefit in easily leveraging those investments for wired access as well. Because tunneled-node provides tighter and simpler unification of wired and wireless access, switches simply tunnel traffic to the MC, which handles the traffic much as it would handle tunneled AP traffic. Employees can connect through their desktop and log in, be assigned a role, and receive customized access. Later, the employee can connect wirelessly using a laptop in a meeting room and again, log in, be assigned the same role and receive the same access.

Note that the tunneled-node feature tunnels wired endpoint traffic to an MC, but it does not encrypt the traffic. If your customer requires encryption across the wire, you should recommend adding MACsec, which is supported in many AOS-Switches such as the 3810 and 5400R Series. For the 5400R, make sure to select modules that support MACsec.

Per-User tunneled-node considerations

When the customer needs tunneled-node, you should typically recommend per-user tunneled-node. This feature gives customers the flexibility to tunnel some traffic, such as employee traffic, to the MC but forward other traffic, such as IP phone traffic, locally for lower latency (see Figure 5-15).

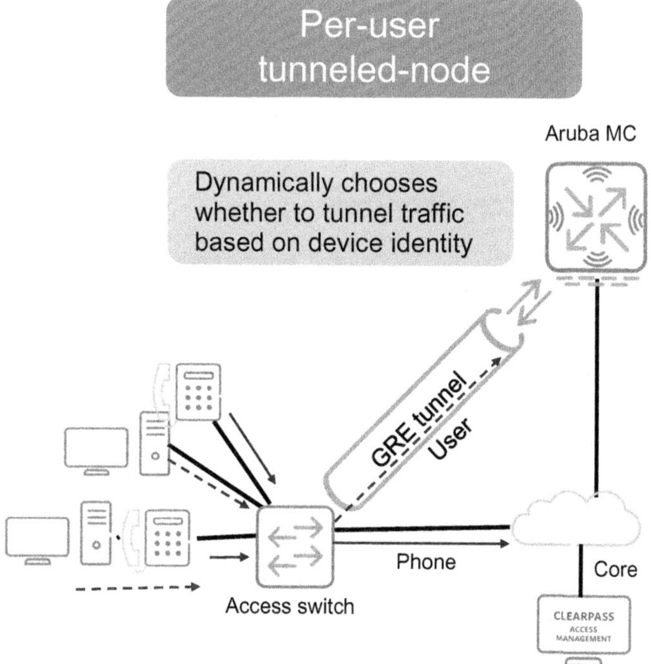

Figure 5-15 Per-user tunneled-node considerations

Per-user tunneled-node works with a standalone MC or an MC that is part of a multi-controller architecture that the Mobility Master (MM) manages. The switch can tunnel traffic to a cluster of MCs, which enhances resiliency and provides better load balancing.

With per-user tunneled node, the switch does not consume licenses on the MC or MM. The MC does require one tunnel per tunneled-node device, and tunneling wired endpoint traffic to the MC does, of course, impose burdens on the MC in terms of users and firewall sessions. Remember to add the wired endpoints to total load when sizing your controllers.

If you are planning a tunneled-node solution for a customer with existing devices, make sure that both MCs and switches meet software requirements.

Aruba hardware MCs require software 8.1 or above to support per-user tunneled-node. A virtual MC (VMC) requires software 8.2 or above. Aruba 2930F, 2930M, 3810, and 5400R zl2 Series switches support per-user tunneled-node as of software KA/KB/WC16.04. The 2540 and 2530 Series do not support tunneled-node. Note that the switch CLI calls per-user tunneled node "role-based" mode.

Aruba devices support a legacy mode for tunneled node called port-based mode. You would need to recommend this mode if the customer could not upgrade the Aruba controllers to version 8.1 for some reason. This mode is described below for your reference.

Port-based tunneled-node

With port-based mode, the switch forwards all traffic that a port receives to the MC. The MC is then responsible for implementing authentication and determining the device role, as well as applying firewall and other policies. MCs require AP and PEF licenses for each tunneled-node switch that operates in port-based mode, much like the switch was an AP. In addition, port-based tunneled-node does not work with MCs in a cluster. Port-based mode is available on Aruba 2930F, 2930M, 3810, and 5400R zl2 Series switches with software XX16.02 and above.

Per-user tunneled-node access control

When you select per-user tunneled node, the switch must enforce a form of authentication such as 802.1X, MAC-Auth, or both. In addition, the switch must use role-based authorization. When a device authenticates, the RADIUS server such as ClearPass determines the device's role and communicates that role to the switch. The role name sent by ClearPass must match a role name configured on the switch. The switch then uses that role name to determine whether to tunnel the traffic. In the example, shown in Figure 5-16, the switch has two roles that match ClearPass roles, Empl-1 and Voice. Employee traffic requires tunneling while voice traffic does not.

The proper settings for a role that requires tunneling include the tunnel redirect action and a secondary role. The secondary role must match a role name on the MC, which the MC uses to apply the proper firewall policies.

In short, when you plan per-user tunneled-node, the switch enforces authentication while the MC is responsible for applying the security policies.

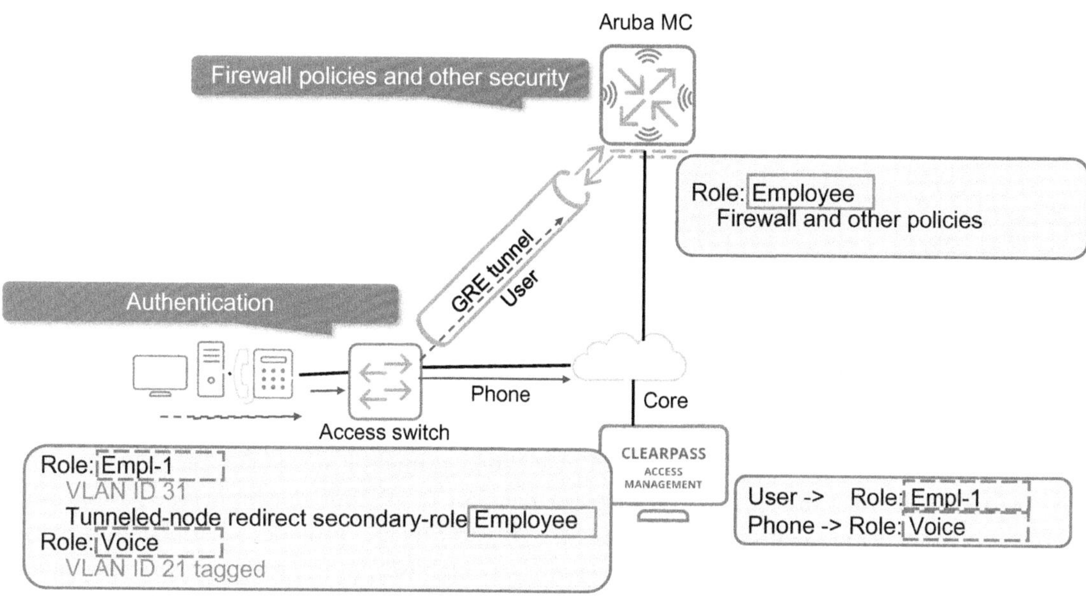

Figure 5-16 Per-user tunneled-node access control

As indicated with the dashed and solid boxes in Figure 5-16, the ClearPass role name must match the role name of the switch. The secondary role name under that role on the switch must match the MC role name.

In addition to the tunnel-redirect setting and secondary role, the switch must define an untagged VLAN ID for any role used for tunneling. This VLAN must exist on the MC.

The switch role used for local forwarding could include additional settings such as a VLAN assignment, but does not have to.

Note that the switch makes the decision whether to tunnel a device's traffic whenever a device authenticates. If the device connected to the port changes, the switch can make a new decision for the new device connected to the port. Sometimes multiple devices can connect to the port because, for example, a VoIP phone and computer have a daisy-chained connection to the same port. The switch can tunnel all traffic from one device and forward all traffic from another device locally, based on the roles assigned to the devices.

Authentication

Although 802.1X is a typical choice, the switch can enforce any form of authentication with per-user tunneled-node. For example, you could support employee and guest authentication with tunneled-node as follows. The switch port enforces 802.1X and MAC-Auth to ClearPass. An employee connects and authenticates with 802.1X. ClearPass sends the employee role, which is set up for tunneling with a secondary role for authorized employees. A guest connects and fails both 802.1X and MAC-Auth. The switch has an initial role that also tunnels traffic to the controller and has a secondary role named something like "unknown." The MC could apply captive portal to the "unknown role" and allow the guest to log in and receive limited access.

Details on tunnel negotiation between the switch and MC

A new process introduced on the mobility controller for the per-user tunneled-node feature is the **"tnld_node_mgr"** process.

The switch and MC can negotiate the use of port-based or per-user based tunneled-node using capability discovery. This feature sends a message from the switch to the MC using the PAPI protocol. If the MC firmware is AOS 8.1 or above, the MC will a respond with a special message, which tells the Aruba switch that the MC is capable of handling per-user tunneling along with per-port tunneling. The Aruba switch will work in per user mode or per port mode, but not both. The Aruba switch establishes the heartbeat GRE tunnel with the MC. The feature also provides additional information if the MCs are in a cluster mode, allowing the switch to establish tunnels with an active and standby anchor controller much like an AP would.

Figure 5-16 shows a single tunnel for simplicity. But when the switch determines that a user's traffic must be tunneled, it can set up a tunnel for the user's traffic. When an MC operates in a cluster, the switch might actually tunnel different users' traffic to different MCs. The switch sends a message to MC, which contains the user-mac and GRE key, port index, user-role, VLAN, and tagged/untagged information. The user entry will be created with the secondary role on the mobility controller. All the policies configured like DPI, firewall, and WebCC for the user-role will be applied to the user.

IDS and WIPS

In this section, you will consider Aruba IDS and WIPS solutions.

Review: Access Points and Air Monitors

Aruba APs can play a dual role, both providing data connectivity for clients and scanning the air to help detect and contain wireless threats. However, the AP prioritizes providing data connectivity for clients, which is its primary responsibility. Aruba also offers Air Monitors (AMs) or Wireless Sensors to scan the air and detect Rogue APs or other IDS events. These AMs do not support clients and are dedicated to providing IDS/WIPS.

An AM can function with only an AP license and, in an MM-based architecture, an MM license. However, it is required that the system has the same number of AP, PEFNG, and RFP licenses, so effectively, you must add an AP license, PEFNG, and RFP license for each AM.

Aruba 8.x provides WIDS/WIPS functionality in the base OS; however, RFP licenses are recommended for more advanced functionality. This functionality includes spectrum analysis that can make AirMatch more accurate. (Note that dedicated spectrum analysis requires APs configured as Spectrum Monitors (SMs), though. AMs monitor for 802.11 threats.) The RFP license also includes detection of more threats, signature-based IDS, tarpit shielding, event correlation, and more.

Advantages of dedicated AMs

Although APs that serve clients can provide some IDS/WIPS capabilities, dedicated AMs offer many advantages. First, they can detect and classify threats much more quickly, often in seconds as opposed to minutes. APs are slower at detecting threats because they also serve clients and can only scan a channel for a brief period. A rogue AP, for example, might not be transmitting at that moment. Therefore, it might not detect a rogue AP until it has scanned the rogue's channel several times.

Some threats do not send much traffic, such as ad hoc networks or wireless bridges; an AM can detect these threats quickly while an AP might not detect the threat at all. AMs also scan rare channels in addition to all regulatory channels.

AMs also provide much better wireless containment. An AP is mostly confined to performing wireless containment on the channel on which it operates. Although it is possible for it to change channels, it can only do so if it is not supporting any clients. A dedicated AM can contain a rogue on any channel without negatively affecting the company's legitimate wireless services.

AMs also permit remote packet captures on any channel, which can be useful for troubleshooting. Finally, if a customer chooses to so configure them, AMs can add to network availability by automatically converting to AP mode if a neighboring AP fails.

In short, dedicated AMs are recommended if the customer wants wireless containment and faster detection of threats.

Wireless threat protection framework

As Figure 5-17 shows, the first phase of the WIDS/WIPS is to Discover which includes continuous RF monitoring of wireless devices, activity, and configuration across all 802.11 channels. Next, the IDS/WIPS performs automatic classification of threats and nonthreats is critical to RF security. Then automatic containment to block any rogue or intruder. Lastly, the IDS/WIPS performs automated logging and reports distribution ensures compliance with wireless security policies and regulations.

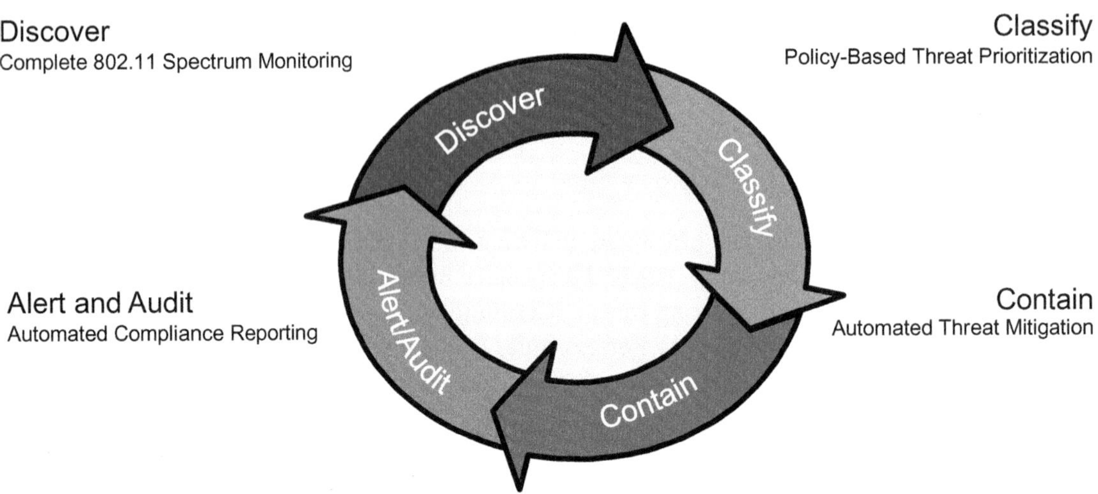

Figure 5-17 Wireless threat protection framework

IDS/WIPS design

If you plan to deploy dedicated AMs, you should place approximately one AM centrally for every four APs. However, this recommendation is dependent on the AP density that you have planned. If you have a higher-density AP design, you will require fewer AMs per AP, perhaps one per 7–10 APs.

For the best rogue detection and to provide wired containment, the AM must be on the same VLAN as the rogue AP. If the customer network has multiple VLANs to which a rogue AP might connect, you should extend each of them to the AMs. To extend multiple VLANs to an AM, you must tag the VLANs on the connected switch port. The same rule applies if you are using APs, rather than dedicated AMs, for the WIDS/WIPS functionality.

The VLAN only needs to be extended to one AM or AP for all AMs or APs to detect rogue APs connected to that VLAN. However, the AM or AP needs to be in the same VLAN as the rogue to implement wired containment, so extending all VLANs to all AMs is recommended.

Activity 5: Design access control and security

In this activity, you will continue to design the network for the customer scenario that was introduced in Activity 1. You will determine the customer requirements for access control, and you will plan a solution to meet the customer's security needs.

To complete this activity, use the information about the scenario provided with the previous activities plus the additional information provided below.

Scenario access control and security

This scenario provides the details that you need to plan an access control and security solution for Corp1.

Corp1 users and devices

Corp1 has 900 employees who work at the main site. The company also has 45 employees who do not work at the main site, but sometimes visit there. The site could have up to 930 users a day, including these workers and guests.

In addition to connecting with their laptops, employees often bring their own tablets and smartphones and connect them as well. The customer expects that users could be logged in to the network on a couple devices

Besides user devices, the main site has about 100 non-user devices such as printers and voice conferencing equipment.

Existing Corp1 authentication components

Corp1 has a Microsoft Active Directory (AD) domain. The company-issued laptops are joined to the domain.

The company's current wireless solution uses 802.1X with PEAP-MSCHAPv2 to Microsoft NPS RADIUS servers. NPS simply authenticates the users and permits network access without customizing their access in any way.

Corp1 security needs

Corp1 is motivated to increase the level of security at their campus. IT security staff have mentioned that they are interested in EAP-TLS; however, they are concerned about the level of management required to deploy certificates to user devices. The company currently has no solution for Bring Your Own Device (BYOD).

The site is badge-controlled, but IT security staff are interested in increasing protections against internal threats. The Aruba controllers' firewall and deep packet inspection are key selling points of the solution to them.

Many employees occasionally take their laptops to other locations and connect them to insecure networks. The company had an issue in the past year with an infected device, which took time to identify and remediate. Security staff are concerned about a repeat of this incident or something worse. They are looking for ways to ensure that devices connected to the network have their firewalls on and comply with other security policies.

The sales representative has brought up the risks of wireless DoS attacks and rogue APs. Corp1 likes the sound of the WIPS capabilities, but is concerned about the cost and complexity of deploying additional devices for the solution.

Finally, the customer wants to offer guest access. Guests should be restricted from internal network access and just receive Internet access after they register in a portal and agree to terms and conditions. Most guests will connect wirelessly, but the customer likes to offer wired access to guests in meeting rooms.

Activity 5.1: Design authentication and access control

Activity 5.1 objectives

You will recommend appropriate security settings for WLANs, based on the customer's requirements. You will also make a recommendation for wired access control.

Activity 5.1 steps

1. Refer to the scenario for this activity. Make a case for why this customer should implement Aruba ClearPass. What advantages and business benefits does it provide?
2. In addition to ClearPass Policy Manager, which components will you recommend for the customer to use (ClearPass Guest, OnGuard, and OnBoard) and why?
3. Make a high-level plan for the wireless authentication and access control, including:
   - The number of WLANs and which users and devices will connect to each WLAN
   - The type of authentication and encryption implemented on the WLAN
   - The general strategy for applying access control
4. Will you recommend implementing authentication and access control at the wired edge? Explain why or why not.
5. If you have recommended implementing access control at the wired edge, make a high-level plan, including:
   - Whether to use per-user tunneled-node
   - The type or types of authentication to implement
   - The general strategy for applying access control

Activity 5.2: Design ClearPass server sizing and licensing

Activity 5.2 objectives

You will plan the number of licenses required for this solution. (Assume that you are proposing ClearPass version 6.7.) You will then recommend a ClearPass appliance or VM.

Activity 5.3 steps

1. Do you have enough information to plan the Access licenses? What should you clarify with the customer?

   After further discussion with the customer, you have clarified that the main site currently has at most 2400 devices connected in a day, based on DHCP logs. You discussed how users might start to connect wireless devices more frequently when the wireless network improves. You have agreed that the solution should support 3000 concurrent devices, which allows three devices per user, including visiting employees and guests, a hundred nonuser devices, and another hundred devices for leeway.

2. How many of the following licenses will you recommend (if you are not planning to recommend a feature, you do not need to plan any licenses for it):
   - Access
   - Onboard
   - OnGuard
3. What would you discuss with the customer to determine whether to propose a ClearPass appliance or VM?

Activity 5.3: Plan certificates

Activity 5.3 objective

You will plan the certificates required for the customer.

Activity 5.3 steps

1. Which devices require certificates?
   - If you are implementing 802.1X, consider the EAP method.
   - If you are implementing Captive Portal authentication, also consider the need for HTTPS certificates.
   - Also consider the need for HTTPS certificates for the MC and ClearPass UIs.
2. Will you recommend that the customer use CA-signed certificates rather than self-signed certificates? What are the advantages of your recommendation?

Activity 5.4: Determine needs for AMs

Activity 5.4 objectives

Determine whether to add AMs to the BOM.

Activity 5.4 steps

1. Refer to the scenario for this activity. Will you recommend the deployment of dedicated AMs or not?
2. If you have not already recommended RPF licenses, will you do so?

Activity 5.5: Add security solutions to the BOM

Activity 5.5 objectives

You will add ClearPass components and licenses to your Bill of Materials (BOM).

Like previous activities, this activity was written to be used with Iris. If you do not have access to Iris, however, you can use the HPE Networking Online Configurator. (Visit **http://hpe.com/networking/configurator** to access this tool.)

Activity 5.5 steps

1. Launch Iris.
2. Open your project from the previous activities (**File > Open**).
3. Click your site and open the workspace (see Figure 5-18).

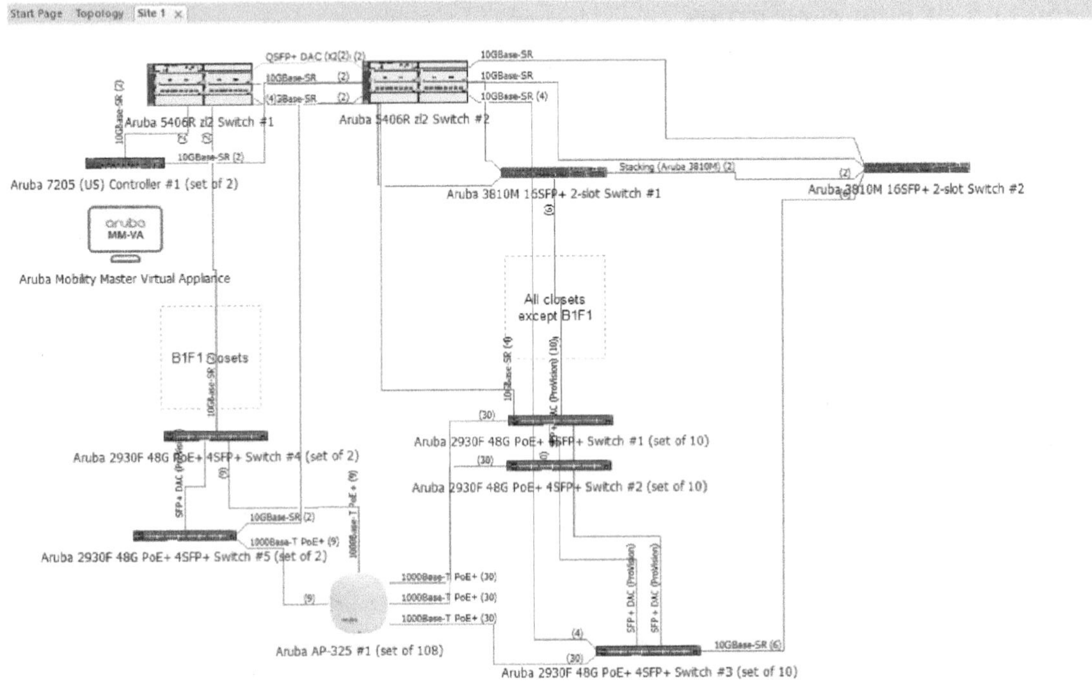

Figure 5-18 Workspace in Iris

4. In the catalog, expand folders as shown and add **ClearPass** (see Figure 5-19).

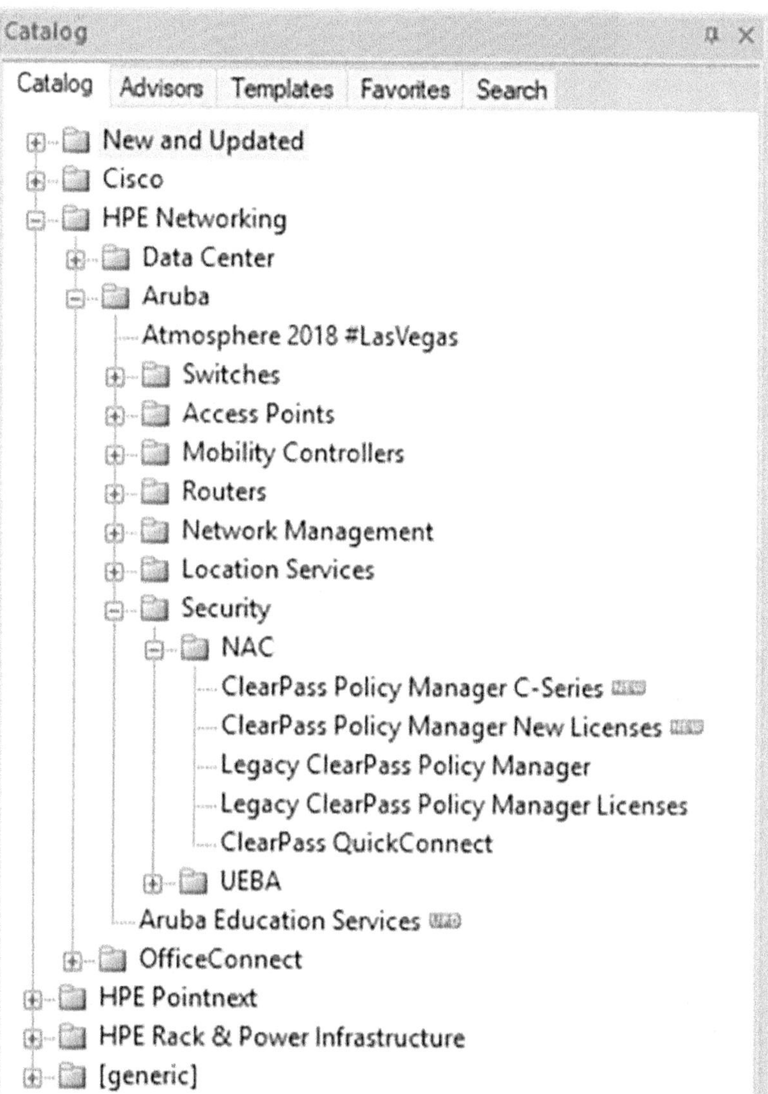

Figure 5-19 Add ClearPass in Iris

5. In the Properties window, you can select the ClearPass model. Assume that you discussed needs with the customer, and the customer has indicated that the data center has a server available that meets the ClearPass sizing requirements. Select a virtual appliance (see Figure 5-20).

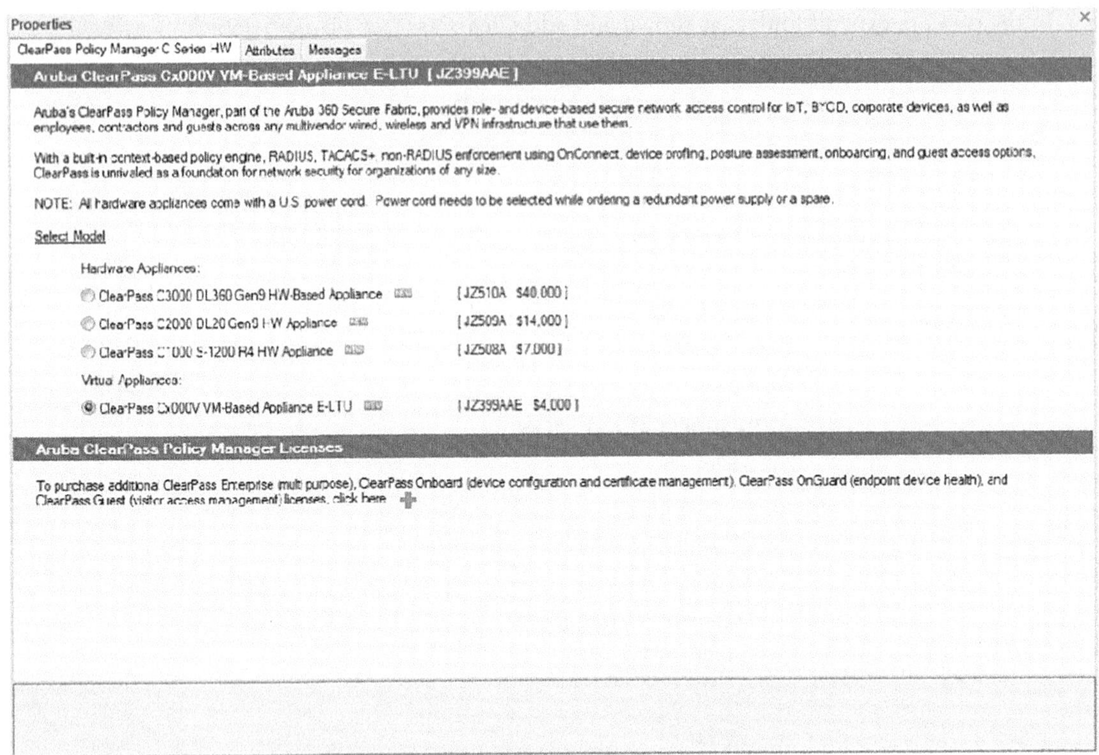

Figure 5-20 Select a ClearPass virtual appliance in Iris

6. Select the + icon to add licenses.
7. In the workspace, select the license icon (see Figure 5-21).

Aruba ClearPass Policy Manager New Licenses

Figure 5-21 Select the license icon in Iris

8. Open the Properties window.
9. For Total number of concurrent endpoints for ClearPass Access, select the number of Access licenses that you planned.
10. Select the other tabs and add the licenses that you planned (see Figure 5-22).

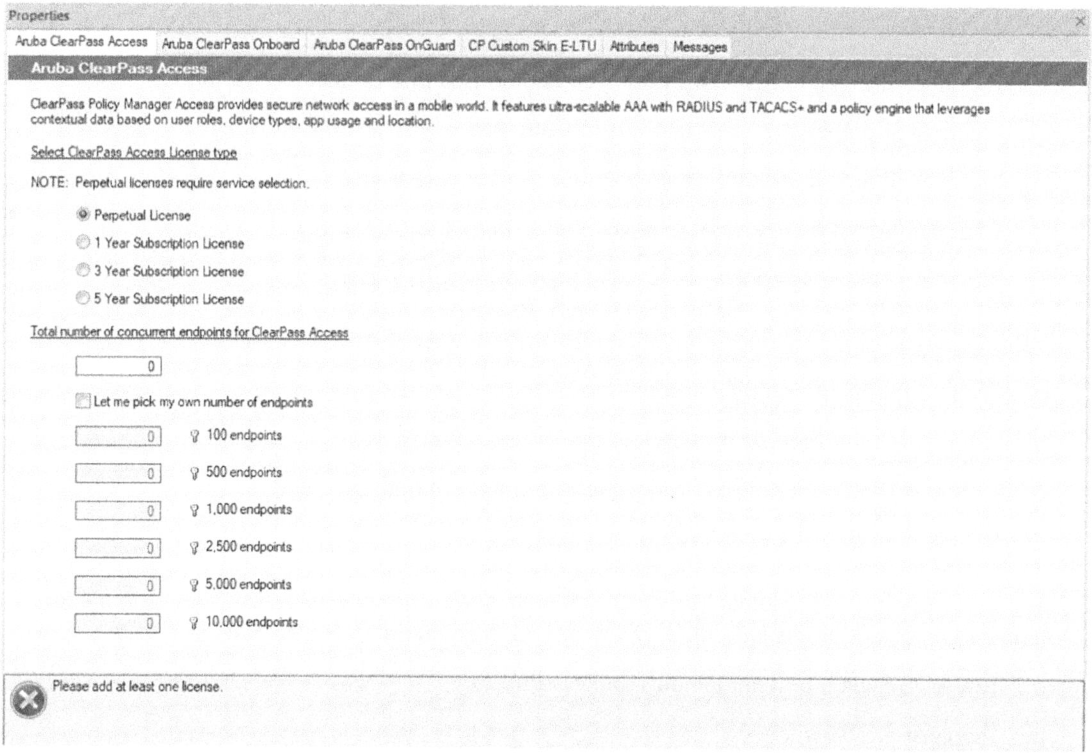

Figure 5-22 Add ClearPass licenses in Iris

11. If you added any APs to act as AMs, remember to add them and their licenses to the BOM.
12. Save the configuration.
13. Select the dollar sign button or select File > Quotation to see your BOM.

Answers to the questions included in activities are provided in "Appendix: Activities and Learning Check Answers."

Summary

In this chapter, you learned to follow best practices for wireless security. You reviewed the capabilities and architecture of Aruba ClearPass and learned how you can use it to control employees using role-based policies. You also learned how you can extend access control to wired users for consistency across the complete unified network. You next explored designing solutions for guest access control.

You then learned how to propose the proper licenses to support your customers' requirements.

Finally, you turned your attention to planning the correct AP modes to support IDS and WIPS.

Learning check

1. What defines the number of ClearPass Access licenses that a customer requires?
   a. Number of employee user accounts in ClearPass
   b. Number of employee and guest user accounts in ClearPass
   c. Total number of devices that ever authenticate to ClearPass
   d. Number of devices concurrently authenticated to ClearPass
2. What is one best practice for implementing 802.1X with PEAP-MSCHAPv2?
   a. Install a certificate on all client devices
   b. Configure client devices to validate the RADIUS certificate
   c. Create a DHCP scope for devices that have not yet authenticated
   d. Supplement 802.1X with WEP encryption
3. You are planning to implement per-user tunneled node on an AOS-Switch and authenticate tunneled-node endpoints with 802.1X. What device will enforce 802.1X?
   a. The AOS-Switch that implements tunneled node
   b. The core AOS-Switch
   c. The MC to which the switch tunnels traffic
   d. The MM

Answers to the learning checks in this study guide are provided in "Appendix: Activities and Learning Check Answers."

6 VLAN Design

EXAM OBJECTIVES

- ✓ Translate a customer's needs into technical requirements.
- ✓ Given a customer's requirements for a single-site campus, design the high-level architecture.
- ✓ Given the customer's requirements for a single-site environment, determine the component details and document the high-level design.
- ✓ Given a customer scenario of a single-site campus environment (less than 1000 employees), design and document the logical and physical network solutions.

Assumed knowledge

- Switching and routing technologies such as VLANs and IP addressing
- 802.11 wireless standards
- Aruba 8.x architecture

Introduction

VLANs define the broadcast domain. Correlating with an IP subnet, a VLAN forms a basic building block for a network. A solid VLAN plan makes a network simpler to deploy and manage. And even though VLAN boundaries can be invisible to users, a good VLAN plan can also improve their experience as they access the network.

This chapter presents some of the challenges for designing VLANs for unified wired and wireless environments.

You will look at the best practices for deploying wired VLANs, which tend to follow more traditional guidelines. You will also learn about planning logical VLAN IDs and subnets and look at some example VLAN designs both with and without tunneled-node, which affects where wired device VLANs must be deployed.

A single VLAN can actually be the best choice for a modern wireless deployment. In addition to learning why, you will learn how to ensure that the single VLAN deployment works well in the customer environment.

Wired VLAN deployment

You will first consider best practices for deploying wired VLANs and learn how to plan logical VLAN IDs and subnets.

Benefits of VLAN segmentation for wired networks

While extending a VLAN as far as possible works well for wireless networks, which need to support roaming, wired networks often benefit from more segmentation. First, create separate VLANs for wired users and devices and wireless networks. From there, you can further divide users and their devices for several reasons. These include supporting access control by putting different types of users and devices in different VLANs. You may want to place specialized wired devices in their own VLANs to easily manage them and ensure the proper QoS.

You can think of each user type or device type as an access category. You should plan at least one VLAN per access category. However, sometimes you may plan multiple VLANs per access category because of the VLANs' other important function: limiting the size of the broadcast domain.

In IPv4, every device in the VLAN must process broadcast traffic, and broadcasts can impose a burden by congesting links and tying up endpoints' processes. The more devices in a VLAN, the more broadcasts that every device must handle. In addition, gratuitous ARP broadcasts can fill up devices' ARP cache. Similarly, multicasts extend through a broadcast domain. While you no longer have to make the broadcast domain as small as in the past, it still makes sense to keep broadcast domains a sensible size.

Placing different types of users and devices in different VLANs can be very important for wired networks. The Aruba mobility controller firewalls are role-based, but wired traffic might pass through a more traditional firewall, which uses the devices' source addresses to decide how to control the traffic. Therefore, placing different types of wired users and devices in different VLANs makes it easier for the firewall to control the different traffic in different ways.

For example, sometimes by looking at the IPv4 address, the firewall can tell several things about a user including the user's group, location, and whether the user is an employee, contractor, or guest.

A few organizations with large public IPv4 networks use public addressing within the campus. If the public IPv4 networks are noncontiguous, limited VLANs might be required to correspond with each noncontiguous segment.

Isolate traffic for access control

Figure 6-1 shows one example of how you can group devices based on access control considerations. A basic design features a VLAN (or VLANs) for users. APs could be on their own VLAN, but they can also be deployed on any VLAN available in the network. If APs are implementing IDS/WIPS, rather than dedicated Air Monitors (AMs), APs will only be able to detect rogues on the VLAN to which it is attached.

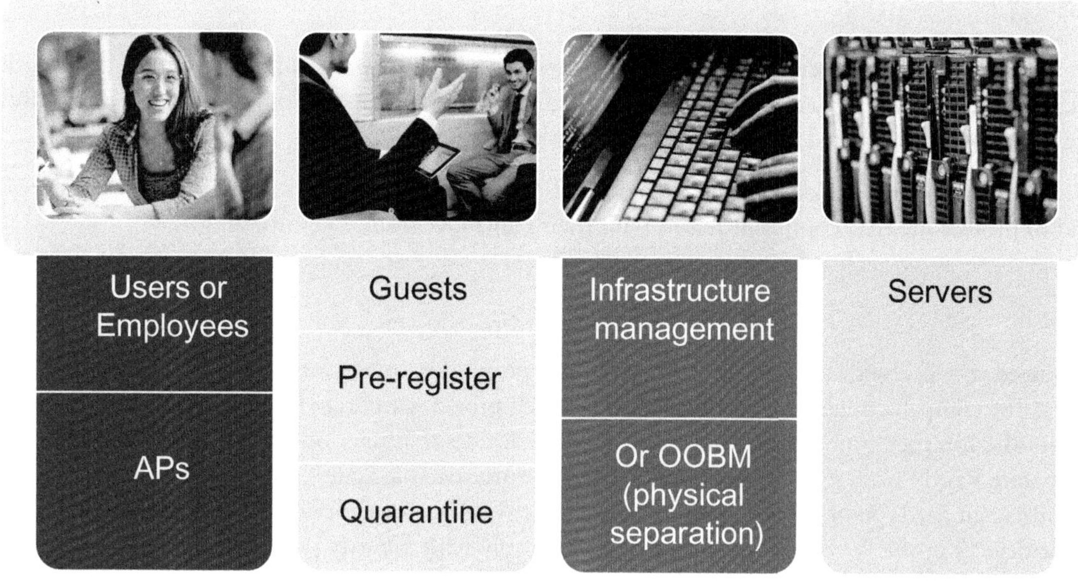

Figure 6-1 Isolate traffic for access control

If the company offers wired access for guests, you should put employees and guests in different VLANs. Depending on the complexity of the guest solution, you might create VLANs for guest devices before the guests have registered and logged in. A company may also need a quarantine VLAN for devices that are detected to have viruses.

You should place switches' management IP addresses in an isolated VLAN as this is the first line of defense against unauthorized access. Alternatively, you can establish an out-of-band-management (OOBM) network if your switches support OOBM ports. In this case, the switches will use separate physical links for their management traffic, so you do not need to create a VLAN on the production links for that purpose.

If the campus has local servers, you would typically place those VLANs in separate VLANs, unless special requirements call for the servers to be in the same broadcast domain as clients.

Users or employees

This category includes the devices through which employees or other users connect to the network. Some companies might have more complicated requirements to place different types of employees in different VLANs, dividing the users by department. However, a simple user or employee category works for many customers. The more granular level of control might require authentication to identify the user and make the assignment. The authentication server could alternatively apply access control policies directly to the user session, allowing multiple types of users to share the same VLAN but have access to different resources.

Access Points (APs) / Air Monitors (AMs)

To support rogue AP detection, you should deploy AMs in the network and trunk/Tag the ports with VLANs that have no APs connected. If only using Aps, then the APs will only be capable of classifying rogues who are located in the same VLAN as the APs.

It is best practice to use Control Plane Security (CPSec) to protect the APs from unauthorized access and to protect the APs' communications with their controller from eavesdropping.

Guest

You need to work with the customer to define the proper treatment for unknown or unauthorized users. The company might want a guest VLAN, which provides access to limited resources or perhaps just to the Internet. Some guest solutions also place devices that have not yet registered for access in a separate VLAN from guest devices that have been authorized for guest access. The customer might require a similar type of VLAN, called a quarantine VLAN. An endpoint integrity solution might place devices in this VLAN if the devices fail to comply with security policies such as requirements for antivirus software. The VLAN typically provides access to limited resources that help the user to remediate the problem.

Management (Network infrastructure devices)

This category includes network infrastructure devices such as switches, routing switches, routers, and sometimes mobility controllers (MCs). It might also include servers that act as network infrastructure devices such as traffic optimizers and load balancers. Routing switches and routers typically have IP addresses on other VLANs as well. However, the devices are managed on the network infrastructure VLAN. For this reason, this VLAN is often called the management VLAN.

Some customers prefer to physically isolate the management traffic rather than simply separating it into a VLAN. You can create this physical isolation with an OOBM network. Some AOS-Switches provide OOBM ports, including the Aruba 2930M, 3810, 5400R zl2, 8320, and 8400 Series switches. The OOBM ports are Ethernet ports that connect only to the switch management plane, not to its data plane, ensuring the complete, secure isolation. If you are planning to use the OOBM ports, remember to make a plan for the OOBM network. You might add one or two switches to the proposal to support the connections. Those switches would then connect to any management stations, or, if you want to provide management access from anywhere, to a core or aggregation level switch. The connection from the OOBM network into the rest of the network should be on a dedicated VLAN, and you should plan the appropriate ACLs to control access to this VLAN. For example, if you are using role-based access control, ACLs for most roles could prevent access to this VLAN, but ACLs for the IT staff role could permit access. Also remember that the switch configuration should set OOBM as the listening port for management protocols such as Telnet and SSH.

Servers

This category includes servers and devices, such as printers, that provide services. A company with fewer servers might connect them directly to a campus LAN core switch, or at least have this switch act as their router. In that case, the campus LAN would need to include one or more VLANs for servers. A true data center, on the other hand, might have a more complicated topology and connect to the campus LAN over routed connections.

Group traffic for specialized devices

As Figure 6-2 shows, there are several types of access categories you may want to create for specialized devices. A common example is Voice over IP (VoIP). Placing voice traffic in a different VLAN from data traffic protects the VoIP phones from other VLAN broadcasts. In addition, the voice traffic requires a higher QoS than traditional data traffic. If you place all of the traffic in its own VLAN (or VLANs) it makes it easier for you ensure that QoS.

For similar reasons, you may want to group devices that handle a great deal of multimedia traffic in its own VLAN, or you could group devices designed to run multicast applications together in order to minimize multicast routing.

You can also group a company's specialized devices according to the application. For example, you could place point of sales (PoS) devices in their own access category. Other examples of such devices include nurse stations, security cameras, and Internet of Things (IoT) devices. These devices might or might not require high-priority handling. However, they do form an access category of devices with similar needs and traffic patterns. You may want to segregate the devices for security reasons as well.

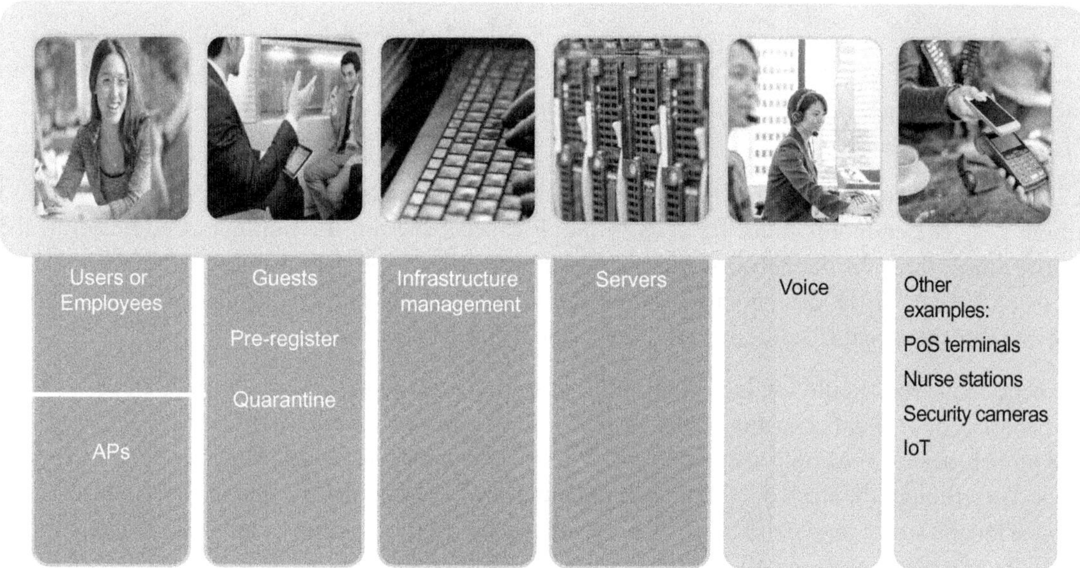

Figure 6-2 Group traffic for specialized devices

Minimize broadcast domain

Once you have a list of access categories, such as employees, guests, and voice, you should determine the estimated number of endpoints for each category. Then you need to decide if you want to divide the access category into multiple VLANs based on broadcast domain limitations. The traditional guideline limits a wired VLAN to a /24 subnet, which provides enough IP addresses for 253 endpoints. For modern networks, you can often use /23 subnets, which support up to 509 endpoints, or /22 subnets, which support about 1021 endpoints.

You should also correlate the VLAN with logical divisions in the access layer, such as a closet or floor. It is also best practice to limit the number of hops over which a VLAN extends. For example, it is usually best to terminate VLANs at the aggregation layer in a three-tier topology.

However, as you make your plans for larger campuses, keep in mind balancing the need to limit the broadcast domain with the need to stay under the maximum number of VLANs, which is theoretically 4094. Of course, you can reuse VLAN IDs across routed boundaries. For example, you generally route all traffic between the campus core and the data center core, so the campus and data center can use overlapping VLAN IDs (but not subnets). More VLANs might come with a cost of their own, and you must balance the advantages of smaller VLANs with disadvantages of more VLANs. For most enterprise networks using core routing, you should aim for a limit of about 200 to 250 VLANs per site. As long as you are not using MSTP, access layer switches generally only need to support the specific VLANs for their endpoints.

A few trends have made it possible for you to plan larger VLANs such as ones associated with /22 subnets. Many endpoints now have more processing power, which reduces the impact of many broadcasts. Operating systems also tend to treat broadcasts more efficiently. Higher bandwidth for connections also reduces the impact of broadcasts.

The need to avoid issues with spanning tree stood behind the guideline for limiting the extent of VLANs. Now VSF and backplane stacking can eliminate spanning tree design as a concern. VSF and backplane stacking also combine many ports into a single virtual switch, so you could extend a VLAN across those ports.

It is best practice to correlate the maximum number of endpoints that connect in a VLAN with a subnet size. The endpoint count for each subnet is calculated from the number of addresses in the subnet minus three for the network address, broadcast address, and default router IP address. If you are using VRRP for the default gateway, you must plan two additional IP addresses for the default gateway (for the actual IP addresses assigned to the redundant gateways).

As mentioned, you should use logical divisions as well. For example, you might have an access layer that consists of several backplane stacks with four members that each have 48-ports. Each stack has 192 ports for users. You could assign one user VLAN, associated with a /24 subnet, to each backplane stack. Even though the stack does not reach the limit of 252 endpoints, it makes more sense to assign two backplane stacks to two different VLANs rather than overlap multiple user VLANs across them. Or if you are planning larger VLANs, you could assign the same user VLAN, associated with a /22 subnet, to five backplane stacks (960 ports). Again, the number of ports is a bit below the maximum, but six backplane stacks would exceed the 1021 limit, and it does not make sense to split the sixth stack between two different VLANs. You might also use a closet or a floor as a logical place for a VLAN division.

Finally, as you plan the VLAN size, keep some additional guidelines in mind. The user count does not always tell you how many ports are required for the Users access category. The company can hire more users. Offices might have multiple Ethernet jacks, and users might connect devices to all of them. Therefore, the maximum port count is about the number of edge ports at the site minus any devices that you plan to put in their own VLANs such as security cameras or VoIP phones that do not offer computer connections. If your count approaches the maximum size that you have selected for the VLAN, you should plan to divide the access category as a buffer for future growth.

Generally, the same considerations apply to other categories. An inventory of devices might not tell you the eventual size of the VLAN.

It is fine to have a VLAN with fewer than 250 endpoints if you have security reasons to create such a VLAN or if the division makes sense in the customer topology (such as a floor with 200 jacks). However, you do not need to divide access categories into VLANs smaller than this simply for the sake of having smaller VLANs.

Example: Assign logical VLAN IDs

Although the VLAN ID does not really matter beyond being unique within the Layer 2 domain, it is often a good idea to choose IDs in an organized fashion. Network administrators can easily remember which VLANs are associated with each function.

In the example provided in Table 6-1, the architect is using the first digit in the VLAN ID to indicate the access category or VLAN type, such as switch management, voice, employees, and guests. If you divide an access category into multiple VLANs in different locations of the network, you can use the second digit to indicate the location. In this example, each floor is associated with one VLAN, and the floor number is the second ID in the VLAN. The company does not support many wired guest devices, so those are all placed in the same VLAN.

Table 6-1 Logical VLAN ID example

Access category	Location	Examples
Switch management = 10		Switch management =VLAN 10
Voice = 2	Floor 1 = 1	VoIP phones on floor 1 = VLAN 21
	...	...
	Floor 5 = 5	VoIP phones on floor 5 = VLAN 25
Employees = 3	Floor 1 = 1	Employees on floor 1 = VLAN 31
	...	...
	Floor 5 = 5	Employees on floor 5 = VLAN 35
Guest = 40		Guest = VLAN 40

Table 6-1 provides just one example of how to assign VLAN IDs. You can adapt similar plans to the particular customer environment.

For example, in a multiple building campus, you can use one digit to indicate the building and another to indicate the floor.

Switches support VLAN IDs from 1 to 4094. VLAN 1 is the default VLAN on AOS-Switches, which means that it is the VLAN to which all ports belong by default. Sometimes administrators use VLAN 1 as the management VLAN. However, using a different VLAN ID provides better security. You may want to use VLAN 1 as a dead-end VLAN, which is not carried on uplinks. Then if someone connects to a port that is not yet set up for the correct VLAN, the device cannot obtain network access.

Assign logical subnet addresses

You must plan an IP addressing scheme for the VLANs. Although some enterprises have public IP address spaces allocated to them, most enterprises use private IP addresses, defined by RFC 1918, for devices in their campus LAN. In a new deployment, you can choose any private network address. However, choosing logical addresses will help network administrators as they manage, maintain, and troubleshoot the network. If you want, you can leverage the meaning embedded into the VLAN ID, by using IP addresses that correspond to the VLAN IDs. For example, the third octet in the address could match the VLAN ID (see Figure 6-3).

Also, try to ensure that every location has contiguous IP address blocks to enable simpler route summarization. Summarization occurs at octet boundaries and binary blocks within an octet.

You may try to plan subnet addresses to support summarized ranges within ACLs and firewall. For example, employee devices have IP addresses between 10.1.16.1 and 10.1.19.254, enabling you to specify that range for any ACL. The ArubaOS firewall supports aliases, which help you easily group together addresses for referencing in a rule. Therefore, this consideration might be less important than route summarization.

Finally, remember that you cannot always balance all of the principles. Sometimes you might choose not to correlate the subnet IP address and the VLAN ID so that you can better create contiguous IP addresses.

Balance guiding principles:

- Correlation with VLAN ID
- Contiguous addresses to:
  - Prepare for route summarization
  - Sometimes prepare for ACLs and firewall rules

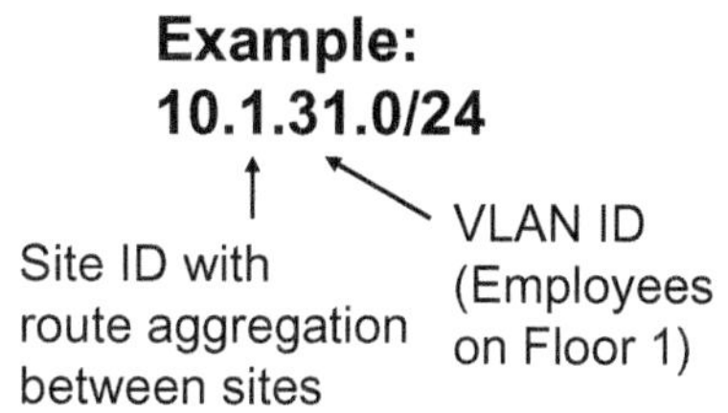

Private IP network	Number of subnets
10.0.0.0/8	65,536 /24 32,768 /23 16,384 /22
172.16.0.0/12	4096 /24 2048 /23 1024 /22
192.168.0.0- 192.168.255.255	256 /24 128 /23 64 /22

Figure 6-3 Assign Logical Subnet Addresses

Keep some additional tips in mind for each of these considerations.

Corresponding with VLAN ID

You could take different approaches for corresponding the IP subnet with the VLAN ID. For example, the architect used XY for the VLAN ID, in which X is the VLAN type and Y is the closet or floor number. The network address can also include this information, using one of these schemes:

10.X.Y.0/24

172.16.XY.0/24

192.168.XY.0/24

If the location is very large and will correspond with an OSPF area, you should put the location first in the VLAN ID to support route summarization. You may also choose to correspond with just part of the VLAN ID. For example, the first part of the VLAN ID might indicate the building number while the second part indicates the third octet in the subnet address.

Preparing for route summarization

You should select a block large enough that, if the company needs to add networks to an area in the future, they can find addresses within that block. If you use the second octet in the 10.0.0.0/8 private address range to indicate the area, the area supports up to 256/24 networks.

Do not only count the subnets currently in use. Also, think about the future. Planning for the future prevents a situation in which you would have to add a noncontiguous block to the network, introducing more route summaries per area. For example, if you think that you might need to add more subnets later, you might plan two /16 blocks for the area such as 10.0.0.0/16 and 10.1.0.0/16 or 10.2.0.0/16 and 10.3.0.0/16. In this case, the area should always start with an even number in the second octet so that you can create one route summary (10.2.0.0/15) on the area border routers (ABRs).

If the area that you need to summarize is smaller, you could divide it into smaller blocks such as 10.1.0.0/17 (128/24 subnets) or 10.1.0.0/18 (64/24 subnets). This would make sense when you need to conserve IP addresses. On the other hand, you should also value logical numbering and ease of management. Choose a system that makes sense to you and to the network administrators who will manage the system.

Summarization occurs at octet boundaries and binary blocks within an octet. For example, you can aggregate routes from 10.1.0.0/24 through 10.1.255.0/24 with 10.1.0.0/16. You can also aggregation routes from 10.1.0.0/24 through 10.1.15.0/24 to 10.1.0.0/20. However, routes from 10.1.0.0/24 through 10.1.19.0/24 require several aggregated routes. You generally want to aggregate routes between sites or between different segments of the network such as the campus and data center. When you use OSPF, area boundaries define where you can aggregate routes. One simple way to promote route aggregation is to assign 10.X to each OSPF area with X corresponding to the area ID.

If you are making the VLAN ID correspond with the subnet address, you could use the same VLAN IDs or different ones between areas. For example, you could use VLANs 1000 to 1255 in area 1 (subnets 10.1.0.0/24 through 10.1.255.0/24) and VLANs 2000 to 2255 in area 2. Or you could reuse VLAN IDs 1-255 in both areas because the areas have layer 3 boundaries between them.

Preparing for ACLs

Grouping devices in a VLAN can help you control a user or device's access rights. Sometimes you can make it easier for administrators to configure the ACLs and firewall rules by using contiguous blocks of addresses for VLANs to which the same rules apply. In addition, simpler ACLs impose less of a burden on the device that is processing them, which could be important for a solution that filters a lot of traffic. However, this consideration might not be as important.

You can also combine both uses. The route summarization block should encompass the blocks that you create for access rights. For example, you could assign blocks of 256/24 subnets to each OSPF area and then divide those blocks into four blocks of 64/24 subnets for various VLAN types.

Using contiguous blocks

As you choose contiguous blocks, remember to think in binary. In other words, blocks of addresses come in sizes that correspond to powers of two. That is, you can have a block of two /24 subnets, a block of four /24 subnets, a block of eight /24 subnets, and so forth.

Guest network

Also note that some customers like to make the guest network as distinctive as possible. You might use an entirely different IP address block for this VLAN. For example, you could use 192.168.1.0/24 for the guest VLAN in a network using the Class A 10.0.0.0/8 block. The guests are often routed directly out to the Internet by a security appliance or firewall, so their traffic does not need to fit within the internal routing scheme. Using a different classful network can help administrators distinguish guest devices and to treat them differently. It might make setting up the guest solution easier when the customer has separate teams. Finally, it can also help to conceal your IP addressing scheme from a hacker.

Fit into an existing solution

Up until now, you have been looking at best practices for planning wired VLANs and sub-netting for a new network. However, often you are updating an existing network. As a general rule, you should follow the current design unless you have a good reason to change. For example, a site might be using wired VLANs associated with /24 subnets. Although it is possible to use larger subnets, there is no pressing reason to change. And you do not want to interfere with the routing and DHCP scopes already in place. On the other hand, if the company has a very large VLAN and is experiencing issues, you could recommend dividing the VLAN.

If you are expanding the number of ports, you may need to add more VLANs, fitting them with the current scheme, where possible. You should also look out for new VLANs that you might want to add. For example, if the company is adding IP security cameras or IoT devices, you could add a VLAN for those specific device types. Or if the customer is not currently following best practices to isolate different types of users and devices—such as guests and employees—you might recommend doing so.

Detect and address issues with large broadcast domains

You may want to investigate whether a large wired VLAN is working well for the customer when the associated subnet uses a larger prefix length. Signs of issues could include slow connections and poor performance. However, many issues can cause such symptoms from inadequate network capacity to lack of quality of service (QoS) measures. One indication that the issues could stem from a large VLAN is an excessive number of broadcasts on links. Generally, you should expect broadcast traffic to be less than about 20 percent of the traffic.

If the network uses spanning tree, then issues with the spanning tree could also indicate that a VLAN extends too far across the topology. Once again excessive broadcasts can indicate a problem. In addition, you can check switch logs for Topology Change Notices (TCNs).

In either case, it is important to take a systematic approach to searching for issues, checking port utilization across the network at many different times of the day. A centralized solution such as AirWave can help you to establish a pattern of behavior over time. You can also set up RMON alerts to trigger SNMP traps when ports experience excessive broadcasts or a switch receives. By tracking these traps on AirWave, you can begin to gain a comprehensive view of the issue.

Once you have determined that a VLAN might be too large, you can begin to address the issue by dividing the VLAN. As you do, make sure to take the proper change management steps with the customer to ensure that the change does not have any unexpected effects.

For example, VLAN 10 might be associated with subnet 10.1.0.0/16 subnet, support 5000 devices, and extend across 100 access layer switches. You might decide to divide this VLAN into 20 VLANs (such as 101 through 120). If you want to re-use the former addressing space, you could each associate these VLANs with subnets 10.1.0.0/24 through 10.1.20.0/24. The customer IT staff would need to reprogram DHCP scopes for the new VLAN. You should also make sure to discuss with the customer whether any devices in the VLAN have static IP addresses, which might need to be updated. You would also need to check routing switches. In addition to adding the VLANs to the routing switches that will act as default router, you must make sure that other switches can reach the new VLANs. If you are dividing up the VLAN gradually rather than all at once, you should use subnets in a new addressing space to prevent overlap.

You would also need to consider how you will assign ports to the new, smaller VLANs. For the simplest approach, you could simply assign the new VLANs to switches based on their location. In this example, you might assign VLAN 101 to five switches in a similar location, VLAN 102 to the next five switches, and so on. However, you should discuss with the customer whether this approach meets their needs. Are there any endpoints that need to be in the same VLAN so that they can use broadcast-based protocols or more easily exchange multicasts? In this case, you must take care to group those devices in the same VLAN.

Example two-tier wired VLAN design without tunneled-node

Figure 6-4 shows a wired VLAN scheme for a two-tier topology. In this topology, VoIP phones are in VLANs 21 and 22. Employees and APs are in VLANs 31 and 32. Figure 6-4 shows only two access layer switches, but the actual network would have several access layer switches using VLANs 21 and 31 and several more using VLANs 22 and 32.

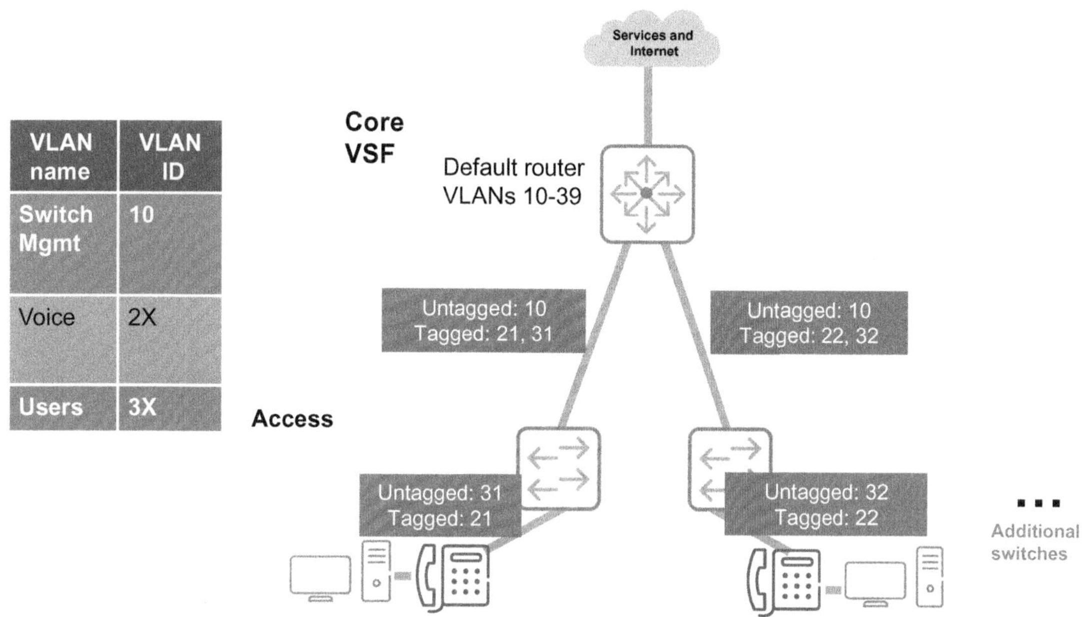

Figure 6-4 Example two-tier wired VLAN design without tunneled-node

The two-tier design makes it simple to extend VLANs across the campus as desired. For example, you can extend VLAN 10 across the site and place the switch IP addresses in it. You can make the switch management VLAN the untagged VLAN on all switch-to-switch links, rather than have VLAN 1 untagged on those links. This dead-ends the default VLAN, preventing users from connecting to it. Or you could remove untagged traffic from the switch-to-switch links if you prefer.

Note that this network topology is focusing on logical connections. The core switch is actually two physical switches in a VSF fabric or backplane stack. And the single links between the various components are actually link aggregations.

Example three-tiered wired VLAN design without tunneled-node

You can create similar schemes in three-tier topologies. However, you should often route between the aggregation layer and core. Therefore, you cannot extend the same VLAN into segments served by different aggregation layer switches. However, you can reuse the VLAN IDs, which can be helpful for management purposes. Multiple access layer switches can share the same configuration file, for example. The VLAN IDs would then be associated with different IP networks on the different aggregation layer switches.

For example, the network shown in Figure 6-5 has an aggregation layer in each building in the campus. Both buildings are using VLANs 10-39.

However, the VLANs terminate at the aggregation backplane stack, which provides routing for those VLANs. In Building 1, VLAN 31 uses IP address 10.1.31.0/24, while in Building 2, VLAN 31 uses IP address 10.2.31.0/24.

Finally, remember that routing switches must have a VLAN for their connections between each other as well. It is best practice to terminate all the VLANs used by endpoints at the default router. Then create a dedicated VLAN for each Layer 3 switch-to-switch link. Using a dedicated link for each separate connection helps to make the routing solution converge more quickly.

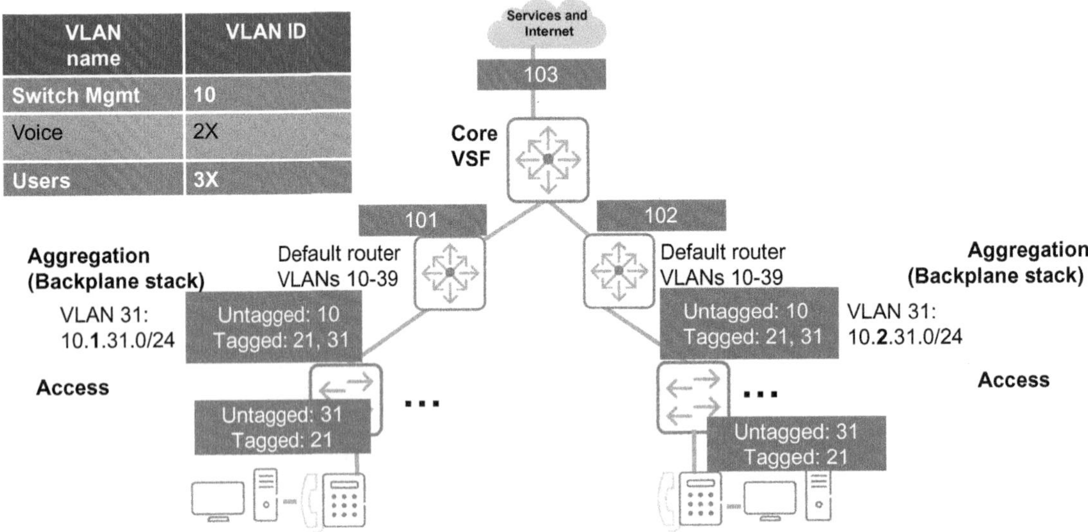

Figure 6-5 Example three-tiered wired VLAN design without tunneled-node

For simplicity, Figure 6-5 does not show all of the access layer devices in each building. However, in the real network, additional access layer switches would exist. If the building had enough switches, they would place devices in VLANs 22, 23, 32, 33, and so on.

Some administrators prefer to use different VLAN IDs across Layer 3 boundaries, which is also a valid approach. For example, in this design, Building 1 could use VLANs 100-139, and Building 2 could use VLANs 200-239.

Example wired VLAN design with tunneled-node

When you use per-user tunneled node, the switch authenticates devices that connect to its port and determines how to handle the traffic based on the role name returned by ClearPass. The role on the switch, in addition to indicating whether the switch tunnels the traffic to the MC or not, also specifies the VLAN ID. This ID must match a VLAN ID on the MC (see Figure 6-6). The MC has these tunneled node user VLANs tagged on its port, and the core Layer 3 switch (or VSF fabric) routes for them. Much like a wireless VLAN, the tunneled node user VLAN does not need to extend across the underlying infrastructure; the switch tunnels the traffic to the MC at the core, where the VLAN "appears."

Because you will typically deploy tunneled-node in networks that also feature wireless APs, an AP is shown in Figure 6-6 as well. It is important to understand that the AP's port should not implement authentication and tunneled-node. Instead, the AP is placed in a VLAN that is switched and routed locally, allowing the AP to reach the MC.

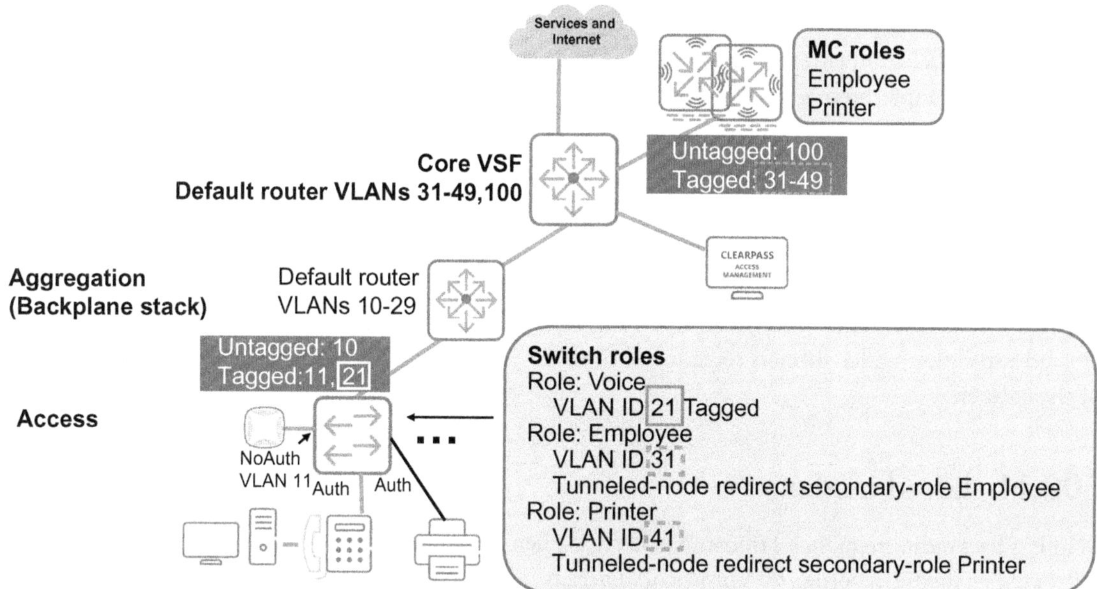

Figure 6-6 Example wired VLAN design with tunneled-node

Administrators can manually configure AP ports to have the correct ID and not to implement authentication. However, Aruba switches can also be configured to detect an Aruba AP using LLDP, authorize it, and apply a device profile with the correct ID.

This network topology in Figure 6-6 shows just one access layer switch that implements tunneled-node, but, of course, in the real world, the network would feature many more access layer switches. You should generally keep the VLANs for tunneled node devices around the same size as traditional wired VLANs. MCs can suppress all broadcasts except DHCP, ARP, and VRRP on a VLAN; however, this can still leave a high number of those broadcasts. Although MCs can convert ARP and DHCP broadcasts to unicasts and perform many other forms optimization, those mechanisms apply to wireless users.

Finally, note that when MCs are deployed in a cluster, any MC could potentially terminate traffic for any tunneled-node device, and the MCs must support the same VLANs.

WLAN VLAN deployment

You will now consider why a single VLAN might be the best choice for a wireless deployment.

Challenges in implementing VLANs in a wireless network

While wired VLAN design rules continue to remain rather traditional, the wireless world has changed. Where we used to assume a user with a single device, we now see users each with three to four devices. Mobility has become important due to the mobility of these new devices. Segregating users in VLANs on a group of APs can cause an issue with roaming, and using VLANs with /24 subnets can be limiting.

The growth of Wi-Fi in enterprise Wireless Local Area Network (WLAN) started more than a decade ago. Initially, people used wired networks as a primary medium to connect to networks, and Wi-Fi was just an option. To support a handful of devices connecting to Wi-Fi, network administrators used to assign separate VLAN for wireless clients, which met the users' needs.

With the launch of mobile devices in 2007, the use of Wi-Fi increased exponentially. Because many new mobile devices do not have wired ports, Wi-Fi becomes an essential medium for connectivity. To support increased numbers of clients on a WLAN, network administrators started adding more VLANs for wireless users, and VLAN pooling became a popular concept. Network administrators created multiple smaller subnets to reduce broadcast multicast traffic rather than increasing the size of the subnet.

802.11 shared medium

Wireless uses radio frequencies transmitted across open air, which is a shared medium (see Figure 6-7). Wi-Fi uses a medium contention protocol. There is no VLAN distinction between wireless users and the AP.

In CSMA/CA, as soon as a device has a packet to send, it checks to be sure the channel is clear and no other devices are transmitting. This includes other devices or the AP, as radio is a unidirectional medium. If the channel is clear, it sends the packet. If the channel is not clear, the device waits for a randomly chosen time and tries again. No user gets priority over others.

Once the AP receives the packets, it places them in a GRE tunnel and sends them to the MC.

The MC will decrypt, firewall and place the packet in the appropriate VLAN and switch or route the packet.

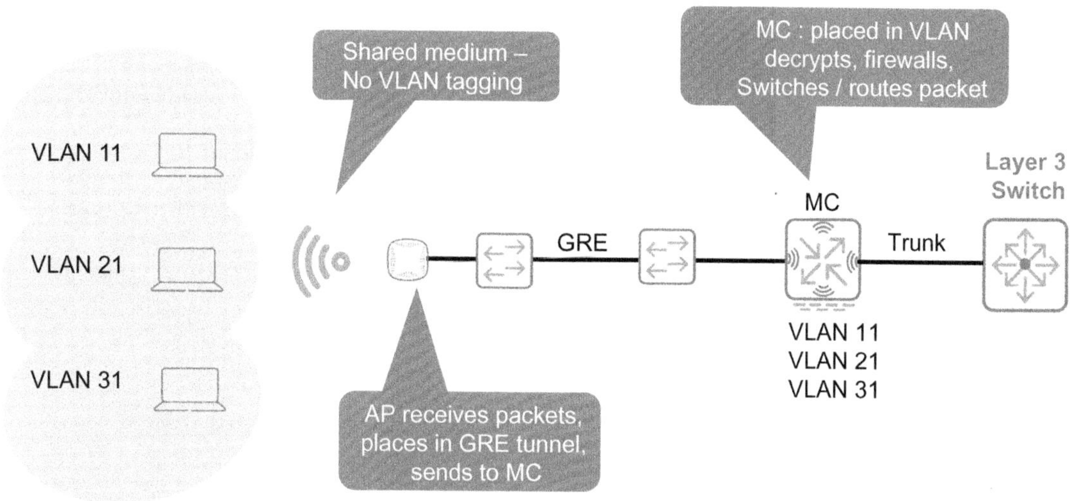

Figure 6-7 802.11 shared medium

VLAN options

In an Aruba network, you can choose how to use VLANs. Small sites with a small group of users can use an AP group with a single VLAN /24. If the company has many sites, each with a group of APs, and each site warrants a VLAN/24, then this is a valid option (see Figure 6-8).

A larger site with a significant number of users could use VLAN pooling. VLAN pooling uses a hashing algorithm or a round-robin system to distribute users into various subnets.

VLAN pooling works well, but still has some issues in IPv6 and roaming. Sometimes there can be an inconsistent use of the VLANs.

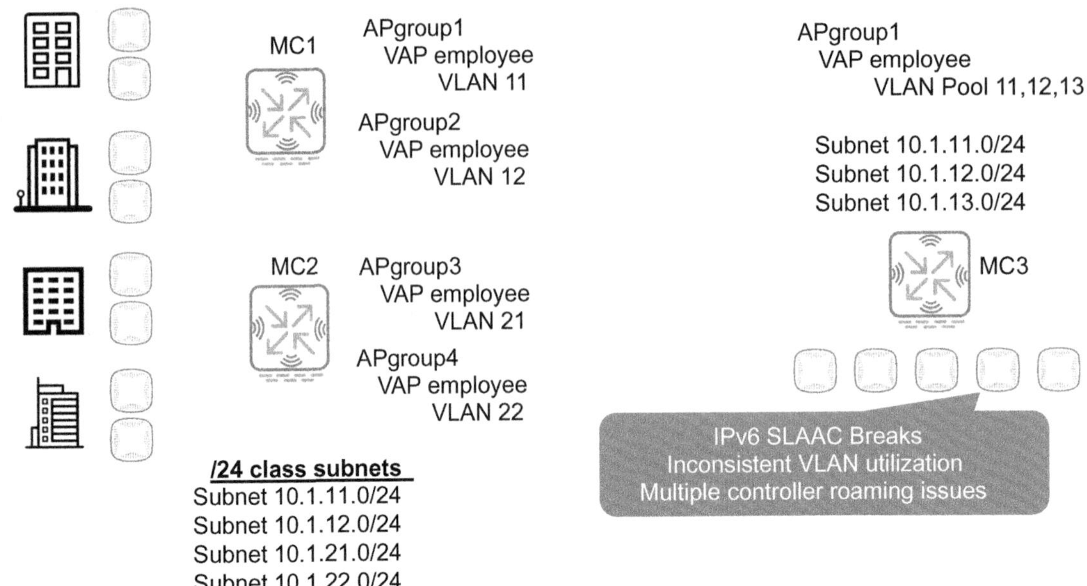

Figure 6-8 VLAN options

You can use several methods to assign a client to a VLAN, in order of precedence.

1. The default VLAN is the VLAN configured for the WLAN.
2. Before client authentication, the VLAN can be derived from rules based on client attributes such as SSID, BSSID, client MAC, location, and encryption type. A rule that derives a specific VLAN takes precedence over a rule that derives a user role that may have a VLAN configured for it.
3. After client authentication, the VLAN can be configured for a default role of an authentication method, such as 802.1X or VPN.
4. After client authentication, the VLAN can be derived from attributes the authentication server returns, via a server-derived rule. A rule that derives a specific VLAN takes precedence over a rule that derives a user role that may have a VLAN configured for it.
5. After client authentication, the VLAN can be derived from Microsoft Tunnel attributes like Tunnel-Type, Tunnel Medium Type, and Tunnel Private Group ID. This does not require a server-derived rule.
6. After a client authenticates, the VLAN can be derived from Vendor-Specific Attributes (VSA) for RADIUS server authentication. This does not require a server-derived rule. If a VSA is present, it overrides any previous VLAN assignment.

Single VLAN campus

As Figure 6-9 shows, the single VLAN design refers to one large subnet to serve all the clients connecting to an SSID in the campus environment with contiguous RF. Controllers and tunneled APs make it possible to scale to larger subnets in the range of /22 to /16.

Clusters of MCs can handle several APs and users, and balance these between the MCs. Take, for example, an MC that has the power to handle 2048 APs and a capacity of 32,768 devices. Putting all the devices in a single /22 subnet would make roaming simpler. A cluster of MCs with multiple APs will distribute the users in a single /16 subnet.

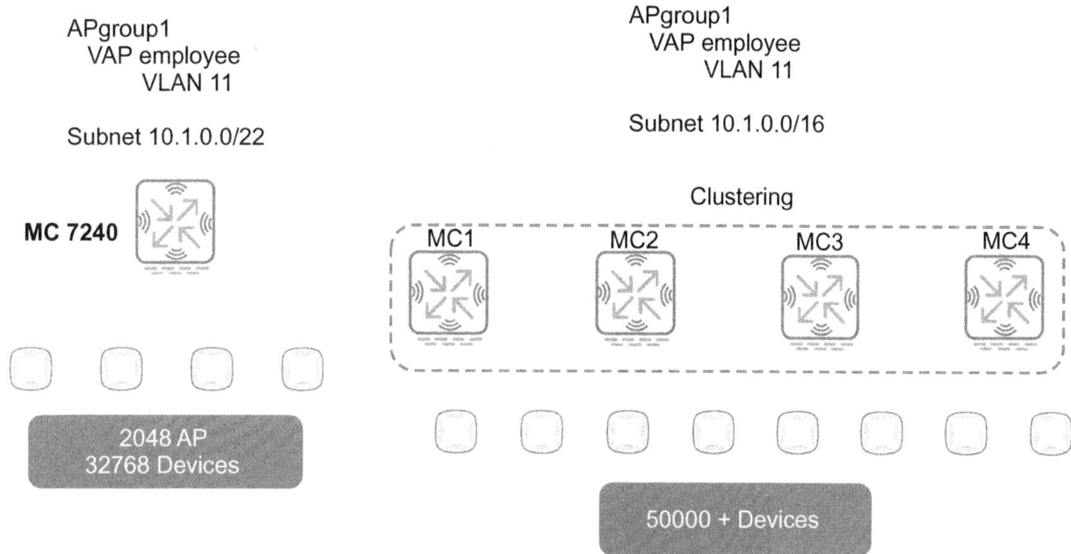

Figure 6-9 Single VLAN campus

The single VLAN design is simple and smart. It can greatly reduce the complexity of the WLAN design, and it addresses the IPv6, DHCP, and roaming challenges seen with VLAN and VLAN pooling. At the same time, you can meet all of the requirements of the wireless LAN using the single VLAN design.

Large universities can use the maximum advantage of this design with thousands of wireless clients across a large campus with contiguous RF followed by large enterprises with multiple buildings at a location. The Single VLAN design recommends using the same VLAN throughout the campus with contiguous RF. If you have multiple buildings in different locations such as school districts or corporate branches spread across different cities or towns, then you should use different VLANs and subnets. The Single VLAN architecture recommends using one large subnet for all the clients connecting to an SSID. However, you can use separate VLANs for clients connecting to separate SSIDs.

Ideally, on the campus WLAN, you should use separate VLANs for employee and guest SSIDs. Lastly, wired and wireless clients should not share the same VLAN. Use separate VLANs for wired clients and, if needed, use multiple smaller subnets to restrict the broadcast domain for wired devices. The Single VLAN architecture is for wireless LAN only, as the controller has a lot of visibility and control over wireless users, but none for wired devices. The obvious problems related to large VLANs on wired networks still apply.

Single controller roaming

Aruba provides a centralized management system to manage roaming, which allows wireless traffic mobility to be seamless. Since APs are simply radios, when a client roams from one AP to another, it only changes radios. The controller maintains the state of authentication and encryption while the client controls mobility.

In the example shown in Figure 6-10, the client hangs onto AP1 as long as it can. Once the client device moves to AP2, from the same MC, the record is updated showing the client is now on AP2. In this case, it does not matter if AP2 is in a different AP group. The MC knows which VLAN this user belongs to. However, the user may need to regenerate keys.

The objective of the network is to retain the user in the same VLAN to keep all sessions clean.

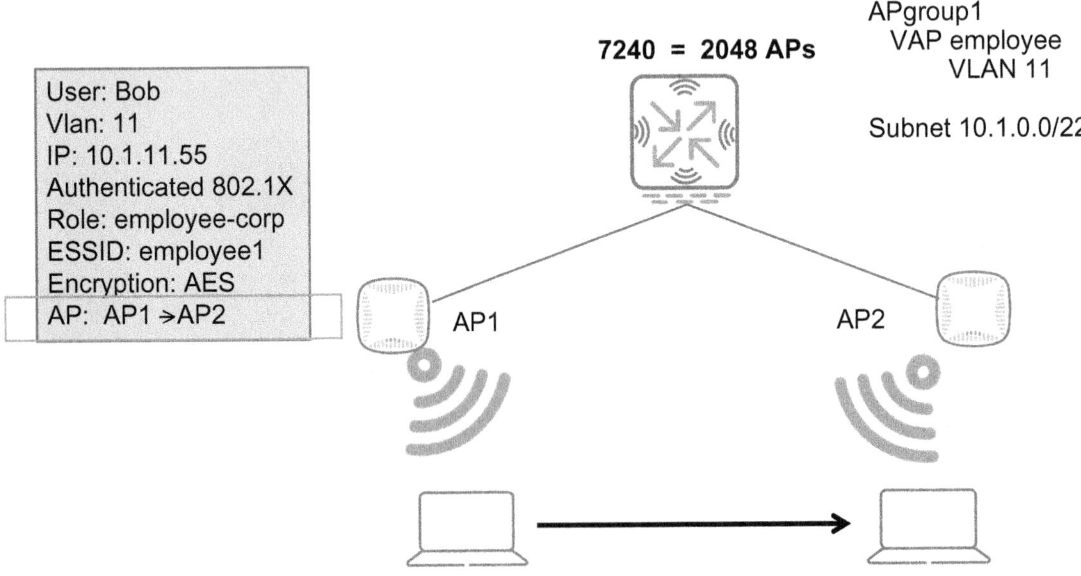

Figure 6-10 Single controller roaming

In the case of dynamic key encryption, using WPA, the keys are derived as a function of both the client's and the AP's MAC addresses. As the client moves from one AP to another, new keys are derived for security purposes. During the time, the client takes to renegotiate its keys, data or voice traffic cannot be exchanged between the infrastructure and the client. This could result in the session being dropped.

WPA2 offers key-caching mechanisms that alleviate these roaming effects by allowing the clients and APs to cache keys. This support is not available in the WPA standard. However, on designated handsets, Aruba controllers have implemented a mechanism for WPA key-caching similar to the Opportunistic Key Caching feature available for WPA2.

The key negotiation protocol in IEEE 802.11i specifies that, for 802.1X based authentication, the client is required to renegotiate its key with the RADIUS or another authentication server supporting Extensible Authentication Protocol (EAP) on every handoff. This is a time-consuming process. The solution is to allow for the part of the key derived from the server to be cached in the wireless network. Then, the cached key can provide the basis for a reasonable number of future connections, avoiding the 802.1X process. Currently, the Opportunistic Key Caching feature, based on 802.11i, performs the same task. 802.11r differs from OKC by fully specifying the key hierarchy.

OKC is a technique not defined by 802.11i, but available for authentication between multiple APs in a network where those APs are under the common administrative control of a single controller.

When using OKC, a station roaming to any AP in the network will not have to complete a full authentication exchange. Instead, it will perform the 4-way handshake to establish transient encryption keys. Although OKC is not a part of the 802.11i standard, several wireless vendors have adopted this technique and have achieved interoperability. Most notably, Microsoft has provided support for OKC.

Single VLAN cluster

An MC cluster with multiple MCs and APs could distribute the users in a single /16 subnet. The AP will have an AP Anchor Controller (AAC) and a Standby AP Anchor controller (S-AAC).

When a user associates to an AP, the AP will look up a bid table and determine which MC will be this user's User Anchor Controller (UAC). The UAC is a role given to cluster members from an individual user perspective.

As Figure 6-11 shows, the bid system may assign another user from the same AP and the same SSID to another MC in the cluster. This MC would be the new user's UAC.

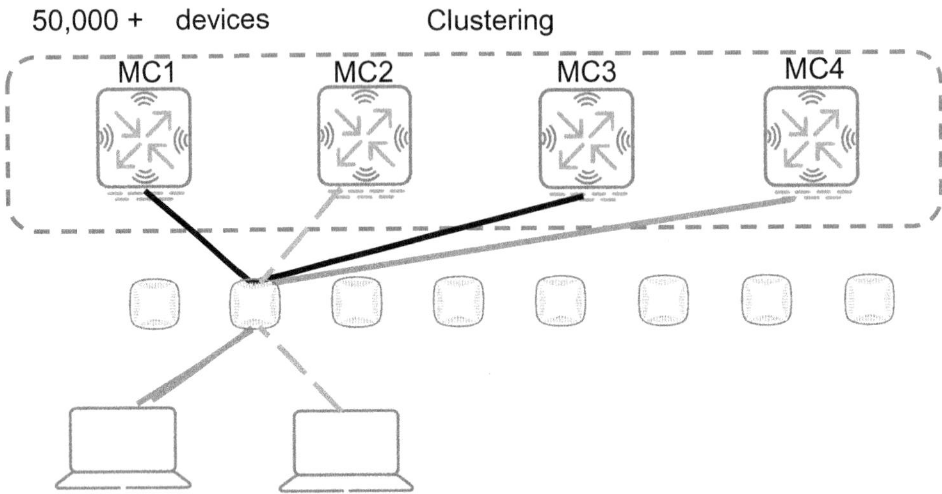

Figure 6-11 Single VLAN cluster

The objective of the cluster is to provide high availability to all the clients and ensure service continuity when a failover occurs. It improves scalability, performance, and redundancy.

The UAC Mobility Controller handles all wireless client traffic including:

- Association/Disassociation notification
- Authentication
- All unicast traffic between the Mobility Controller and its clients
- Roaming clients among APs

Upon associating to the AP, a client UAC Mobility Controller is determined and the AP uses the existing GRE tunnel, to push the client traffic to its UAC.

If no GRE tunnel exists, the AP will create a dynamic tunnel to that client's UAC. When that client roams to a different AP, the original AP tears down the dynamic tunnel, if no other clients are using it.

The new AP that the client roamed to will follow the same tunneling guidelines as above to push traffic to the same UAC.

Cluster roaming

When a client associates to the AP, a user anchor point is selected based on a bucket map. This will be one of the MCs in the cluster.

As Figure 6-12 shows, if the user roams to another AP, the AP will look up its bucket map and select the same MC. Then the AP will build a GRE tunnel to the MC and forward the user's traffic.

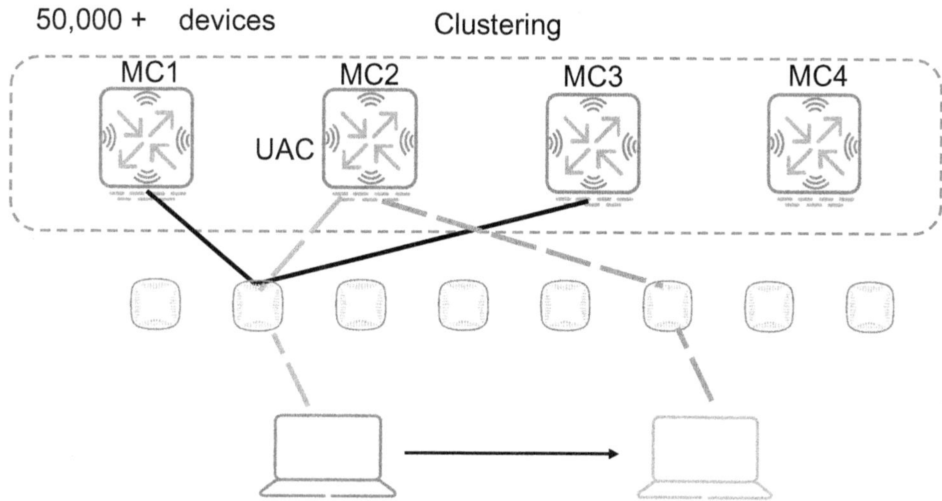

Figure 6-12 Cluster roaming

Clustering is based on keeping client processing, including signaling and traffic, anchored to a managed device regardless of which AP the client roams to, as long as it is within the cluster's control scope.

When a client associates to an AP, its Wi-Fi MAC address goes through a hashing algorithm to produce a decimal number [0..255]. That decimal number is used as an index to a mapping table that provides the UAC controller for that client. This mapping table is also known as a Bucket-Map. The cluster leader computes the bucket-map on a per ESSID basis and pushes it to all the APs in the clustering domain.

At the same time that a UAC is assigned to a given user, a standby UAC gets assigned to that user if Redundancy is enabled.

Advantages of single VLAN

There are many advantages of a single flat VLAN. The design is simple with one scope and easy to support with just one large subnet for all the wireless clients connecting to an SSID. The network administrator does not need to configure multiple VLANs, DHCP scopes, extend VLANs across multiple devices, or configure redundancy for default gateways in each VLAN. On top of that, the network administrator does not need to configure a VLAN mobility or IP mobility scheme for clients roaming across different controllers. Configuration of Layer 2 or Layer 3 roaming on the controller is simply no longer necessary. Since the client VLAN remains the same, the client keeps using the same IP address no matter where the client roams, either on a different AP on the same controller or an AP on a different controller. This solves the roaming complexity for the network administrator.

The simplicity of a wireless LAN design makes it very easy for network administrators to support it. If issues arise, it is very easy to troubleshoot them. Network administrators do not need to determine which VLAN the client falls into, or try to isolate an issue related to that VLAN or network-wide.

The single VLAN design solves Ipv6 SLAAC challenges. Since multiple RAs do not come from IPv6 routers, there is no issue of the client getting the wrong v6 address.

Single VLAN gives the administrator address efficiency. With one VLAN, there is no issue of some VLANs being completely exhausted while other VLANs stay less utilized. In large campus environments with thousands of users, this helps to avoid the issue of clients not getting IP addresses due to some VLANs being drained.

Recommendation for single VLAN

A campus with complete RF coverage would be a good place for a single VLAN as it will simplify roaming for users. Hospitals and large university campuses are locations where simplified roaming is an advantage.

Aruba recommends separate VLANs for customers who have several buildings with noncontiguous RF coverage.

Aruba does not recommend using derived VLANs in a single VLAN design. The single VLAN should be in the same VAP with the same SSID. You cannot simply set up a single VLAN on the Aruba switch. You must make sure the core can handle the large VLAN as well.

The single VLAN design is intended for wireless clients connecting to APs in the same campus with contiguous RF. If RF is not contiguous, then you should use different VLANs/subnets for different RF domains. If the client can roam from one building to another in the campus without dropping off from the SSID, you should use a single VLAN.

The typical use cases for using the single VLAN design include university campuses, large enterprise headquarters, and hospitals with multiple buildings. If the customer has multiple buildings and offices spread across different geographic locations, then RF is not going to be contiguous. Thus, they should use separate VLANs. The typical use case is branch offices or home offices.

Customers must not use VLAN derivation by user role or server-rule. If clients, connected to the same SSID, are in different VLANs, the behavior is the same as VLAN pooling. Thus, all the concerns with VLAN pooling will apply to the design. The large subnet is for wireless clients connected to the same SSID only. Use different VLANs for different SSIDs (for example, employee vs. guest SSID).

If customers have multiple controllers serving a campus network with contiguous RF, they should extend the client VLAN all the way back to their core switch, router, or firewall where the default gateway exists. Customers using the single VLAN design must ensure that their uplink switches, router, or firewalls can support the required number of ARP and MAC address table entries depending on the subnet size. You should ensure that the DHCP server can support a DHCP scope as large as the size of the client VLAN.

Key considerations

To set up a single VLAN, there are key considerations you need to verify (see Figure 6-13). Is your DHCP server capable of creating a large scope and distributing these addresses? Many Linux-based servers are not capable of this process. Is your router capable of handling large Address Resolution Protocol (ARP) tables? Is your switch capable of large bridge and MAC tables?

Upstream firewalls will likely be configured based on VLAN rules. Also, there is the problem of broadcast and multicast. However, Aruba has the wireless solutions for these problems.

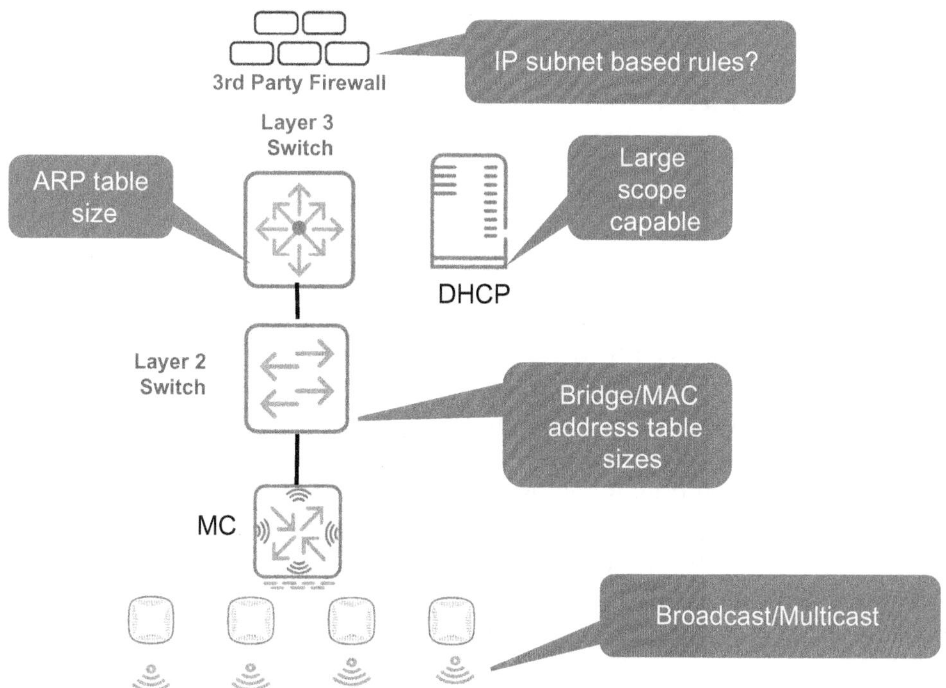

Figure 6-13 Key considerations

To support a large VLAN, other devices on the network also need to be capable of handling the workload. Most of the time the wireless client VLAN is configured as Layer 2 on the Aruba controller and uplink switch. The uplink router is configured as a Layer 3 interface working as a default gateway for that VLAN. While using large VLANs, the ARP table on the router can also be sizeable depending on how many clients are connected to the WLAN. Thus, network administrators should find out any limitations on the router and address them accordingly.

The same goes for switches. Although the controller uplink switch does not need to handle ARP entries, the bridge/MAC address table entries can be very large. The network administrator should consider this. In the case of single VLAN design, the DHCP server needs to be able to support large DHCP scopes, as big as /16, which many Linux-based DHCP servers are not capable of supporting. Thus, the network administrator needs to ensure that the DHCP server can handle a large DHCP scope. At the same time, the DHCP server needs to be powerful enough to lease out IP addresses quickly, which is particularly essential in a failover situation. For most of the enterprise networks, existing firewall rules are based on IP subnets. There are many network administrators still entrenched in this way of thinking. Even inside the network, administrators may configure routing, access control, and QoS based on the VLANs and IP subnet through routers and legacy firewalls.

Obviously, next-generation firewalls are user and application-centric and are reasonably deployed at the Internet gateway and data centers for large enterprise customers. With the Single VLAN Design, all the traffic of wireless clients is in one subnet only. Hence, user and application-centric firewalls are necessary to apply proper firewall rules. Another option is to use the Aruba Controller's built-in stateful firewall to manage firewall policies for wireless clients.

Aruba wireless broadcast/multicast solutions

Aruba can limit broadcast and multicast by converting the necessary traffic to unicast and removing the unnecessary broadcast and multicast traffic, which can also solve many IPv6 issues. Many of these features are configured by default.

To enact the broadcast solutions, first convert broadcast to ARP unicast. When a router sends out an ARP broadcast, there is no need to send this frame to every AP. The MC knows both the user and the AP associated to the user. The MC simply changes the ARP request to a unicast and sends it directly to the user. If this feature is disabled, then the MC does no conversion and the ARP broadcast would transmit to all APs with a user in the subnet, causing degradation of the radios. This unicast feature is on by default. The MC will suppress any ARP that does not correspond to any user in the MC list, which saves the unnecessary broadcast and use of the radio. The MC will broadcast on any of its wired ports or any wired AP port, as well as on any VAP in split tunnel mode (split tunnel mode is commonly used on RAPs). This feature is on by default. Optimizing duplicate address detection controls the flood of gratuitous ARP (GARP) and IPv6 DAD frames. If a user sends a GARP, the MC will forward this frame to the router, but will not flood the APs.

The MC will forward GARPs or DADs coming from the router because this is important information for the clients. For example, if you have redundant gateways and lose the primary gateway, the secondary gateways notifies the client that it is now owner of the IP address. This feature is on by default. Once ARP and DHCP packets have been converted to unicast, the MC will block all unnecessary broadcast and multicast. By default, this feature is disabled, but Aruba recommends enabling the feature. IGMP snooping keeps track of clients subscribed to multicast stream and converts multicast packets to unicast before sending it to wireless clients. This feature is also disabled by default but should be enabled since multicast streaming is required.

Single VLAN optimizes IPv6 traffic

Clients IPv6 addresses can be auto configured by listening to RA frames or sending RA frames to the router. All RA and RS message are sent as specific multicast addresses. This could affect the wireless performance.

A periodic RA sent in a flat VLAN is not a major issue. However, an issue could develop with VLAN pooling when users are on the same AP but in different VLANs, because RAs would be advertised on each VLAN. To avoid this issue, the MC will intercept the RAs and unicast them to the clients.

A client will forward DAD packets to inform others of its IP address. If all clients in a flat VLAN send out DAD frames, then each client would need to retain this information in its ARP table. Client devices have a limited ARP table so this could be an issue. Aruba MCs stop the forwarding of DAD frames to clients. However, DAD frames from the router are still forwarded.

Identity-based Aruba firewall

Role derivation will determine what role and rights the user will receive, regardless of the AP, SSID or subnet the user falls into. For example, in the diagram shown in Figure 6-14 Pedro, Robert, Frank, and Susan are on the same SSID and are on the same subnet, but all have different roles. The IP address they have is irrelevant to what role they receive.

Figure 6-14 Identity-based Aruba firewall

Roles can be derived locally on the MC or specified by a RADIUS server or ClearPass. If there is no locally derived role and no server derived role, then the user falls into the authentication default role.

Note

You can write user rules that say if you are on this AP or if you are in this VLAN, you are assigned a role.

With Aruba networks, not only can you gain contextual insights about users, devices, and application states, but you can also influence the way the infrastructure behaves. For example, if there is a need to update the policy enforcement for a specific user's device, you can program the network to route traffic differently. This will improve data security and change the way physical spaces interact with mobile devices based on their location or identity.

Activity 6: Recommended VLANs for new network

In this activity, you will determine the customer requirements in preparation for designing a new network: Use the scenario information from previous activities plus the additional information below to complete the activity.

Scenario VLANs

This addition to the main scenario will explain the customer's current VLAN structure.

Current VLAN network structure

Corp1 has two network closets in the main corridor of every floor (you have planned an upgrade for these closets). Each closet was assigned a /24 subnet.

Building 1

Building 1 Floor 1 closet 1: 10.2.1.0/24 VLAN 101

Building 1 Floor 1 closet 2: 10.2.2.0/24 VLAN 102

Building 1 Floor 2 closet 1: 10.2.3.0/24 VLAN 103

Building 1 Floor 2 closet 2: 10.2.4.0/24 VLAN 104

Building 1 Floor 3 closet 1: 10.2.5.0/24 VLAN 105

Building 1 Floor 3 closet 2: 10.2.6.0/24 VLAN 106

Building 2

Building 2 Floor 1 closet 1: 10.2.7.0/24 VLAN 201

Building 2 Floor 1 closet 2: 10.2.8.0/24 VLAN 202

Building 2 Floor 2 closet 1: 10.2.9.0/24 VLAN 203

Building 2 Floor 2 closet 2: 10.2.10.0/24 VLAN 204

Building 2 Floor 3 closet 1: 10.2.11.0/24 VLAN 205

Building 2 Floor 3 closet 2:10.2.12.0/24 VLAN 206

The old aggregation layer switches routed traffic and connected to the data center core on VLAN 1000, subnet 10.1.0.0/24.

The data center has been assigned the 10.1.0.0/16 space. The data center divides this into /24 subnets as required. Currently, 10.1.1.0/24 to 10.1.25.0/24 are used.

VLANs 500 to 550 have also been assigned to the data center as needed.

The Corp1 office network connects to the data center, which connects to the Internet. Two gateway firewalls sit between the data center and the Internet.

Corp1 VLAN requirements

You do not need to plan any changes for the data center. You should assess the VLANs assigned to the closets and determine if they meet the needs.

For the new Wi-Fi network the IT department will create new subnets as follows:

Building 1

Building 1 Floor 1 10.3.1.0/24 VLAN 301

Building 1 Floor 2 10.3.2.0/24 VLAN 302

Building 1 Floor 3 10.3.3.0/24 VLAN 303

Building 2

Building 2 Floor 1 10.3.4.0/24 VLAN 304

Building 2 Floor 2 10.3.5.0/24 VLAN 305

Building 2 Floor 3 10.3.6.0/24 VLAN 306

There are approximately 900 employees at corporate, so IT feels this should be enough IP addresses.

Activity 6.1: Wired VLAN recommendations

Activity 6.1 objectives

You will make recommendations for the VLANs in the updated wired infrastructure.

Activity 6.1 steps

1. What recommendations do you have for the wired infrastructure? Consider these questions:
   - Will you change any VLANs? If so, explain your changes.
   - Will you add any VLANs? If so, plan the VLAN ID and associated subnet.
   - Which devices will act as the default routers for each VLAN? And where do VLANs need to extend? (Keep in mind your security choices from Activity 5.)
   - What VLAN or VLANs will you use between the aggregation layer, if present, and campus core? What VLAN or VLANs will you use between the campus core and data center core?

2. Draw out the logical design using Figures 6-15 and 6-16.
   - In each box, indicate the VLANs configured on switches deployed in that closet or at that layer.
   - For the links between layers, indicate the VLANs carried on the links between switches and whether those VLANs are tagged or untagged.
   - Ignore the Mobility Controllers boxes for now. You will fill out these in the Activity 6.

Use the Network diagrams provided in Figures 6-15 and 6-16.

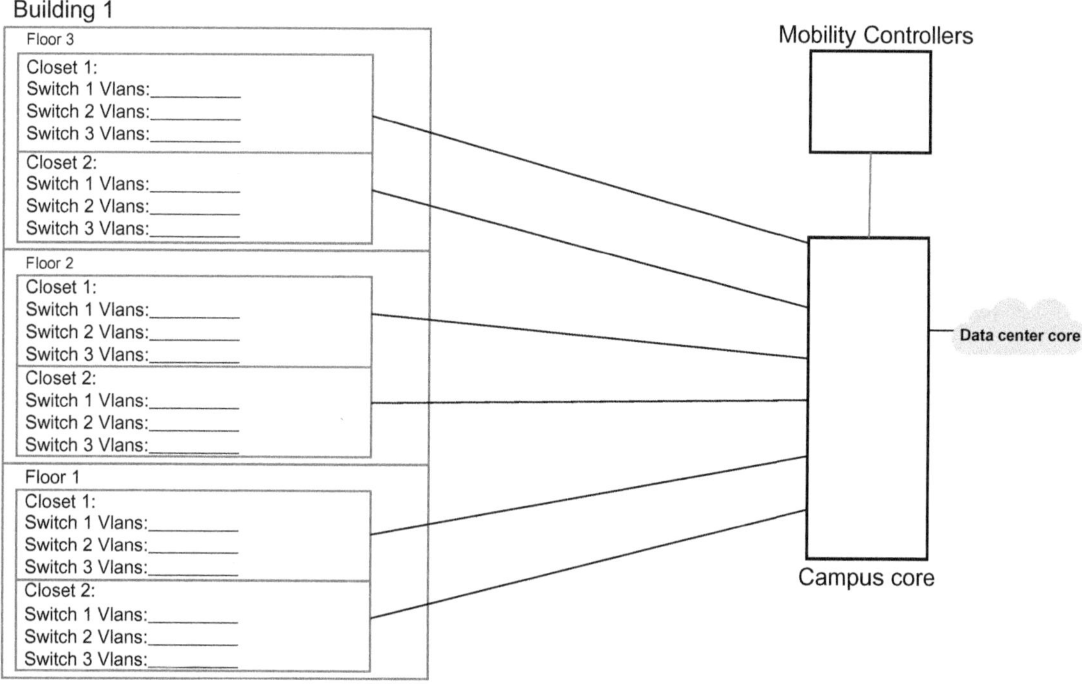

Figure 6-15 Network diagram

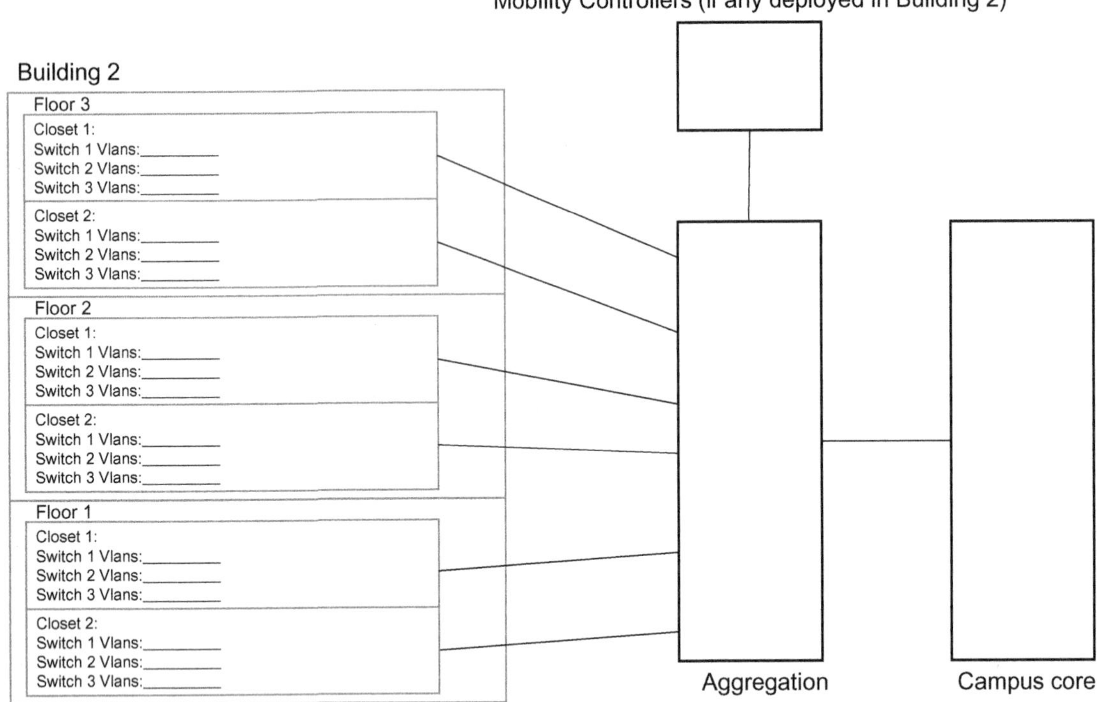

Figure 6-16 Network diagram

Activity 6.2: Wireless VLAN Recommendations

Activity 6.2 Objective

You will make recommendations for the VLANs needed to support the new WI-FI infrastructure.

Activity 6.2 Steps

1. What VLANs have been assigned for the Wi-Fi structure:
   a. Building 1 Floor 1: ____________
   b. Building 1 Floor 2: ____________
   c. Building 1 Floor 3: ____________
   d. Building 2 Floor 1: ____________
   e. Building 2 Floor 2: ____________
   f. Building 2 Floor 3: ____________

2. Do the VLANs and subnets assigned meet your requirements (Y/N)? __________
3. What recommendation do you have for the Wi-Fi infrastructure?
4. List the AP group(s) that you will recommend, along with what VLANs will be used in each AP group VAP.

You can use Figures 6-15 and 6-16- to list your AP groups and VLANs if you want.

Answers to the questions included in activities are provided in "Appendix: Activities and Learning Check Answers."

Summary

You have learned how to follow best practices to deal with the challenges of VLAN planning for both wireless and wired devices. You understand why you often want to create just one VLAN for wireless users, to simplify and enhance roaming, but you should generally create multiple VLANs for wired devices. You can plan the subnets associated with VLANs and also make your plan fit within the legacy environment.

You also understand that when you use tunneled-node, you must match the VLAN assigned to tunneled-node endpoints by the switch with a VLAN on the MC.

Learning check

1. Which best practice should you follow when planning VLANs for wired user devices?
   a. Place wired devices in the same VLANs used by wireless devices.
   b. Extend the wired VLANs as far as possible up to a /16 subnet.
   c. Make sure that VoIP phones and the computers connected to them use the same VLAN.
   d. Keep the wired VLANs at between about 250 and 1000 devices.
2. What is an AP group?
   a. A group of MCs
   b. APs of the same type
   c. A group of APs with the same configuration
   d. AP grouped in a location

3. What are key considerations when planning a single VLAN in a campus?
    a. Contiguous RF
    b. APs of the same type
    c. MAC and ARP tables size
    d. DHCP server capabilities
    e. Third-party firewall

Answers to the learning checks in this study guide are provided in "Appendix: Activities and Learning Check Answers."

7 Redundancy

EXAM OBJECTIVES

- ✓ Evaluate a customer's needs for a single-site campus, identify gaps, and recommend components.
- ✓ Translate a customer's needs into technical requirements.
- ✓ Given the customer's requirements for a single-site environment, determine the component details and document the high-level design.
- ✓ Given a customer scenario for a single-site campus, choose the appropriate components that should be included in the BOM.
- ✓ Given the customer's requirements for a single-site environment, determine and document the logic and physical networks.
- ✓ Given the customer's requirements, explain and justify the recommended solution.

Assumed knowledge

- Aruba 8.x architecture
- Basics of Aruba clustering
- Layer 2 and Layer 3 in the Open Systems Interconnection (OSI) model
- Virtual Router Redundancy Protocol (VRRP)
- Spanning tree protocols
- Link Aggregation Control Protocol (LACP)
- Aruba Virtual Switching Framework (VSF) and backplane stacking

Introduction

As more companies move toward wireless networks, the need for Aruba Mobility Master (MM) and Access Point (AP) redundancy increases. Providing customers with a design that allows for sufficient coverage and redundancy requires a deeper understanding of MM redundancy and how to implement it into your customer's network.

First, you will review the MM responsibilities and how MM redundancy functions in both Layer 2 and Layer 3 mode.

Next, you will learn how you can back up APs and establish Mobility Controller (MC) and AP redundancy.

Finally, you will learn the various types of wired redundancy as well as the differences between redundancy and resiliency in a wired network and why both are essential for wired network redundancy.

MM redundancy

In this section, you will review the MM responsibilities. You will also consider MM redundancy for Layer 2 and Layer 3.

MM responsibilities

In the network, the MM acts as a single point of configuration for global policies such as firewall policies, authentication parameters, and RF configuration (see Figure 7-1). The MM synchronizes all the configurations with the MCs and is responsible for many higher-level functions, including centralized licensing, Unified Communications and Collaboration (UCC) integration, and Aruba AirMatch.

In the event of an MM outage, MCs continue to use their existing licenses for 30 days, and the network remains up. However, you cannot change the configuration items in the controller network. In addition, the MM manages many of the higher level functions, which can cause issues in the network if the MM fails. Aruba recommends a backup MM for these situations. From an economic point of view, a backup MM for a VMM is free. If a customer has a hardware MM, the customer must purchase a backup hardware MM.

You will lose many functions during an MM outage. The MM is the license server and is responsible for the distribution of licenses in the network. With dashboard and graphs, the MM has visibility into all the MCs, APs, and users in the network.

The MM coordinates intrusion detection and protection and uses AirMatch to select power levels and channels for all APs in the network. The MM collects information that AirMatch will calculate and download to every AP in their selected power levels and channels. This process is done every 24 hours when needed.

UCC handles most of the voice, video, conferencing, and desktop-sharing features. In AOS 8.X, the UCC runs as a service on the MM. The Mobility Master's UCC app will receive the XML message information from the SDN API, as well as metadata from the MC in heuristics mode. The MM correlates this information to confirm the application.

The Mobility Controller Web Classification for AppRF can interact with WEBroot through the MM.

The MM maintains the whitelist for all APs within the network.

The MC forwards AMON messages to the MM. The MM executes the client match module and builds the Virtual Beacon Report (VBR) table for each AP and sends it back to the MC and the AP. Client Match will coordinate the client's steering and load balancing.

For AirGroup, the MM maintains a cache table of mDNS records for all the mDNS service advertisements that it sees on the network. The MM only maintains cache records for services and VLANs that AirGroup allows.

MM is responsible for:

- Centralized licensing
- Centralized Visibility, Monitoring and configuring platform
- WIDS
- AirMatch
- User database
- SDN functions
- UCC function
- WebCC proxy
- AP Whitelist DB
- RBCM (Rule Based Client Match)
- AirGroup classification

When the Mobility Master is unavailable:

- MM functions are lost
- WLAN itself will continue to function

The Backup Mobility Master:

- It is free of charge (VM)
- It is a hot standby

Figure 7-1 Mobility Master (MM) responsibilities

Because the MM manages all of these functions, it is important to have a backup in case of network failure. A backup MM will take over the primary MM's responsibilities during an outage. If the MM becomes unavailable, the network continues to run without interruption, but higher level functions will fail. Also, any changes in the network topology or configuration will require the availability of the MM.

MM redundancy overview

There are two types of redundancy: MM redundancy on a site and MM site redundancy (see Figure 7-2). A primary and backup MM, connected to the same broadcast domain using the same Virtual Router Redundancy Protocol (VRRP), can share a single set of licenses.

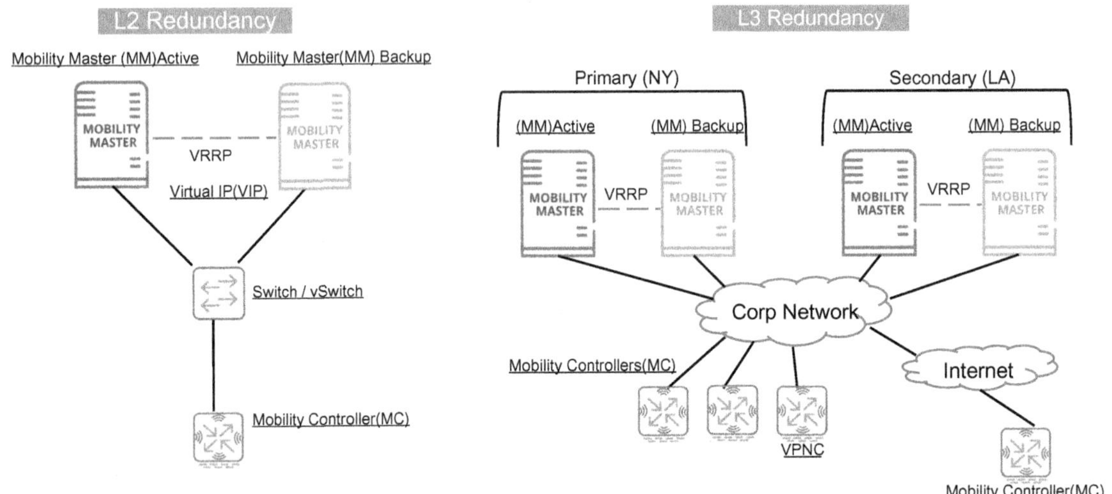

Figure 7-2 MM redundancy overview

Managed devices on the network connect to the MM using the VRRP virtual IP address configured for that set of servers. The primary MM uses the configured virtual IP address by default. However, if the primary MM becomes unavailable, the secondary MM will take ownership of the virtual IP address, allowing managed devices to retain seamless connectivity to an MM device. You can only define one backup MM for each primary MM.

The Layer 3 redundancy is primarily intended for customers who want to handle a complete data center failure during natural disasters or other catastrophic events (see Figure 7-2). This form of redundancy requires the purchase of two MMs, each of which can have a backup.

Layer 3 redundancy will prevent a scenario where an MM acts as a single point of failure if the link to the MM goes down or if a colocated standby MM VRRP controller pair fails due to a network failure or local natural disaster. Configuration and databases are synced automatically from the primary to the secondary data center. Managed devices detect a failure in the primary data center and automatically switch to the secondary data center.

MCs are configured with secondary master IP using full-setup dialog, mini-dialog, ZTP, or as partial configuration on the MM. Each MC will interface with the Health Check Manager (HCM), which will provide the reachability information of both the primary and secondary MM. The MCs that directly terminate on the MM will probe the primary and secondary MM IP to detect the primary failure. If the MCs connects to the MM via VPNC, then MCs probe the primary and secondary VPNC IP to detect primary failure. When the MC detects that it cannot reach the primary MM for 15 minutes, it triggers Layer 3 switchover.

When MCs detect that the primary is down and secondary is up, the MC will tear down the tunnel with the primary and attempt to establish IPsec tunnel with its secondary. The secondary MM will accept MCs only if it detects its tunnel with the primary MM is down.

The MCs can stay in the secondary MM as long as the primary is down. As soon as the primary is up, the tunnels with the MCs will be torn down and the MCs will switch back to primary MM.

MC AP redundancy

You will now focus on how you can back up APs and establish MC and AP redundancy.

Coverage redundancy

If a switch fails, several APs will become unavailable as well. The AP depends on the switch for Power over Ethernet (PoE) power and wired access.

If losing an AP is not an option, you can dual home the APs, as shown in Figure 7-3. Dual-homing, for APs with two wired ports or more, allows you to connect one port to a switch and the second port to another switch. In case of a switch outage, the AP can use its other wired port to send its traffic. The AP will use wired port 0 first as the highest priority. If it is not available, then it will use wired port 1. Without dual-homing, an AP would be connected to a single switch, and any switch outage, cable issue, interface issue, and so forth, would cause an AP coverage outage.

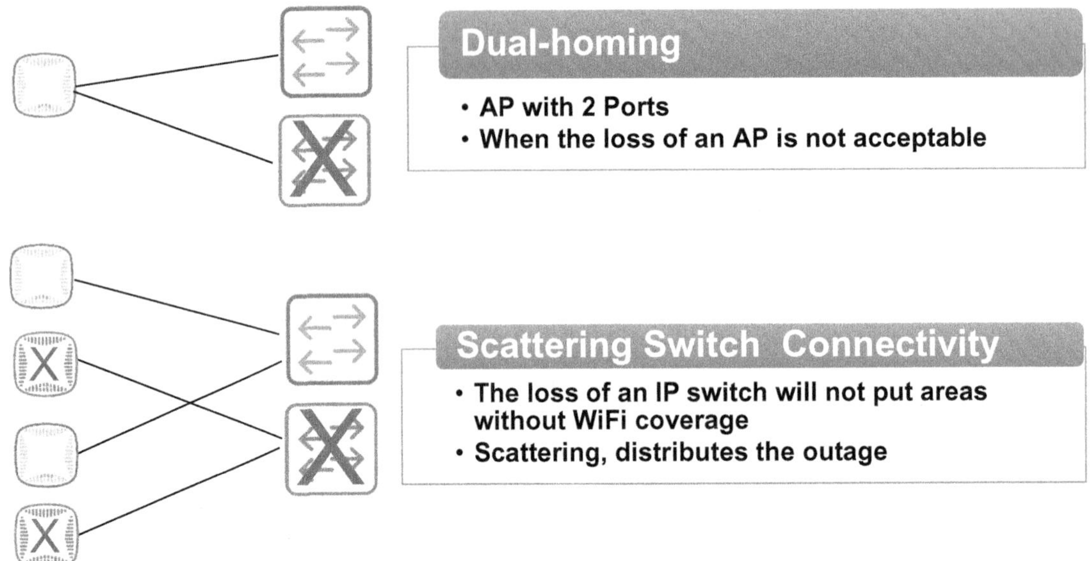

Dual-homing
• AP with 2 Ports
• When the loss of an AP is not acceptable
Scattering Switch Connectivity
• The loss of an IP switch will not put areas without WiFi coverage
• Scattering, distributes the outage

Figure 7-3 Coverage redundancy

If dual homing is not economically feasible, consider staggering your APs between switches (see Figure 7-3). This way, if a switch fails, some APs will remain active.

Scattering switch connectivity is especially useful if you are deploying APs in a tunnel or in a linear fashion. Connecting alternate APs to the same switch staggers the APs. In the case of a switch outage, you would not lose coverage completely in any one area. On the other hand, if you did not stagger the connections, instead connected all APs in the area to the same switch, a switch outage will cause coverage to be lost in that area.

LMS backup redundancy

You can configure LMS backup redundancy for both Campus APs (CAPs) and Remote APs (RAPs). An LMS backup can be more important for RAPs, however, because it is more complicated to connect RAPs to clustered MCs. You can configure a RAP with the LMS IP address of the MC. This could be the IP address of an MC or VRRP VIP between two MCs. You can then configure the RAP with a backup LMS IP address (see Figure 7-4).

If the RAP loses contact with the LMS IP, it will attempt a connection to the backup LMS IP.

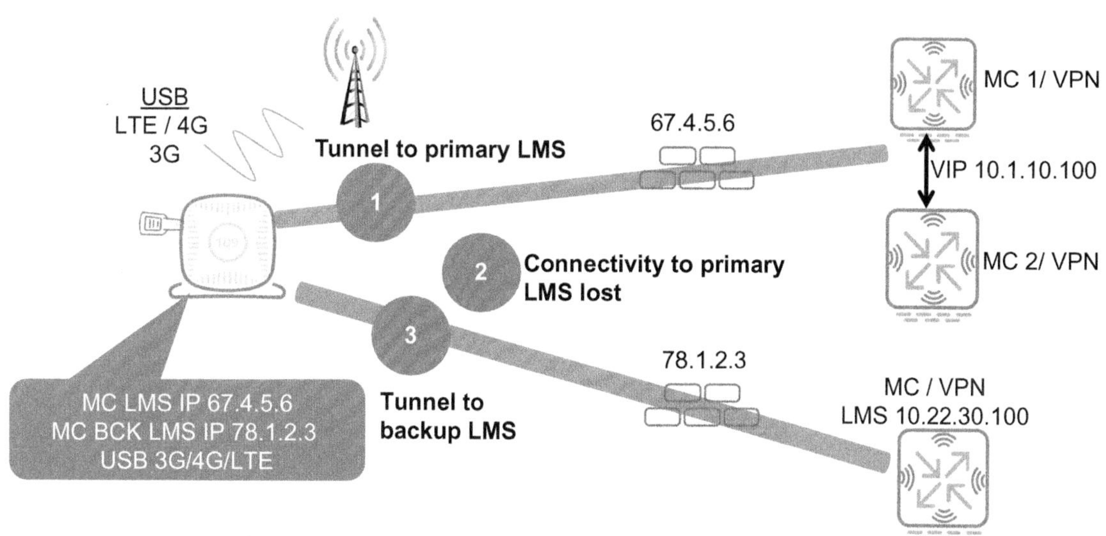

Figure 7-4 LMS backup redundancy

If both the LMS IP and the backup LMS IP are unreachable, the RAP will attempt the USB 3G/4G stick, if you implemented it.

The RAP must terminate the VPN tunnel on an MC. The MC can be behind a gateway firewall. The gateway firewall will use Network Address Translation (NAT) to translate the IP address into the MC IP address.

If the RAP loses its local connectivity then RAP, if configured, will attempt to use the USB 3G/4G/LTE stick. If the USB supports multiple standards, then it will try the highest first.

MC HA failover

AOS6.4 introduced AP fast failover, also known as High Availability (HA). In AP fast failover, the AP simultaneously establishes GRE tunnels to both the primary and backup controllers, as shown in Figure 7-5. In case of a controller outage, the AP can quickly failover to the backup controller since it already established the tunnels.

Note that the standby APs' GRE tunnels to the backup controllers do not consume licenses.

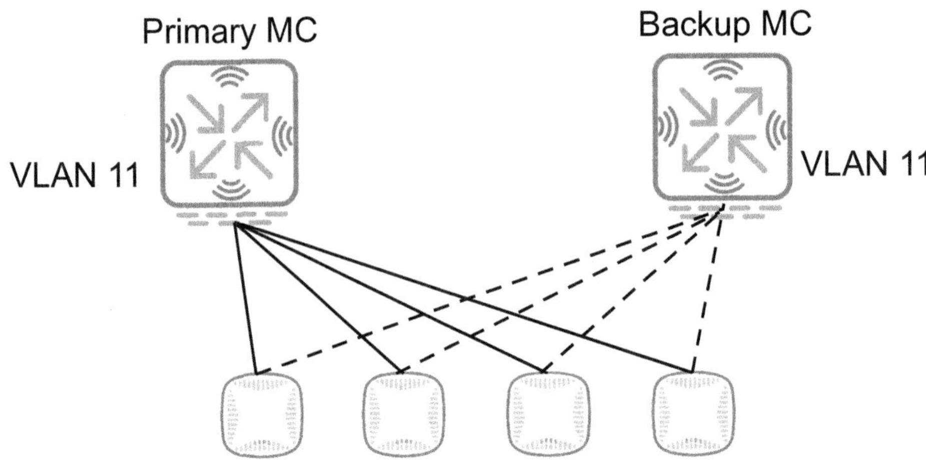

Figure 7-5 MC HA Failover

You can configure HA several ways including:

- Active-Active where both MCs terminates APs as well as act as standby for the other MC.
- Active Standby where the primary MC can server APs, but cannot act as standby for APs.
- Standby where the backup MC does not terminate APs. It only has standby APs with standby tunnels open to it.

When APs failover, users need to see the same VLANs on the other MC.

In VRRP and the backup LMS IP, the AP does not open a redundant connection to the backup controller until it fails over. This causes extra delays for the AP to establish GRE tunnels to the backup controller. Also, in VRRP and backup LMS IP redundancy, the AP does a bootstrap reset of its radios, before failing over, which causes a coverage outage. In comparison, the AP does not have to do a bootstrap in AP fast failover.

When the primary fails, the APs will failover to the backup controller. The licenses from the primary will be returned to the license server. When the AP uses the backup MC, the backup MC will request licenses from the license server.

With APs failover, users need to see the same VLANs on both controllers. An AP is linked to an AP group. The AP group has a WLAN that is configured to place users in a specific VLAN. If the AP fails overs, the AP remains in the same AP group and therefore must place the users of the WLAN in the same VLAN. This VLAN does not need to be in the same subnet, but if not the user will need to request a new IP address, and this is problematic for many devices.

What is clustering?

Clustering is a combination of multiple managed devices working together to provide HA to all clients to ensure the service continuity when a failover occurs (see Figure 7-6).

Clients remain anchored to a single controller throughout their roaming on campus, no matter which AP they connect to, makes their roaming experience seamless since their Layer 2/Layer 3 information and sessions remain on the same MC.

Clients are automatically load balanced within the cluster. When the clients are moved among cluster members, the move is done in a stateful manner.

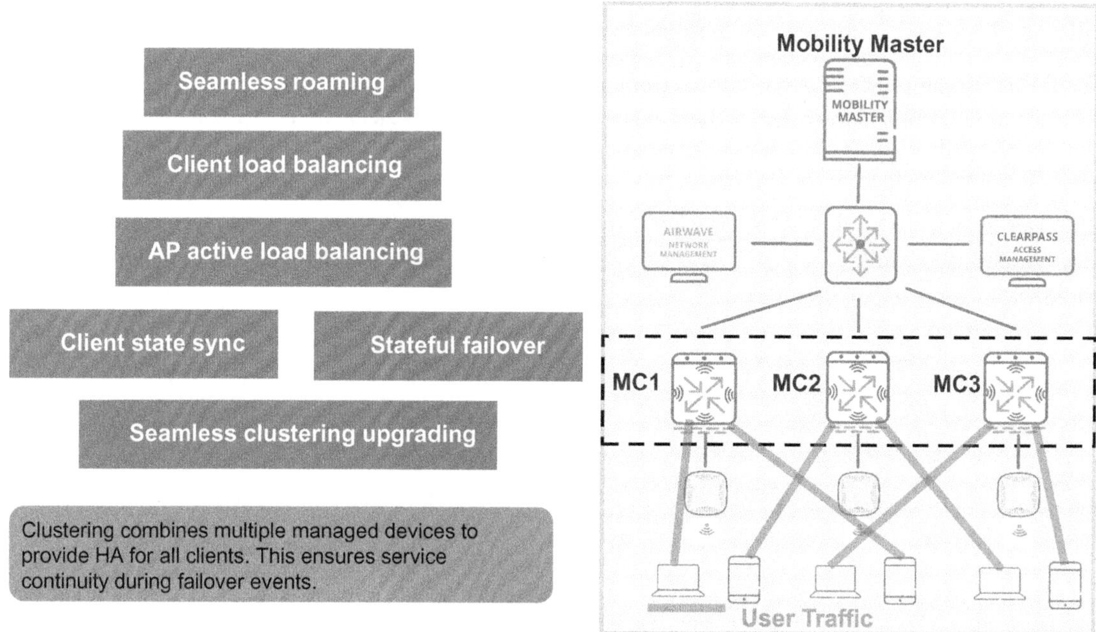

Figure 7-6 What is clustering?

APs are automatically load balanced across cluster members.

Client information is anchored on two controllers in the cluster. Thanks to the full redundancy, in the event of a cluster member failure, the connected clients are failed over to a redundant cluster controller without disruption to their wireless connectivity, or their existing high-value sessions.

The In-service upgrade feature will upgrade controllers and APs in a cluster. There is zero downtime, minimal RF impact and it avoids manual intervention during the upgrade.

As soon as Mobility Controllers are configured as the members of the same cluster, they start handshaking among each other to determine cluster reachability and eligibility. This action ensures all cluster members see each other and that the cluster is fully meshed.

For stateful failover, the cluster needs to be Layer 2 connected. The Live Upgrades feature allows the managed devices and APs in a cluster to automatically upgrade the software from AOS 8.1.x versions to higher AOS versions. Customers can now upgrade their local controller cluster from AOS 8.x to 8.y to benefit from some of the newer features or bug fixes in AOS 8.y without having to manually upgrade each controller in the cluster.

Cluster redundancy

Aruba designed the AOS 8 clustering feature primarily for mission-critical networks. As Figure 7-7 shows, its goal is to provide full redundancy to APs and Wi-Fi clients alike in case of a malfunction of one or more of its cluster members. Clustering supports seamless campus roaming, client stateful failover, client load balancing, AP load balancing, and seamless upgrades.

For AP redundancy, the AP has connections to the Active AP Anchor Controller (A-AAC) and the Standby AP Anchor Controller (S-AAC). The cluster leader will elect these controllers.

User redundancy in a cluster can survive an MC failure.

The user associated to the AP is associated to the User Anchor Controller (UAC) and the Standby User Anchor Controller (S-UAC). You can select these by using a Bucket map table downloaded to all APs.

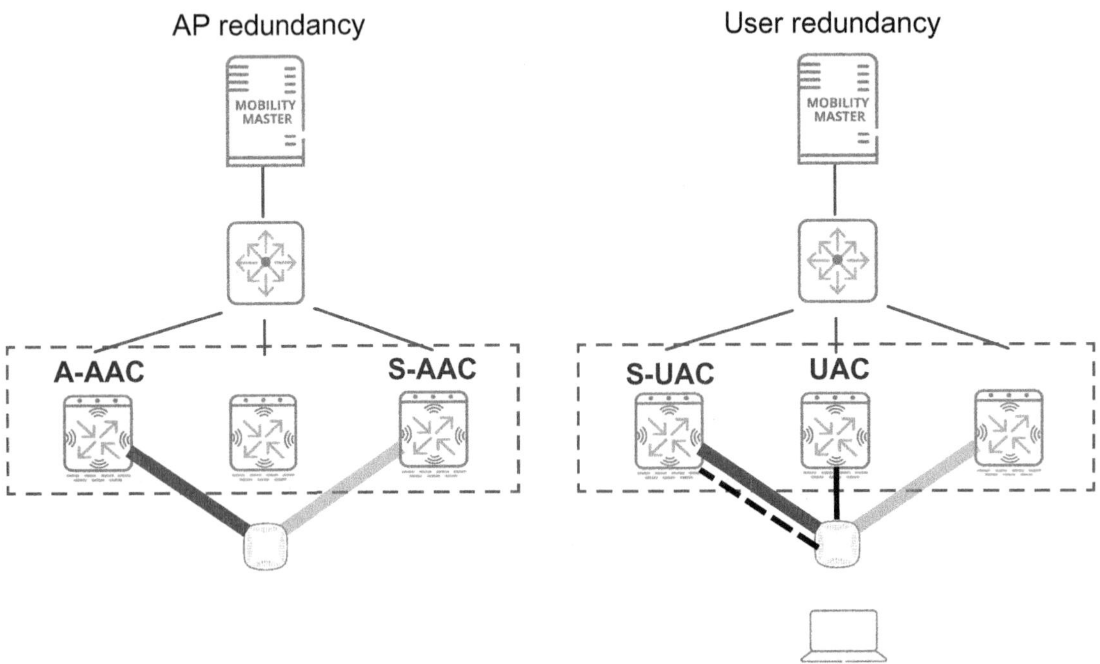

Figure 7-7 Cluster redundancy

The UAC is a role that is allocated to cluster members on an individual user basis, and the S-UAC is its standby. Any cluster controller can be the UAC or S-UAC for a client. The bucket map table helps distribute clients across the cluster controllers.

Upon associating to the AP, a client's UAC Mobility Controller is determined, and the AP uses the existing GRE tunnel to push the client traffic to its UAC. If no GRE tunnel exists, the AP will create a dynamic tunnel to that client's UAC. When that client roams to a different AP, the original AP tears down the dynamic tunnel if no other clients are using it.

If a user roams to a new AP, the AP will follow the same tunneling guidelines as above to push traffic to the same UAC.

A load-balancing scheme makes use of the client load as a percentage of the platform capacity (current client load divided by total platform capacity).

There are three load-balancing thresholds:

- Active Client Rebalance Threshold (default 50%)—the active client load threshold on any node
- Standby Client Rebalance Threshold (default 75%)—the standby client load threshold on any node
- Unbalance Threshold (default 5%)—the minimum difference in load percentage between the max loaded and minimum loaded cluster members

New APs are distributed among cluster members by following the 1% platform capacity logic. The first 1% goes to the active-master/LMS controller, then the master/LMS controller distributes the second 1% to another member, third 1% to another member, and so forth.

Once all APs are connected to the cluster, the load-balancing thresholds will kick in and adjust as needed.

The APs are notified of their AAC via the node list. The node list sent to each AP will depend on the 1% rule.

If a cluster member reaches 50% AP load capacity and the difference between most loaded and least loaded managed devices is 5%, the AP will be redistributed to another cluster controller.

Note that, if you are implementing per-user tunneled-node on Aruba switches, the switch can use a cluster as their tunneled-node server. Much like an AP, the switch establishes tunnels to its active anchor controller and standby controller. The switch also has a bucket map table to help it distribute tunneled-node users to an A-UAC and an S-UAC.

Wired redundancy

You will now focus on wired redundancy. In addition to the various types of wired redundancy, you will learn how redundancy and resiliency differ and why both are essential for wired network redundancy.

Types of wired network redundancy

You can enhance network availability with several general types of wired network redundancy (see Figure 7-8).

Redundant links ensure multiple physical paths between endpoints on a network. In campus LANs, link redundancy generally extends to switch-to-switch uplinks, but rarely goes beyond. Each user's computer has only one connection.

Device redundancy ensures that even if an entire switch fails, network connectivity continues uninterrupted. Device redundancy is often coupled with link redundancy. For example, rather than the access layer switch having two links to the same core switch, it has one link to two different core switches. To provide high availability, implement device redundancy at the aggregation and core layers. If you have link redundancy but no device redundancy, it is best practice to at least provide as much redundancy as possible by connecting the links to different modules on a modular switch. The same best practice extends to VSF links between modular 5400R switches. Assign at least two physical links on different modules to the VSF link.

Redundant hardware components allow a switch to continue to function even if a specific piece of hardware fails. For example, modular switches such as the 5400R and 8400 series switches might have two management modules or multiple power supplies. You can partially offset the vulnerability of a device deployed without redundancy by adding redundant components. This approach can improve availability at the access layer for endpoints deployed without multiple links. You can also use redundant components to improve redundancy for switches deployed as part of a redundant pair, but this might be less important unless the customer has the highest availability requirements.

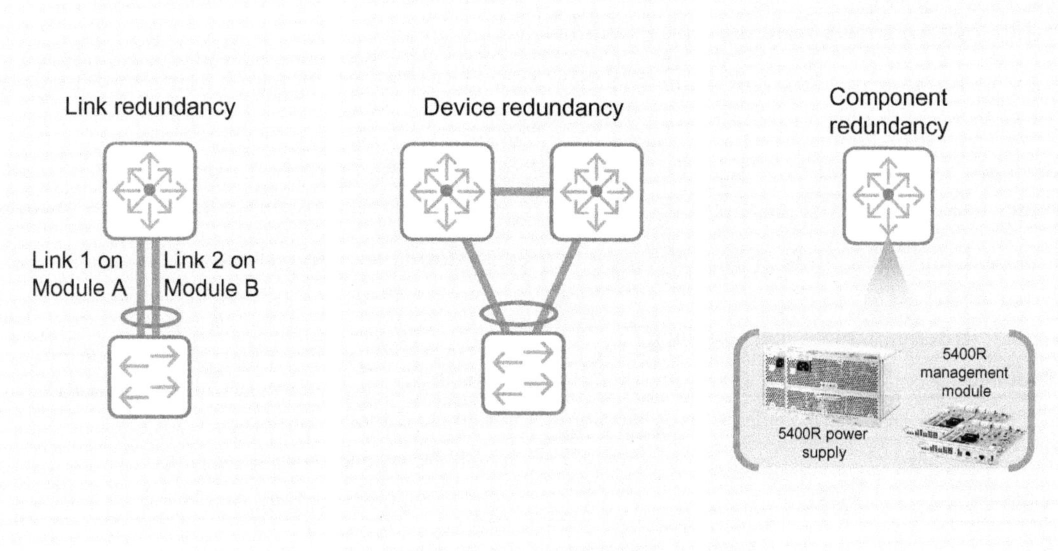

Figure 7-8 Types of wired network redundancy

Installing two power supplies does not always deliver adequate resilience. The switch might require multiple power supplies simply to support the number of modules that you have installed. You should refer to the switch's technical specifications to determine power requirements. To provide redundancy, exceed the required number of power supplies by one.

For Aruba 8400 Series switches, you select fabric modules. The number of fabric modules required to maintain adequate performance depends on the types of interface modules and the amount of traffic that the switch supports. Calculate these requirements and then add one additional fabric module to provide for redundancy.

You can also encourage the company to store backup components so that it can quickly replace failed components. A well-designed solution uses consistent devices across the line so that the same components can back up many devices.

Redundancy versus resiliency

To ensure high availability, you must confirm that the network not only provides redundancy but also resiliency.

Redundancy provides multiple components that fulfill the same function, but resiliency enables the switch to take advantage of those components. Resiliency is the ability to quickly adapt to change and to recover from link and hardware failures. The two concepts are closely entwined. Redundancy provides the foundation for resiliency, while resiliency automatically adapts to failures, ensuring that the redundant components do not go to waste.

Redundant links become resilient when the proper technologies enable switches to use multiple links and to quickly failover from one link to another.

These technologies include LACP and MSTP at Layer 2. At Layer 3, dynamic routing protocols such as OSPF can also play a role in making links resilient.

Similarly, redundant devices become resilient when you configure technologies such as VSF, backplane stacking, or VRRP to enable the switches to act together to provide services such as routing.

Table 7-1 highlights some of the differences between redundancy and resiliency.

Table 7-1 Redundancy versus resiliency

Redundancy	Resiliency
Multiple components that fulfill the same function	Ability to adapt to changes
Redundant links	LACP, MSTP, and dynamic routing
Redundant core switches	VSF, backplane stacking, or VRRP

The result of any failure is service failure. End users, be they company employees, partners, or customers visiting the company's web site, fail to receive the desired service. Users do not care whether the service is unavailable because the only physical link in a network path has failed or because a failed link between routers caused a routing loop that is misdirecting traffic. To users, the result is the same.

A resilient network will protect against service outages and ensure uptime, not only in network links but also in network services and business-critical applications. Resiliency can include Layer 4 and Layer 5 and beyond. Rapid failover can help to support applications that are more sensitive to lost traffic, but the network infrastructure solution is mostly concerned with sustaining availability through Layer 3. Server clustering and load-balancing solutions take over primary responsibility for ensuring service and application availability. However, these considerations are outside the scope of this study guide.

Ensuring resiliency in a traditional design

A traditional network requires Rapid Spanning Tree Protocol (RSTP) or Multiple Spanning Tree Protocol (MSTP) to block redundant links to eliminate loops, as well as to handle failover to a redundant link if the active link fails (see Figure 7-9).

If the customer requires a degree of load-balancing across the links, you should deploy MSTP. In this example, you are planning a two-tier network, so the MSTP region extends across the network. You should assign some VLANs to one instance and other VLANs to another instance. Then set priorities on the redundant core switches such that one is the root for one instance and the other is the root for the other instance.

The core switches must implement VRRP so that they can work together to provide default router services for VLANs. You should align the MSTP roles and the VRRP roles. Use VRRP priorities to ensure each core switch is the VRRP master in its VLANs. Finally, the core switches should use a dynamic routing protocol to communicate with devices in other segments such as the data center. (Redundant static routes could provide an alternative.)

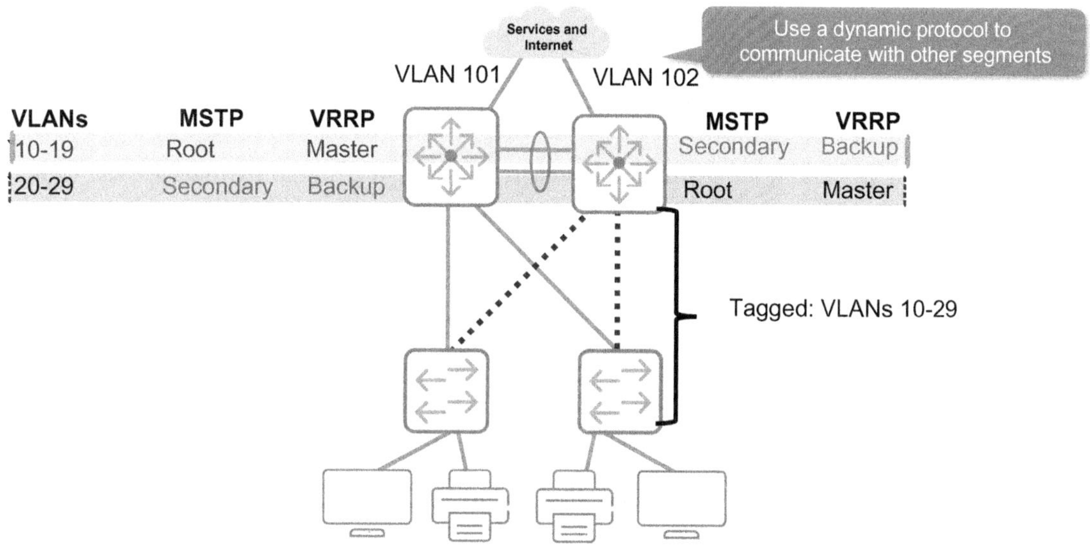

Figure 7-9 Ensuring resiliency in a traditional design

More complete recommendations for designing spanning tree and VRRP follow. These are provided for your reference if you need to deploy a traditional network. In most cases, though, you will use the more modern VRF or backplane stacking-based design.

Spanning tree

AOS-Switches support several different spanning tree protocols, including Rapid Per-VLAN Spanning Tree Plus (RPVST+), RSTP, and MSTP. RPVST+ is a proprietary Cisco protocol and is recommended on AOS-Switches only when you need to integrate those switches into a heterogeneous network that already uses RPVST+ or PVST+. RPVST+ imposes a high-processing burden on switches than the standard spanning tree protocols because it uses a separate Bridge Protocol Data Unit (BPDU) and creates a separate tree for each VLAN.

To choose between RSTP and MSTP, consider whether the customer truly requires active use of multiple redundant links. If so, you should use MSTP. If not, RSTP can be simpler to design.

Then use the following recommendations to avoid some of the most common problems:

1. Keep the topology simple. Ideally, limit the hops between each switch and the root to two. Also minimize loops within loops. Flattening the architecture from three-tiers to two-tiers helps in this endeavor. If you are working with a three-tier topology, implement routing at the aggregation layer. The pair of aggregation switches act as the MSTP roots. Avoid implementing STP across multiple segments in a WAN; route traffic between the segments instead.

2. Explicitly set the priority of the primary root to 0 and the priority of the secondary root to 1 (4096). You might want to set which switch has priority 2, but this is not necessary.

3. Make sure that all devices use the same standard for setting costs, ideally the 802.1t standard.

4. Make sure that the failure of a link never isolates a VLAN. Redundant switch-to-switch links should always carry the same VLANs. This prevents issues if one link fails. Also avoid issues if the link fails between the MSTP root switches. If some VLANs are missing on the new path between the root switches, the root switches, which typically also implement VRRP, take over as VRRP master in those VLANs. The resulting IP address conflict can disrupt connectivity throughout the network. In the simplest and typically recommended setup, all switch-to-switch links in the MSTP region carry *all* VLANs.

   You should assess whether this setup will cause issues in your environment. If the network has many VLANs, it might congest the switch-to-switch links with excessive broadcasts. In that case, you can try another solution (ideally, you can implement VSF or backplane stacking and avoid the issue entirely, but these guidelines are for circumstances when you cannot do so). You can explicitly define the backup path for each instance when a core-to-core link fails by configuring a low cost for the desired instances on the port or link aggregation that connects to the root for that instance. You may also set a favorable priority for this instance on the link in case another switch somehow offers an equal-cost root path. Then make sure that the backup path supports all VLANs in this instance. You could also establish a highly resilient link between the MSTP root switches.

5. If you are concerned about uneven link usage (one link cannot handle the full traffic load on its own), follow these steps:

   a. Implement MSTP.

   b. Minimize the number of MSTP instances. There is no reason to create a new instance in an MSTP region unless you plan to have a new switch act as the root. Usually, two instances meet the need. Then you have two switches, each acting as the primary root for one MSTP instance and as the secondary root for the other MSTP instance.

   c. Make sure that all switches use the same MSTP mappings so that they establish the same MSTP region. This means that if you add a VLAN to one switch, you must add it to all switches and map it to the same instance, even if some switches' uplinks do not carry the VLAN.

   d. Assign VLANs to instances based on the expected traffic load.

6. Understand that instance 0, or the IST, is the instance that switches within an MSTP region use to connect to the Common Spanning Tree (CST). The CST establishes a tree between MSTP regions and between switches that implement different versions of STP (including Cisco's PVST+ or RPVST+). Even if you are not planning to carry traffic on instance 0, you should set the priorities for the switches that you want to be root and secondary root in this instance. This practice provides better behavior in case of misconfiguration of MSTP region settings.

VRRP

VRRP enables clients to use a single IP address for their default gateway address. This address is either a virtual IP address shared by two (or more) VRRP routers, or an address owned by one of the routers, but used by the standby router in case of the owner's failure. Aruba recommends that you use the former setup: each VRRP switch has its own IP address separate from the virtual IP address. Although VRRP enables two (or more) routing switches to present a single virtual IP address for the purposes of routing a VLAN's traffic, the switches act as independent devices with their own control planes and routing processes.

When you need to design VRRP, follow these best practices:

1. Align the VRRP master role with the MSTP primary root bridge and the VRRP back role with the MSTP secondary root bridge.
2. Use preempt mode so that, after a failed default router is restored, that router takes over part of the routing burden again. However, set a preempt delay so that the restored device can discover routes before assuming the master role. Typically, a preempt delay of 15 to 20 seconds is more than adequate, depending on the number of OSPF neighbors that the device has and the complexity of the network.
3. Think about the best way to set up the routing protocol so that failover occurs seamlessly. You should connect each routing switch in the VRRP group to the upstream router in a different VLAN that is dedicated to that link. These VLANs should not run VRRP. If the network is using OSPF, each VRRP switch will establish an OSPF adjacency with the upstream router.
4. Allowing the upstream router to use OSPF ECMP to send traffic to both members of the VRRP group might cause issues for some applications (the traffic sent to the standby member would need to take a longer route, flowing back over the link between the VRRP members, assuming that MSTP roles are also aligned.) Test the solution with real applications, and, if necessary, set costs on VLAN interfaces so that the upstream router prefers the VRRP master of each subnet as the next hop.
5. Rather than use OSPF on both VRRP members, some administrators prefer to avoid these complexities by configuring floating static routes on the upstream router or routers. One route to each VRRP subnet has a lower administrative cost for the master and another route to that subnet has a higher administrative cost for the standby router.

6. You can also configure a track on the VRRP switches. The track monitors connectivity with a key upstream resource and lowers the VRRP priority if connectivity is lost. Then, if the VRRP master loses upstream connectivity, but still has connectivity on the VLANs for which it is master, it concedes the master role to the standby router. The traffic then continues to reach its destination.

Ensuring resiliency in a VSF or backplane stack-based design

VSF and backplane stacking provide the best foundation for all redundant architectures. When you use these technologies, spanning tree and VRRP are no longer required to provide resiliency. To ensure link resiliency in these architectures, you simply need to create link aggregations, keeping in mind that you can combine links on different members in the same link aggregation. The fabric or stack also automatically ensures resiliency at the device level as long as you follow the best practices.

For example, if you configure a VSF fabric at the core to act as the default router for VLANs as shown in Figure 7-10, it automatically continues to provide these functions if the commander fails. It also acts as a single entity in connecting to upstream devices. In the example shown in Figure 7-10, the core VSF fabric uses a link aggregation to connect to the data center core. Failure of a link here does not require routing failover; instead, failover occurs as a simple failover of a link within an aggregation.

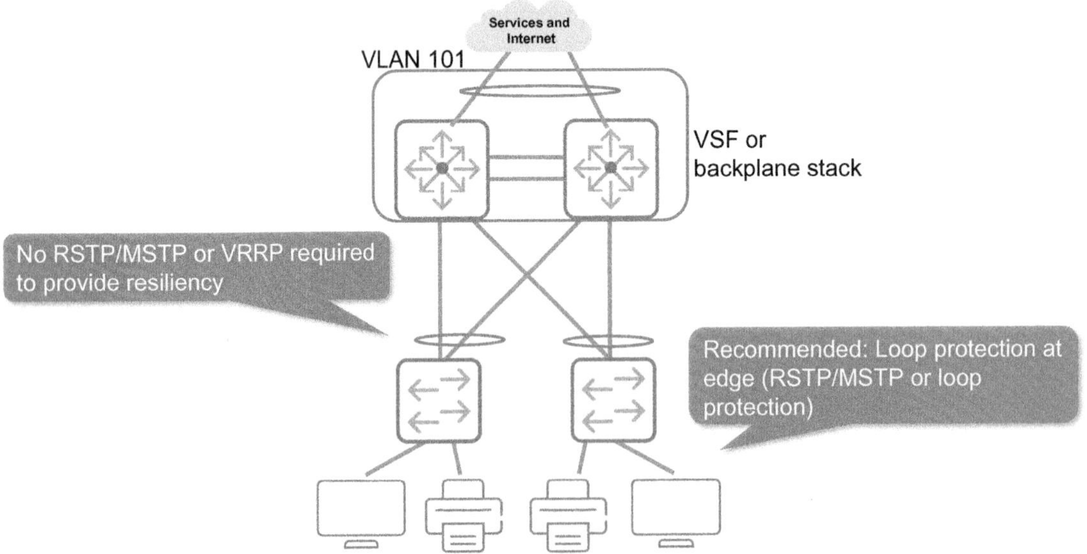

Figure 7-10 Ensuring resiliency in a VSF or backplane stack-based design

Keep in mind that you should enable some form of loop protection even though you generally have a loopless topology.

Otherwise, the network might be brought down if someone connects an unauthorized switch or another misconfiguration creates a loop. You can enable spanning tree, loop protection, or both. The loop protection mechanisms will not actually block any links during normal operation in this topology. They are simply preventing accidental loops on edge ports.

Although spanning tree and loop protection can fulfill a similar function in this topology, the technologies do differ. Spanning tree takes the approach that new ports represent potential loops and block traffic until it has detected that no loop will occur (the port is selected as a root or designated port). Loop protection takes the approach that the switch can forward traffic on a port unless it has detected a loop on it. Each approach has its advantages, and you can select the technology that you prefer.

Comparing the designs

Take a moment to consider how the high availability VSF or backplane stacking provide is superior to the high availability VRRP and MSTP provide. You are specifically considering a situation in which the default router at the core (or aggregation layer) uses VSF or backplane stacking. However, the access layer could be using these technologies as well. Table 7-2 highlights some of the factors you should consider.

Table 7-2 Comparing VSF/backplane stacking design with VRRP + MSTP design

	VSF or backplane stacking on default router	VRRP and MSTP
Is traffic load-shared over redundant switch-to-switch links?	Yes	To a degree (on a VLAN basis)
Is the routing burden distributed across redundant default routers?	Yes	To a degree (on a VLAN basis)
How many IP addresses are used by the default router per subnet?	One	Three—One per VRRP member + one virtual
How do you set up routing protocols?	Setup similar to single routing device (plus graceful restart)	More complicated setup to ensure proper routing and fast failover
How quickly does failover occur?	Less than a second	Seconds

VSF and backplane stacking provide better load-sharing. Because these technologies enable a link aggregation-based design, traffic is load-shared over redundant links on a per-conversation basis. Similarly, VSF and backplane stacking allow both members in a redundant pair to actively route traffic (even though only one of the members builds the routing table, both can use it). VRRP and MSTP only permit load sharing and distributing of routing functions at a per-VLAN basis.

Setting up the VSF fabric or backplane stack to support routing redundancy is simple. The fabric or stack has one IP address on the subnet. Administrators configure IP addresses and routing features in exactly the same way that they would set up a single routing switch. Some routing protocols do require graceful restart, though. With VRRP, each member typically has its own IP address in addition to the virtual IP address. In addition, deploying VRRP involves planning carefully to avoid issues. The routing design is much more complex because each VRRP switch must participate in the routing solution.

Finally, an architecture that uses VSF and backplane stacking provides much faster failover. The network can recover in less than a second from a link failure or a complete device failure. With VRRP and MSTP, link failover occurs in about a second. The VRRP backup router must wait for messages to time out before it takes over as router; default settings failover occurs in about three seconds.

In summary, if available, VSF or backplane stacking is the best way to implement wired network redundancy in terms of resiliency, simplicity, and performance.

Another key advantage of the VSF or backplane stacking-based architecture is that it is more stable than an architecture based on spanning tree.

Troubleshooting STP, RSTP, and MSTP can be very difficult. A loop can introduce all sorts of connectivity issues, and administrators might not immediately identify the cause as a loop. This is because the problem spreads throughout the wired network and can be difficult to trace back to the source. Administrators do not always understand that RSTP and MSTP establish topologies independently of VLAN assignments. This sometimes makes it difficult for them to understand the network's behavior and why one link is or is not blocked. In addition, sometimes a loop occurs intermittently, making troubleshooting even more difficult. For example, someone has introduced a nonenterprise switch, which does not properly prioritize BPDUs. As congestion occurs, the switch loses BPDUs, causing it to unblock a link and create a loop. This problem occurs intermittently and is difficult to trace.

Ensuring routing resiliency

A redundant physical and Layer 2 topology goes far toward supporting routing resiliency. Figure 7-11 shows a network that uses VSF or backplane stacking at the aggregation layer as well as link aggregations, indicated by the links surrounded by circles. However, you also need to ensure that the Layer 3 solution is resilient. In a network with multiple paths, dynamic routing with OSPF provides the best convergence and failover capabilities.

As one best practice, enable OSPF graceful restart on any device that might continue to function when its primary control plane has failed. Such devices include modular switches with redundant management modules and VSF fabrics. Also, make sure to set the OSPF restart timer longer than the time required for the switch to re-establish the routes but lower than the switch's redundancy switchover timer.

Even in a well-designed topology, a link failure can cause a slight disruption. To minimize convergence time, you should minimize the time that it takes OSPF to detect a failed link. You can do so by designing direct, point-to-point wired connections between routing devices. In other words, each link or link aggregation between routing devices has a VLAN and subnet dedicated to that link alone. If the link fails, the neighbor relationship fails, and routes reconverge. Segmenting each link between the Layer 3 switches into its own VLAN also eliminates the need to run spanning tree over the links if you have redundant links, such as the link assigned to VLAN 103 in this example.

If you must extend a VLAN over multiple devices, consider using OSPF Bidirectional Forwarding Direction (BFD) in echo mode. Echo mode BFD enables OSPF neighbors to detect when they lose connectivity with a neighbor in under one second. The Aruba 3810, 5400R, and 8400 switches, which are targeted for the aggregation and core layers, support this feature.

- Dynamic routing with OSPF for networks with multiple paths
- OSPF graceful restart:
  - Implement on switches with redundant management modules, on VSF fabrics, or on backplane stacks
  - Adjust restart timer and redundancy switchover timers
- A dedicated VLAN for each link or link aggregation between routing devices
- OSPF BFD if VLANs must extend over multiple devices

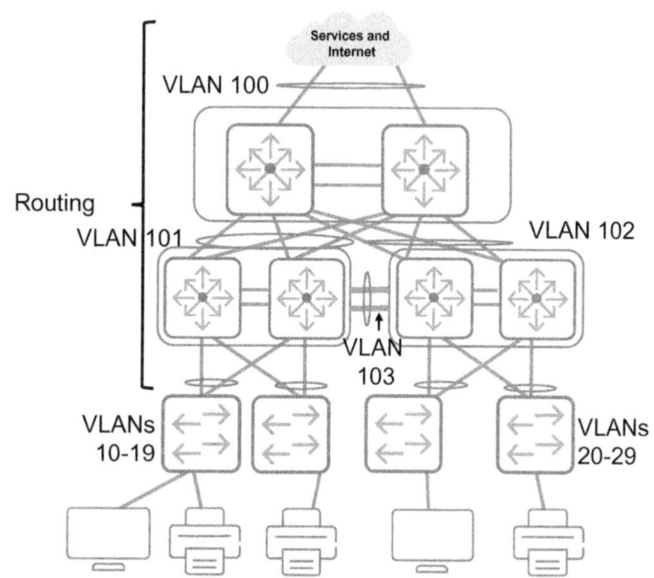

Figure 7-11 Ensuring Routing Resiliency

Supplemental information on OSPF graceful restart

OSPF graceful restart ensures nonstop routing when a VSF or backplane stacking master fails. The same holds true for a switch with a standby management module, but this section refers only to the VSF or backplane stacking master.

The master maintains routing tables and routing protocol information in its control plane. The master proxies this information to other members' control planes, but if the master fails, the routing processes must restart on the new master.

The active VSF members have routes that are proxied to the Forwarding Information Databases (FIBs) in their control planes. They can continue to route traffic with these state routes until the new master builds the routes in its own control plane—and, when the VSF virtual switch is using static routes or Routing Information Protocol (RIP) routes, it does.

However, standard OSPF does not allow OSPF neighbors to entirely resynchronize the LSDBs without tearing down the old adjacency relationship. Therefore, when the new master sends hellos to OSPF neighbors and tries to re-establish adjacency, the VSF virtual switch's neighbors remove the VSF virtual switch's LSAs from their LSDB and notify other OSPF routers to do the same. Routes reconverge away from the VSF virtual switch—even though the switch is perfectly capable of routing the traffic—causing unnecessary disruption to the traffic flow and possible interruption to service.

OSPF graceful restart, on the other hand, enables the VSF virtual switch to ask its neighbors to maintain the neighbor and/or adjacency relationship undisturbed while the new master restarts the OSPF process. Thus traffic continues toward its destination undisturbed.

AOS-Switches always act as restart helpers. You must enable graceful restart manually though by enabling nonstop routing for OSPF. Then adjust the OSPF restart timer to exceed the time required to re-establish adjacency with the neighbors, run the Shortest Path First (SPF) calculation, and repopulate the routing table. Also make sure that the redundancy switchover timer is higher than the restart timer. The OSPF restart time indicates when the switch ends the restart period and sets the newly calculated routes as active. The redundancy switchover timer sets when the switch flushes stale routes. Therefore, if the restart timer is higher than the switchover timer, the switch is still using stale routes and connectivity could be interrupted when the switch flushes them. Set the redundancy switchover timer to about 50% more than the restart time (including the time to run the Shortest Path First (SPF) calculation and repopulate the routing table).

Supplemental information about minimizing link failure detection time

Support for BFD was added in for the 3810 and 5400R Series in KB16.01 software. 5400R switches require v3 modules to support the feature.

When you cannot use BFD, you can still obtain faster than default convergence by setting the OSPF hello timer as low as possible, which is one second. Also make sure to adjust the dead time accordingly. It should be four times the hello timer setting.

Activity 7: Recommended Wi-Fi design and redundancy strategy for the new wireless and wired network

You will now compete a design activity in which you apply what you have learned in this chapter. This activity builds on the scenario introduced in earlier chapters.

In this activity, you will determine the customer requirements for redundancy in the new network and plan to meet those requirements.

Scenario redundancy

You will first learn about the redundancy in the customer's current network.

Current redundancy

On the wireless side, Corp1 has no redundancy. If the previous Wi-Fi controller fails, it takes down a major portion of the wireless network. In the past, this was not an issue, but today, this is a critical failure.

Corp1 has two network closets in the main corridor of every floor. The closets are equipped with three 48-port switches, which you have planned to upgrade. There is no redundancy on these switches.

Current wired resiliency technologies

The company currently uses MSTP and VRRP to provide Layer 2 and Layer 3 resiliency. The company currently routes between the aggregation layer and the core, and it uses OSPF as its dynamic routing protocol.

Corp1's redundancy requirements

Corp 1 requires redundancy. If a controller should fail, there should be no loss of APs or client connectivity.

The corporation understands that if one of the new access layer switches fails, then all APs physically connected to that switch will also fail. The new design should attempt to minimize that failure.

The company requires switch-level redundancy at the campus core and the aggregation layer, if used. The network should continue to function without interruption if a switch at any of these layers fail.

Activity 7.1: Controller Wi-Fi redundancy

Activity 7.1 objective

You will plan MC redundancy.

Activity 7.1 steps

What are your design recommendations for controller failure?

Activity 7.2: AP Wi-Fi redundancy

Activity 7.2 objective

You will plan AP redundancy.

Activity 7.2 Steps

What are your design recommendations for an AP or POE switch failure affecting APs?

Activity 7.3: Wired redundancy

Activity 7.3 objective

You will plan to meet customers' requirements for redundancy within the wired network.

Activity 7.3 steps

1. Assess the plan that you have created for the updated wired infrastructure. Does it meet the customer redundancy requirements?

   If it does not meet the requirements, you will need to adjust the plan in Iris. You can copy and paste a switch to quickly add it the BOM, or you can adjust the Quantity Multiplier.

2. Will you recommend redundant components for any of the proposed modular switches? Will you recommend any spare components?
3. What technologies will you recommend to ensure Layer 2 and Layer 3 resiliency, and how will you implement these technologies? How will you explain the benefits of your plan to the customer?
4. If your plan for Activity 7.1, 7.2, or 7.23 involves adding anything to the BOM, access your project in Iris and make the changes now. If you do not have access to Iris, you will use the HPE Networking Online Configurator. (Visit http://hpe.com/networking/configurator to access this tool.)
5. Save your project.
6. Select the dollar sign button or select File > Quotation to see your BOM.

Answers to the questions included in activities are listed in "Appendix: Activities and Learning Check Answers."

Summary

Congratulations, you now have the tools and knowledge to implement MM, MC, AP, and wired redundancy within your customer's network.

Learning check

1. What are the two modes of HA redundancy?
   a. Active Backup
   b. Active Active
   c. Active Standby
   d. Active Primary
   e. Active Secondary
2. A user is associated to an AP. The User anchor controller must be the same as the AP's anchor controller.
   a. True
   b. False
3. You plan to deploy a pair of Aruba 5406R switches at the core of a two-tier topology. Which technology should you plan to provide the simplest setup and fastest failover for default router services?
   a. VRRP
   b. MSTP
   c. Backplane stacking
   d. VSF

Answers to the learning checks in this study guide are in "Appendix: Activities and Learning Check Answers."

8 Quality of Service (QoS)

EXAM OBJECTIVES

- ✓ Given an outline of a customer's needs for a single-site campus environment or subsystems of an enterprise-wide network, determine the information required to create a solution.
- ✓ Given a scenario, translate the business needs of a single-site campus environment or subsystems of an enterprise-wide environment into technical customer requirements.
- ✓ Given a scenario, explain how a specific technology or solution would meet the customer's requirements.
- ✓ Given the customer's requirements, explain and justify the recommended solution.

Assumed knowledge

- Switching and routing technologies such as the following:
  - VLANs
  - Link Aggregation Control Protocol (LACP)
  - IP routing, including static IP routing and Open Shortest Path First
  - Link Layer Detection Protocol (LLDP)
  - Power over Ethernet (PoE) and PoE+
- Basic wireless network concepts, such as the 802.11 standards
- Frequencies and channels used by 802.11 and the basics of cochannel interference

Introduction

Quality of Service (QoS) design is critical to delivering high-performance enterprise WLANs for voice and video applications. You need to understand the applications in your customer's environment and the special requirements that those applications might have.

Traffic prioritization based on standard QoS technologies helps to ensure good QoS for time-sensitive applications such as voice and video, as well as for other high-priority and mission-critical traffic.

A good experience for users also rests of solid capacity planning and optimization, particularly for the wireless network. The wireless solution must also provide roaming optimization for mobile devices running time-sensitive application such as voice. You will learn about technologies for optimizing wireless capacity and roaming in this chapter, leading into the discussion of high-density RF designs in the next chapter.

Application requirements

You will begin by learning about how to classify applications and determine the requirements specific to that application.

Application classification

When you collect information about a customer's requirements, it is very important to use this information to assess the customer's applications. Network control traffic should always have high priority, but generally devices automatically prioritize this traffic so you do not need to plan to prioritize it. You do need to identify real-time applications, such as voice and video, which require special handling to ensure a good QoS.

You might also need to classify traffic based on how critical it is to the company's mission.

Figure 8-1 shows several common classes listed in the order of priority, with the highest priority at top. You can also see the types of applications associated with these classes.

Many customers do not require every traffic classes. Often you simply need to make a plan to provide the proper QoS to voice and interactive video traffic, as well as perhaps streaming video. Everything else receives best effort forwarding.

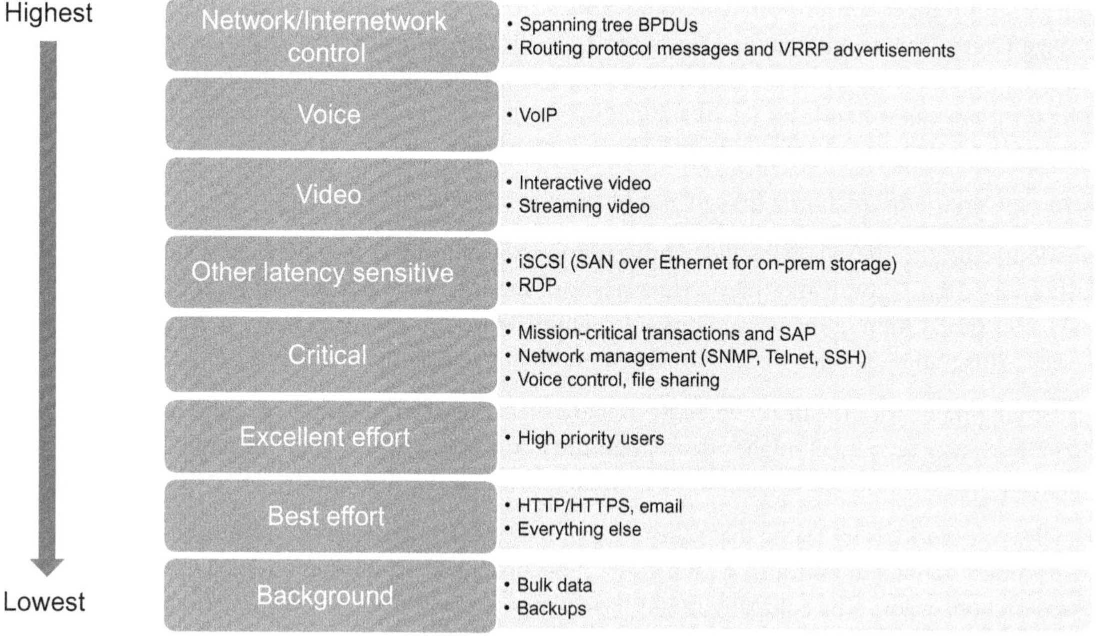

Figure 8-1 Application Classification

Eventually, you will associate each traffic class with a forwarding queue. In the wireless world, devices use one or four queues depending on their capabilities. AOS-Switches typically use two, four, or eight queues. The advantage of using fewer classes, and therefore queues, is both simplicity and, for wired switches, greater buffering space per queue. In addition, the simpler the design, the easier the solution is to deploy and maintain.

For customers who do require more classes, though, you know a bit more about classifying and prioritizing different types of traffic.

As you see, network and internetwork control traffic should have the highest priority. This traffic includes control protocols transmitted by network infrastructure devices themselves. For example, internetwork control traffic includes Open Shortest Path First (OSPF) and other routing protocol messages. Virtual Router Redundancy Protocol (VRRP) advertisements provide another example of internetwork control traffic. The BDPUs used by spanning tree protocol (STP) are network control traffic. All of this traffic require the highest priority because the network cannot ensure delivery for any traffic if these protocols stop functioning correctly.

Multimedia applications, such as voice and video, require the next highest priority. Voice typically requires higher priority than video because it occurs in real-time. You can further divide video into interactive video and streaming video. Interactive video might require higher priority because real-time communications require timely delivery.

The iSCSI protocol carries SCSI communications between an initiator—typically, a server—and a storage target. This technology enables a server to mount volumes from disks hosted on a remote storage array. While you will most often find iSCSI deployed within data centers, some customers do have on-prem storage and run iSCSI traffic over Aruba 5400R and 3810 switches. The iSCSI communications require high priority to prevent excessive latency or packet loss.

Remote Desktop Protocol (RDP) can also be latency sensitive.

Mission-critical traffic might include applications associated with transactional databases as well as SAP applications. Companies may define network management traffic such as SNMP, Telnet, and SSH as critical. Often nonvoice and nonvideo Unified Communications (UC) traffic fits in this class as well. This traffic includes voice control (the traffic for setting up a call) and file sharing.

Classifying traffic into the next two classes depends less on the application type and more on the scenario. For example, a hotel might offer a premium and a best effort service. All traffic from users who purchase the premium service should be granted critical or excellent effort.

Finally, you can identify traffic that takes a lot of bandwidth but does not require timely delivery as background traffic. For example, FTP uses TCP, which can tolerate loss and speed up and slow down based on congestion. The company might have periodic backups that need to execute but can complete anytime during the day without negative effects. You could also define applications that the customer does not care about as background traffic.

Voice requirements for low loss, latency, and jitter

Now that you have classified the customers' applications, you can begin to consider the needs (see Figure 8-2). You need to take steps to protect the higher-priority classes of traffic from the ill effects of congestion: loss, latency, and jitter. Since voice applications tend to be the most sensitive to all three of these issues, you will learn about how these issues affect voice as an example.

Packet loss occurs when congestion prevents the network from delivering all packets. While voice applications can tolerate a few lost packets, too many lost packets will degrade quality and eventually drop the call.

Latency is the amount of time that passes between the sending of a transmission and its arrival at the receiving station. Within a campus network, significant latency typically occurs, not due to the transmission time for signals, but because congestion causes a packet to be buffered before transmission. Latency can also occur if an endpoint's connection is disrupted by roaming or re-authentication. Latency can cause issues for voice because it introduces a slight delay after a person begins to speak. This delay can cause people to talk over each other and sometimes create an echo. The International Telecommunications Union (ITU) recommends no more than 150 ms end-to-end delay for voice traffic.

Because congestion typically occurs in unpredictable bursts, the amount of latency can vary from packet to packet within a stream. This effect is called jitter. Jitter can have a much worse effect on call quality than simple latency because real-time applications are designed to have a consistent packet rate, and jitter causes smaller and greater gaps between packets. It can sometimes even cause packets to arrive out of order. For instance, a VoIP telephone sends a very small packet every 20–30 milliseconds (ms). If the interval between packets grows to 50 ms due to jitter, the transmission will be unsatisfactory for many users.

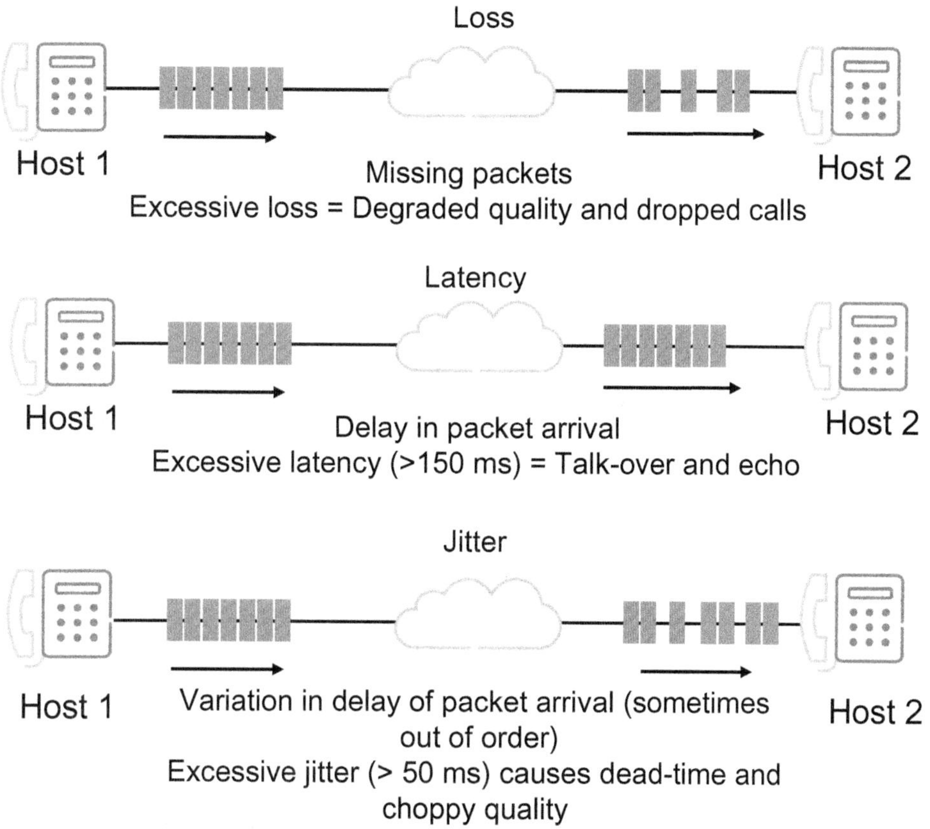

Figure 8-2 Voice requirements for low loss, latency, and jitter

VoIP is particularly sensitive to many of these issues because it is a real-time application that uses UDP. TCP-based applications can recover from packet loss because they use acknowledgments. However, resending the packets does slow down the application. UDP applications do not use acknowledgments and cannot recover lost packets. In a real-time conversation, by the time the packet could be recovered, it is too late for it to be useful.

You should also be aware that some latency is inevitable. While the speed of light in a vacuum is 186,000 miles per second (267,000 km per second), signals move through fiber or copper cable at approximately 125,000 miles per second (200,000 km per second). Consequently, a fiber network stretching halfway around the Earth introduces a delay of approximately 70 milliseconds under the best of circumstances with additional delay introduced by the physical transport medium. However, within a campus, the latency introduced by the physical media is negligible. Instead, you need to worry about the latency introduced by congestion.

If you are planning to support wireless real-time applications such as voice over WLAN (VoWLAN), you should also consider the latency during the re-authentication that occurs during roaming.

Summary of application requirements by class

Table 8-1 summarizes the QoS needs for the common classes of network.

Table 8-1 QoS needs for common classes of network

Common classes	Needs
Network/internetwork control	Low loss, low latency
Voice	Low loss, low latency, low jitter
Video	Low jitter, guaranteed bandwidth
Critical	Low loss, low latency
Excellent effort	Varying needs, high priority
Best effort	Varying needs, normal priority
Background	Varying needs, low priority

Network control protocols require low loss to work correctly. Most protocols are built to withstand one or two lost packets, but not many. For example, if a Standby VRRP member misses three VRRP advertisements, it assumes that the Master is down and takes over as Master. Most network control protocols can tolerate more latency than voice, but most still work best with latency under one second.

Video traffic, like voice traffic, is susceptible to jitter, which can cause a choppy quality. Streaming video can tolerate some latency due to buffering mechanisms. However, excessive latency can negatively affect interactive video due to the same reasons as voice. Video traffic also requires guaranteed levels of relatively bandwidth, which you should take into account when planning capacity.

Critical applications also require low loss and low latency (of course, the applications we mentioned earlier can also be critical. Here, you are examining applications of other types that are critical due to the purpose for which the business uses them).

The last three classes listed in Table 8-1 typically include a wide array of applications, depending on the customer scenario. The main thing that you should remember is that the network should deliver excellent effort traffic before best effort traffic, which should be delivered before background traffic.

If the network does use iSCSI keep in mind that iSCSI is a TCP application, which can recover lost packets and respond to congestion. Nonetheless, the underlying SCSI communication were designed for direct links dedicated to data transfer, and iSCSI operates best when the application experiences low latency, low loss, and low jitter. In addition, iSCSI can consume a high amount of bandwidth, so you should provision links accordingly.

Meet application requirements

As you have seen, in a campus network, congestion causes excessive packet loss, latency, and jitter. Therefore, the first step toward ensuring good QoS is minimizing congestion by planning adequate capacity.

Nonetheless, even a well-designed network features oversubscription and the possibility of momentary bursts of congestion, which can cause excessive damage for time-sensitive solutions. Therefore, you will also need to implement QoS mechanisms to protect this sensitive traffic and ensure that users have a positive experience running all necessary applications (see Figure 8-3).

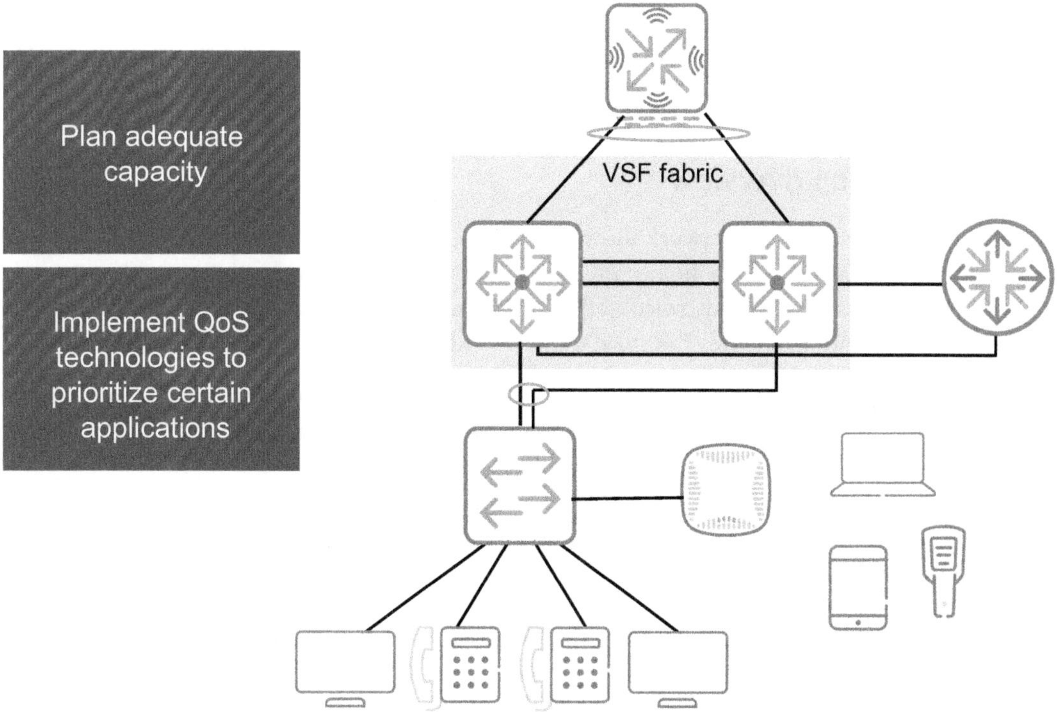

Figure 8-3 Meet application requirements

Switch-to-switch links are often oversubscribed and thus provide less bandwidth for outgoing traffic than the total amount of traffic that could be incoming from other devices—often many times less. Usually, endpoints need to send and receive only a small fraction of their bandwidth, so oversubscription makes sense. However, on a moment-to-moment basis, fluctuating traffic patterns and bursts of traffic can cause congestion. The lower-bandwidth link acts as a bottleneck, and if traffic continues to arrive more quickly than the port can forward it, the port's buffers fill, and it drops traffic.

In other words, even when infrastructure links typically provide ample bandwidth, implementing QoS technologies can be important because even brief periods of congestion can degrade the quality for time-sensitive applications. QoS helps to protect high-priority traffic during such periods of network congestion. For example, the wired infrastructure might provide enough bandwidth all the time except at 9 a.m. when all the employees arrive.

Implementing the proper technologies can also help to improve user experience over wireless connections. These can be subject to greater unpredictability than wired connections due to the shared medium and nature of wireless signals.

Traffic prioritization

Once you have planned proper capacity across the campus, you should add prioritization for the most important traffic to ensure timely delivery for this traffic even if temporary congestion occurs.

Traffic prioritization overview

Traffic prioritization enables the network infrastructure to place different types of traffic in different queues and then forward traffic in higher-priority queues preferentially. In the example shown in Figure 8-4, one wireless device sends voice traffic, and another sends data traffic. Each infrastructure device in the traffic's path prioritizes the voice traffic in order to minimize loss, latency, and jitter.

You will learn about several technologies that help network devices to recognize traffic's priority and communicate that priority to other devices. These technologies include 802.1p, Differentiated Services (DiffServ), and Wi-Fi Multimedia (WMM).

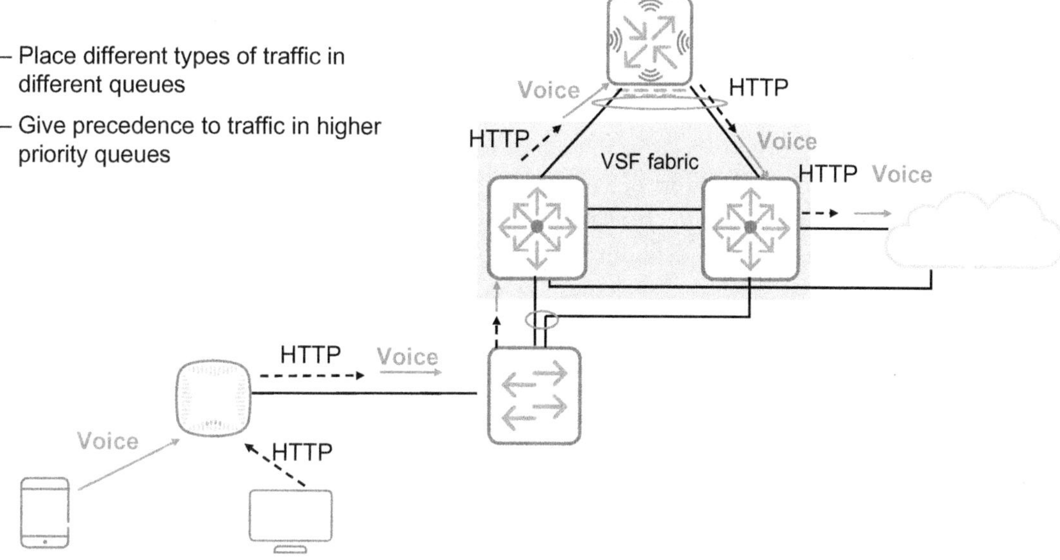

Figure 8-4 Traffic prioritization

You can break down the communications into several segments. It is important to consider how traffic is classified and handled at each segment of its journey to ensure the proper QoS.

First, you need to consider QoS between the wireless device and the AP or a wired endpoint and its switch. If the wireless devices themselves lack support for WMM or another QoS technology, the traffic cannot receive priority handling on this segment of the journey. This segment of the journey is less important for wired devices such as the computers connected to the access layer switch in this example. Each of these devices has its own full-duplex link.

Next, you need to consider the AP's link to the switch and the switch-to-switch links across the network. Congestion can occur on these links because they are typically oversubscribed with multiple downstream devices sharing the links' bandwidth. You should remember that every device in the path can affect the QoS by either correctly or incorrectly prioritizing the traffic.

Finally, remember to consider traffic in both directions of a conversation and ensure that return traffic is properly classified and prioritized as well.

Ethernet prioritization: 802.1p and DiffServ

The 802.1p standard uses a three-bit field in the 802.1Q VLAN tag to indicate priority for any wired traffic, both IP and non-IP. The standard defines eight values between 0 and 7, with 7 as the highest priority value and 0 as the default priority value. Based on the 802.1D standard, values 1 and 2 actually indicate lower than best effort forwarding.

The 802.1p values only persist over wired connections with VLAN tags. Therefore, networks often use the Type of Service (ToS) field in the IP header to communicate traffic priority. Differentiated Services (DiffServ) uses the first six bits in this field to define how to handle the traffic. A DiffServ value is called a DiffServ Code Point (DSCP).

Instead of simply defining priority relative to each other, the 64 DSCPs were intended to define distinct forwarding behaviors, or Per Hop Behaviors (PHBs). However, as implemented across campus networks, a DSCP primarily serves to correlate to an 802.1p value and determine the priority queue for the packet. Figure 8-5 shows the default mappings on AOS-Switches. DSCP value 0 is the default value.

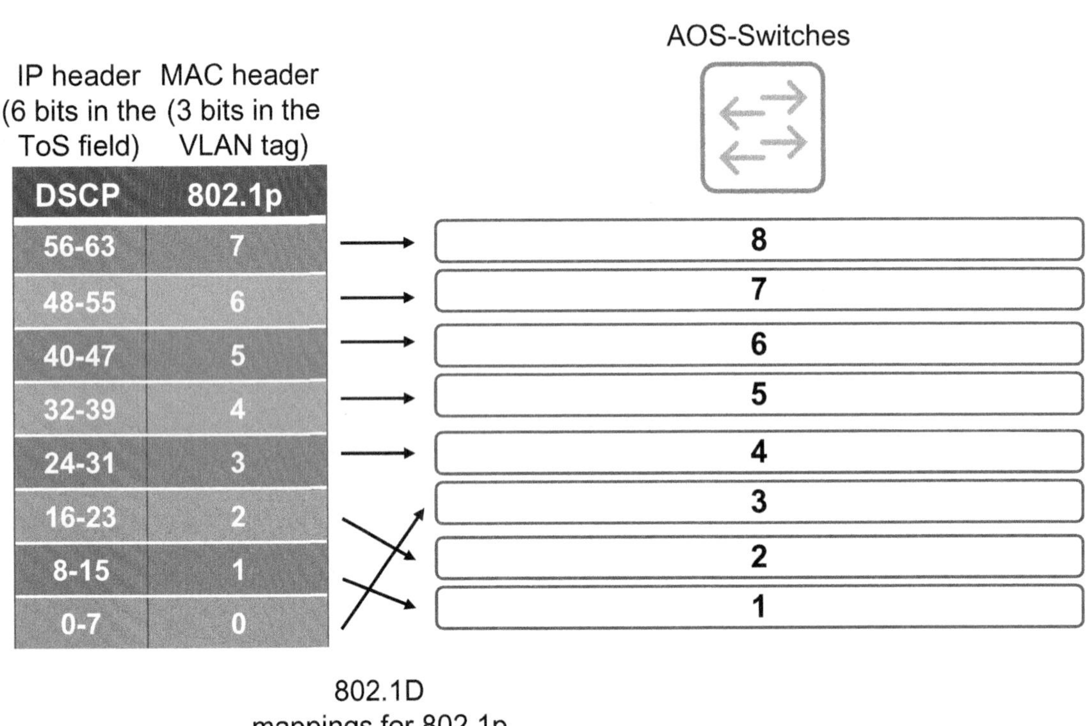

Figure 8-5 Ethernet prioritization: 802.1p and DiffServ

AOS-Switches correlate 802.1p values to queues based on the recommendations in 802.1D, the standard that originally introduced 802.1p. The 802.1Q-2014 standard contains the most recent recommendations, which are similar but assign value 2 a *higher* priority than 0. The 802.1D recommendations match well with the way that wireless devices generally prioritize traffic, though.

All AOS-Switches allow you to adjust the DSCP mappings but only some, such as the 2930F/2930M switches, allow you to adjust the 802.1p to queue mappings. Because applications often use DSCP rather than 802.1p to indicate the value, though, adjusting the DSCP maps can have the intended effect.

Wireless prioritization: WMM

Wi-Fi Multimedia (WMM), which is based on the Enhanced Distributed Channel Access (EDCA) portions of the 802.11e standard, allows wireless devices to divide their traffic into different Access Categories (ACs), or queues. The highest queue is intended for voice traffic, and the next highest for video traffic. The next queue is best effort traffic, and the lowest queue, for background traffic. WMM prioritizes traffic for wireless transmission by using advantageous settings for contending for the medium. The higher the WMM queue, the better the chance a device has to transmit.

Clients and APs can use traffic's DSCP to determine in which AC traffic belongs. Wireless NICs map DSCP values to WMM queues as shown in Table 8-2. For APs and clients within a WLAN to use WMM, the WLAN must have WMM enabled on it. The APs' MC should also define DSCP maps. Recommended ones are shown in Table 8-2, although some additional considerations apply, which you will consider below.

Table 8-2 Mapping of DSCP values to WMM queues on wireless NICs

DSCP typical client mapping	Recommended DSCP map for Aruba solution	WMM queue
48-63	56, 48	AC_VO (voice)
32-47	40, 32	AC_VI (video)
0-7, 24-31	24, 0	AC_BE (best effort)
8-16	16, 8	AC_BK (background)

If the AP connects to devices that do not support WMM, it can use a simpler prioritization method that gives voice traffic higher priority and all other traffic best effort delivery.

WMM dictates the same general behavior as Distribution Coordination Function (DCF), the mechanism by which 802.11 devices ensure that they only transmit when the medium is free. Before transmitting a first frame, stations listen for an IFS. If they detect contention, they must select a random value between 0 and the contention window (CW) and multiply that value by the time slot. After the medium becomes free for an IFS, they wait for the backoff time, possibly stopping the timer and deferring again if they detect contention.

However, EDCA specifies a different set of values for the parameters used by each AC during this process. The values for the parameters are configurable, but you would usually use the default ones, which ensure that AC_VO traffic can transmit before AC_VI, which can transmit before AC_BE, and so on. The sections below describe the parameters.

Arbitrary IFS (AIFS)

Instead of using the DIFS as the time to listen for contention before transmitting, WMM-capable station's use the AIFS for the AC to which the traffic belongs. The AIFS equals the SIFS plus the AIFSN times the slot time. When the AIFSN is set at 2, an AC has an AIFS equal to the non-WMM DIFS. The lower the AIFSN, the better the chance the device has a being able to transmit first.

CWmin

The CWmin defines the value for the CW the first time that a device backs off before it has detected any retransmissions. The lower the CWmin, the better the chance that the station will select a low multiplier for the backoff time, increasing that station's chances of transmitting first when multiple devices contend for the medium. By default, the AC_BE uses the same CWmin value as DCF (non-WMM traffic), while higher priority queues have lower values, and the AC_BK has a higher value.

On Aruba devices, the CWmin is set as an exponential value (ecwmin), which determines the CWmin with this formula: 2n – 1 = CWmin. For example, if you set the ecwmin to 4, the CWmin is 15 (24 – 1 = 15).

CWmax

Similarly, the higher priority ACs have a lower CWmax, which ensures that these queues continue to have a higher chance to transmit even if collisions occur. In fact, the difference between the queues' CW becomes even more marked as the CW approaches the CWmax. Therefore, voice and video traffic has a much better chance of being transmitted first as congestion increases.

The CWmax is also set as an exponential value (ecwmax).

Transmit opportunity (TXOP)

When a device wins control of the medium, it can request a TXOP in the frame. This value determines the amount of time over which the device can transmit without contention. Often VoIP frames are small. The transmit opportunity allows devices that are transmitting smaller frames to transmit more frames, allocating bandwidth more evenly between devices.

Options for where priority is applied

It is important that you understand which devices are responsible for applying traffic with its priority, both so that you can integrate with the customer's QoS schemes and so that you can understand the responsibilities of the solution you are delivering (see Figure 8-6).

Applications on endpoints can mark traffic for a specific DSCP, as can Windows policies for endpoints that are part of a Microsoft domain. If the endpoint is wireless, a WMM-capable wireless NIC will automatically place the traffic in the WMM queue that correlates to that DSCP.

Specialized devices such as VoIP phones can also mark their traffic with 802.1p and DSCP values.

If the endpoints do not have the capability, or are not trusted, to assign the correct QoS values to their traffic, the switches and MCs can classify the traffic. Switches would be responsible for marking wired devices' traffic, while the MC handles wireless traffic. Although these policies can be very sophisticated, particularly on MCs, they cannot take effect until the traffic flows through the device in question.

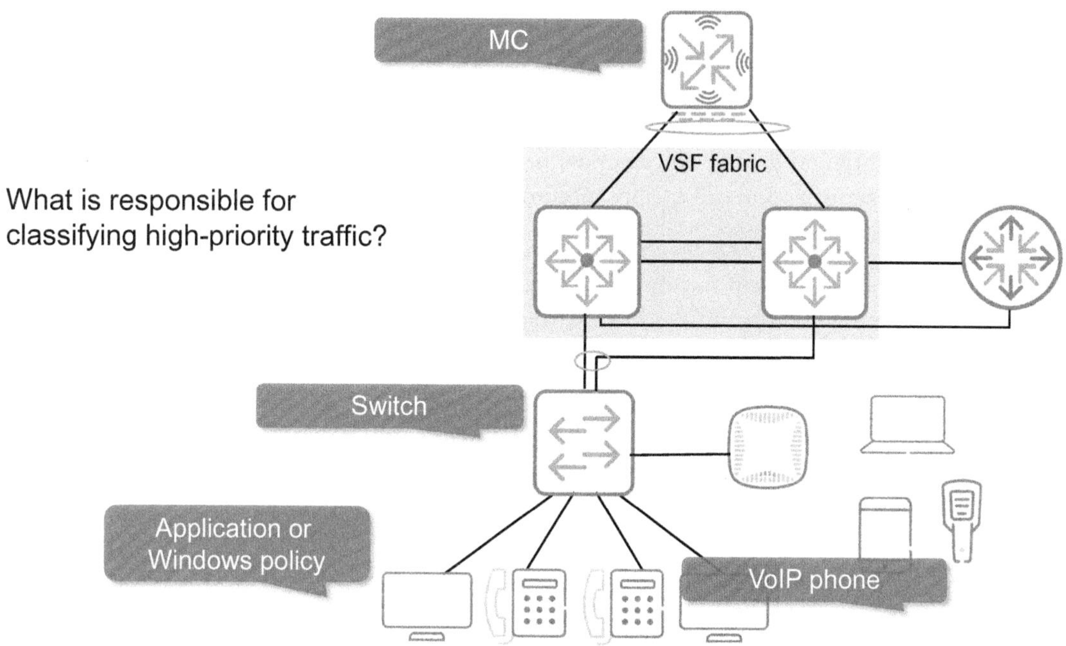

Figure 8-6 Options for Where Priority Is Applied

You can also use a mix of strategies. For example, some applications on trusted servers might mark their own traffic, while the wireless infrastructure classifies traffic from wireless endpoints based on important business priorities as interpreted by QoS policies.

Prioritizing traffic at the endpoint can be particularly crucial for traffic from wireless clients. An Aruba AP will automatically mark traffic with the DSCP that corresponds to the incoming WMM queue and transmit the traffic with the appropriate prioritization onto the wired network. If the traffic arrives in the best effort WMM queue or the wireless client is not WMM-capable, the AP forwards the traffic with the default priority. Although the MC can remark the traffic with the desired priority, the traffic must first travel to the MC at the default priority.

Aruba wireless QoS features

Whether the customer needs the Aruba solution to honor priorities applied elsewhere or to intelligently prioritize traffic, a wide array of features in ArubaOS 8.x improve quality for time-sensitive and mission-critical applications.

In addition to APs' ability to honor traffic's existing priority with WMM, MCs can classify traffic with a variety of sophisticated techniques. Aruba MCs, both standalone or deployed with an MM, support a special voice role to which you can assign users and devices that need to run voice applications. This role comes preconfigured with a variety of firewall policies that ALGs and heuristics to determine precisely which traffic should be prioritized. Customers can also take those predefined firewall policies and apply them to roles of their choice.

For an even more nuanced approach, you can use the SDN capabilities in the Unified Communications Modules (UCM). The SDN controller communicates with the customer's UC application and automatically configures the wireless infrastructure devices to apply the correct priority to traffic. In an MM-based architecture, the MM runs an SDN controller as well as all the UCC-related ALGs, providing better scalability and greater efficiency with less overhead on MCs. UCM on an MM also provides centralized statistics and call quality monitoring. Make sure to recommend an MM to give the customer these features.

Aruba wireless solutions also help to improve performance for wireless voice and video with voice- and video-aware roaming and scanning. Finally, the broadcast and multicast optimization features about which you have already learned help to speed up multicast applications and improve performance in wireless cells generally.

MCs can also enforce bandwidth contracts, which limit bandwidth per-role or per-user. These help to ensure that less important users, such as guests, take only a fair share of their bandwidth.

AOS-Switch QoS features

For the most part, you can be sure that any AOS-Switch that you propose supports the full suite of QoS features.

In addition to using incoming 802.1p and DSCP values to prioritize traffic, the switches can classify traffic on their own based on incoming port, incoming VLAN, RADIUS-assigned role, or Layer 2 to Layer 4 information (see Table 8-3.) Classifier-based policies provide a great deal of flexibility.

AOS-Switches also support ingress and egress rate limiting of traffic, and most switches support class-based rate limiting. They also support LLDP-MED for automatic provisioning of LLDP-MED-capable VoIP phones with the correct VLAN ID, QoS priority values, and location information.

The Aruba 2540 and 2530 Series do have a slightly less powerful feature set. Limitations include no class-based policies. These switches can still classify traffic based on Layer 2 to 4 information; however, they cannot do so as flexibly as the other switches. You might also reconsider the 2530 if the customer needs better congestion management. This switch supports fewer queues and much less deep buffers than the others.

Table 8-3 QoS features on AOS-Switches

Features	5400R, 3810M, 2930M, 2930F	2540	2530
Max queues	8 per-port	8 per-port	4 per-port
Buffers	Deeper	Deeper	Less deep
Classification	• Port and VLAN-based • Dynamic RADIUS or role-based • Application-based using Layer 2 to 4 information and class-based policies	• Port and VLAN-based • Dynamic RADIUS or role-based • Application-based using Layer 2 to 4 information	• Port and VLAN-based • Dynamic RADIUS or role-based • Application-based using Layer 2 to 4 information
Queue scheduling	Strict priority with GMB	Strict priority with GMB	WRR
Rate limiting	Ingress, egress, and class-based	Ingress and egress	Ingress and egress
Other features	LLDP-MED for auto VoIP phone provisioning		

The 5400R, 3810M, 2930M, and 2930F have packet buffers between about 12 and 13.5 MB. Refer to the latest switch datasheets for details.

All AOS-Switches let you adjust the number of queues lower than the maximum. On switches that support eight queues, you can change the number to two or four. On 2530 switches, which support four queues, you can lower the number to two. Sometimes a customer does not require the granularity of eight queues for eight different types of traffic. Lowering the number of queues provides more buffering capacity per queue, helping to prevent dropped traffic if congestion occurs.

Different switches also implement queue scheduling differently. The 5400R, 3810, 2930M, 2930F, and 2540 Series switches support strict priority queuing with a Guaranteed Minimum Bandwidth (GMB) for each queue. Over each service period, the switch ensures that each queue receives at least its GMB, which is a specific percentage of the total bandwidth. By default, the GMBs weighted such that highest priority queues receive more bandwidth and the best effort queue receives sufficient bandwidth. After all queues receive their GMB, some bandwidth might be unused because a queue did not need all of its bandwidth. The switch then allocates unused bandwidth based on strict priority: it serves each queue, starting at the highest, until that queue is empty or the bandwidth is consumed.

The 2530 Series uses Weighted Round Robin (WRR) queuing. In each service period, WRR forwards packets from each queue in turn, but weighting the number of packets that it forwards from each queue based on priority.

If the customer has VoIP phones that support LLDP-MED, you can explain other benefits of LLDP-MED. With LLDP-MED, the switch can send information such as the phone's physical location, which can help to support Emergency Call Services (ECS). When phones receive Power over Ethernet

(PoE) from the switch, LLDP-MED can help the switch allocate and deliver exactly the power that the phone needs. LLDP-MED can also help in tracking and troubleshooting VoIP phones through a central management platform. SNMP servers can read detailed VoIP endpoint data inventory, PoE status, and information to help troubleshoot the IP telephone network and call quality issues.

Considerations when the application marks the traffic: AP to client

In addition to understanding the QoS features available on the products that you are proposing, you need to collect some information from the customer about whether and how their applications mark traffic for high priority. This will let you know to adjust settings on MCs or recommend that the customer use new values.

First, you will examine a scenario in which two users have a video conference using Microsoft Skype for Business. As Figure 8-7 shows, one user has a wireless connection, and the other user has a wired connection.

WMM is enabled on the SSID, as recommended to support voice. And the wireless implementer has configured DSCP mappings as recommended by Aruba. However, no firewall policies or SDN integration are configured on the MC, and the scheme does not match the customer scheme.

The laptops have Windows group policies that tell them to transmit audio traffic with DSCP 46, a commonly recommended DSCP. However, as you remember, DSCP 46 corresponds to the video AC. The DSCP recommendations and the WMM recommendations are simply out of sync.

With no mapping for DSCP 46, the AP passes on the traffic with the DSCP for the Video AC.

Worse, because the MC has not been set up to remark voice traffic, the voice traffic arrives with DSCP 46 on the AP. The AP does not recognize its priority and sends it in the best effort AC.

DSCP values 16 through 23 have a similar issue. DSCP classifies this traffic as excellent effort, while WMM puts it in the background queue.

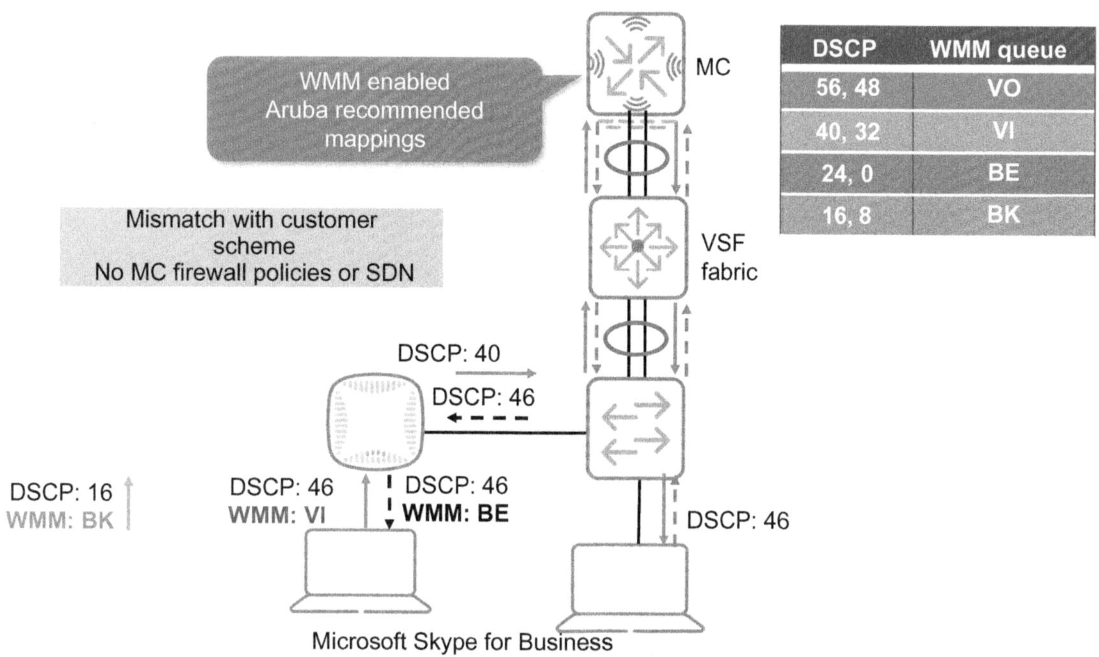

Figure 8-7 Considerations when the application marks the traffic: AP to client

AOS-Switches, by default, also map DSCP values 16 to 23 to a lower priority queue than best effort.

Keep in mind that similar issues can occur with other mismatches in the solution from the customer requirements. If WMM was not enabled, neither clients nor APs could send traffic in a higher priority WMM queue. If WMM was enabled but no mappings configured on the MC, the AP would keep the incoming DSCP from the wireless client. But it would not forward traffic with DSCP 46 with a high priority to wireless clients.

Solution 1: Work with the customer's scheme

You will now look at recommendations for ensuring that your solution works with the customers' solution.

In the first approach shown in Figure 8-8, the MC has mappings that correspond to the DSCP values that the customer is already using. It also has firewall policies to assign voice traffic to the high-priority queue.

However, the client itself continues to send the traffic in the video queue; the Aruba solution settings cannot change this behavior. The AP sees that the traffic arrives in the video queue, and its DSCP does not match a mapping for this queue. Therefore, the AP remarks the traffic to DSCP 40. When the traffic passes through the MC, though, the MC can use its firewall policies and SDN integration to remark the traffic for DSCP 46 and the high-priority queue.

When traffic marked with DSCP 46 and destined to a wireless client arrives on the AP, the AP forwards the traffic in the correct voice queue.

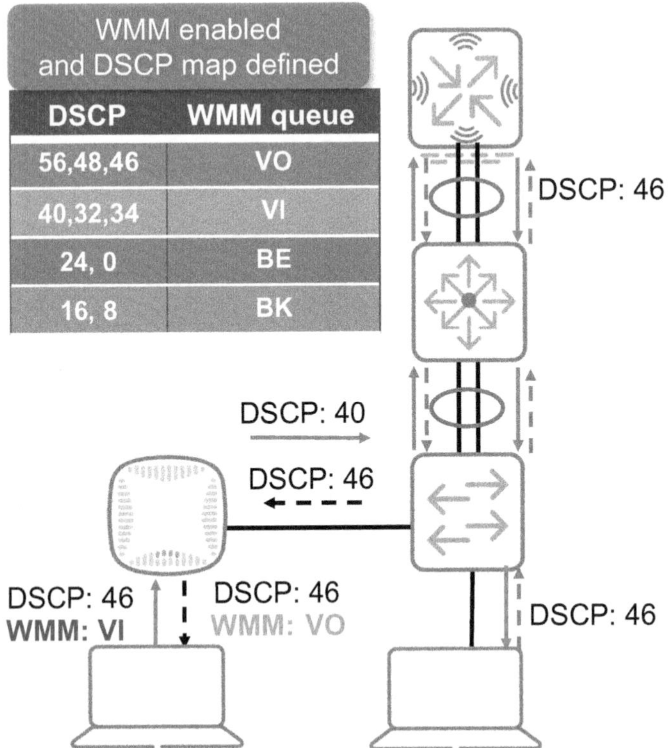

Figure 8-8 Solution 1: Work with the customer's scheme

With this approach, the voice traffic has the wrong priority on portion of the path from the client to MC. The MC then intelligently selects the traffic and remarks it. While it is not ideal for the voice traffic to have the wrong priority at any point, you often have to work with the scheme that the customer uses, and the segment of the journey most likely to cause—the air between the client and the AP—is outside of your control. Once the traffic arrives on the wired network, DSCP 40 might be acceptable. For example, if an AOS-Switch is using its default DSCP mappings, it will forward voice traffic with either DSCP 40 or 46 in queue 6, and it will forward video traffic with DSCP 34 with a slightly lower priority in queue 5.

To make the AP retain DSCP 46 when it forwards wireless client traffic that arrives in the video queue, you would have to map DSCP 46 to the video queue in the MC settings. Then the MC would need to use firewall policies to remark voice traffic with a different DSCP such as 56 or 48, which is mapped to the voice queue.

Solution 2: Adjust the customer's scheme

If possible, you should recommend that the customer changes their policies to mark voice traffic with DSCP 56. (Or if the customer prefers to reserve 56 for network control traffic, mark it with DSCP 48.) In either case, the client will now place the voice traffic in the voice queue. The Aruba wireless solution can now use the recommended DSCP to WMM mappings shown in Figure 8-9.

- Customer adjusts voice DSCP
- Voice DSCP assigned to high priority queue
- Firewall policies and SDN assign voice traffic to high priority queue

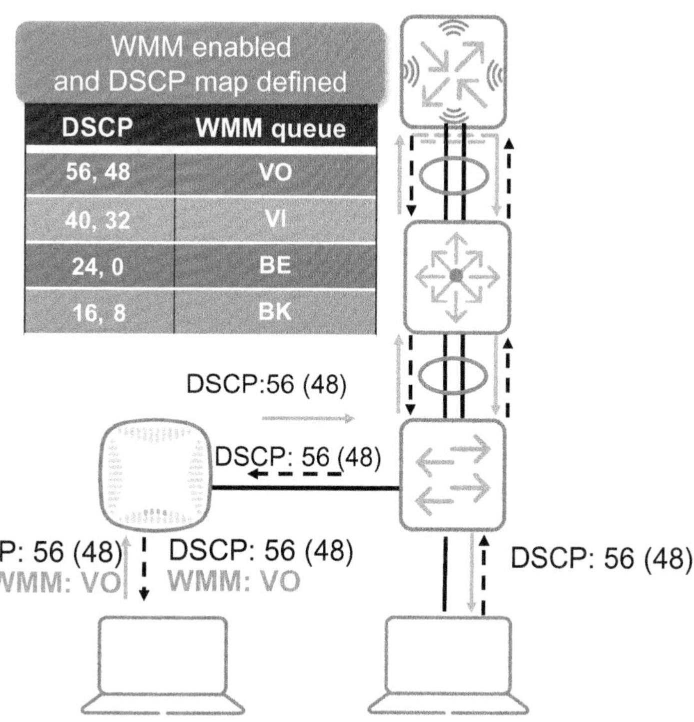

Figure 8-9 Solution 2: Adjust the customer's scheme

You can also recommend using DSCP 40 for video. If the customer continues using 34, the implementer should make sure to add 34 to the map for the VI queue.

Similarly, you should recommend that the customer avoid using DSCPs 16-23 for excellent effort traffic because WMM maps this traffic to the background queue, which receives less than best effort priority. DSCPs 24-31 will better serve the purpose.

Additional considerations for the path from AP and client to switch

The next point to consider is what happens when the traffic flows from the AP to its switch.

By default, AOS-Switches do not use incoming DSCPs to prioritize traffic, although they do use 802.1p values. 802.1p values require tagging, though. The AP is sending untagged traffic, as is the wired device. That means that the switch keeps the incoming traffic's current DSCP, but does not prioritize the traffic, instead sending it in the best effort queue with all traffic at default priority (see Figure 8-10).

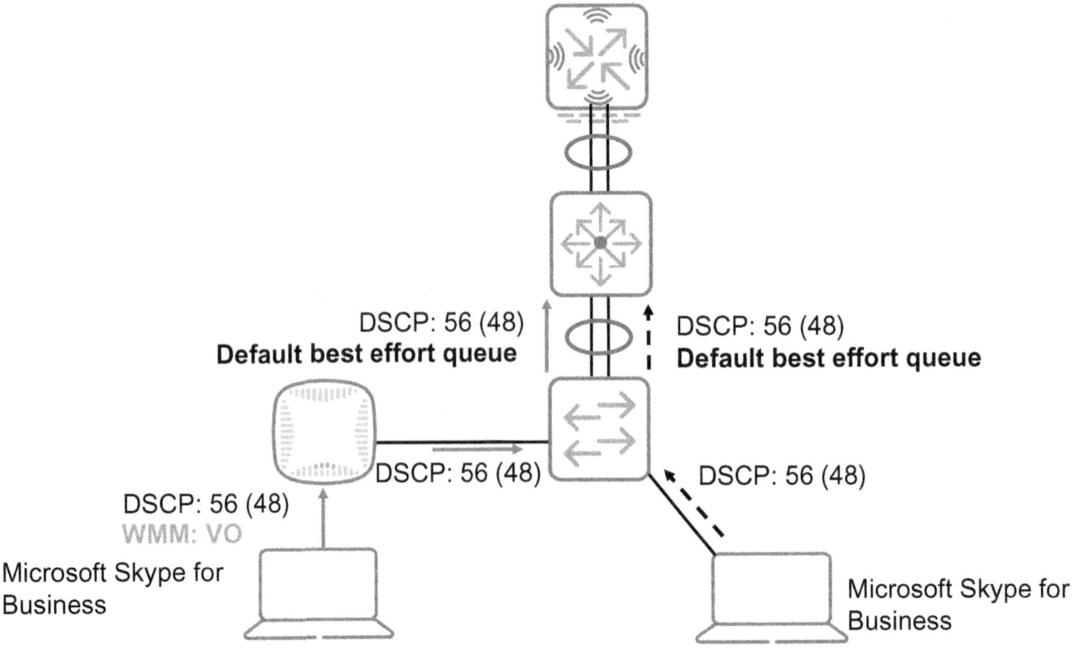

Figure 8-10 Additional considerations for the path from AP and client to switch

Switch recommendations

To ensure that the traffic continues to receive high-priority across its entire path, plan a DSCP trust policy for the Aruba switches. As Figure 8-11 shows, options include enabling trust on specific ports or globally, which enables trust on all ports. Make sure that ports trust DSCP when they connect to endpoints configured to mark their own traffic or to Aruba APs and MCs.

The switch then forwards traffic with a DSCP in the correct queue.

Ports that connect to untrusted endpoints should not trust DSCP. If you are not sure which edge ports will connect to APs and which will connect to untrusted endpoints, plan the "device aruba-ap" trust setting, which enables the switch to trust DSCP on the port only if LLDP indicates that an Aruba AP is connected to it.

The switch's default DCSP mappings match the recommend Aruba ones, so if you are having the customer create policies to assign voice traffic to 48 or 56, the switch already prioritizes the traffic correctly.

If the customer needs to use DSCP 46, you might want to change the switches to map DSCP 46 to 802.1p 6 (queue 7) for better consistency.

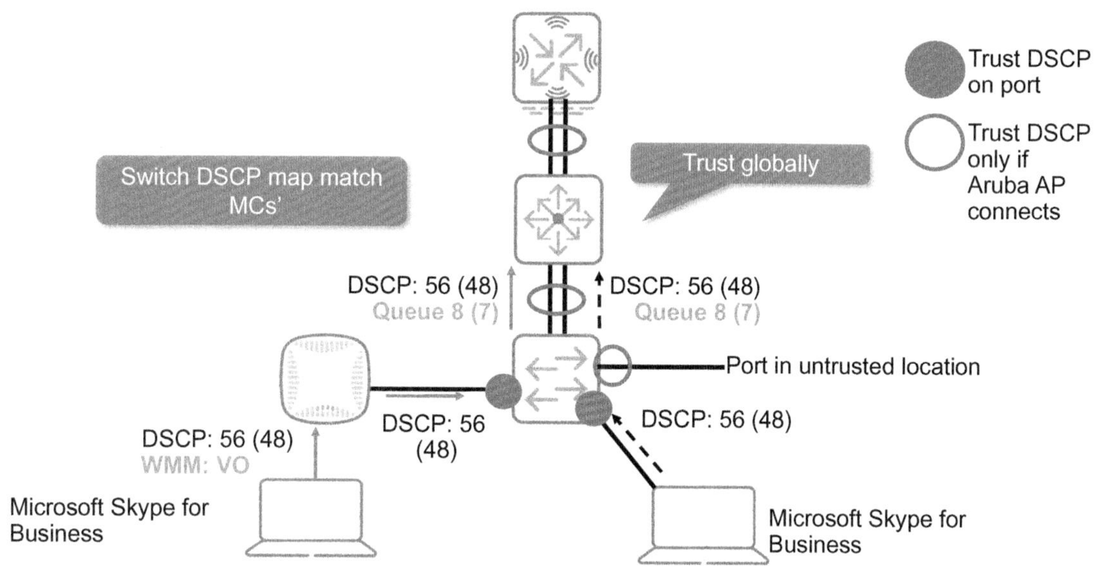

Figure 8-11 Switch recommendations

Capacity planning

You will next look at capacity planning in more detail.

Capacity planning overview

It is important to include capacity planning and strategies to meet your customers' needs (see Figure 8-12). Most modern campuses should feature Gigabit to the wired edge. You should plan appropriate oversubscription on switch-to-switch links. Oversubscribing the links too much can lead to poor performance.

Wireless capacity planning depends a great deal on a proper RF design. You need to ensure that clients can operate at high data rates by providing a good signal across the campus. In high-density environments, you also need to deploy enough APs such that the bandwidth is shared among the right number of clients. It is recommended to have 40–60 users per radio in most typical enterprise indoor deployments. However, user density per AP radio also depends on the AP model.

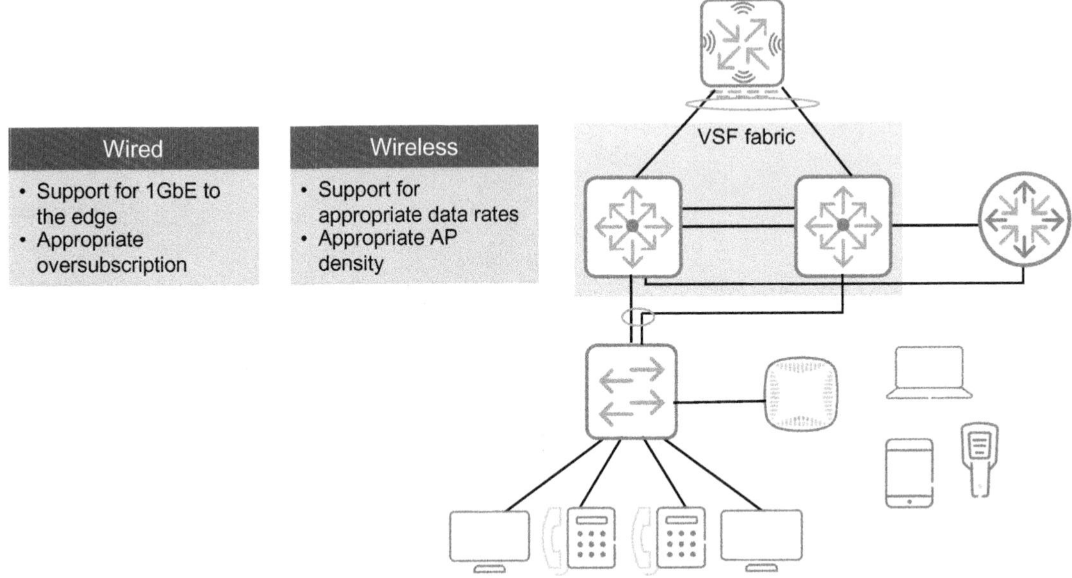

Figure 8-12 Capacity planning overview

When you make a capacity-based AP plan, it is also important to remove support for the lower data rates, helping to ensure that clients connect to an AP that offers a stronger signal and higher data rate.

For the wired network, the definition of appropriate oversubscription for varies based on customer circumstances and application usage. However, appropriate oversubscription at the wired edge tends to range from about 10:1 for higher bandwidth applications to about 20:1 or 24:1 for light users. In a three-tier topology, up to 4:1 oversubscription is appropriate between the aggregation layer and the core. The core should have no oversubscription. In a two-tier topology, the core might introduce up to 4:1 in connections to other network segments such as the data center.

Issues that prevent efficient use of wireless capacity

Even when you have planned AP locations and density well, various issues can prevent full use of the available wireless capacity.

Each RF channel is a shared medium. If two nearby APs operate on the same or overlapping channel, the system offers the same capacity as it would with a single AP. If the APs are just far enough apart that some of their clients cannot hear each other, the situation can be worse. The network will experience a large number of collisions and retransmissions that slow down communications. Other sources of non-RF interference can also take up airtime or cause errors in transmissions.

It can also be difficult to determine the maximum capacity of a wireless cell because all clients within the cell rarely use the same data rate. Slower clients reduce the capacity for the entire cell. Since the wireless medium is shared, every device must wait while the slow client transmits, effectively shifting the cell capacity toward the lowest common denominator.

Finally, unpredictable client distributions can stand in the way of efficient usage of the capacity. If one AP supports just a few clients while another AP supports many, the extra capacity that the first AP is contributing goes to waste.

Different clients support different data rates due to varying signal levels, and their data rates may vary as they move. A sticky client remains connected to an AP even after roaming relatively far away and dropping to a low data rate. In addition, clients have widely varying capabilities. 802.11n clients do not support the highest 802.11ac rates, while legacy 802.11a/g clients operate at much lower speeds. Furthermore, many 802.11n and even 802.11ac clients have limited antennas so they do not support the higher data rates available through more MIMO spatial streams.

Features to optimize wireless capacity

The Aruba wireless solutions offer several features to help overcome these issues and optimize the capacity offered by the complete wireless solution.

As Figure 8-13 shows, AirMatch helps to minimize co-channel interference between APs and maximize the capacity. This feature is available in Mobility Master (MM)-based architectures with ArubaOS 8. Wireless solutions that have no MM or that run on legacy software use Adaptive Radio Management (ARM) to limit co-channel interference. AirMatch is the preferred solution because ARM tends to be less stable and relies on each AP to adjust its channel rather than a centralized plan. Therefore, if customers are relying on the automatic channel and transmit power plan capabilities make sure to recommend an MM.

ClientMatch provides intelligent load balancing and client steering to help even out the client distribution on APs as much as possible while maintaining good signals. This feature can also help to eliminate sticky clients, increasing the number of clients that operate at a high data rate. Again, an MM provides a more effective version of the feature.

The MM-based solutions are more effective because they consolidate data from APs across the network. AirMatch then uses centralized intelligence to make an optimal channel and transmit power plan for every AP radio. ClientMatch also uses centralized intelligence to help APs continuously steer clients to the best AP.

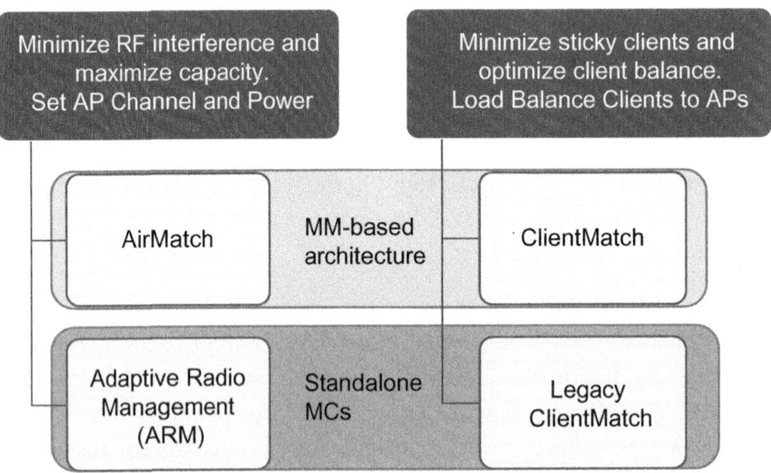

Figure 8-13 Features to optimize wireless capacity

AirMatch/ARM recommendations

To efficiently control the RF characteristics of each band and implement the recommendations included in this guide, create separate ARM profiles and assign them to their individual radio profiles (see Table 8-4).

Note that the "Default Value" column refers to the 802.11a and 802.11g radio settings. These settings are actually for the 5 GHz and 2.4 GHz radios, but 802.11a and 802.11g are the names within the profile.

Table 8-4 AirMatch/ARM recommendations

Feature	Default value	Recommended value	Comments
Transmit Power (dbm)	**802.11 A (5 GHz) radio**: Min 12/Max 18 **802.11 G (2.4 GHz) radio**: Min 6/Max 9	**Open office**: 5 GHz: Min 12/Max 15 2.4 GHz: Min 6/Max 9 **Walled office or classroom**: 5 GHz: Min 15/Max 18 2.4 GHz: Min 6/Max 9	• The difference between minimum and maximum Tx power on the same radio should not be more than 6 dbm. • Tx power of 5 GHz radio should be 6 dbm higher than 2.4 GHz radio.
Channels	• 80 MHz channels enabled. • ISM, U-NII-1, and U-NII-3.*	• 80 MHz channels can be used in Greenfield deployments. • U-NII-2 and U-NII-2e (DFS) channels must be used when operating on 80 MHz channels. • Remove channel 144 from list. • Consider using 40 MHz or 20 MHz channels for better channel separation.	• Enable DFS channels if you are not close to an airport or military installation. • Enabling DFS channels could create coverage holes for clients who do not support it. • Most of the clients do not scan DFS channels initially. This makes roaming more inconsistent when using these channels. • Very few clients support channel 144. • 20 MHz or 40 MHz channel width will help in reducing channel utilization in high-density open air environment.

* Assess the customer environment for legacy devices and sources of radar (such as airports and military)

Additional recommendations

AirMatch, as well as the legacy ARM, includes support for a feature called Airtime Fairness. This feature provides more transmission opportunities to clients that support higher data rates and fewer to legacy clients, preventing the legacy clients' slow transmissions from taking up all of the airtime. This maximizes the overall capacity that the AP radio offers.

As listed in Table 8-5, the recommended setting, Fair Access, in the QoS Traffic Management Profile, enables Airtime Fairness. This setting improve performance for environments with clients of varying abilities.

Table 8-5 Recommended setting for environments with mixed clients

Feature	Default value	Recommended value	Comments
QoS Traffic Management Profile (traffic shaping)	Default access	Fair access	Provides equal airtime to all the clients.

Table 8-6 lists another recommended setting to ensure good support for voice devices. Some VoIP handsets do not support channel 165, so it should be disabled. You should check the voice device type with the customer.

Table 8-6 Recommended setting to support voice devices

Features	Default value	Recommended value	Comments
Channels in ARM profile	• U-NII-2 and U-NII-2e disabled. • Channel 165 enabled.	• Disabled. • Disable this channel if the VoIP client doesn't support it.	• Voice devices do not scan many channels. • Some VoIP handsets do not support channel 165. Check the device manual for support details.

You already learned about other recommendations for voice devices such as enabling WMM and setting up the DSCP maps.

It is also recommended that the Aruba solution implements multicast and broadcast optimization, as is also recommended for a single VLAN.

ClientMatch

Figure 8-14 shows how ClientMatch monitors each client's capabilities and connection on a WLAN using probe requests and data frames the client sends.

Each AP forms a client probe and data report, which includes a list of all the clients that an AP can hear, including the SNR.

By default, an AP sends out this information to its MC every 30 seconds. The MC then sends the information to the MM. The MM can then consolidate all of the information from the system and create a Virtual Beacon Report (VBR), which maps each client to all the radios that can hear the client. The MM sends the VBR of each client to the MC and then the AP with which that client is associated.

Based on the information received in the VBR, an AP may decide to initiate band steering or sticky move for the clients associated to it. However, the controller will decide whether to dynamically load balance the clients or not.

When a ClientMatch is initiated to move a client to the desired radio, all the radios in the RF vicinity (except the one you select) blacklist the client for a short duration (default: 10 sec). This ensures that the client moves to the desired radio.

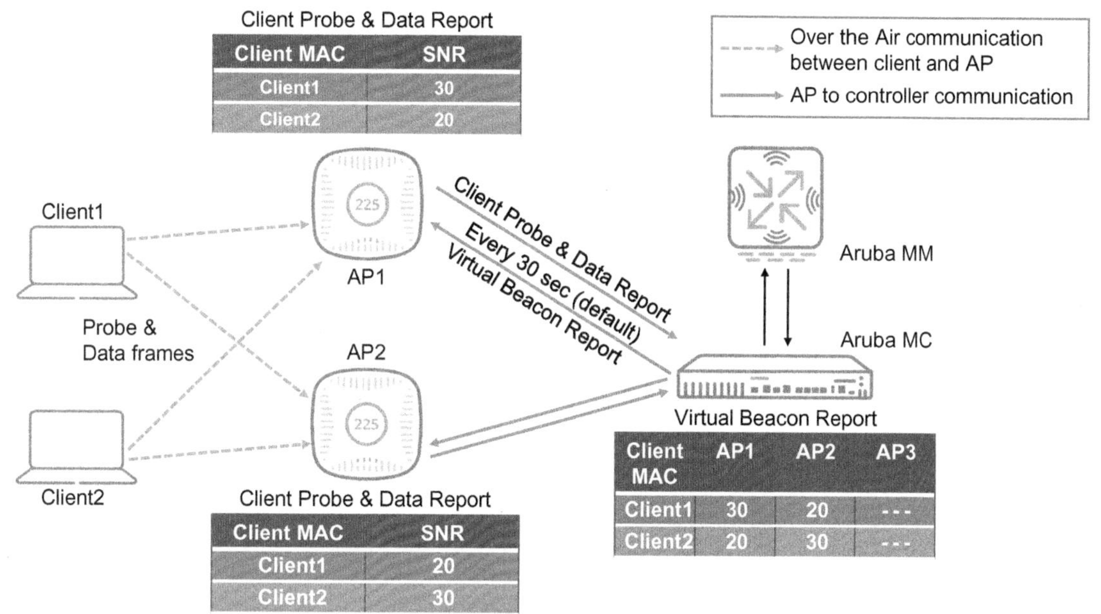

Figure 8-14 ClientMatch

Roaming optimization

You will now learn about technologies for optimizing wireless capacity and roaming.

Roaming optimization recommendations

When you use a capacity-based approach to deploying APs, clients can often receive a fairly good signal from multiple APs. You can help clients connect to the best AP by removing some of the lower data rates. But to help the implementers follow Aruba recommendations, collect information about clients in the environment. Some specialized legacy clients or Internet of Things (IoT) clients might require that they leave the lower data rates.

For your reference, Table 8-7 provides recommended values to optimize roaming.

Table 8-7 Recommended settings to optimize roaming

Features	Default value	Recommended value	Comments
Data rates (Mbps)	**802.11a:** Basic rates: 6, 12, 24 Transmit Rates: 6, 9, 12, 18, 24, 36, 48, 56 **802.11g:** Basic rates: 1, 2 Transmit Rates: 1, 2, 5, 6, 9, 11, 12, 18, 24, 36, 48, 56	**802.11 a & g:** Basic rates: 12, 14 **802.11 a & g:** Transmit rates: 12, 24, 36, 48, 56	• May affect client performance, depending on AP density and client type. Contact Aruba SE or TAC before adjusting the values. • If you have gaming or IoT devices connecting to the network, then add data rates 5, 6, 9, and 11 Mbps to the G radio basic and transmit rates.
Beacon Rate (Mbps)	By default lowest configured basic rate.	For both 802.11a and g radio, use 12 or 24.	Sends out beacon at rate configured and not at the default lowest basic rate.
Local Prob Req Threshold (db)	0	0	AP stops responding to client probe request if SNR is less than 15 db.

Fast roaming recommendations

Voice and other real-time applications require fast roaming. If a client roams within an MC cluster, the roam should occur seamlessly because the client remains connected to its User Anchor Controller (UAC), which has all of its association information.

Table 8-8 provides recommendations for features that support fast roaming. For example, Opportunistic Key Caching (OKC) is recommended to speed roaming.

Due to inconsistent client support, 802.11r is not recommended. Because 802.11k and 802.11v can help clients roam more quickly, implementers should use the recommended settings for these features.

Table 8-8 Recommendations for features that support fast roaming

Features	Default value	Recommended value	Comments
Opportunistic Key Caching (OKC)	Enable	Enable	Avoids full 802.1X key exchange during roaming by caching the opportunistic key. NOTE: Mac OS and iOS devices do not support OKC.
Validate Pairwise Master Key ID (PMKID)	Enable	Enable	Matches PMKID sent by client with the PMKID stored in the Aruba controller before using OKC.
Extensible Authentication Protocol over LAN (EAPOL) Rate Optimization	Enable	Enable	Sends EAP packets at lowest configured transmit rate.
802.11r	Disable	Disable	Enable this feature only after testing it in the lab with different types of devices connecting to the network.
802.11k	Disable	• Enable with beacon report set to **Active Channel Report**. • Disable **Quiet Information Element** parameter from the **Radio Resource Management** profile.	Helps clients make a quicker decision to roam.
802.11v Fast BSS Transition	Disabled	When 802.11k is enabled, 802.11v **Fast BSS Transition** also gets enabled.	Helps clients to roam faster.

Activity 8: Design QoS

You will now compete a design activity in which you apply what you have learned in this chapter. This activity builds on the scenario introduced in earlier chapters.

In this activity, you will use the customer scenario to determine the requirements for the customer's applications. You will plan how to meet these applications' needs for QoS.

Scenario QoS

Discussions have revealed a significant new requirement. Corp1 has a Microsoft Skype for Business Unified Communications solution. The company uses Skype for Business primarily for the voice and video conferencing capabilities. One of the reasons that Corp1 is upgrading the network is due to poor user experiences with these applications, even though admins set up Skype for Business to use the recommended QoS settings. The customer has emphasized that the new solution must improve the quality.

The customer has explained that most video conferencing occurs with the equipment in the meeting rooms in Building 1, Floors 2 and 3 and the board rooms in Building 1, Floor 3. However, users also place calls from their desks. Users need to be able run Skype for Business anywhere, whether they are on a wireless or wired connection, and receive the same quality of service.

Based on logs in the Skype for Business, the customer tells you that up to 300 users are in calls concurrently.

The customer has mentioned these other applications in use within the campus:

- HTTP/HTTPS
- Email
- FTP
- Network file sharing
- Printing
- Telnet and SSH
- SNMP

Other applications are used within the data center, but you do not need to plan that area of the network.

Activity 8.1: Classify applications

Activity 8.1 objectives

You will classify the customer's applications and identify applications for which you need to guarantee QoS.

Note that the customers' new requirements might lead you to recommend a higher capacity AP deployment. However, you will wait to make this plan until Activity 9.

Activity 8.1 steps

1. Refer to the scenario for this activity. Which applications might need high priority in order to provide a good experience to users running those applications or protect critical services?
2. What else would you discuss with the customer to further classify applications and identify high-priority ones?

Activity 8.2: Plan QoS measures

Activity 8.2 objectives

Create a plan for providing the proper QoS to sensitive and critical applications.

Activity 8.2 steps

Assume that after further discussion with the customer, you have determined that Corp1 does not want to prioritize the traffic of one employee group over another. The customer does want to prevent guests from monopolizing bandwidth by limiting guests to 10 Mbps.

1. What do you need to plan to ensure that wired traffic receives the correct priority?
2. What do you need to plan to ensure that wireless traffic receives the correct priority? Include ensuring that the wired infrastructure supports the plan.
3. What additional technologies do Aruba solution provide to help ensure a good QoS for critical and time-sensitive traffic?

Activity 8.3: Describe Aruba benefits

Activity 8.3 objective

Explain to customers the benefits that Aruba solutions provide for modern networks with a diversity of multimedia traffic.

Activity 8.3 steps

1. Plan a brief (1 or 2 minute) presentation on benefits that an Aruba solution provides for the customer's sensitive applications such as voice. How will the solution enhance users' experience?

Answers to the questions included in activities are provided in "Appendix: Activities and Learning Check Answers."

Summary

In this chapter, you learned how to assess customers' applications, to classify them, and to take measures to ensure that receive the proper handling. The foundation for QoS is good capacity planning. However, to deal with congestion that can still occur, you should implement traffic prioritization technologies that ensure that the most important and time-sensitive traffic is forwarded first.

You considered extra measures for ensuring that wireless users have a good experience with voice applications. These measures include fast roaming and voice-aware 802.11 rekeying as well as MC features for voice management.

Learning check

1. What describes the effect jitter can have on voice and video applications?
   a. Jitter generally has fewer negative effects on these applications than latency.
   b. Jitter triggers the UDP transport protocol to slow down transmissions.
   c. Jitter can cause a choppy quality, worse than that caused by a consistent delay.
   d. Jitter, which refers to the percentage of packets dropped, causes ill effects at about 5%.
2. What is one advantage that an MM provides for customers with Microsoft Skype for Business over a standalone MC?
   a. Role- and firewall-based prioritization
   b. Bandwidth contracts
   c. Support for WMM
   d. Better scalability for SDN integration
3. What is one way that an MM-based architecture helps to optimize wireless capacity better than a legacy controller architecture?
   a. MM delivers ARM, a technology that optimizes broadcast and multicast traffic.
   b. MM offers AirMatch, which can make channel plans based on centralized intelligence.
   c. MM enables support for eight queues within WMM rather than just four.
   d. MM replaces legacy ClientMatch with AirMatch, which helps match clients to the best APs on the fly.

Answers to the learning checks in this study guide are provided in "Appendix: Activities and Learning Check Answers."

9 High-Density Design

EXAM OBJECTIVES

- ✓ Evaluate a customer's needs for a single-site campus or subsystems of an enterprise network, identify gaps and recommend components.
- ✓ Given a customer's requirements for a single-site campus, design the high-level architecture.
- ✓ Given a scenario, explain how a specific technology or solution would meet the customer's requirements.
- ✓ Given a scenario for a single-site campus, choose the appropriate components to be included in a Bill of Materials (BOM).

Assumed knowledge

- Basic wireless network concepts, such as the 802.11 standards
- Frequencies and channels used by 802.11 and the basics of co-channel interference
- Familiarity with the Aruba Mobile First Architecture, specifically the 8.x architecture
- Basic network design principles

Introduction

Very High Density (VHD) enterprise networks have unique requirements and design criteria. In this chapter, you will learn how to take these requirements and plan a VHD deployment by calculating the number of Access Points (APs) and Mobility Controllers (MCs) necessary for a VHD WLAN. This is a complete process and requires you to factor in many variables such as Associated Device Capacity (ADC), take rate, and address design. You will also see some examples of AP placements for VHD designs and how to plan your own VHD design.

RF channel design

You will first focus on RF channel design.

What is an RF channel? It is a collision domain

The channel entity is an 802.11 collision domain, which is an independent block of capacity in an 802.11 system. A collision domain is a physical area in which 802.11 devices, which attempt to send on the same channel, can decode one another's frame preambles. It is also a moment in time. Two nearby stations on the same channel do not collide if they send at different times.

Collision domains are also dynamic regions that constantly move in space and time, based on which devices are transmitting.

The concept of a collision domain is specific to the 802.11 MAC layer. All radio systems can interfere with one another if two transmitters attempt to send at the same time on the same frequency. However, 802.11-based technologies are unique because they apply Carrier-Sense Multiple-Access with Collision Avoidance (CSMA/CA). As you probably know, the collision avoidance mechanism uses a virtual carrier-sensing mechanism as well as a physical energy detection mechanism. What you may not be aware of is the role that frame preambles play in the virtual carrier sense and, therefore, the true shape of the collision domain in both space and time.

You should not use the word "cell" as a synonym for collision domain. Cells are typically engineered areas where the signal-to-interference-plus-noise ratio (SINR) or received signal strength indicator (RSSI) exceeds a specific target value. The so-called "cell edge" is the radial distance from an AP at which this value is hit. However, the collision domain extends until the SINR goes below the Preamble Detection (PD) threshold. The area of the cell is far smaller than the area of the collision domain.

Data rate efficiency

As a general rule, every transmission in a VHD collision domain should use the maximum possible rate for all three 802.11 frame types: data, control, and management.

Figure 9-1 shows a 2D slice of the model, which focuses on the data rate and distance axes. The vertical axis includes the 802.11 legacy and MCS rates grouped by the modulation they share in common. The horizontal axis shows distance from the AP.

The chart in Figure 9-1 shows several important points:

- The average PHY data rate that is used for data frames between any station (STA) and an AP should follow the rate curve shown in green. If the rate seen on air is less than expected, this indicates an operational problem or an issue with the system design.
- The average PHY data rate used for control frames should be pushed as high as it will reliably go. Do not accept the default values in VHD environments. Figure 9-1 shows a dotted blue line for the default 6 Mbps setting and a solid blue line for a 24 Mbps setting (16-QAM modulation with 1/2 coding).
- The average PHY data rate that is used for management frames should likewise be pushed much higher than the defaults for the same reasons.

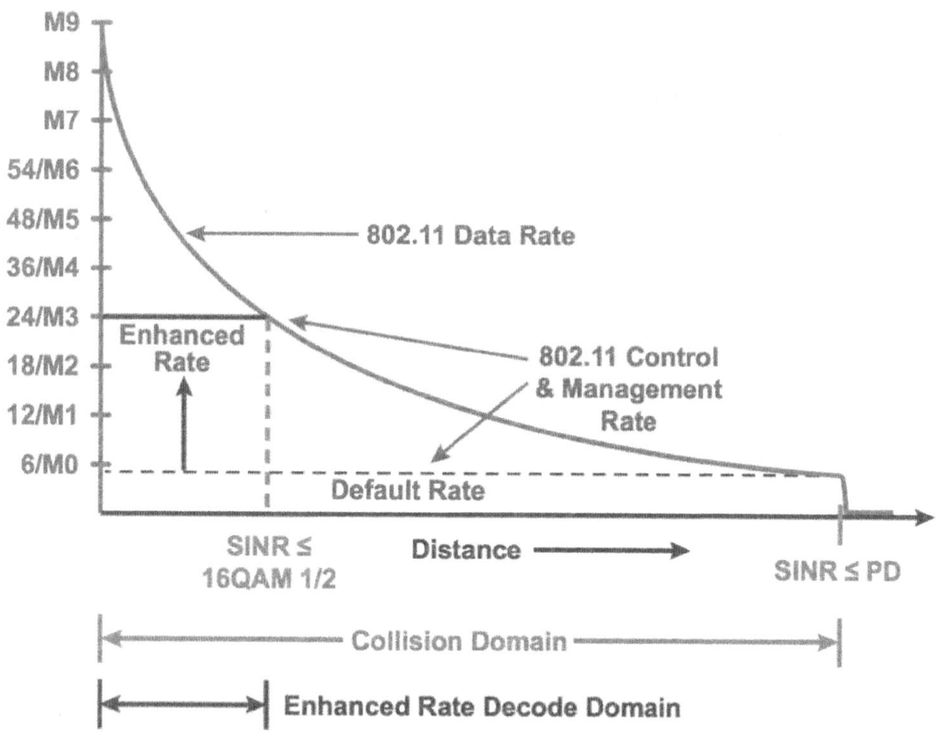

Figure 9-1 Data rate efficiency

As you think about your SSID rate configurations, you always want to push the rate used as high as possible toward the allowable limit on the curve. "Chapter EC-3: Airtime Management" of the Aruba *Very High- Density 802.11ac Networks Engineering and Con⊠guration Guide* discusses this issue in depth across many different types of 802.11 transmissions.

Data rate efficiency also corrects a common misunderstanding among WLAN engineers and architects.

The conventional wisdom is that as the data rate for control and management frames is increased, the cell size "shrinks." A higher SINR is required to decode the faster rate so that payload is not decodable beyond a specific point. This same thinking is behind the common practice of "trimming out" low OFDM data rates.

However, the chart in Figure 9-1 clearly shows that if the payload rate is changed, that change does not alter the interference range of the legacy preamble detection. Those preambles must be sent using Binary Phase-Shift Keying (BPSK), and they cannot be changed. As a result, the collision domain size is unaffected. Distant STAs that decode the preamble still mark the channel as busy for the full duration of the frame even if the payload cannot be recovered.

Trimming out low control and data rates does have several practical benefits, which are discussed at length in "Chapter EC-3: Airtime Management" of the Aruba *Very High-Density 802.11ac Networks Engineering and Con⊠guration Guide*. However, trimming those rates does not change the size of the collision domain.

What is airtime?

In order to manage airtime, you need to have a clear understanding of what airtime is and how it works.

In a high-level overview, you can think of 802.11 airtime as a continuous series of alternating idle and busy periods on a given channel (or collision domain). You measure the length of each period in a unit of time, such as milliseconds (ms) or microseconds (⊠s). When you view this at a high level, the time each period requires varies constantly.

In the example shown in Figure 9-2, you have alternating idle and busy periods on three adjacent, nonoverlapping channels.

In the time period on the diagram in Figure 9-2, channel X is in the middle of a transmission on the left, and ends up in an idle state. Channel Y begins in idle and ends busy. Channel Z is idle except for periodically repeating transmissions, such as an access point (AP) beacon.

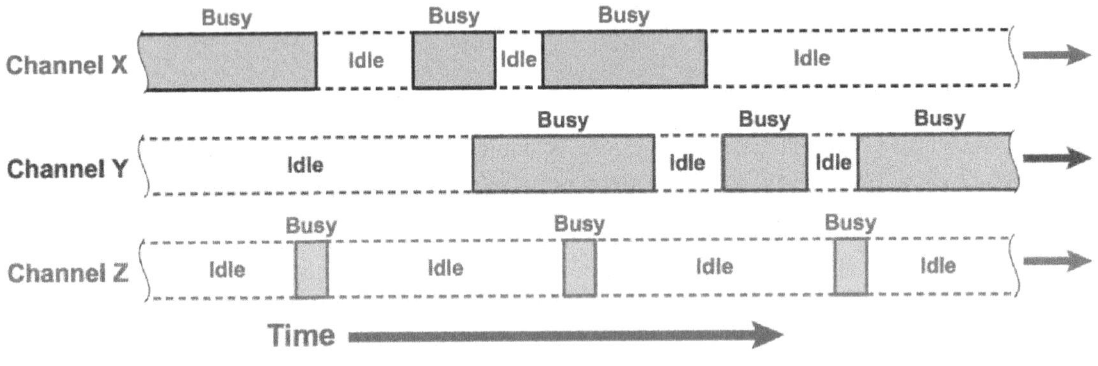

Figure 9-2 Data Rate Efficiency

Very high-density RF coverage

You will now consider how to design coverage for very high-density (VHD) environments.

RF coverage strategies

You can use radio coverage in three ways, regardless of the type of area you want to serve (see Figure 9-3).

In Overhead Coverage, you can place APs on a ceiling, catwalk, roof, or other mounting surface directly above the users to be served. You should always use the integrated antenna APs for ceilings of 10m (33 feet) or less.

With Side Coverage, APs are mounted to walls, beams, columns, or other structural supports that exist in the space to be covered.

In a Floor Coverage design, picocells are created using APs mounted in, under, or just above the floor of the coverage area. APs with integrated antennas are used for any VHD area of under 5000 seats.

Overhead Coverage: APs are placed on a ceiling, catwalk, roof or other mounting surface directly above the users to be served.

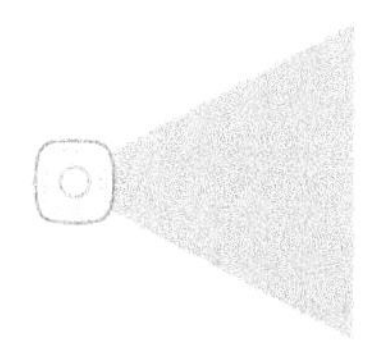

SideCoverage: APs are mounted to walls, beams, columns or other structural supports that exist in the space to be covered.

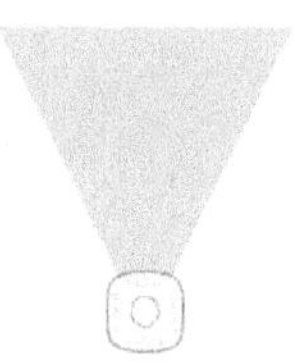

Floor Coverage: This design creates picocells using APs mounted in, under or just above the floor of the coverage area.

Figure 9-3 RF Coverage strategies

Overhead coverage

Overhead coverage is a good choice when the customer wants uniform signal everywhere in the room. Overhead coverage does not allow for RF spatial reuse because of the wide antenna pattern and multipath reflections.

In the example shown in Figure 9-4, note the 20-MHz channel width, and that no channel number is used more than once.

Figure 9-4 is an example of a static, nonrepeating channel plan a wireless architect intentionally chose. Overhead coverage requires access to the ceiling with minimal difficulty or expense to pull cables and install equipment.

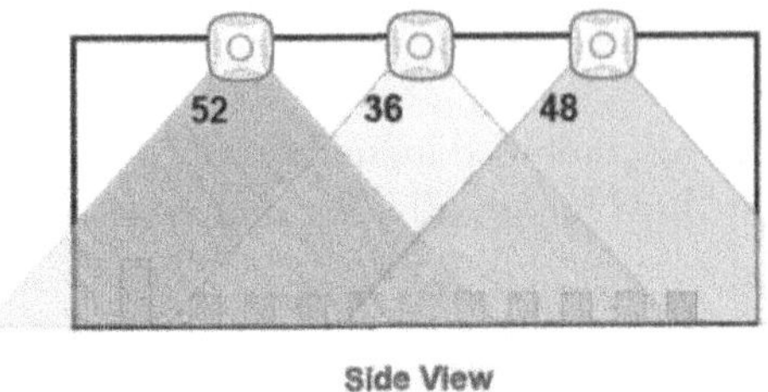

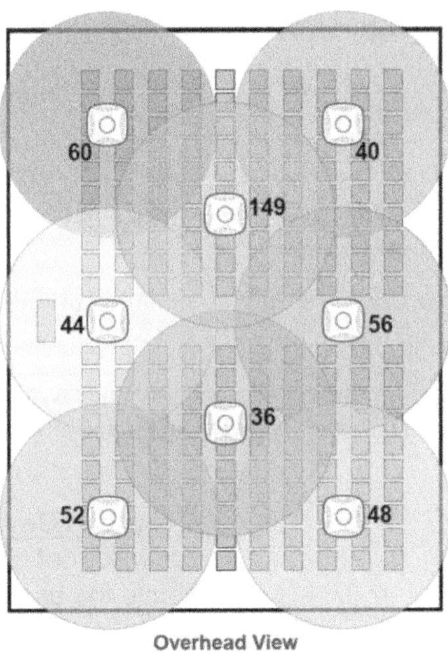

Figure 9-4 Overhead coverage

Examples—Overhead coverage #1

Figure 9-5 shows an example of overhead coverage APs provide in an auditorium. Note that the APs are placed in an overhead mounting design.

Figure 9-5 Examples—Overhead Coverage #1

Examples—Overhead coverage #2

Figure 9-6 shows another example of APs providing overhead coverage by mounting on a steel catwalk on the ceiling of an auditorium.

Figure 9-6 Examples—Overhead coverage #2

Side coverage

Side coverage is very common in VHD areas with wall, beam, and column installations (see Figure 9-7). Some ceilings are too difficult to reach, others have costly finishings that cannot be touched, or there may be no ceiling such as open-air atriums. Side coverage does not allow for RF spatial reuse in indoor environments when mounting to walls or pillars. In addition, 50% of the wall-mounted AP signals are lost to the next room (and 75% of the signal in the corners).

Note that adjacent APs on the same wall always skip at least one channel number.

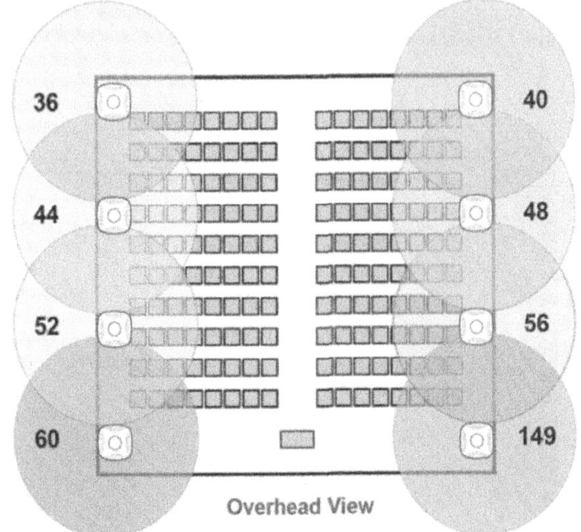

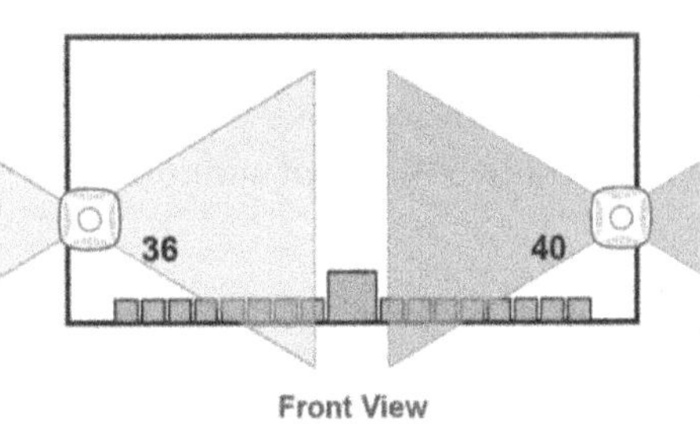

Figure 9-7 Side coverage

No RF spatial reuse

No RF spatial reuse occurs when every AP can be heard everywhere in the room (see Figure 9-8). This may happen in high-density environments such as stadium bowls due to the RF propagation.

In Figure 9-8, the lighter shading indicates where the signal is at least -55dBm, -60dBm, or -65dBm.

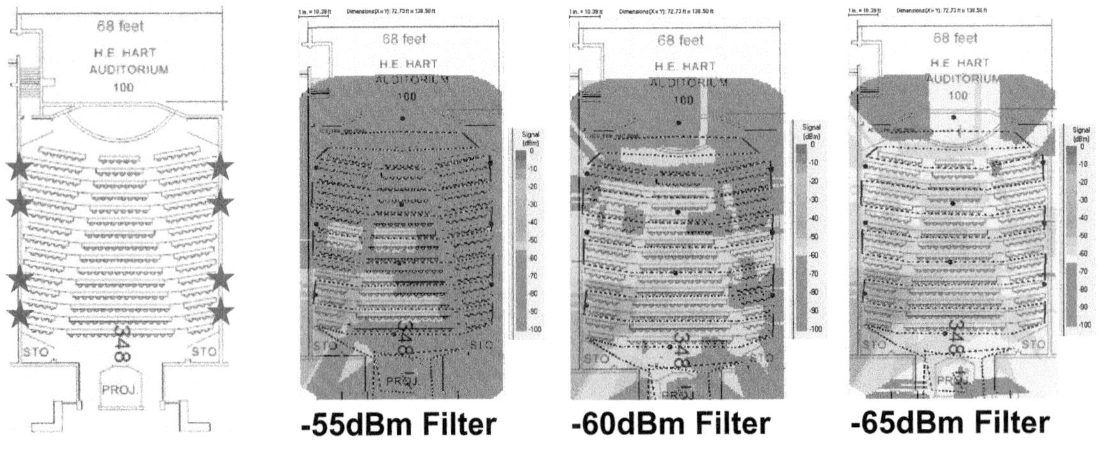

Figure 9-8 No RF spatial reuse

Examples—Side coverage

Figure 9-9 shows an example of APs mounted on the outside walls of a lecture hall to provide side coverage.

Figure 9-9 Examples—Side coverage

Floor coverage

Venues with less than 10,000 seats should always use overhead or side coverage rather than floor coverage (see Figure 9-10). Larger venue can use a picocell design. The "picocell" option has been proven to deliver significant capacity increases as the density of picocell can be much higher than an overhead or side coverage. The picocell design leverages absorption that occurs to RF signals as they pass through a crowd (also known as "crowd loss"). However, the cost and complexity of picocells may not always justify the extra capacity it generates.

- Venues <= 10K seats should always use overhead or side coverage.
- Above > 10K seats "picocell" can be used.
- Density of picocell can be much higher
- Picocell design leverages absorption
- Cost and complexity of picocells may not always justify the extra capacity generated.

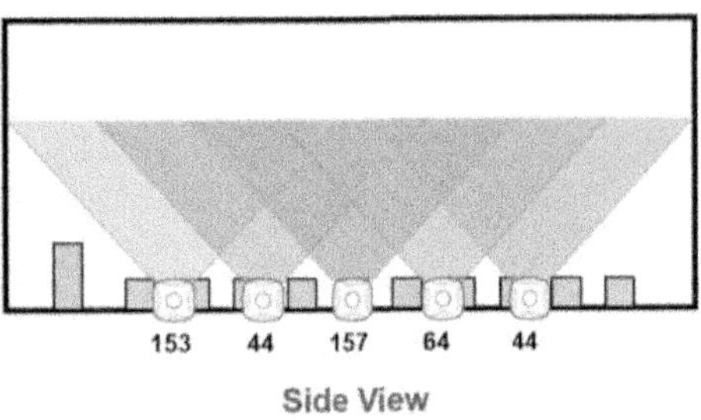

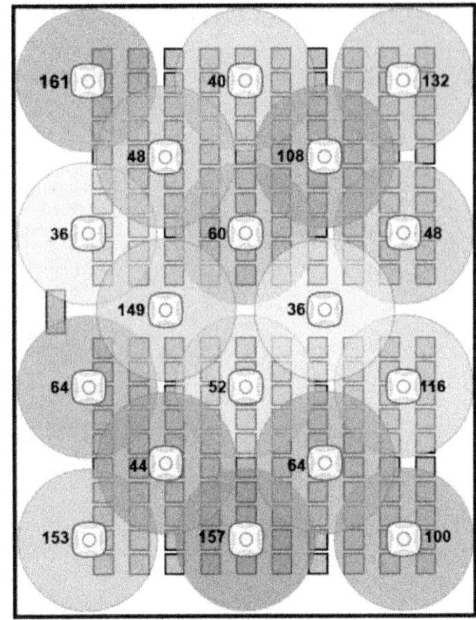

Figure 9-10 Floor coverage

Examples—Picocell

Figure 9-11 shows some examples of APs mounted on the floor underneath stadium seats to provide picocell coverage.

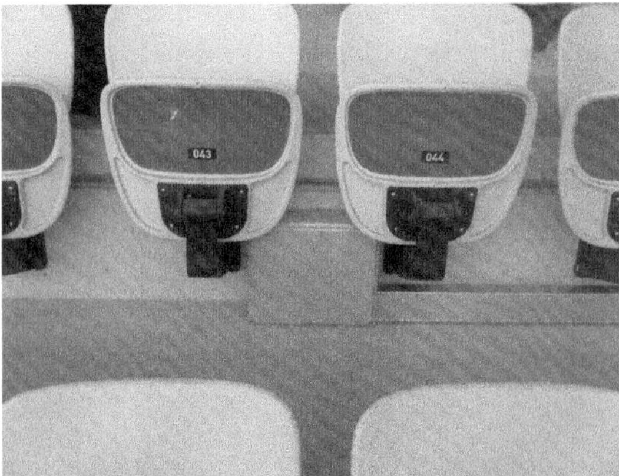

Figure 9-11 Examples—Picocell

Choosing AP model for VHD areas

For high-density and VHD areas, favor the APs that are designed to support more spatial streams for both non-MU-MIMO and MU-MIMO clients. For example, as of the publication of this study guide, the Aruba AP-330 and AP-340 Series provide the best performance for these environments. Select the AP-330 Series to provide coverage in the 2.4 GHz range or the AP-340 Series if you are confident all devices support 5 GHz.

Because these APs support more spatial streams, they can deliver consistently higher throughput for clients that also support more spatial streams. However, in many environments, most clients support no more than two spatial streams. In this case, the greater performance of the AP-330 and 340 Series becomes more apparent as the client density increases. As more clients begin to support MU-MIMO, these APs will be able to take advantage of their multiple spatial streams more often, transmitting two streams to one client and two streams to another, for example. In other words, deploying the APs future proofs the network.

You can use the model number as a general guide. The AP-330 and AP-340 Series have higher series numbers than the AP-310 and AP-320 Series and offer support for higher-density environments. However, you need to consider the AP-360 Series and AP-370 Series separately; these are outdoor APs while the others are indoor APs.

AP placement for VHD areas

When APs are mounted in VHD areas, they should be staggered in horizontal placement, but all facing the same direction, as shown in Figure 9-12. In side-mounting scenarios, you can use the rear lobe of RF coverage to your advantage to provide coverage in the opposite direction of the APs antenna.

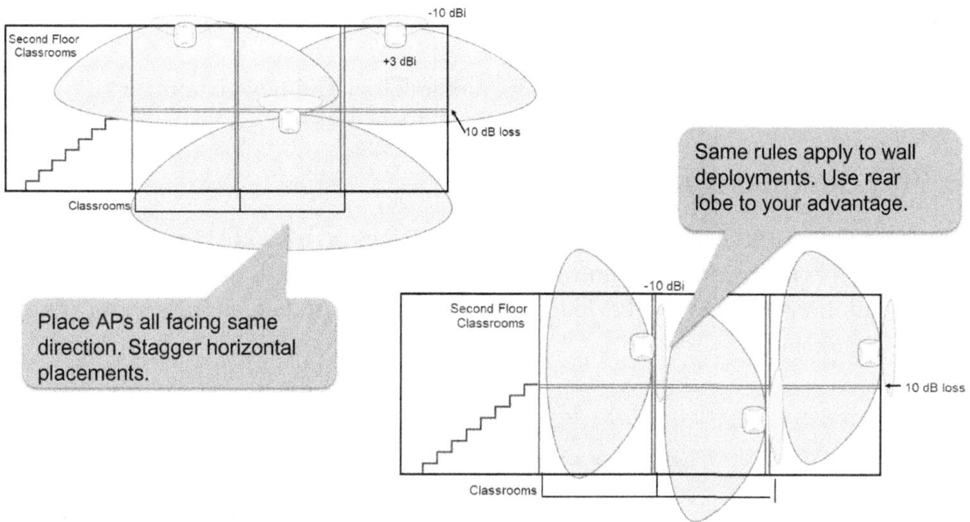

Figure 9-12 AP Placement for VHD areas

Back-to-back APs on same wall

You should always mount APs back to back on different channels to prevent cochannel interference, as shown in Figure 9-13. You can place APs on adjacent 5 GHz channels if they are staggered to ensure physical spatial separation.

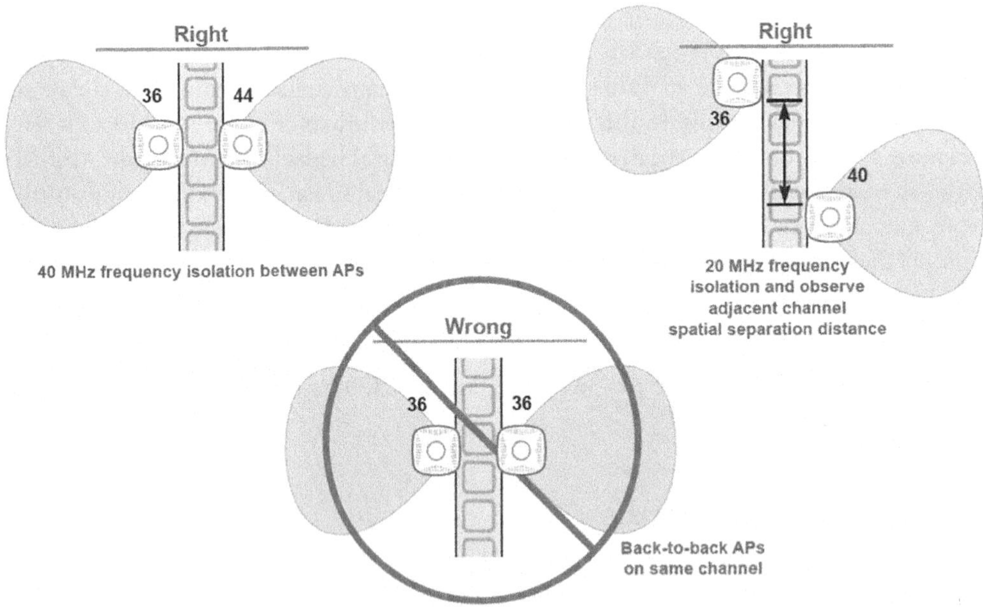

Figure 9-13 Back-to-back APs on same wall

VHD design

You will now learn about the Aruba VDD design methodology and how to apply it.

Aruba VHD design methodology

As Figure 9-14 shows, the Aruba VHD design methodology starts with Device Dimensioning to evaluate the WLAN clients and determine how many the solution needs to support. It then moves to WLAN Dimensioning to size the APs, followed by Infrastructure Dimensioning to determine the infrastructure requirements.

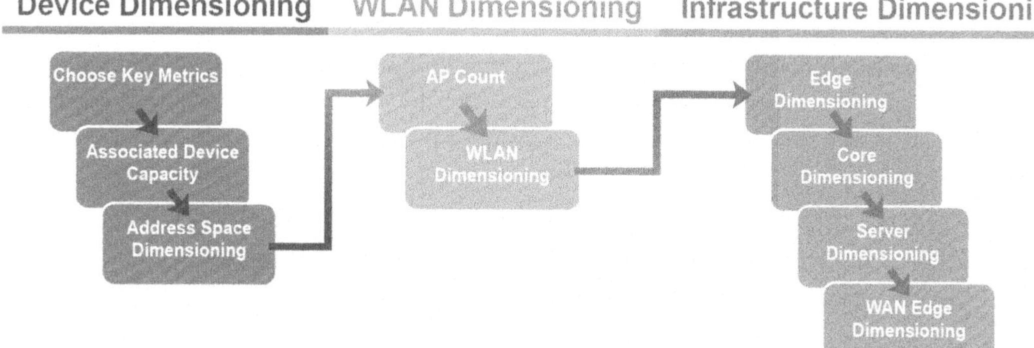

Figure 9-14 Aruba VHD design methodology

Step 1—Key design criteria for typical VHD WLAN

To determine the device dimensioning of a VHD WLAN, first assign a value such as seating capacity, average devices per person, or take rate (which means what percentage of users will have wireless devices). From these basic metrics, you can calculate additional values such as seats or area covered per AP as well as associated devices per radio (see Table 9-1).

Table 9-1 Metrics for device dimensioning

Metric	Definition	Typical Value
Seating capacity	Number of people the facility can hold.	Varies
Average devices per person	Typical number of discrete Wi-Fi enabled devices carried by a person visiting the VHD facility.	1–5
Take rate	Percentage of seating capacity with an active Wi-Fi device.	50%–100%
Associated device capacity	Take rate multiplied by the average number of Wi-Fi enabled devices per person.	Varies
Seats or area covered per AP	How many square meters (square feet) or seats each AP must serve—essentially the physical size of a radio cell.	Varies
Associated devices per radio	The design target of how many associated devices should be served by each radio on an AP.	150
Average single-user goodput	What is the minimum allowable per-user bandwidth when multiple users are attempting to use the same AP?	512 Kbps–2 Mbps
5 GHz vs. 2.4 GHz split	Distribution of clients across the two bands.	5 GHz: 75% 2.4 GHz: 25%

Step 2—Estimate associated device capacity

As Figure 9-15 shows, estimating Associated Device Capacity starts with the seating/standing capacity of the VHD area you want to cover. You then estimate the take rate (50% is a common minimum). Next, you choose the number of devices you expect per person. This varies by venue type. It might be lower in a stadium and higher in a university lecture hall or convention center salon.

For example, 50% of a 70,000-seat stadium would be 35,000 devices, assuming each user has a single device. With 100% of a 1000-seat lecture hall where every student has an average of 2.5 devices, the Associated Device Capacity would be equal to 2500.

More users should be on 5 GHz than 2.4 GHz. You should compute Associated Device Capacity by frequency band. In general, you should target a ratio of 75%/25%. You should assume association demand to be evenly distributed throughout the coverage space.

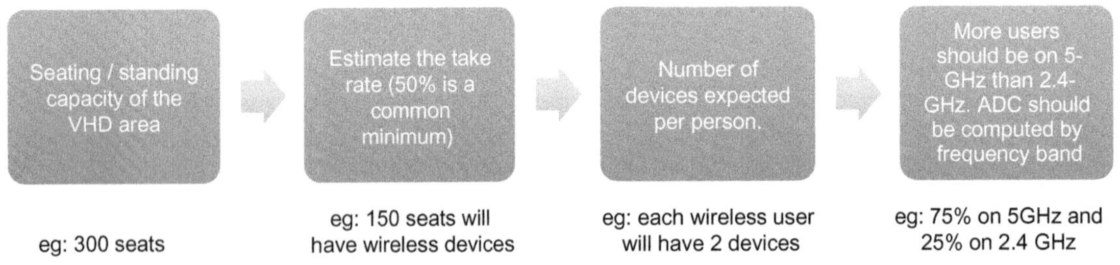

Figure 9-15 Step 2—Estimate associated device capacity

Step 3—Address design

As you learned earlier, Aruba recommends large flat VLANs for WLAN

Table 9-2 provides a few Associated Device Capacity examples and address space estimates for an indoor 20,000-seat arena. Notice the plan is for Guest users to have a take rate of 25% now and 50% in the future with 75% of the clients on 5 GHz and only 25% on 2.4 GHz.

Table 9-2 Associated Device Capacity examples and address space estimates for indoor arena

User group	Devices (Now)	Devices (Future)	% 5 GHz	% 2.4 GHz	Minimum subnet size
Guest / Fan	5000 (25% take rate)	10,000 (50% take rate)	75%	25%	/18
Staff	100	300	100%	0%	/23
Ticketing	50	100	100%	0%	/24
POS	50	200	100%	0%	/24
Team	15	100	100%	0%	/24
TOTAL	**5,215**	**10,700**	**8,200**	**2,500**	-

Table 9-3 shows the average Associated Device Capacity and address space estimates for example university lecture halls.

Table 9-3 Associated Device Capacity examples and address space estimates for university lecture halls

User group	Devices (Now)	Devices (Future)	% 5 GHz	% 2.4 GHz	Minimum subnet size
Student	20,000	45,000	75%	25%	/16
Faculty	2000	4000	100%	0%	/20
TOTAL	**22,000**	**49,000**	**37,750**	**11,250**	-

Step 4—Estimate AP count

In the high-density design, you should plan for 150 associations per radio and 300 per AP (see Figure 9-16). Although ArubaOS supports up to 255 associations per radio, Aruba recommends you plan for only 150 associations per radio, so 60% loading with 40% headroom since all VHD areas experience inrush/outrush. Planning for extra headroom allows for user "breathing" when users are initially connecting. Also remember to increase max users, which is 64 by default in the SSID profile.

150 associations per radio, 300 per AP

ArubaOS supports up to 255 per radio

All VHD areas experience inrush/outrush

Remember to increase max users in the SSID profile

$$\textit{AP Count} = \textit{5-GHz Radio Count} = \frac{\textit{Associated Device Capacity (5 GHz)}}{\textit{Max Associations Per Radio}}$$

Figure 9-16 Step 4—Estimate AP count

Step 5—Dimension controllers that terminate APs

For high-density designs that include up to 32,000 users, platform size must be greater than or equal to the Associated Device Capacity. For this number of users, redundant power supply units (PSUs) are critical so you should not use a 7205 or 70XX series controller. In addition, active/active redundancy is critical for this large number of users (see Table 9-4).

Performance metrics could change with new AOS versions.

Table 9-4 Aruba 7210, 7220, 7240, and 7280 Series Controller

<table>
<tr><th>Model</th><th>7210</th><th>7220</th><th>7240</th><th>7280</th></tr>
<tr><td>Maximum number of LAN-connected APs</td><td>512</td><td>1024</td><td>2048</td><td>2048</td></tr>
<tr><td>Maximum number of users</td><td>16,384</td><td>24,576</td><td>32,768</td><td>32,768</td></tr>
<tr><td>Active firewall sessions</td><td colspan="4">2,015,291</td></tr>
<tr><td>Firewall throughput</td><td>20 Gbps</td><td>40 Gbps</td><td>40 Gbps</td><td>100 Gbps</td></tr>
<tr><td>AES encrypted firewall throughput</td><td>6 Gbps</td><td>20 Gbps</td><td>40 Gbps</td><td>80 Gbps</td></tr>
<tr><td>MAC addresses per VLAN</td><td>64000</td><td>128000</td><td>128000</td><td></td></tr>
<tr><td>FW Session Creation Rate (1000 sessions/sec)</td><td>249</td><td>326</td><td>481</td><td>481</td></tr>
<tr><td>802.1X Auth Rate (transation/sec), EAP ON</td><td>169</td><td>219</td><td>297</td><td>297</td></tr>
<tr><td>802.1X Auth Rate (transaction/sec), EAP OFF</td><td>115</td><td>185</td><td>220</td><td>220</td></tr>
<tr><td>Captive Portals (transactions/sec), 2K bit</td><td>81</td><td>114</td><td>132</td><td>132</td></tr>
<tr><td>100Base-T ports</td><td colspan="4">2</td></tr>
<tr><td>10 GbE ports (SFP+)</td><td colspan="3">4</td><td>8</td></tr>
<tr><td>40 GbE ports (QSFP+)</td><td colspan="3">NA</td><td>2</td></tr>
<tr><td>Redundant PSUs</td><td colspan="4">Yes</td></tr>
</table>

Step 6—Edge design

For the edge switch design in a high-density environment, Aruba recommends having full nonblocking 1 GbE ports downstream to APs, full 802.3 at PoE with 30W on all ports, and Cat-6A cabling. You should connect the APs with 2 10 GbE uplinks for redundant core connections.

Step 7—Core design

For the switch core design, verify that Address Resolution Protocol (ARP) cache and forwarding tables in core switches are large enough to handle a big flat user VLAN and the controller-to-core uplinks are sized at 2X the WLAN throughput computed in the capacity plan. A 1–2 Gbps over-the-air value becomes 2–4 GbE on the controller uplink. The controller should not be the default gateway for the WLAN clients as the first hop redundancy is critical.

Step 8—Server design

For the DHCP/DNS servers in a high-density design, the key metric is transaction time, which should be <= 5ms. This is much more critical than transaction rate and should be modeled at 5% of seating capacity over 5 minutes. For example in an 18,000 seat arena × 5%/300 seconds = 3 discovers per second.

Aruba strongly recommends a carrier-grade DHCP/DNS servers such as Infoblox. Lease times should be 2 times the duration of the event (8 hours suggested).

You should model DNS at 1 request/device/second, and the captive portal rate = DHCP arrival rate. The RADIUS loads depend on whether devices are using 802.1X.

Step 9—WAN edge design

The WLAN uplink bandwidth is estimated using the Aruba Total System Throughput process. Minimum bandwidth is dual load-balanced 1 Gbps links for a country with 20+ channels in 5 GHz using all/most DFS channels. Any VHD area with 20+ APs should easily be able to generate 1 Gbps of load.

You may require WLAN uplinks greater than 2 Gbps if you are attempting RF spatial reuse.

In addition, all edge equipment must be fully redundant.

Deployment example auditorium

In this section, you will explore a deployment example.

Example scenario: Typical multi-auditorium

This example could be a hotel conference center or a university building with multiple adjacent auditoriums. Figure 9-17 shows the physical layout, and Table 9-5 lists metrics for the Associated Device Capacity.

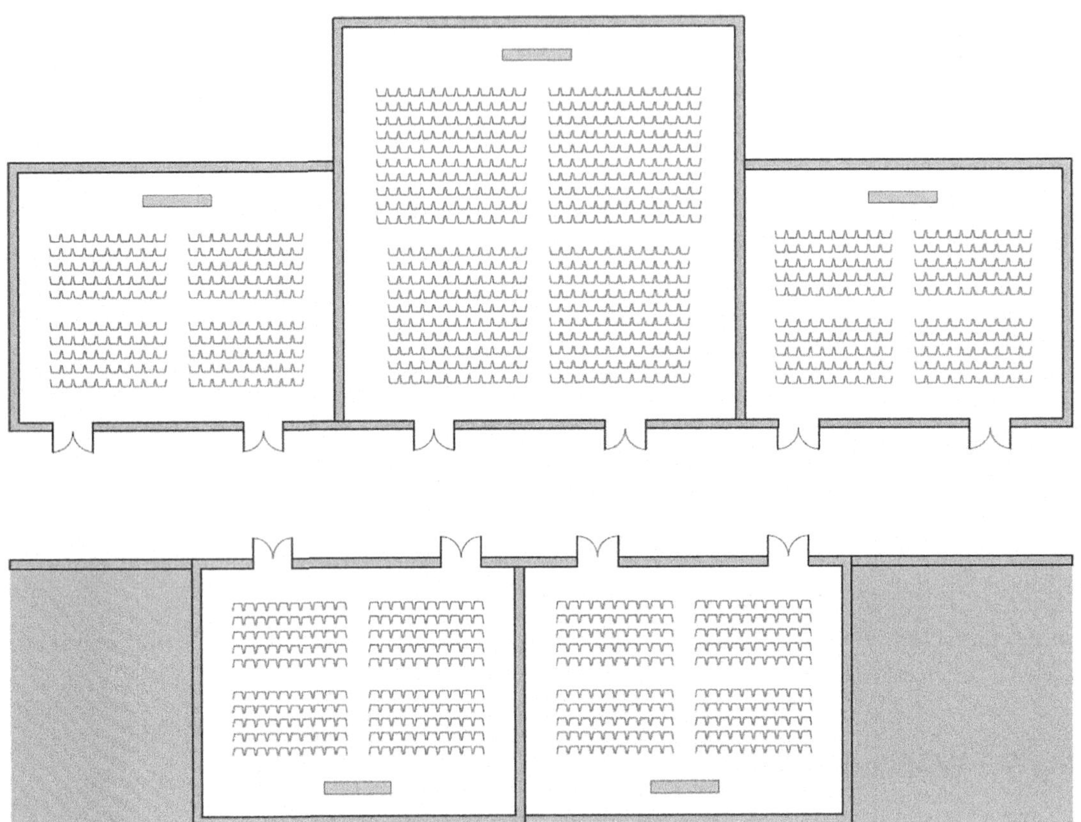

Figure 9-17 Physcial layout

Table 9-5 Associated Device Capacity for the example scenario

Metric	Target
Take Rate	100%
Average devices per person	Work/study—3–5 (future) Fan/guest—1–2 (future)
Associated devices per radio	150
Average single-user goodput	1 Mbps
5 GHz vs. 2.5 Ghz split	5 GHz: 75% 2.4 GHz: 25%

Step 1—Understanding load in auditoriums

You can use Table 9-6 and Table 9-7 to understand the load in auditoriums.

Table 9-7 will show information concerning:

- Common apps for web browsing, email, and office collaboration.
- Class presentation and exam software that are bursting with high concurrent usage.
- Cloud service latency that is not visible to users.

Table 9-6 VHD spatial stream blend lookup table

VHD Usage Profile	Devices/Person (Now)	Devices/Person (Future)	1SS (%)	2SS (%)	3SS (%)
Work/Study	3	5	30%	60%	10%
Fan/Guest	1	2	50%	50%	0%

Table 9-7 Network characteristics of common auditorium applications

User Category	Application	Bandwidth	Latency	Duty Cycle
Work/Study	Play courseware (non video)	500 Kbps	Medium	Medium
	Play courseware (video streaming)	1 Mbps+	Low	High
	Test/exam/quiz	Under 250 Kbps	Real-time	Synchronized bursts
Fan/Guest	General Internet usage	500 Kbps	Medium	Low
	Email	Under 250 Kbps	High	Low
	Social media (text)	500 Kbps	Medium	Low
	Photo/video cloud sync	1 Mbps+	High	Low

Step 2/3—Estimate associated device capacity

Use the formulas provided earlier in this chapter to calculate the Associated Device Capacity:

- Start with the seating capacity.
- Multiply the capacity by the number of devices per user, as well as the take rate, to estimate the Associated Device Capacity.

  In the example that you have been exploring, guest, faculty, and staff all have different numbers of devices, so you would multiply the number of people in each group by the per-user device number and take rate for that group. You would then total the results.

- Determine the percentage of devices on each frequency band. Multiply the Associated Device Capacity by these percentages to determine the 2.4 GHz Associated Device Capacity and the 5 GHz Associated Device Capacity.
- Determine the required address space.

  If you have different types of users such as guests, faculty, and staff, remember that you might need to place the users in different subnets. You would then need to determine the address space for each.

Table 9-8 provides some examples.

Table 9-8 Estimate Associated Device Capacity

Room Number	Seats	ADC (Now)	ADC (Future)	5-GHz ADC (Future)	2.4-GHz ADC (Future)	Minimum Subnet Size
Room A	200	600	1000	750	250	/22
Room B	200	600	1000	750	250	/22
Room C	500	1500	2500	1875	625	/20
Room D	200	600	1000	750	250	/22
Room E	200	600	1000	750	250	/22
Staff/House	–	25	75	75	0	/24
GUEST ADC	**1300**	**3900**	**6500**	**4875**	**1625**	**/19**
STAFF ADC	**–**	**25**	**75**	**75**	**0**	**/24**
TOTAL ADC	**1300**	**3925**	**6575**	**4950**	**1625**	

Step 4—Estimate the AP count

Next, estimate the AP count. (See Figure 9-18.)

- Begin with the 5 GHz Associated Device Capacity (assuming that a higher percentage of devices connect to 5 GHz).
- Divide by the desired maximum associations per radio, which is recommend at 150 for VHD deployments.

$$\text{AP Count} = \text{5-GHz Radio Count} = \frac{\text{Active Device Capacity (5 GHz)}}{\text{Max Associations Per Radio}}$$

Figure 9-18 Step 4—Estimate the AP count

Table 9-9 provides some examples of estimating AP counts.

Table 9-9 Examples of estimating AP count

Room Number	5-GHz Guest	5-GHz Staff	Total 5-GHz Devices	Devices per Radio	AP Count
Room A	750	15	765	150	6
Room B	750	15	765	150	6
Room C	1875	15	1890	150	13
Room D	750	15	765	150	6
Room E	750	15	765	150	6
Hallway	500	15	515	150	4
TOTAL	**5375**	**90**	**5465**		**41**

Step 5—Calculate system throughput

Finally, determine the system throughput calculations, excluding co-channel interference (CCI). Convert the number of APs to the number of channels that they offer. In the example deployment, the environment permits the use of DFS channels. Within each VHD, each AP represents one 5 GHz channel because the 5 GHz band offers many nonoverlapping channels. However, each area only has three nonoverlapping 2.4 GHz channels. Therefore, all AP will advertise a 5 GHz channel, but some APs may not advertise a 2.4 GHz channel due to the density of the deployment.

You can then multiple the number of channels by the estimated channel capacity to calculate an aggregate bandwidth (see Table 9-10).

Table 9-10 Calculate system throughput

Room Number	AP Count	Channels (USA DFS)	Avg. Channel Bandwidth	Aggregate Bandwidth
Room A	6	9	67 Mbps	603 Mbps
Room B	6	9	67 Mbps	603 Mbps
Room C	13	16	67 Mbps	1072 Mbps
Room D	6	9	67 Mbps	603 Mbps
Room E	6	9	67 Mbps	603 Mbps
Hallway	4	7	67 Mbps	469 Mbps
TOTAL	**41**			**3,953 Mbps**

Activity 9: Meet customers' high-density requirements

You will now continue designing a network for the customer scenario that was introduced in Activity 1. In this activity, you will use the customer scenario to determine the customer requirements for solving the Wi-Fi issues in the meeting area. You will then make a plan to meet those requirements. You will also plan a higher- capacity design.

Scenario high-density deployment

This addition to the main scenario explains the need for a high-density deployment.

Current Meeting Area

Corp1 has monthly all hands meeting in Building 1 in the open floor space. Every month a stage and 900 chairs are placed in this area. The CEO, CTO, and CFO all make presentations that last 2 hours. Breakfast is served to attract all the employees.

In the past, they did a few slide presentations and speeches. Of late, the CEO encourages all the employees to logon and take a look at online sales and financial information. On a company APP, they ask questions and attendees can win prizes.

The Data Center server is always ready for this hit that occurs every month, but the previous wireless was not capable of handling the load. Many employees complain that they cannot see the online information or enter the quiz contest.

Corp1 High-density requirement

The CEO has mandated a solution to this problem. The Wi-Fi network should be able to handle 900 employees in the open space of Building 1, Floor 1.

Activity 9.1: Plan a solution for the meeting room

Activity 9.1 objectives

You will plan new AP count and layout for meeting area.

Activity 9.1 steps

1. What is your new AP count for the meeting area?
2. What is your recommendation for the layout of these APs in the meeting room?

Activity 9.2: Plan higher capacity across the site

Activity 9.2 objectives

Based on new customer requirements, such as the need to support voice and video, you will redesign the RF coverage to support higher capacity. You will plan a new AP count and consider the implications for your design. An SE returning from the customer site reports that the site has high cubicles with metal cabinets on the top.

Activity 9.2 steps

1. To support higher speeds across the wireless network, you might need to deploy APs more densely. You will now assess your plan.
2. Return to VisualRF and view your plan for Building 1, Floor 2. This floor is representative of the floors for business users.
3. Think about the square footage of the floor, as well as its physical characteristics. Adjust the AP plan as required to provide the proper capacity. You can use the heat maps, speed, and voice views in VisualRF to help.
4. After you make your adjustments, how many APs are you planning for the business user floor? Explain your plan.
5. Multiply this number by 4 for the total number of APs on the business floors (Building 1, Floor 2 and all floors in Building 2).
6. Similarly assess your plan for Building 1, Floor 3, the executive floor. How many APs will you deploy on this floor to support voice and video? Explain your plan.
7. How many APs will you plan for Building 1, Floor 1, including the ones that you planned for the meeting area?
8. How many APs are you now recommending for the site? (Add the numbers for steps 4, 5, and 6).
9. What implications does this change have on the rest of your plan? Think about as many of the potential effects as you can, going over all the topics that you have learned about up until now.
10. Return to Iris and adjust your plan. Change the Quantity multiplier for APs. Add new licenses and make any other required changes.

    If you do not have access to Iris, you can use the HPE Networking Online Configurator. (Visit http://hpe.com/networking/configurator to access this tool.)
11. The AP icon will turn red when you add more APs to the quantity because these APs have no power. Connect the new APs to switches. If you do not have enough ports available on the switches, add switches or modules to support the APs.
12. Save your project.

Answers to the questions included in activities are provided in "Appendix: Activities and Learning Check Answers."

Summary

You now have the knowledge necessary for completing a VHD RF design. You know how to plan appropriate VHD RF coverage for large deployments.

Learning check

1. Which is true of side coverage established by APs with omnidirectional antennas?
   a. It is required for very high density deployments in large venues.
   b. Picocells are required, increasing the number of APs in the design.
   c. Spatial reuse increases, distorting the signal.
   d. 50% of AP signals are lost to the next room.
2. As a general guideline how many associations should you plan for a high-density design?
   a. 150 per radio
   b. 170 per radio
   c. 210 per radio
   d. 255 per radio

Answers to the learning checks in this study guide are provided in "Appendix: Activities and Learning Check Answers."

10 Branch Deployments

EXAM OBJECTIVES

✓ Given a scenario, translate the business needs of a single-site campus environment or subsystems of an enterprise-wide environment into technical customer requirements.

✓ Given customer requirements for a single-site campus environment or subsystems of an enterprise-wide environment, determine the component details and document the high-level design.

Assumed knowledge

- Aruba product portfolio, including Remote Access Points (RAPs)
- Aruba 8.x architecture

Introduction

As companies grow, they often look for ways to expand their business through remote branch locations, home office workers, and on-the-road employees. Companies need cost-efficient and secure ways to give their remote staff corporate access.

In this chapter, you will review the different branch deployment options Aruba offers and their requirements. You will also explore the different options for corporate access, including Aruba Remote Access Points (RAPs), Virtual Intranet Access (VIA), and Instant AP (IAP) Virtual Private Networks (VPNs). Finally, you will see how a branch office controller works.

Remote deployment options

You will begin by reviewing Aruba branch deployment options.

Remote branches, home workers, and road warriors

As Figure 10-1 shows, many companies have remote branch locations, home office workers, and employees on the road. All these locations need corporate access. Costly WAN links are no longer an effective solution for most sites. However, using inexpensive local Internet access does provide a cost-efficient solution.

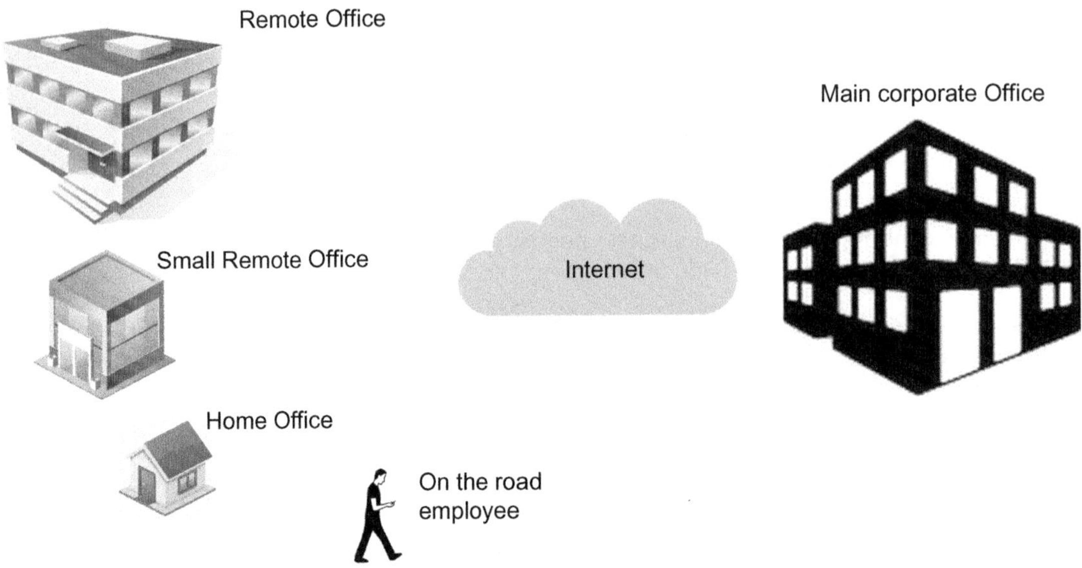

Figure 10-1 Remote branches, home workers, and road warriors

Employees who work remotely can easily install an Aruba AP to securely access the corporate network. Once employees are connected, they have the same experience they would have if they were at the corporate office—thanks to a zero-touch VPN link to an Aruba controller in the data center.

If an organization's branches or small-to-medium-sized businesses (SMBs) need Wi-Fi or switching, deploy Aruba IAPs and Aruba switches with optional cloud management to simplify the enterprise network.

The RAP and VIA, a VPN client, are solutions to securely extend corporate access to small branches, temporary sites, and home offices with simple plug-and-play.

Remote access options

As Figure 10-2 shows, a RAP is a single AP that sets up a VPN connection to a Mobility Controller (MC) that is configured with a VPN service. RAPs are ideal solutions for home or small offices. You can convert any AP into a RAP.

If you require a larger number of APs at the branch office, then a branch office controller could be your solution.

Another remote solution is a cluster of IAPs. The Virtual Controller (VC) in the cluster can set up a VPN connection to an MC that is configured with VPN service. You can use Activate to automate this service.

Road warriors can download a VIA application on their laptops or smartphones, which gives them user access to the corporate network.

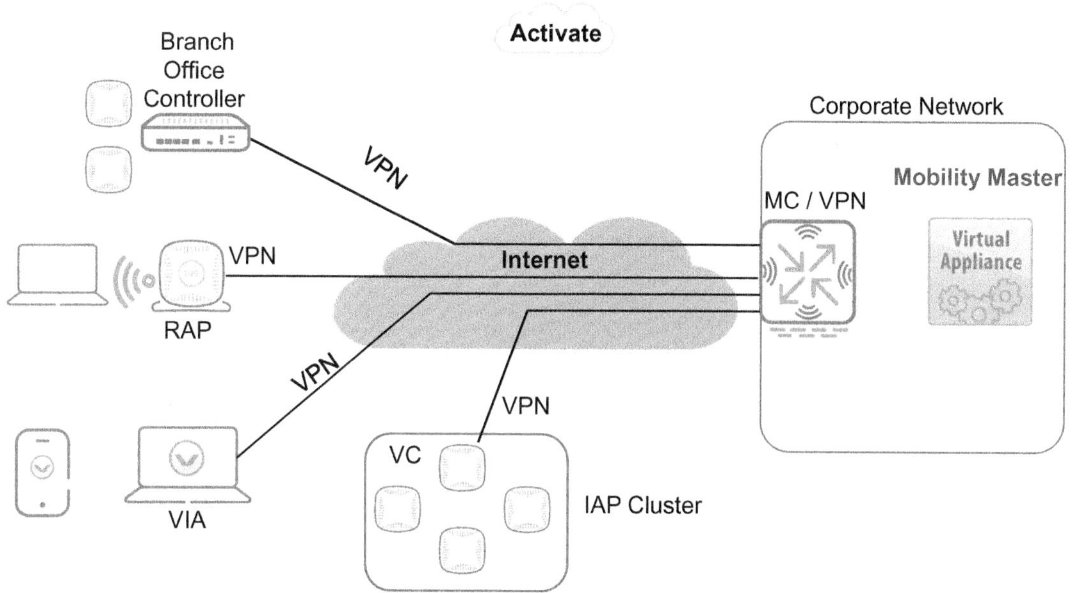

Figure 10-2 Remote access options

Remote Access Points (RAPs)

You will now take a closer look at Aruba RAPs.

RAP overview

RAPs are very popular because they simplify the user experience. A user associating to an SSID in the corporate network would find the same SSID, encryption type, and access rights on a RAP (see Figure 10-3). Therefore, associating to the RAP is the same experience as associating to a Campus AP (CAP). You do not need to enable third-party VPN clients. Because all user association, authentication, and access rights are the same, troubleshooting a user issue is in the same place for a RAP and CAP.

RAP is perfect for home offices or small remote branches that need Wi-Fi and access to the corporate network.

Aruba Activate can communicate the location of the MC or VPN controller to the RAPs, reducing the cost and time to deploy wireless in remote branches and home offices.

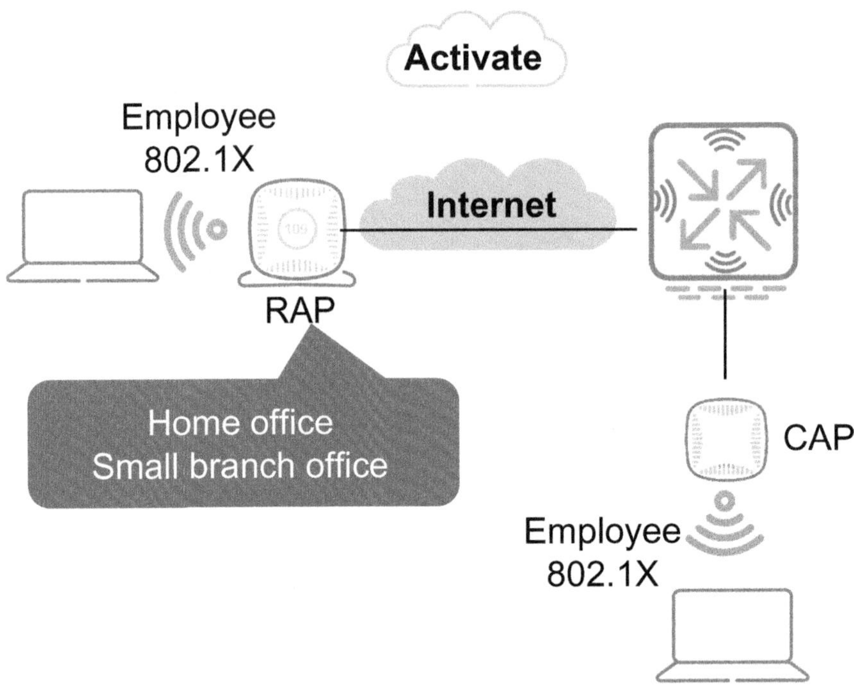

Figure 10-3 RAP overview

Deployment of RAPs uses zero touch with the aid of the cloud-based Activate server. You can ship a new RAP to a location. At the location, a user needs to give the RAP power and a wired connection with Internet access. The Remote AP will boot up as an IAP, or newer versions, as a Unified AP (UAP).

The RAP series of APs offers a wide range of enterprise-class features, including role-based network access, policy-based forwarding, and Adaptive Radio Management (ARM), which optimizes Wi-Fi client performance and ensures that Aruba APs stay clear of interference. With a wireless data rate up to 300 Mbps in the 2.4-GHz and 5-GHz radio bands, the RAP-100 Series has one 10/100/1000BASE-T wired uplink, one 10/100BASE-T local wired port and one USB port to connect to 3G and 4G networks.

RAP deployment options

RAPs must communicate with a controller. If you are deploying a few RAPs, you can have the RAPs go through a gateway firewall to gain access to a controller inside the corporate network. Aruba recommends you place a controller or controllers in the DMZ, if you are deploying several RAPs for corporate use.

For large installations across multiple sites, the Aruba Activate service significantly reduces deployment times by automating device provisioning, firmware upgrades, and inventory management. With Aruba Activate, IAPs and UAPs can easily reconfigure themselves as RAPs, lowering the cost and time to deploy wireless in remote branches and home offices. An IAP will attempt to communicate with the Activate server. A UAP will first attempt to find a controller. If the UAP does not find a controller, it will attempt to communicate with the Activate server.

The Activate server will look up a list of devices, and if it finds this IAP or UAP in the list, Activate will apply the device's rule. This rule could be to convert the IAP to a RAP and direct it to an MC with VPN services. Once connected to the MC with VPN, the RAP will receive its configuration and start advertising the WLAN.

If you want to convert CAPs into RAPs, you can complete this process manually.

The RAP and its controller communicate with Protocol Application Programming Interface (PAPI). This PAPI control traffic is encrypted in the L2TP/IPsec tunnel. The gateway firewall will need to use Network Address Translation (NAT) on the traffic sent back to the controller. RAPs will use NAT-Traversal (NAT-T) to maintain the VPN traffic across gateways that implement NAT functions.

The gateway firewall must allow User Datagram Port (UDP) port 4500 for the Internet Protocol security (IPsec) connection and NAT-T. The UDP port 500 is used for the VPN ISAKMP (see Figure 10-4).

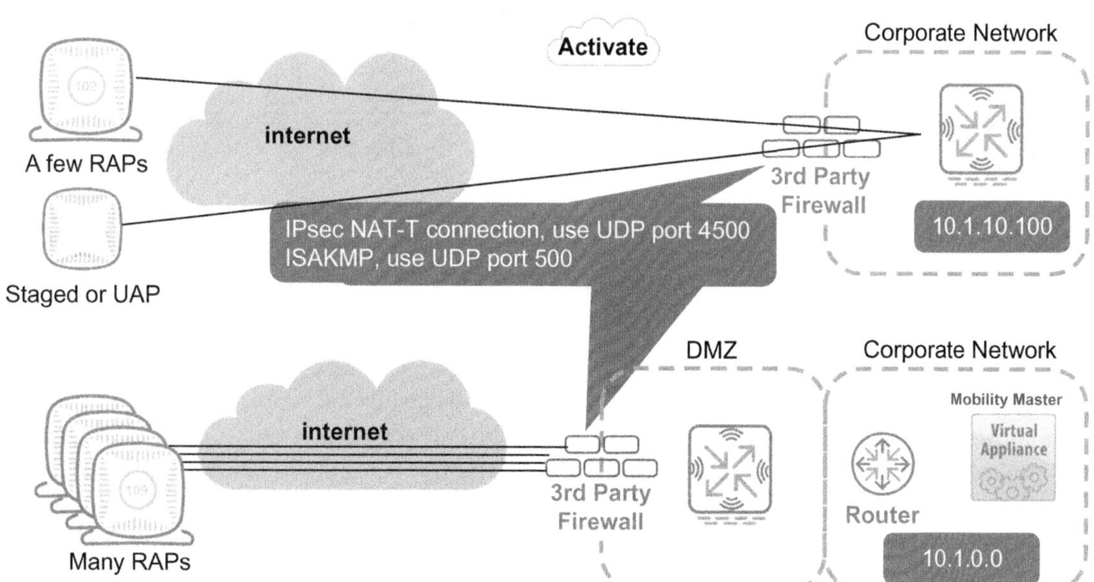

Figure 10-4 RAP deployment options

RAP WLAN forwarding modes

As Figure 10-5 shows, you can configure SSIDs in three modes for RAPs. You can have one SSID in Tunnel mode, another SSID in split-tunnel mode, and still another in Bridge mode.

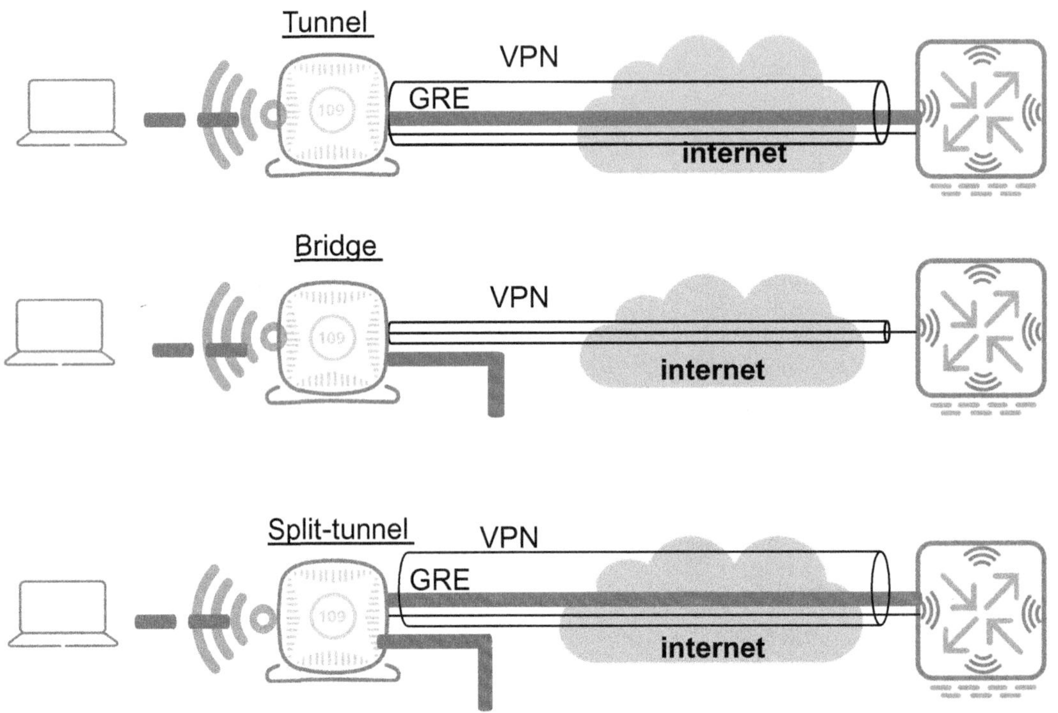

Figure 10-5 RAP WLAN forwarding modes

In Tunnel mode, the RAP sends data traffic inside a GRE tunnel to the controller. It still encrypts its own control traffic, as mentioned before, but does not encrypt client traffic (The GRE tunnel is encapsulated in IPsec, but not encrypted by IPsec). Instead, it is assumed that the WLAN enforces WPA2-Enterprise, the user passes 802.1X authentication, and the user traffic is encrypted with WPA2 AES. This user data is encrypted and unencrypted at the controller, which also firewalls, switches, and routes the traffic. However, you can configure a double encryption option to have the RAP encrypt all traffic. In either case, clients obtain a corporate IP address, based on the VLAN configured in the controller.

You can use the tunnel option when the network requires all traffic to be sent to corporate. For example, network administrators might want all branch traffic to be firewalled and pass via the corporate content server. Alternatively, the branch might have an IP voice solution, and the Session Initiation Protocol (SIP) server is at the corporate office.

In Bridge mode, the RAP will encrypt and decrypt the traffic, but all traffic is bridged locally. The RAP removes the 802.11 frames and bridges the packet into the local wired LAN. The client acquires its IP address locally. You will typically use this option when you prefer local access instead of corporate access. No firewall rules will be applied.

You can use Split-tunnel mode for employees that may need to tunnel their traffic to corporate, but also have local requirements. The RAP, with the aid of a firewall policy, will decide if traffic must be tunneled to the controller or bridged locally. The 802.11 frames are either tunneled or bridged, while the role and policy determines destination. The client obtains an IP address from a VLAN on the controller in the corporate network. This is a popular choice when users not only need corporate access but also require some local traffic. For example, an employee needs access to corporate servers but needs to print locally.

Note that if you select the "double encrypt" option mentioned earlier, the RAP encrypts the user traffic (already encrypted or not) in the L2TP/IPsec tunnel. L2TP is a tunneling protocol that offers no protection. IPsec provides per-packet data origin authentication, data integrity, replay protection, and data encryption. IPsec performs authentication with a certificate or a pre-shared key. Once the IPsec SAs are successfully created, then the L2TP connection performs a user-level authentication.

Virtual Intranet Access (VIA)

In the next section, you will focus on VIA.

VIA overview

VIA is a part of the Aruba remote networks solution intended for teleworkers and mobile users. A software application for laptops or smartphones, VIA is integrated into the Aruba solution. VIA enables remote clients to connect over a VPN tunnel to an MC, which firewalls and forwards the clients' traffic. The devices are also displayed as clients in the Mobility Master (MM).

When a client with VIA connects to a network, VIA detects that the network environment, whether trusted or untrusted. A trusted network refers to a protected office network that allows users to access the corporate intranet directly. Untrusted networks are public Wi-Fi hotspots, such as airports, cafes, or home networks. If VIA senses that the client has connected to an untrusted network, it will automatically attempt to establish a VPN connection to the MC VPN server. If IPsec is blocked, for whatever reason, VIA will fallback and attempt a Secure Shell (SSL) connection. You can use the same mobile device credentials that authenticate users to the WLANs to authenticate the VIA user.

Two-factor authentication (TFA, T-FA, or 2FA) is an approach to authentication that requires the presentation of two or more of the three authentication factors. Two popular forms of two-factor authentication include RSA SecurID and other token card products and smart cards. RSA SecurID is a hardware and software-based authentication mechanism that generates unique authentication codes at a specified interval using an RSA SecurID token. Smart cards provide two-factor authentication using a certificate and personal identification number (PIN).

The Duo application can integrate with ClearPass and support two-factor authentication for VIA on mobile devices. For classified or highly sensitive network deployments, VIA has been enhanced to support RFC 4869, Suite B Cryptographic Suites for IPsec. VIA with Suite B is enabled with the optional ArubaOS ACR module. (Suite B is a set of cryptographic algorithms developed by the National Security Agency as part of its Cryptographic Modernization Program. It is designed to serve as an interoperable cryptographic base for both unclassified information and most classified information.)

Duo includes some clever features. An admin can use an app on their phone (Duo Mobile) to generate a one-time-use PIN, or the admin can have Duo (in the cloud via the RADIUS proxy) send users' phone a push notification. That push notification opens up the app and allows users to opt to Approve or Deny access. The only caveat you may encounter with the push notification is that the ClearPass Policy Manager admin user interface times out after 10 seconds, which might not give sufficient time for the push notification to reach the phone and for the user to unlock the phone, launch the app, and press "Approve."

VIA deployment

You can install VIA on several different types of devices from laptops to smartphones (see Figure 10-6). Laptops can install the VIA client locally or download the VIA client from the controller. The smartphone will download the VIA as an app.

VIA will automatically determine if it is on a public or corporate network. When on the corporate network, VIA will disable itself.

You can configure VIA clients to tunnel all traffic or to work in split tunnel mode. This allows local traffic to remain local and corporate traffic to be sent across the VPN tunnel.

For VIA implementations, the solution requires either VIA licenses or PEFV licenses. The VIA licenses are per-user. They can be installed on an MM, pooled, and deployed to any MC that needs to terminate a tunnel dynamically. A PEFV license, on the other hand, is bound to a particular MC and enables it to terminate VPN tunnels for VIA users.

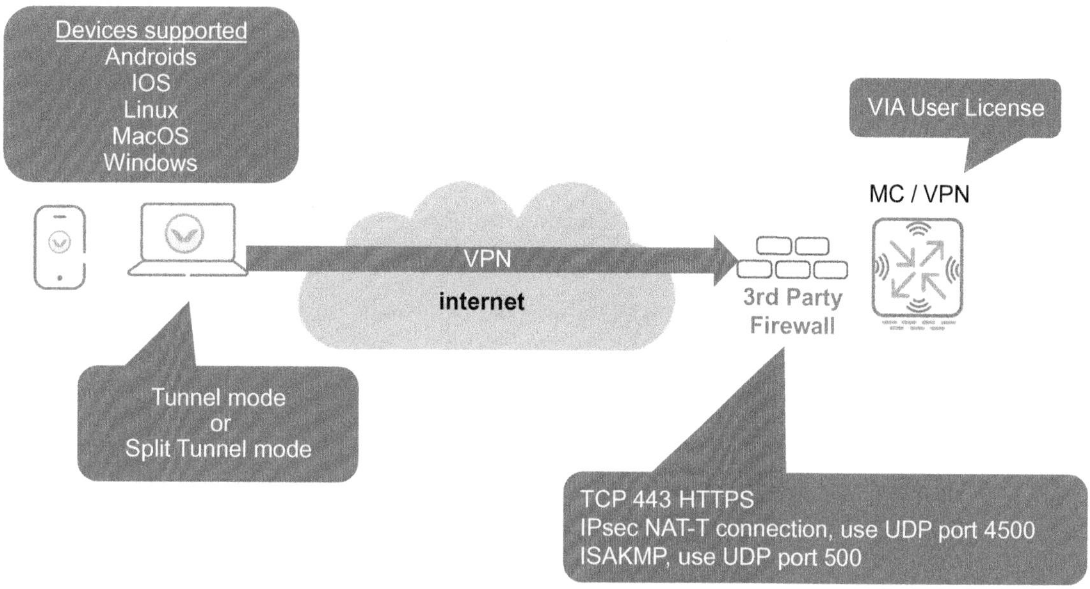

Figure 10-6 VIA Deployment

These are the ports needed for VIA:

- TCP 443 is during the initialization phase. VIA uses HTTPS connections to perform trusted network and Captive Portal checks. It is mandatory that you enable port 443 on your network to allow VIA to perform these checks.
- You can use the UDP port 4500 for the IPsec connection and NAT-T.
- You can also use the UDP port 500 for the VPN ISAKMP.

For traveling employees VIA is a better solution. VIA is an Aruba VPN client that is totally integrated in the Aruba solution. VIA is dormant, on your laptop or smartphone when the client is on the corporate network, but is automatically activated when the client is on a foreign network.

End-users use a web authentication profile to login to the VIA download page using https://<corporateVIA>/via. If they are setup, they can download a VIA client. If more than one VIA authentication profile is configured, users can view this list and select a profile during client login.

In the IPhone, you can download the VIA in the app store by looking for Aruba VIA. You can do the same on an Android device, using the Play store or the Google Play store.

Instant Access Point (IAP) Virtual Private Networks (VPNs)

You will now focus on IAP VPNs.

Aruba IAPs

Aruba Instant is a simple, easy to deploy turnkey WLAN solution consisting of one or more IAPs. An Ethernet port with routable connectivity to the Internet or a self-enclosed network is used for deploying an Instant wireless network. An Instant AP can be installed at a single site or deployed across multiple geographically dispersed locations. Designed specifically for easy deployment and proactive management of networks, Instant is ideal for small customers or remote locations without requiring any on-site IT administrator.

With Aruba Instant, traffic is locally bridged, but it still provides the flexibility of secure VPN tunnel to Mobility Controllers.

IAP VPN deployment

For remote locations that need enterprise-grade Wi-Fi that is simple to deploy and manage, a great option is controller-less Aruba IAPs that provide robust, resilient, and secure branch network services.

The IAP cluster can be autonomous, but some customers use them in branch office solutions and still require corporate access. As Figure 10-7 shows, the IAP can set up a VPN connection back to an MC, providing access to corporate resources. The IAP cluster is independent and has a VPN to access the corporate network.

When you install new IAP clusters, the IAP will communicate with the Activate server and find the location of the MC. The IAPs have a management interface that offers visibility into a single Instant network. You can easily manage multiple Instant networks with cloud-based Aruba Central or on-site with the Aruba AirWave management system.

Future growth and changes in the network may better be served by a controller. You can convert Instant APs to controller-managed APs, which will protect your AP hardware investment.

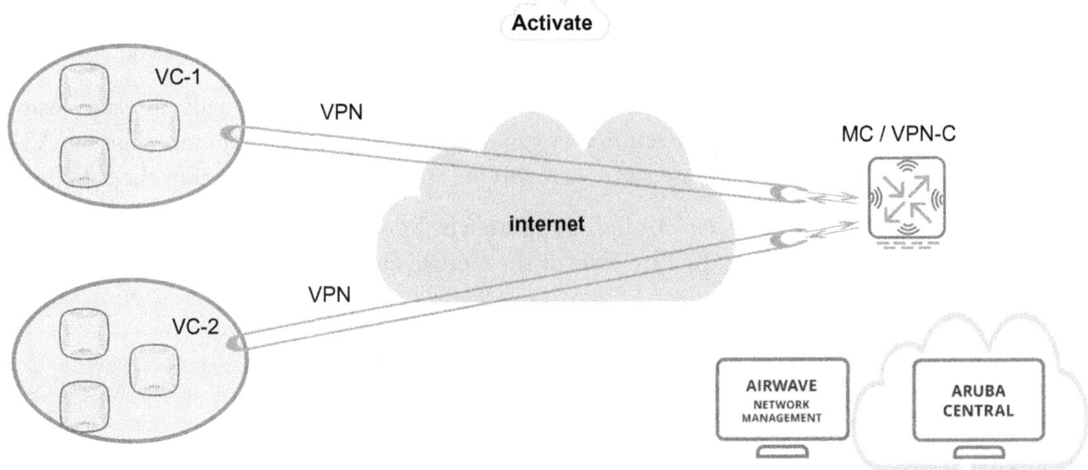

Figure 10-7 IAP VPN deployment

There are four types of VPN setups for the IAPs. Aruba IPsec and Aruba GRE must terminate on an MC. Manual GRE and L2TPv3E can terminate on a VPN server. By preference, Aruba IPsec is used.

Aruba IPsec tunnel

An Aruba IPsec tunnel is configured to ensure that the data flow between the networks is encrypted. When configured, the IPsec tunnel to the controller secures corporate data. Characteristics and benefits of the IPsec tunnel include the following:

- Only supported with an Aruba Controller.
- Can only configure a whitelist and VPN pool on the controller.
- Supports all modes of operation including Layer 2, Layer 3, and local.
- Encrypts both AP and client traffic.
- Can traverse NAT boundary with UDP 4500.

Aruba GRE

Aruba GRE does not require any configuration on the Aruba Mobility Controller because it will act as a GRE endpoint. Characteristics of the Aruba GRE tunnel include the following:

- It is only supported with an Aruba Controller.
- IAP will set up an IPsec tunnel to program the controller with a datapath, which, in turn, will set up the GRE tunnels. No manual GRE configuration is needed on the controller.
- IAP has a peer GRE tunnel used to transport client traffic.
- Only Layer 2 modes of operation are supported.
- GRE failover is supported.
- It cannot traverse NAT boundary.

Manual GRE

A manual GRE requires GRE tunnels to be explicitly configured on the GRE-endpoint, which can be an Aruba Mobility Controller or any device that supports GRE termination.

L2TPv3E

L2TPv3E Layer 2 tunnel protocol allows an IAP to act as a L2TP Access Concentrator (LAC) and tunnel all wireless clients with Layer 2 traffic from an AP to a L2TP Network Server (LNS).

Table 10-1 provides IAP IPsec throughput for IAP VPNs. If the network needs more throughput, then a branch controller is recommended.

Table 10-1 IAP IPsec throughput for IAP VPNs

AP Model	Tunnel	Throughput
IAP 325	IPsec	153 Mbps
IAP 315	IPsec	120 Mbps
IAP 335	IPsec	90 Mbps
IAP 303H	IPsec	31 Mbps
IAP 203R	IPsec	20 Mbps
IAP 203H	IPsec	20 Mbps
IAP 203RP	IPsec	19 Mbps
IAP 207	IPsec	19 Mbps

Aruba Activate

Aruba Activate is a cloud-based, zero-touch provisioning system. Aruba Activate provides plug-and-play capability to an Aruba Instant cluster, which allows rapid deployment of the clusters with minimal or no IT expertise.

An IAP or UAP will communicate with the cloud-based Activate server. If the Activate server has a setup configuration for this AP, it sends a message back to the AP. This message will include its setup instructions.

The AP may be directed to a Central cloud-based management system to receive its configuration and after that be managed, or the AP may be directed to the AirWave management system. From here, it will receive its configuration and be managed.

The AP can be converted into a Campus AP (CAP), where it will be directed to an MC.

If you need remote access, you can convert the AP to a RAP. It will then be directed to an MC with VPN services.

UAPs

UAPs have a new setup procedure. First, the UAP searches for a controller with the five standard methods, using the static configuration or option 43 in a DHCP response. If there is no static information or an option 43, then the UAP will attempt to reach the MC by a broadcast, a multicast, and a DNS query.

If the AP cannot discover a controller, it will listen for IAP master beacons. If it receives a master beacon, the AP will join that Instant cluster.

If the AP cannot discover a controller or locate an Instant cluster, the AP tries to connect to an AirWave server (if one was provided via DHCP/DNS options). The AP receives its image and configuration from AirWave and joins the appropriate Instant cluster.

If an AirWave server is not found, the AP connects to Activate and looks for a provisioning rule on Activate. If the AP finds a provisioning rule to connect to an AirWave server, it will attempt connection to AirWave and receive its image and configuration accordingly. If the AP finds a provisioning rule to connect to Central, it will communicate with Central and receive its image and configuration. If the AP finds a provisioning rule to connect to a controller as a CAP (or RAP), it will attempt to communicate with the controller and receive its image and configuration.

Even if no provisioning rules are configured on Activate, the UAP will still upgrade its image to the latest available full Instant image. After a reboot, if nothing was found during the discovery process, the UAP becomes the VC and forms its own Instant cluster. It broadcasts an SSID called SetMeUp-xx:xx:xx (where xx:xx:xx are the last hex octets of the UAP MAC address).

After it broadcasts the SSID, if there is no keyboard input or an active WebUI session for 15 minutes, the UAP will reboot and start the discovery process again.

Upon connecting to this SSID, you are presented with the Instant web UI where you can perform the usual Instant-related tasks. Other UAPs that are subsequently brought up will discover this Instant cluster and join it. This ends the discovery process.

There might be scenarios where the UAP does not have Internet connectivity and cannot reach Activate to upgrade its manufacturing image to a full Instant image. In this case, when no controller-based or controller-less networks are discovered, the UAP broadcasts the SSID SetMeUp-xx:xx:xx. However, connecting to this SSID will give you a provisioning web UI and not the Instant web UI. Using this UI, you can upgrade the UAP image to AOS or InstantOS, or perform maintenance and debugging tasks.

After the SSID is broadcast, if there is no keyboard input or an active WebUI session for 15 minutes, the UAP will reboot and start the discovery process again.

IAP DHCP modes when using VPN

When using VPN, the IAP DHCP mode decides how the IP address will be passed to the Wi-Fi clients. This could be local addresses or corporate addresses. The IAP DHCP mode also determines if traffic is switched or routed. As Figure 10-8 shows, there are Local modes, Centralized L2/L3 modes, and Distributed L2/L3 modes.

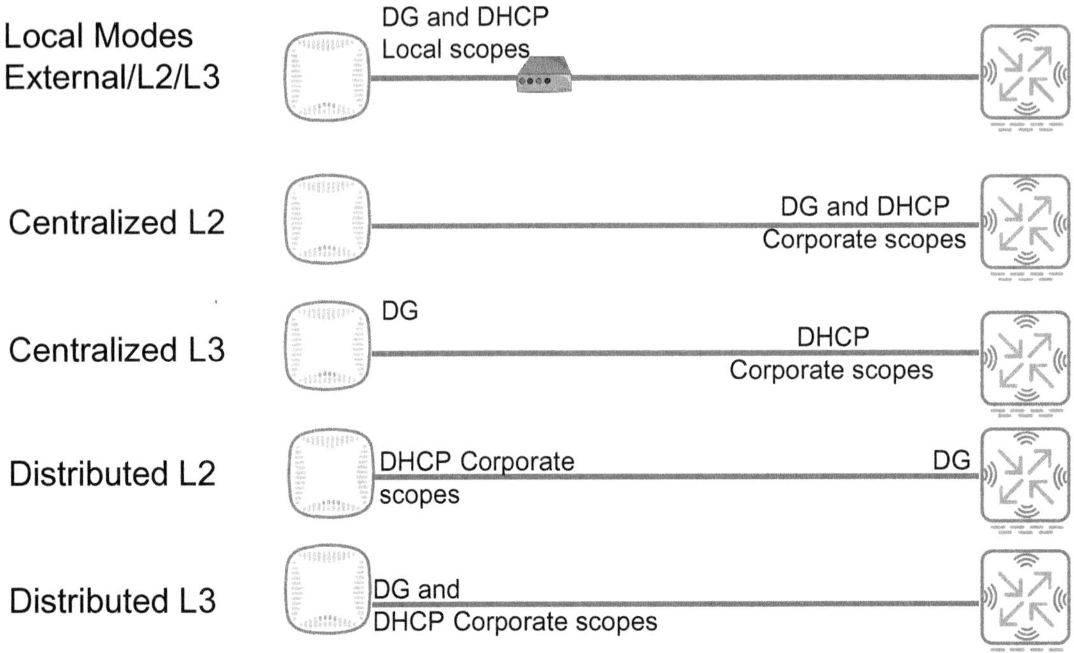

Figure 10-8 IAP DHCP modes when using VPN

The Local Mode is similar to the local network of a home wireless router, but is not an L2 or L3 extension of your corporate network. The DHCP and default gateway (DG) are local. You will use this option when you only require local access, but the Captive Portal is generated from a central location.

In Centralized L2 Mode, the DHCP server is centralized and the branch is an L2 extension. In this mode, the DHCP server and the gateway for the clients reside in the corporate network and the IAP cluster is essentially extending the corporate network VLAN to the branch office or your home office.

In Centralized L3 Mode, as the name implies, the DHCP services are centralized in the corporate network, but the branch subnet is an L3 extension of your corporate address space. The DG is the IAP VC with a corporate IP address.

Distributed L2 means your DHCP server is distributed, and your branch is a L2 extension of your corporate subnet. The DHCP server is distributed, and it is, therefore, the VC's responsibility to run the DHCP server and give out valid corporate addresses. In this mode, the IAP acts as a DHCP server for a corporate subnet.

With Distributed L3 Mode, the subnet is part of your overall corporate IP address space. The master IAP in the branch acts as the default gateway and the DHCP.

The Local Mode is similar to the local network of a home wireless router, but with VPN capabilities and enterprise-grade features. In local mode, the master IAP of the cluster has a local subnet, for example, 192.168.0.0/24, which is not an L2 or L3 extension of your corporate network. The master AP in the cluster is also the DHCP server and the gateway for clients. Local L2 and Local L3 are all local services, and the DHCP and DG are always local. You can use this option when you only require local access, but the Captive Portal is generated from a central location.

IAP DHCP mode recommendations

In local mode, the IAP cluster at the branch has a local subnet, and the master AP of the cluster functions as the DHCP server and gateway for clients. The local mode is well suited for branch guest networks that use a Captive Portal sever located in the data center for guest authentication.

Aruba recommends the centralized L2 mode only if Layer 2 extension is mandatory for branches. The mode is well suited for organizations that stream multicast videos to remote branches. The DHCP server and the gateway for the clients reside in the corporate network.

Centralized L3 moves the DG into the VC cluster, but remains part of the corporate network. All DHCP requests are sent to the corporate network. If the VPN were to fail, the users with IP addresses would still have some form of local access.

Aruba recommends distributed L3 mode for organizations that do not require Layer 2 extensions. The network is autonomous; Distributed L3 mode contains all broadcast and multicast traffic to a branch and eliminates any WAN bandwidth consumption. If you still require multicast and broadcast to corporate then use the Distributed L2 mode.

Figure 10-9 summarizes these recommendations.

Local External/L2/L3	Local access only
	Recommended for guest networks with a centralized captive portal servers.
Centralized L2	Recommended only if multicast traffic to a branch is required.
	If it is not required, use L3 modes.
Centralized L3	Recommended for deployments that require the use of a centralized DHCP server.
Distributed L2	Recommended only if multicast traffic to a branch is required.
	If it is not required, use L3 modes.
Distributed L3	Recommended for all deployments for site survivability.

Figure 10-9 IAP DHCP mode recommendations

Branch office controller

Finally, you will consider when to recommend a branch office controller.

Branch office controller

Aruba 7000 Series controllers offer integrated wireless, switching, and hybrid WAN services for distributed enterprises that need to rapidly deploy and manage an all-in-one customer-premise equipment solution for each branch or remote location. The 7000 series controllers can support between 16 and 64 APs and 4 to 24 ports, depending on the model you select (see Figure 10-10).

First, connect the 7000 series to an Internet connection and let it download its configuration from a centralized controller. The 7000 series will communicate with Activate to find the corporate centralized controller. This is a simple plug in and go solution.

The MM will manage the branch office controller like any other MC. You can also use AirWave to monitor the branch controllers.

A branch deployment can use backup solutions, such as 4G and LTE technologies.

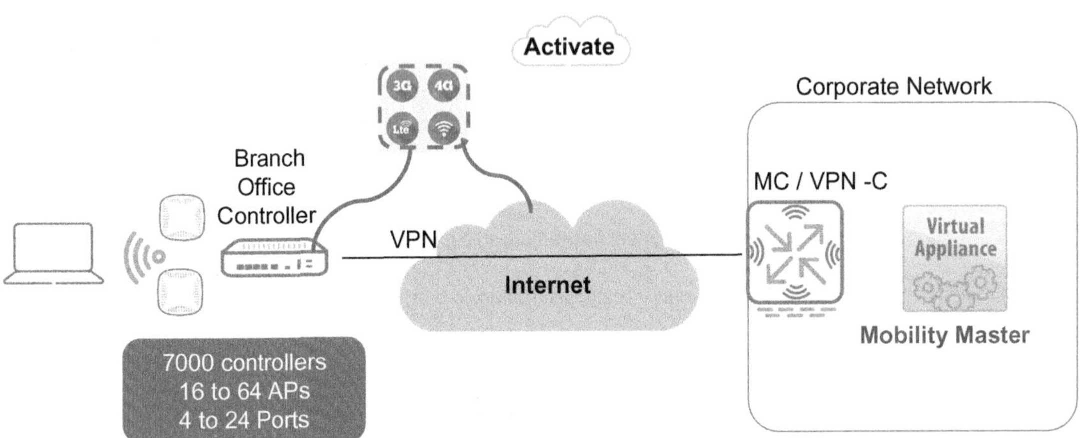

Figure 10-10 Branch office controller

Organizations can also extend advanced policy management, guest services, and employee self-service to the branch with the Aruba ClearPass Policy Management system.

Because deploying two branch office controllers can be complicated, if the customer has strict survivability requirements, IAPs might be the better choice. Other considerations apply as well, though. If you need to decide between a branch office controller or an IAP cluster with VPN, refer back to

"Chapter 3: Aruba Campus Design," which provides comparison between controller-based and IAP-based networks.

Starting with Aruba 8.2.0.0, by default, the DHCP lease limits for the 7000 Series Controllers are increased to those of the user limits (see Table 10-2). Also, a new CLI command, **ip dhcp increase-lease-limit**, is introduced in Aruba 8.2.0.0 for additional DHCP scope.

Table 10-2 DHCP lease limits for the Aruba 7000 Series Controller

Platform	DHCP Lease Limit in Previous OS	DHCP Lease Limit in Aruba 8.2.0.0	Additional DHCP Scope with CLI Option Enabled
7005 Controller	512	1024	2048
7008 Controller	512	1024	2048
7010 Controller	1024	2048	4096
7024 Controller	1024	2048	2048
7030 Controller	2048	4096	4096

Activity 10: Remote access design

In this activity, you will continue designing a network for the customer scenario introduced in Activity 1 and expanded in the activities that followed. In this activity, you will decide which remote solution will best solve the customer's remote access issues. Use the scenario information from previous activities plus the additional information provided below to complete this activity.

Scenario branch deployments

Determine the requirements for remote access.

Current remote access

Corp1 has 45 sales employees around the country. Their major complaint has been poor access to main campus servers. The sales personnel often find themselves at customer sites like offices, warehouses, dock yards, and truck yards, with no Wi-Fi access. They need access to corporate to gain access to servers with critical information. The company has created several apps to help the sales people, but without Internet access to corporate these apps are unusable.

Corp1 new remote access requirements

Corp1 wants to give their sales staff full access to corporate servers even when they are on the road.

A second issue is the lack of office space at corporate. Corp1 is looking into work from home solutions. Home employees would need corporate access and also have corporate phones installed in their home office. Corp1 expects up to 100 home office workers by the end of the year.

Activity 10.1: Sale employees

Activity 10.1 objectives

You will recommend a solution for the 45 sales employees that need constant access to corporate servers.

Activity 10.1 steps

1. What remote solution do you recommend?
2. Describe the advantages to your solution.
3. Does your solution have any network requirements?

Activity 10.2: Work from home employees

Activity 10.2 objective

You will recommend a solution for work from home employees.

Activity 10.2 steps

1. What remote solution do you recommend?
2. Describe the advantages to your solution.
3. Does your solution have any network requirements?

Activity 10.3: Add the remote solutions to the BOM

You will now adjust your BOM in Iris to show your plan. If you do not have access to Iris, you can use the HPE Networking Online Configurator. (Visit http://hpe.com/networking/configurator to access this tool.)

1. Return to your Iris application and, if necessary, open your project.
2. Expand the catalog and add any MCs that you planned in Activity 10.1 and 10.2. (See Figure 10-11.)

Figure 10-11 Add MCs in Iris

3. In the Properties window for the MCs, select the appropriate model.
4. Also add any RPs. Again, select the appropriate model in the Properties window. Remember that you can click the Attributes tab and use the Quantity Multiplier. (See Figure 10-12.)

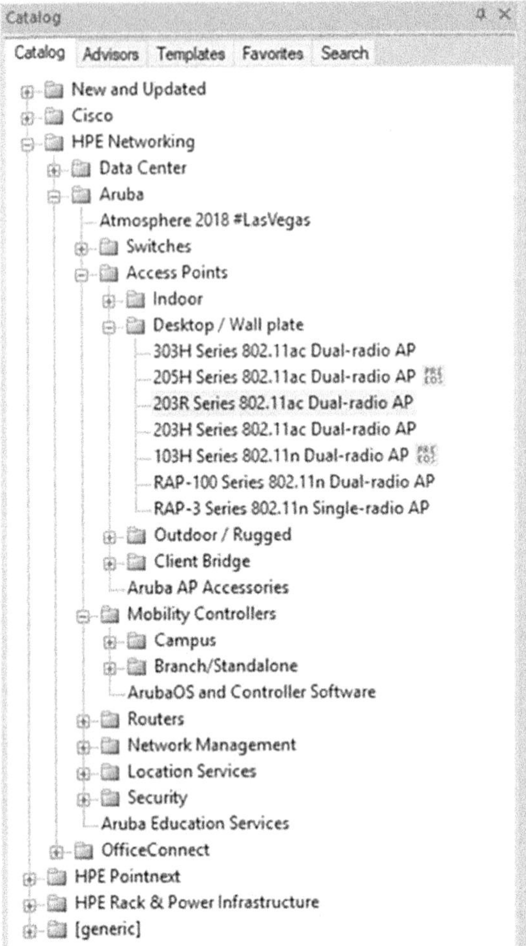

Figure 10-12 Add RPs in Iris

5. Remember to add the appropriate licenses for the new APs and controllers on the MM (or on standalone controllers if you are not using an MM).

   You can find these in the MM Properties window > Mobility Master Appliance tab.

6. Remember that the remote user solution requires additional licenses. If you opt for VIA licenses, add them to the MM. You can find these in the MM Properties window > ArubaOS tab.

   If you choose a PEFV license instead, it must be added to an MC.

7. Save your project.

Answers to the questions included in activities are provided in "Appendix: Activities and Learning Check Answers."

Summary

You now have a better understanding of the various options Aruba offers for remote network access. You can provide cost-efficient and secure solutions to your customers that will best suit their needs and help secure their corporate network.

Learning check

1. What APs can be deployed as RAPs
   a. AP 300 series
   b. RAP 100 series
   c. AP 200 series
   d. 205H series
2. VIA can be installed on what devices?
   a. Linux
   b. MacOS
   c. IOS
   d. Windows
   e. Androids
3. What is an IAP's DHCP Distributed L3 mode?
   a. DG on IAP and DHCP at corporate
   b. DG and DHCP with corporate scopes on IAP
   c. DG and DHCP with local scopes on IAP
   d. DG at corporate and DHCP on IAP

Answers to the learning checks in this study guide are provided in "Appendix: Activities and Learning Check Answers."

11 Network Management

EXAM OBJECTIVES

- ✓ Given the customer's requirements for a single-site environment, determine and document a detailed network management solution.
- ✓ Given the customer scenario, determine and document licensing requirements.
- ✓ Given the customer's requirements, explain and justify recommended solution.

Assumed knowledge

- Networking protocols such as Simple Network Management Protocol (SNMP) and Internet Control Message Protocol (ICMP)
- Aruba product portfolio
- Aruba 8.x architecture

Introduction

Aruba recommends that every customer have a network management system that meets their needs and provides them features to streamline their network processes. Aruba offers several choices for network management including Aruba AirWave and Central. AirWave provides customers a site management system while Central provides customers a cloud-based management system. In this chapter, you will take a closer look at AirWave and Central and their license requirements.

You will learn how to provide your customers recommendations for either AirWave or Central based on their system requirements and the features they need to manage their networks effectively.

Network management introduction

This section introduces you to Aruba management solutions.

Network management options

Aruba AirWave is an easy-to-use network operations system that manages wired and wireless infrastructure from Aruba and a wide range of third-party manufacturers. It also includes visibility and controls that let you optimize how devices and apps perform on your network. You can install AirWave in a VM environment or purchase as an appliance (see Figure 11-1).

Aruba Central is a cloud-based management system that offers a simple, secure, and cost-effective way to manage and monitor Aruba Instant APs, switches and branch gateways. It does not require VM or installation.

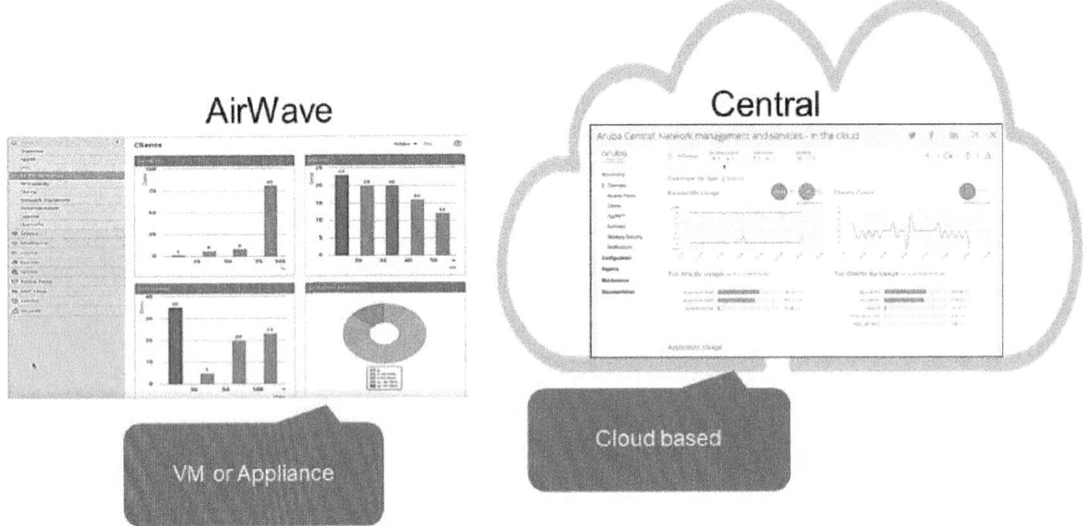

Figure 11-1 Network management options

Ease of deployment

One of the key challenges in deployments that include multiple interconnected buildings is the deployment of APs or switches. Aruba addresses this challenge with Aruba Activate, which is a cloud-based, zero-touch provisioning system. Aruba Activate provides plug-and-play capability to an Aruba Instant cluster, which allows rapid deployment of Aruba Instant clusters with minimal or no IT expertise.

The cloud-based Activate server will direct the IAP cluster or Aruba switches to one of the management systems, as Figure 11-2 shows.

- If the devices are sent to Central, then Central will download the devices' configuration.
- If the devices are sent to AirWave, then AirWave will download the devices' configuration.

Both Central and AirWave are capable of sending a configuration down to the devices. Once a device receives its configuration, it is ready to provide service and be monitored by the AirWave or Central management device.

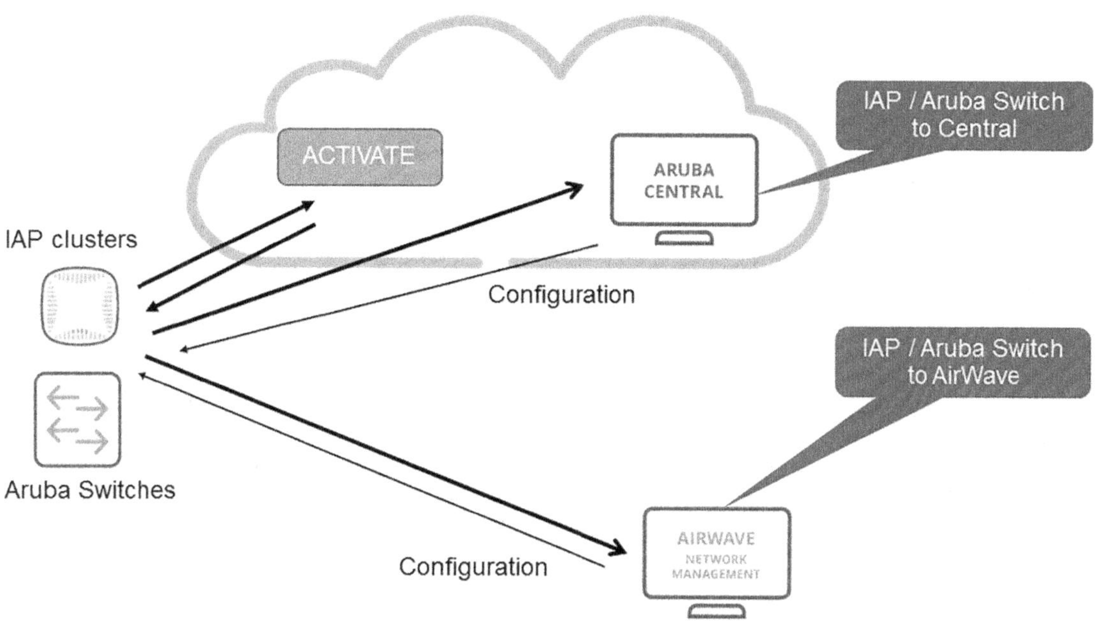

Figure 11-2 Ease of deployment

When an AP in factory default mode powers on and connects to the Internet, it automatically checks with Aruba Activate. Activate supplies provisioning data to the Instant AP or Aruba switch, which then communicates with its configuration master—in this case the Central or AirWave management platforms. The AP's configuration master pushes the configuration and any required firmware update to the IAP. If the requirements for these devices ever change and you need to re-provision an IAP, move the IAP to another folder within Activate and then set the IAP back to its factory-default setting by pushing the factory reset button. The IAP will repeat the provisioning process, applying the provisioning rules in its new folder.

Devices are grouped into folders that can contain a set of provisioning and email notification rules. If you want to group your devices by unique provisioning requirements, you can create separate folders that each contain different provisioning rules. You can also group devices by location or device type by creating folders or subfolders for each device type or installation site.

Aruba AirWave

You will now take a closer look at AirWave.

AirWave architecture

The AirWave Wireless Management Suite (AWMS) uses a simple wireless network management architecture. The AWMS monitors and manages the wired/wireless network as long as it has routable access.

The AirWave Management Platform (AMP) can manage up to 4000 devices, but this depends on the hardware platform. Anything above 4000 devices requires another AMP server.

AWMS provides failover capability for high-availability environments. As Figure 11-3 shows, the AirWave Glass aggregates all the information from multiple AMP servers in the network.

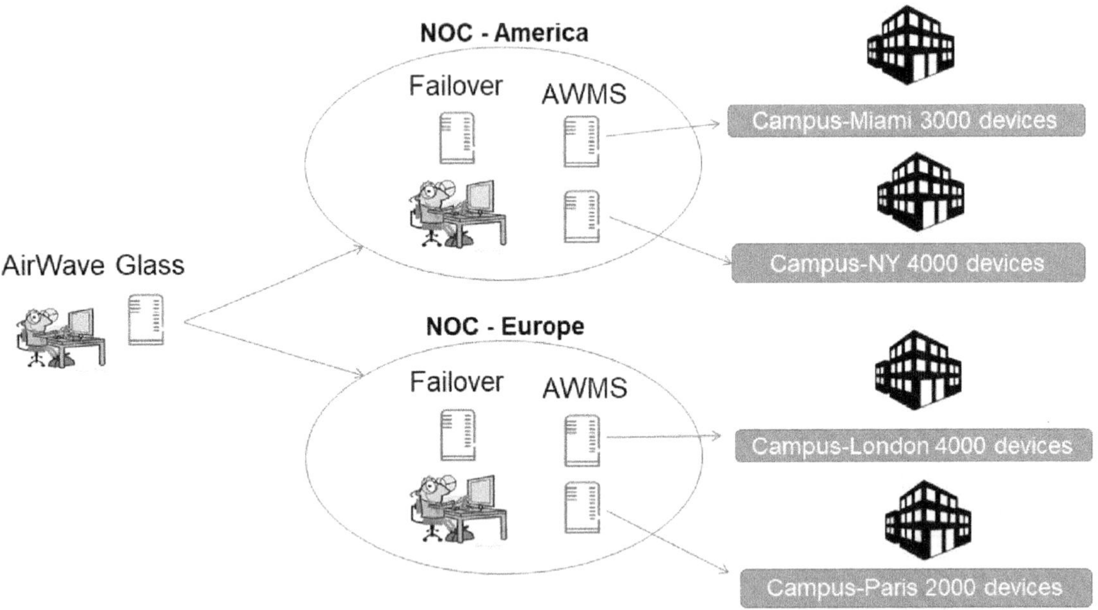

Figure 11-3 AirWave Architecture

AirWave Glass provides customers running multiple AirWave instances, a single console that clearly displays network infrastructure data for visibility, reporting, and troubleshooting. An advanced search engine enables IT administrators to efficiently get to the data they need without suffering through a "click" marathon. Built-in single sign-on (SSO) login security, lets IT staff safely go from Glass to individual AirWave consoles for efficient administration and privileges control. The same role-based access controls that are used in AirWave also manage who does what. Glass is available in physical and virtual appliance options.

One failover server can be the backup for many other AMP servers. If an AirWave server fails, the failover server will take over management operations.

Note that once failover takes over from one AMP server that failed, it effectively stops being a failover server and does not act as a failover for the other AMPs anymore. You can have a many-to-one failover setup, but remember that failover can only fail over one other AMP at a time.

Once the failover has loaded the backup file of the failed AMP server, it basically becomes that server. For example, you have AMP-NY and a failover server. AMP-NY fails and the failover server loads the

AMP-NY backup files. The failover server then reboots and becomes AMP-NY. The only difference will be the servers IP address.

AirWave capabilities

AirWave gives you the ability to monitor controllers, APs, IAPs, switches, and routers in real time as well as troubleshoot and diagnose Wi-Fi clients and dveices. It displays network health through several graphs including RF performance, RF capacity and network deviation.

The AirWave clarity function diagnoses clients association, authentication, DHCP request, and DNS times and averages. This helps point the administrative staff to probable server issues instead of Wi-Fi issues.

AppRF and Visual RF help administrators understand network traffic through heatmaps of the network, location tracking, web reputations, web categories, applications, and destinations.

AirWave also includes RAPIDS, a rule-based intrusion detection system to help secure your network. AirWave can also do firmware upgrades that save time and help maintain a consistent OS level on many devices.

There are multiple types of reports that you can customize and generate daily, weekly, monthly, or yearly to help you monitor your network.

Figure 11-4 provides a list of AirWave capabilities. It is important to note that AirWave only supports monitoring functions for Aruba 8.x; for example, it does not support configuration management or firmware upgrades.

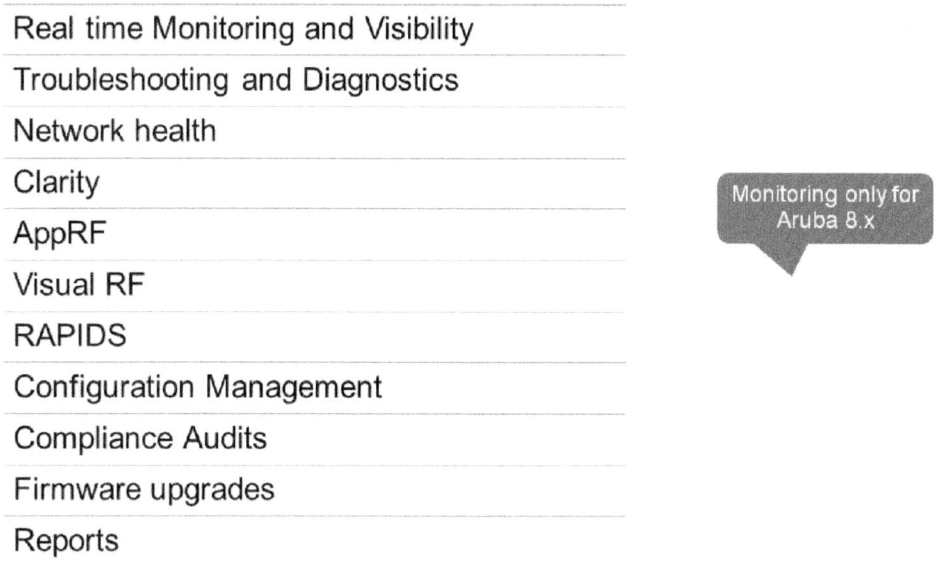

Real time Monitoring and Visibility
Troubleshooting and Diagnostics
Network health
Clarity
AppRF
Visual RF
RAPIDS
Configuration Management
Compliance Audits
Firmware upgrades
Reports

Monitoring only for Aruba 8.x

Figure 11-4 AirWave capabilities

Monitor/manage device communication

AirWave can manage most leading brands and models of wireless infrastructure and obtains monitoring information from any device, including switches, routers, controllers and APs. It also includes universal device support, which enables monitoring of many less commonly used devices (see Figure 11-5).

AirWave uses several different protocols to communicate with the various devices depending on their capabilities. You also have the option to enter SNMP credentials. If you do not enter SNMP credentials, AirWave will provide Internet Control Message Protocol (ICMP) monitoring for the universal devices.

You can monitor key elements of the wired network infrastructure including upstream switches, RADIUS servers, and other devices. Once you are monitoring devices, you have all the functions of AirWave at your disposal such as monitoring, firmware, upgrades, reports, and audits.

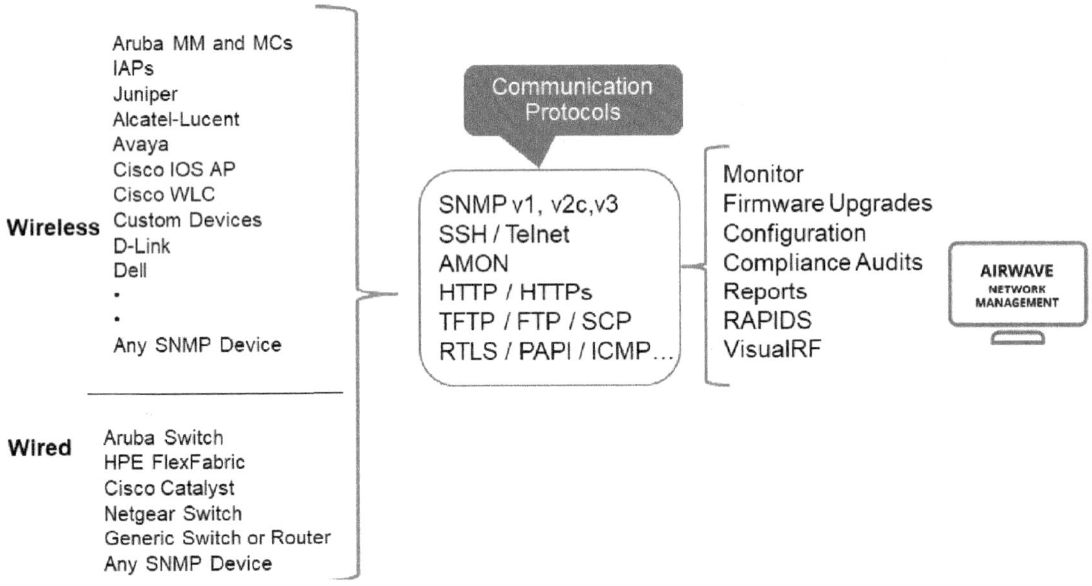

Figure 11-5 Monitor/manage device communication

Additional functions of AirWave include:

- Automatically see every user and device wireless and remote on the network
- Measure response times and failure rates for client association with Wi-Fi radio
- Authentication with a RADIUS server
- Gathering IP addresses through DHCP and resolving names for DNS services

- Monitor wired infrastructure that connects wireless controllers and APs
- View radio errors, including noise floor and channel utilization information, frequent causes of connectivity problems
- Drill down from network-wide to device-level monitoring views
- Store and view RF performance, capacity and application-level statistics, web traffic and network deviations over a 40-week period

Tested hardware platforms

The single AirWave server can manage up to 4000 devices. These include controllers, APs, router, and switches. Keep in mind that clients are not devices. The more devices you manage, the more hardware resources you will need.

If you have less than 1500 devices, use the AW_HW630-Pro configuration. If you have more than 1500 devices use the 4000 AW-HW630-ENT configuration. A device is a switch/router controller or AP. Clients associating to AP are not considered devices in AirWave.

You can get an AirWave OVA file and implement AirWave on a virtual machine (VM). You should use the same specs as listed in the image with a 20% overhead. On a dedicated platform, the entire server is dedicated to AirWave. When AirWave is installed as a VM, this could be a shared environment. Therefore, a 20% overhead is added.

Table 11-1 lists the recommended specifications for optimum AirWave server performance. You need to be sure the disk subsystem can withstand random write rates. Sustained sequential write rates will not help, because AirWave writes are primarily random.

Table 11-1 Recommended specifications for optimum AirWave server performance

Model	AW-HW630-PRO	AW-HW630-ENT
Maximum Devices	1500	4000
CPU	1 x Intel Xeon E5-2640, 2.5 GHz	Dual Xeon E5-2640 v3 2.6 GHz
Physical Cores	6	12
Passmark Score	9761	14846
RAM	48 GB	96 GB
Disks (RAID10 only)	6 x 146 GB 15k rpm SAS	8 x 146 GB 15k rpm SAS
Storage Capacity	800 GB	1.1 TB
Maximum Input/Output Operations Per Second (IOPs)	2132	2842

AirWave includes a 64-bit CentOS operating system based on Red Hat Enterprise Linux (RHEL) and is installed by default.

Choosing the CPU

For most AirWave installations, Aruba recommends selecting high-performance Xeon, because Aruba performs scalability testing using Intel-based hardware.

Choosing the memory

AirWave's memory recommendations scale linearly with the managed device count. The recommended memory usually results in best overall performance, especially when AirWave servers are running at full load on device counts with large floorplans. Aruba defines best performance by minimal disk reads due to sufficient caching, thus allowing for maximum disk write performance and minimal CPU I/O wait time

Factors that affect AirWave performance

If you do not properly configure AirWave, its performance will be affected. Several factors can affect the AirWave performance in areas such as CPU, memory, and IOPS.

How often your clients roam impacts the AirWave server's disk write demands and storage requirements. The number of clients AirWave serves has a direct and significant impact on hardware sizing requirements as well. AirWave requires more processing resources to identify new clients from existing clients. Switches also demand more processing resources from AirWave.

Additional factors that can affect AirWave performance include:

- Traps consume processing resources, and trap rates are associated with client mobility.
- AMON allows AirWave to collect enhanced data from Aruba devices on certain firmware versions. AMON rates affect CPU and memory.
- More aggressive or frequent AirWave SNMP polling would require a server with increased CPU and IOPS capacity to handle the increased workload.
- The processing of rogue APs consumes CPU and disk resources on the AMP.
- The numbers of floor plans and campuses AirWave manages directly impacts the amount of memory the VisualRF feature uses.
- Long retention periods increase the bulk of data AirWave has to manage. This may result in larger disk capacity needs.
- The type, frequency, and scope of reports can have a large impact on the AirWave.
- IOPS and disk capacity can also be affected.

Figure 11-6 summarizes the other factors that can affect the performance of AirWave.

Client Mobility	Impacts disk writes and storage
Number of Clients per AP	Hardware sizing requirements
Ratio of New clients	More processing resources
Wired Switch percentage	More processing resources
Trap Rate	Increases CPU and IOPS capacity
SNMP Pool periods	Increases CPU and IOPS capacity
AMON	CPU and memory resources
RAPIDS	CPU and Disk recourses
VisualRF	Impacts Memory
Data Retention Periods	Disk capacity
Reports	IOPS and disk capacity

Figure 11-6 Factors that affect AirWave performance

Aruba Central

You will now learn more about Aruba Central.

Central modes

Central is a cloud-based management system that you can use for your own company management needs or by a service provider offering management services to other companies.

As Figure 11-7 shows, the Standard Enterprise interface is intended for customers who manage their respective accounts end to end. In the Standard Enterprise mode, the customers have complete access to their accounts. They can also provision and manage the respective accounts.

Central offers the Managed Service Portal for managed service providers who need to manage multiple customer networks. With Managed Service Portal, the MSP administrators can provision customer accounts, allocate devices, assign licenses, and monitor customer accounts and their networks. The administrators can also drill down to a specific end customer account and perform administration and configuration tasks. The end customers can only access their respective accounts, and those features and application services to which they have subscribed.

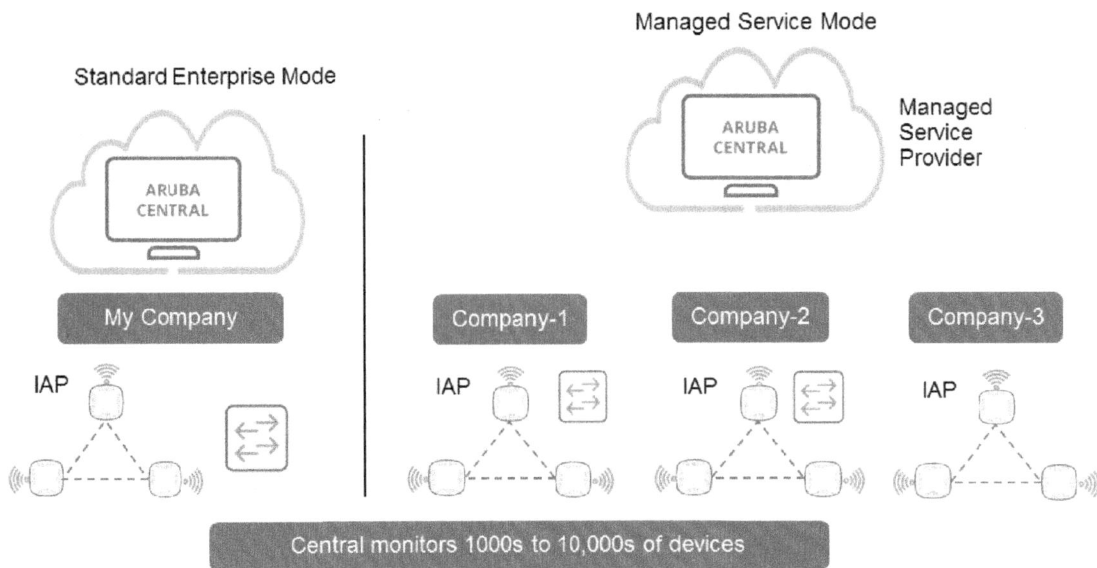

Figure 11-7 Central modes

Central capabilities

Aruba Central offers a simple, secure, and cost-effective way to manage and monitor your Aruba Instant APs, switches, and branch gateways. With Aruba Central, you can get your network up and running in minutes with intelligent Zero Touch Provisioning. The intuitive dashboards along with reporting, maintenance, and firmware management make monitoring and troubleshooting easy– no technical expertise required.

Get an at-a-glance summary of your entire network health via an intuitive dashboard, designed to let you drill-down for more information with just a few simple clicks. Using this single dashboard, you can view metrics for instant access points, switches, and clients, along with flagged alerts and location mapping.

- Central also has advanced capabilities such as customizable guest Wi-Fi and presence analytics for smarter decision-making.
- Clarity diagnoses clients association, authentication, DHCP request, and DNS times and averages. This points the admin staff to probable server issues and not Wi-Fi issues.

Presence analytics is an APP that gives you the capability to gather insight into customer patterns. Know how many customers passed by a location, entered the location, stayed on, and for how long.

Central also offers the application visibility to display network traffic giving the administrator information on web reputations, categories, application, and destination. Central can enable guest sponsors, such as receptionists, event coordinators, and other non-IT staff, to create temporary guest accounts or allow self-registration.

Central can perform the following functions:

- Identify rogue devices on the network with Wireless Intrusion Detection System (WIDS)
- Function as the focal point for configuring multiple IAP clusters or switches
- Perform firmware upgrades and schedule the upgrades
- Store your management data indefinitely so you can create reports containing historical data if needed
- Presence analytics give you the capability to gather insight into customer patterns
- Identify how many customers passed by a location, entered the location, stayed on, and for how long

The standard enterprise mode provides a complete view of the devices that Central monitors and manages for a specific customer. The Managed Service Portal provides a consolidated view of the networks of various customers the service provider manages. Because Central is a cloud-based service, it is always on and gathering information.

Figure 11-8 summarizes Central's main capabilities.

Network Health Monitoring and Troubleshooting
Clarity
Application visibility
Cloud Guest
WIDS
Configuration and Firmware Management of devices
Reporting
Presence Analytics
Standard enterprise mode or Managed service mode
Always on and API accessible

Figure 11-8 Central

Central communication

Most of the communication between devices on the remote site and the Central server in the cloud is carried out through HTTPS (TCP443) (see Figure 11-9). However, you may need to configure TCP 80 for firmware upgrades. Network time protocol, UDP 123, is used to configure time zones when the IAP cluster is up for the first time. TCP port 2083 is used for RADIUS authentication for guest management, but if port 2083 is blocked, then the HTTPS protocol is used.

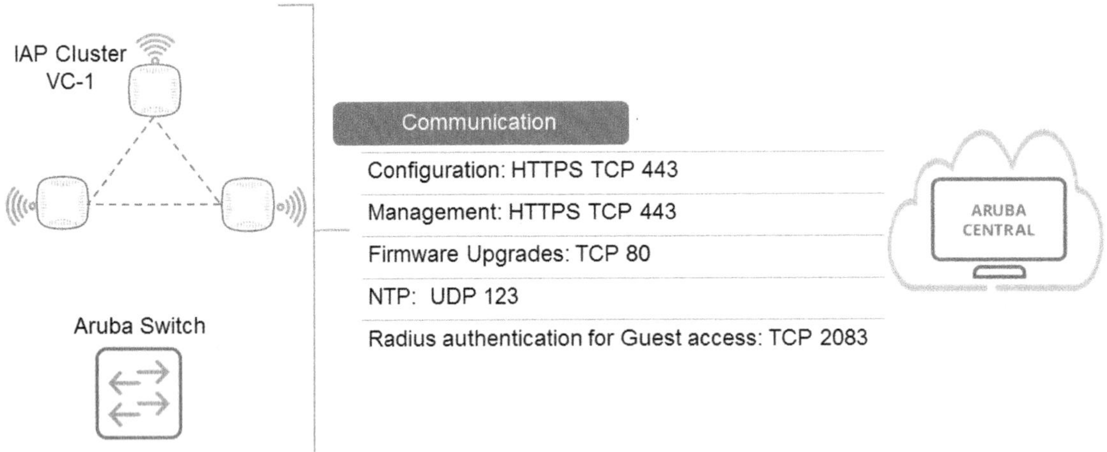

Figure 11-9 Central communication

Aruba switch required software

As of the publication of this course, Aruba Central can manage these switches with this software:

- Aruba 2930M Switch Series WC.16.04.0004 or later
- Aruba 2920 Switch Series WB.16.02.0012 or later
- Aruba 2930F Switch Series WC.16.02.0012 or later
- Aruba 3810 Switch Series KB 16.03.0003 or later
- Aruba 5400R Switch Series KB.16.04.0008 or later
- Aruba 2530 Switch Series YA/YB 16.04.0008 or later
- Aruba 2540 Switch Series YC.16.02.0012 or later
- Aruba Mobility Access Switch Series S1500-12P, S1500-24P, S2500-24P, S3500-24T, ArubaOS 7.3.2.6, 7.4.0.3, or 7.4.1.4

Licenses

Finally, you will focus on AirWave and Central licensing.

Licenses

The AirWave server has a device number license. Every device AirWave monitors will use one license. This does not include client devices (see Figure 11-10).

To create a failover server, you need the failover license. For Glass, you need the Glass license. To evaluate Airwave, the sales team can generate an evaluation license.

For Central, you can start with a device management subscription if you only need network management and monitoring capabilities. This subscription includes all capabilities except Clarity, Cloud Guest, and Presence Analytics. You a services subscription to take advantage of these features. Both types of licenses are based on the device count.

To evaluate Central, your sales team can generate a 90-day evaluation license.

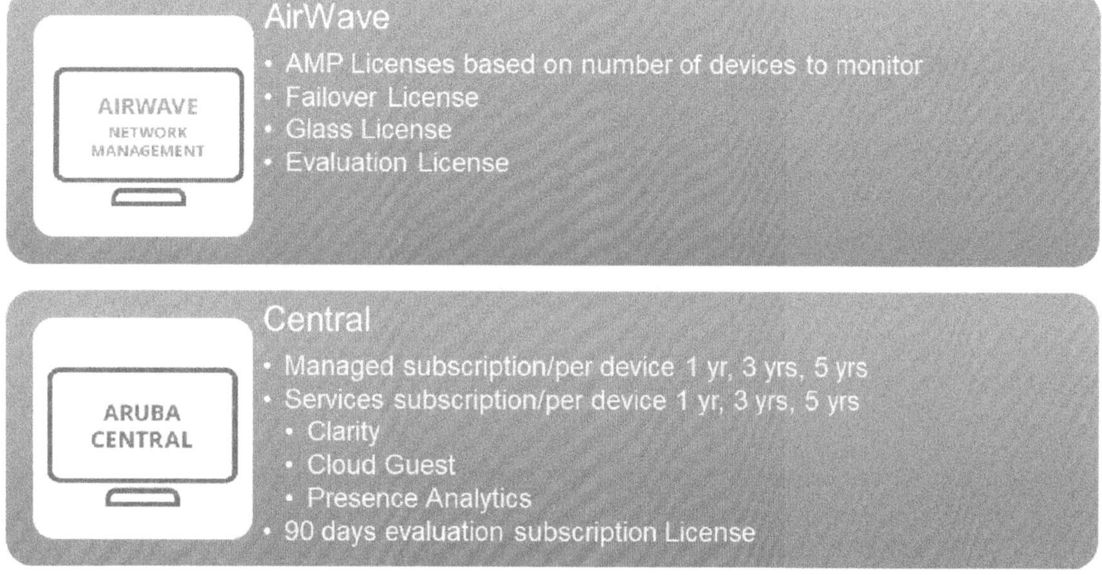

AIRWAVE
NETWORK MANAGEMENT
AirWave
• AMP Licenses based on number of devices to monitor
• Failover License
• Glass License
• Evaluation License
ARUBA CENTRAL
Central
• Managed subscription/per device 1 yr, 3 yrs, 5 yrs
• Services subscription/per device 1 yr, 3 yrs, 5 yrs
• Clarity
• Cloud Guest
• Presence Analytics
• 90 days evaluation subscription License

Figure 11-10 Licenses

When a subscription assigned to a device expires or is canceled, Central checks the inventory for the available subscription tokens for the device and verifies if the subscription has adequate license tokens. If the subscription has adequate capacity, Central automatically assigns the longest available subscription token to the device. If not, Central ensures that the subscriptions are utilized to the full capacity by assigning as many devices as possible. Central does not support automatic assignment of subscriptions in the Managed Service Mode.

Support services and training

Support services are an important part of every proposal. IRIS automatically adds basic support services to accompany products added to the BOM, but some customers may need more. You can extend the coverage period from one-year to three-years, for example. You can also select higher levels of support such as proactive consulting rather than only reactive support for issues such as failed hardware.

In addition, some customers do not have the expertise they need in house, so they might need help implementing and deploying their IT solutions. You can also add training to the BOM because most customers are more satisfied when they are taught how to implement and manage their solutions.

Activity 11: Management design

In this activity, you will complete the network design for the customer scenario introduced in Activity 1. To complete your design, you will decide what management platform to use to solve the customer's network management issues. You will use the scenario information from previous activities plus the additional information provided below to determine which solution to recommend.

Scenario management

This addition to the main scenario will explain the current customer's network management system and future requirements.

Current network management

Corp1 was managing their closet switches with a third-party management system that relies on SNMP and has poor graphs and diagnostic tools by today's standards.

Corp1 new management requirements

Corp1 is looking for a management system that can monitor both the wireless and wired network. Corp1 would also like the management system to monitor the data center network equipment, which includes HPE FlexFabric switches so that staff can see whether links are up or down.

Activity 11.1: Management design

Activity 11.1 Objective

You will now recommend the best management system for the customer requirements.

Activity 11.1 steps

1. What management platform do you recommend?
2. Justify your recommendation.

Activity 11.2: Licenses

Activity 11.2 objectives

You will now recommend the licenses you will need for this customer.

Activity 11.2 steps

1. Based on your chosen management platform list the required license(s).
2. Explain why you need these licenses and what they are used for:

Activity 11.3: Add management components to the BOM

You will now add your recommended management solution and its licenses to your BOM in Iris. If you do not have access to Iris, you can use the HPE Networking Online Configurator. (Visit http://hpe.com/networking/configurator to access this tool.)

1. Return to your Iris application and, if necessary, open your project.
2. Expand the catalog and add your selected management solution.
   - If you chose AirWave, add either:
     - A hardware appliance and licenses
     - Licenses only for a VM
   - If you chose Central, add it.
3. Click the component or components that you added and fill out your plan. Make sure to specify the correct number of licenses.

   Now that you have completed your BOM, you will check the services that IRIS has added automatically.
4. Click the Addon Services and Licenses icon in the top bar (see Figure 11-11).

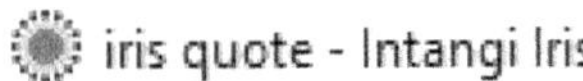

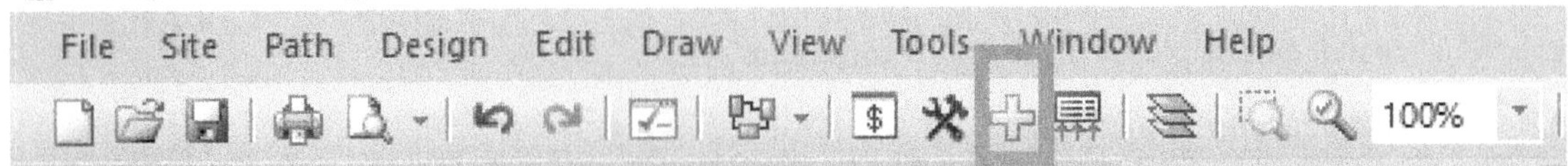

Figure 11-11 Addon services and licenses icon in Iris

5. As you see, Iris has automatically added hardware and software support services to the BOM.
6. You can talk with the customer and adjust services if you like.

   For example, you can extend all of the services to a three-year term. To change the service type for many products at the same time, *first* click the column for the new service that you want in any row. Then click the Select all items icon (see Figure 11-12).

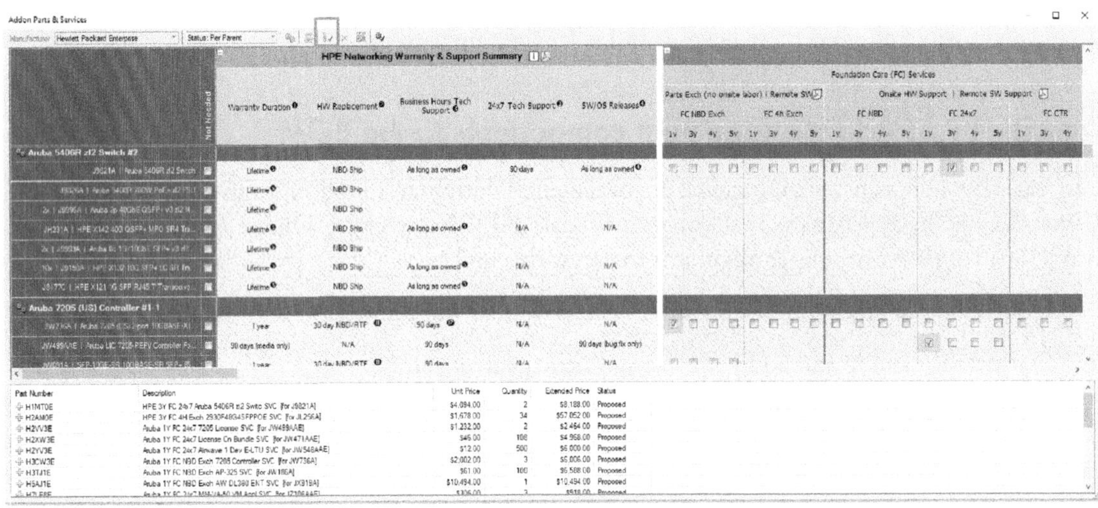

Figure 11-12 Add services in Iris

7. When you are done examining or adjusting services, close the window.

   You will now add training.
8. Double-click any of the products in the Iris workspace.
9. In the Properties window, click the Messages tab.
10. See that Aruba recommends training to accompany all Aruba hardware products. Click the Add button (see Figure 11-13).

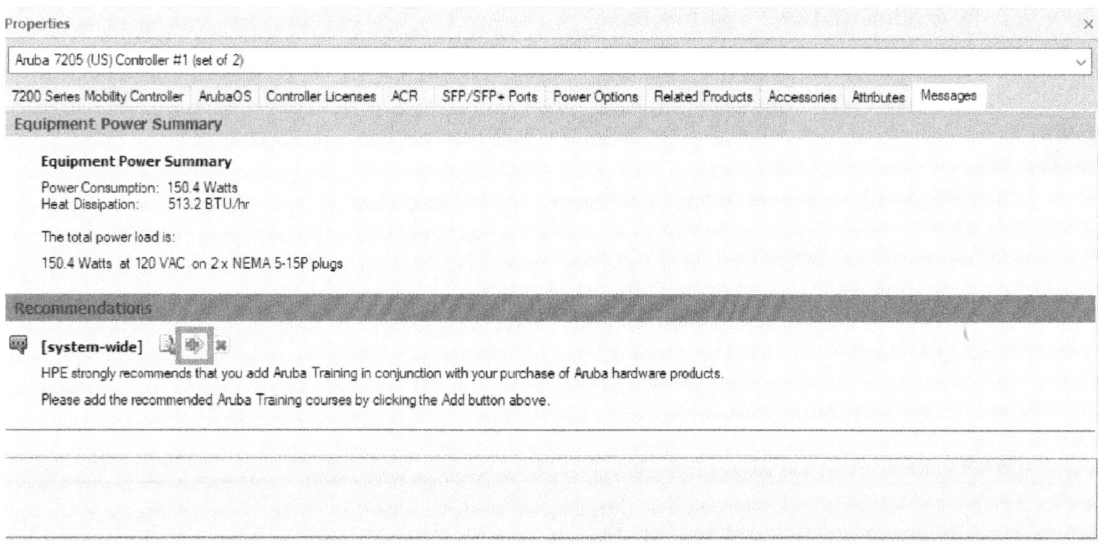

Figure 11-13 Add training in Iris

11. Education Services are added to the BOM and its Properties window is displayed.
12. Select appropriate training and set the number of students to 2. You can use the Type of training/exam filter to narrow your choices based on the hardware (see Figure 11-14).

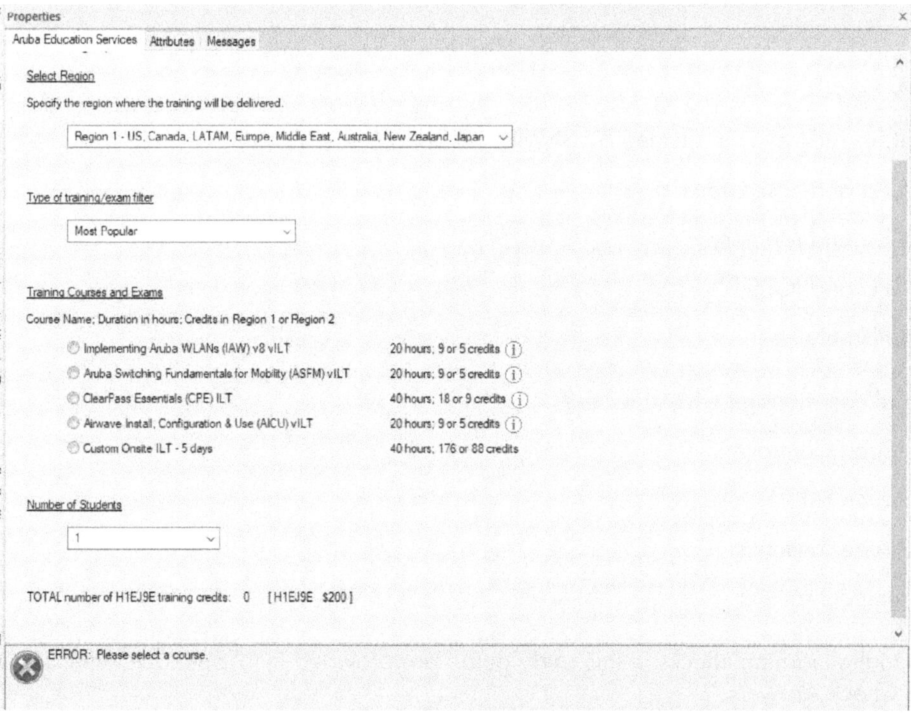

Figure 11-14 Select appropriate training courses

13. Close the window and save your project.
14. You have completed your BOM. Click the Quotation icon and find your final proposal amount.

Note

There should be no errors in your IRIS BOM when completed.

There should be no errors on the following:

POE budget.

Power Supplies.

Transceiver distance mismatch.

Licenses.

Answers to the questions included in activities are provided in "Appendix: Activities and Learning Check Answers."

Summary

In this chapter, you learned how to design the correct management platform for your client. You learned the requirements of implementing Aruba's two management platforms.

Learning check

1. What, in AirWave, is affected by SNMP Poll periods ?
   a. CPU and Disk capacity
   b. CPU and Memory
   c. CPU and IOPS
   d. CPU only
2. Central can manage what devices?
   a. Aruba Controllers
   b. Aruba IAPs
   c. Aruba Switches
   d. Junipers routers

Answers to the learning checks in this study guide are provided in "Appendix: Activities and Learning Check Answers."

12 Aruba Campus Design

EXAM OBJECTIVES

✓ Given a customer's needs for a single site, determine the information required to create a solution.

✓ Evaluate a customer's needs for a single-site campus, identify gaps and recommend components.

✓ Select the appropriate products based on a customer's needs for a single-site campus.

✓ Given a customer's requirements for a single-site campus, design the high-level architecture.

✓ Given a scenario for a single-site campus, choose the appropriate components to be included in a Bill of Materials (BOM).

Assumed knowledge

- Wired and wireless network design principles
- Aruba Mobility Master and Controller architecture
- Basic knowledge of 802.11 and radio frequency (RF) concepts
- Basic knowledge of switching and routing

Introduction

Throughout this study guide, you have learned the process for designing a network. In this chapter, you have the opportunity to design a network for a new customer. You will be given a new customer scenario and will then design the solution for the customer.

You can compare your solution to the example solution included in "Appendix: Activities and Learning Check Answers."

Customer's goals and site information

You will first focus on understanding the customer's goals and learning about their site.

You will design a network for a customer who needs a wireless and wired network upgrade. As you will learn in the sections that follow, this customer wants to improve performance and provide ubiquitous wireless coverage. On the wired side, the customer wants to upgrade the bandwidth across the campus and between the campus and the datacenter. Furthermore, the company wants to provide better wireless and virtual private network (VPN) access for remote users.

The customer also wants to improve security for both wired and wireless access and needs a management solution that provides insight to both the wireless and wired network. They want to reduce issues and improve their ability to resolve any issues that occur.

Overview of CorpXYZ's goals

CorpXYZ has an existing wireless network. Originally installed to provide secondary access for employees and the company's few guests, the wireless network was designed to provide coverage and support 802.11 a/bg.

As more employees began to use the wireless network, the company wanted to improve performance. They decided to replace some of the Access Points (APs) with APs that support 802.11n but did not hire a network architect to redesign the RF coverage. Several areas now have 802.11n APs, including the areas where software developers and the company executives work.

The site has ample meeting rooms and lobbies, and the customer is pushing employees to work together more. However, employees complain that they cannot connect to the wireless network consistently in these areas or run the applications they need.

On the wired side, the customer has shifted to centralized and cloud-based solutions over the years. Users have started to complain about slow connections, and much of their work requires access to services and files in the data center.

The customer lacks insight into the wireless and wired networks and does not know how to address problems as they occur. And problems are occurring more and more often.

CorpXYZ requires a new 802.11ac network that provides ubiquitous coverage, solves the performance issues, and helps the customer avoid another upgrade for several years. It should allow employees to connect more devices in more locations and work wirelessly as easily as on a wired workstation. The customer also wants to upgrade wired bandwidth across the campus and to the data center.

In addition, the company wants to provide better wireless and Virtual Private Network (VPN) access for their growing number of remote users, who work at home.

The customer would like three-year support services. The partner will include their own services for installing and setting up the network, so you do not need to include those Aruba services.

Users and devices

CorpXYZ has 250 employees working on-site. Approximately 165 of these users are using laptops that have docking stations, which are daisy-chained to Voice over IP (VoIP) phones. The remaining users are software developers who require more powerful workstations, which are also daisy-chained to VoIP phones. Most developers are also using tablets.

In fact, almost all users have smartphones and tablets, which they connect to the wireless network.

All the on-site employees use Web browsers, e-mail, Microsoft Skype for Business, and printing. They also access and store files in a network file system. In addition, they have Web-based access to a time card application, project management software, Customer Relationship Management (CRM) application, and other business applications in the datacenter.

Software developers currently do a lot of their work locally, but they sometimes use Secure Shell (SSH) and Remote Desktop Protocol (RDP) to access platforms in the data center. They also occasionally push Docker containers, which are typically 200 MB–1 GB, to the data center.

VoIP phones use Microsoft Skype for Business. They support 802.1X, Link Layer Discovery Protocol–Media Endpoint Discovery (LLDP-MED), and Power over Ethernet (PoE). They can draw up to 8.4 W.

CorpXYZ currently has 214 remote users, who work from home. Their job is to contact potential and existing customers, using call center software. The company provides these users with laptops and VoIP phones. In addition to the call center software, these users access the company's time card application and email system. These users are also using smartphones, and about one-fourth of them have Wi-Fi printers because they prefer to print out the scripts that they use when calling customers.

Currently, these users are responsible for providing their own wireless network, which is all too often spotty and unreliable. Understanding that these issues sometimes disrupt calls, which reflects poorly on the company, CorpXYZ wants to provide a standard wireless solution. The company also wants to provide a standard VPN solution for each remote user, so they can ensure users' access to the company's network is secured.

CorpXYZ has a small number of guests. Most guests are visiting the company's executives and are only allowed in the conferences rooms near the executives' offices. The maximum number of guests never exceeds 10 guests on a given day.

Each building floor also includes four printers, which have wired connections. The company recently purchased a new color printer for each floor. These printers are Wi-Fi capable, and the company would like to move the printers to a new location, which does not have a RJ45 jack.

Site information

CorpXYZ leases three floors in a five-story building in a business park. Access to all floors is controlled by badges.

All floors use an identical floorplan, which is shown in Figure 12-1, and similar density and distribution of users. For simplicity in this activity, make the plan for one floor and multiply by three. The floor dimensions are

- Dimensions: 300 x 215 feet (91 m x 65 m) at the widest points
- Total floor area (including center): Due to an irregular shape, 48,450 square feet (5915 square m)
- Excluded area: 12,300 square foot (1143 square m) central area that includes elevators, stairs, and other spaces not leased to CorpXYZ
- Ceiling height: 15 feet high (4.5 m) with a drop ceiling at 10 feet

Note that you will not plan wireless for the central area that includes elevators, stairs, and other spaces not leased to the customer.

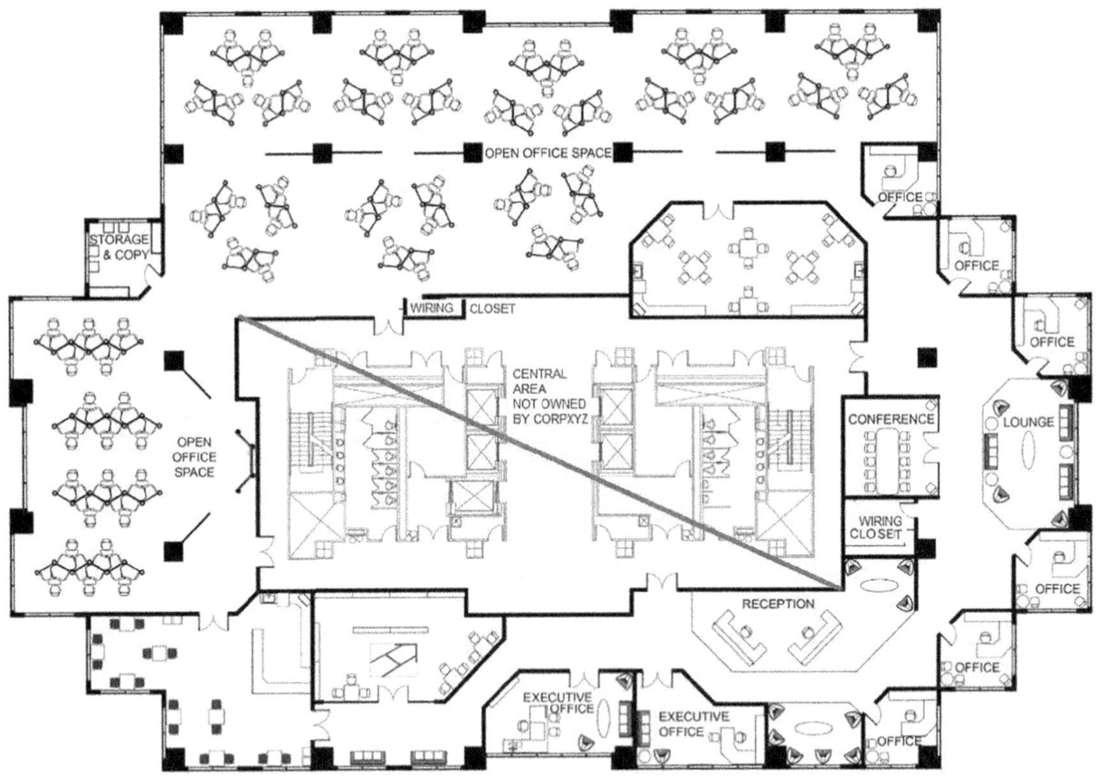

Figure 12-1 CorpXYZ floor 1

Roaming

CorpXYZ wants to allow employees to roam within a particular floor but not between floors because they are sharing the building with other companies.

Security

Although the CorpXYZ executives and IT manager understand the benefits of wireless access, they are concerned about the related security risks. The company currently uses a freeware RADIUS server to implement 802.1X with WPA2 for wireless access. However, they want to implement additional security measures. They are interested in a solution that offers deep packet inspection and firewall protection.

Relying on the fact that employees require a badge to gain physical access to each floor, the company has not yet imposed 802.1X security for wired connections. Recently, however, a guest wandered into an area where nonemployees are restricted from accessing. The company now sees the need to protect against internal threats. CorpXYZ is also interested in providing the same experience for users, no matter if they access the network through a wired or wireless connection.

Existing network

Many years ago, IBEWIFI vendor supplied the original 24 802.11 a/bg APs, seven per floor. Over the years, the customer has gradually replaced some APs with 802.11n APs and added other APs. To save money, the company purchased 802.11n APs from another vendor. The customer now has 36 APs. Users can receive coverage in most areas, but the network is slow. Users have noted several places where performance seems particularly slow.

CAT6a cable extends to each desk, which has one wired jack. Reception areas also have a few live ports. The total number of drops on each floor is 112.

The company has one wiring closets per floor. Each closet has three switches, which provide:

- 48 10/100/1000 Mbps PoE (802.1af) ports
- Two 1 GbE SFP SX transceivers
- Layer 2 forwarding

These access layer switches do not provide authentication capabilities.

Each access layer switch has two 1 GbE fiber connections to one aggregation switch, which is located in the wiring closet on Floor 1. (Fiber runs between the closets on Floor 1 and Floor 2 are 6 m and between the closets on Floor 1 and Floor 3 are 13 m.) The aggregation switch has four 1 GbE links to the data center, which is located in the basement of the building. The fiber connecting to the data center also terminates in this closet: eight strands of OM4 MM cable (85 m run).

The aggregation switch is not redundant and is quite outdated. It does not support 10 GbE.

Existing logical network

The current network has one VLAN at the campus. Wired and wireless users are in the same VLANs. The VLAN uses subnet 10.50.0.0/22. The customer says that this should work for the new deployment, but you can use 10.62.0.0/22 also if you want.

Future growth

The company plans to add 20 to 25 remote users per year over the next 5 years. They also plan to add 10–15 software developers per year and 5–10 Marketing and Sales personnel per year over the next 5 years. In 3 years, they want to lease half of Floor 4 in the building, but until then they will add new employees as remote users and provide cubicles on-site for them so they can work onsite occasionally.

Activity12.1 new network design

You will now plan the upgrade for CorpXYZ and use Iris to create a Bill of Materials (BOM). If you do not have access to Iris, you can use the HPE Networking Online Configurator to create a BOM. (Visit http://hpe.com/networking/configurator to access this tool.)

A floorplan for this customer scenario was included in the email message you received from HPE Press when you purchased this study guide. You can use this floorplan to plan coverage in Aruba VisualRF Plan.

Make sure that your plan meets customers' management, redundancy, and security, as well as connectivity and performance requirements. Be prepared to explain your plan to your colleagues.

Below you will find hints to help you complete your design.

Hints

- If you plan coverage in VisualRF, remember to exclude the central area. You can use the floorplan boundary feature.
- Remember to multiply your plan for one floor by three.
- Remember to add all necessary licenses (MM, AP, PEFNG, RFP, ClearPass Access, and so on).
- Remember to make a plan to provide wireless and wired redundancy.
- If you are planning VSF or backplane stacking, remember to add the equipment required for the VSF or stacking links.
- Remember to check that your switches have a sufficient PoE budget to power the APs and VoIP phones.

- Remember to consider whether the customer might benefit from a management solution.
- Does your solution for the remote users meet all of their needs? Remember that they need wireless coverage, a VPN connection to the main site

Summary

Congratulations! You have successfully designed a network for the new customer. (Remember that you can compare your solution to the example solution included in "Appendix: Activities and Learning Check Answers.")

13 Practice Exam

Introduction

The Designing Aruba Solutions exam tests your ability to design best practice Aruba solutions to meet a variety of customer needs for sites with less than 1000 users. To pass the exam, you must demonstrate that you can collect the information relevant to designing the solution, analyze this information to determine customer requirements, and create a plan to meet those requirements. The exam tests you on all components of the design from selecting the wired and wireless products; designing the physical and logical topology; creating the security, QoS, and management plans; and delivering the BOM.

The intent of this study guide is to set expectations about the context of the exam and to help candidates prepare for it. Recommended training to prepare for this exam can be found at the HPE Certification and Learning website (https://certification-learning.hpe.com). It is important to note that, although training is recommended for exam preparation, successful completion of the training alone does not guarantee that you will pass the exam. In addition to training, exam items are based on knowledge gained from on-the-job experience and application and on other supplemental reference material that may be specified in this guide.

Minimum qualifications

Typical candidates for this exam are networking IT professionals who have architect experience with Aruba wireless and wired switching solutions. They have relevant field experience focused on interpreting architectures and customer requirements to design Aruba subsystems or single campus network solutions.

HPE6-A47 exam details

Below are details about the HPE6-A47 exam.

- Number of items: 60
- Exam time: 1 hour 30 minutes
- Passing score: 68%
- Item types: Multiple choice (single-response), multiple choice (multiple-response), and multiple choice with scenarios
- Reference material: No online or hard copy reference material is allowed at the testing site.

HPE6-A47 exam objectives

This exam validates that you can successfully perform the following objectives. Each main objective is given a weighting, indicating the emphasis this item has in the exam.

- Gather and analyze data, and document customer requirements. Weighting: **10%**
  - Given an outline of a customer's needs for a single-site campus environment (less than 1000 employees) or subsystems of an enterprise-wide network, determine the information required to create a solution.
- Evaluate the requirements, and select the wired and wireless networking technologies for the design. Weighting: **18%**
  - Given a scenario, evaluate the customer requirements for a single-site campus environment (less than 1000 employees) or for subsystems of an enterprise-wide network to identify gaps per a gap analysis, and select components based on the analysis results.
  - Given a scenario, translate the business needs of a single-site campus environment (less than 1000 employees) or subsystems of an enterprise-wide environment into technical customer requirements.
- Plan and design an Aruba solution per the customer requirements. Weighting: **31%**
  - Given a scenario, select the appropriate products based on the customer's technical requirements for a single-site campus environment (less than 1000 employees) or subsystems of an enterprise-wide environment.
  - Given the customer requirements for a single-site campus environment (less than 1000 employees) or subsystems within an enterprise-wide environment, design the high-level architecture.
  - Given a customer scenario, explain how a specific technology or solution would meet the customer's requirements.
- Produce a detailed design specification document. Weighting: **33%**
  - Given a customer scenario for a single-site campus environment (less than 1000 employees), choose the appropriate components that should be included in the BOM.
  - Given the customer requirements for a single-site campus environment (less than 1000 employees), determine the component details and document the high-level design.
  - Given a customer scenario for a single-site campus environment (less than 1000 employees), determine and document a detailed network management design.
  - Given a customer scenario for a single-site campus environment (less than 1000 employees), design and document a detailed network security solution.

  - Given a customer scenario of a single-site campus environment (less than 1000 employees), design and document the logical and physical network solutions.
  - Given the customer scenario and service-level agreements, document the licensing and maintenance requirements.
- Recommend the solution to the customer. Weighting: **8%**
  - Given the customer's requirements, explain and justify the recommended solution.

Test preparation questions and answers

The following questions help you to measure your understanding of the material presented in this study guide and determine if you are prepared to take the HPE6-A47 exam. Read all of the choices for each question carefully before selecting the correct answer. Remember to select all correct answers for each question.

You can check your responses in the "Answers" section at the end of this chapter.

Questions

The first questions in this practice exam relate to a scenario. Read the scenario and then answer questions 1–4 with reference to this scenario.

Scenario

A customer wants to upgrade to 802.11ac. The customer has an eight-year-old office building with three floors. The company has a total of 350 employees with about 100–120 employees per floor. Each employee has a company-issued laptop. The employees can also connect to the network with their personal devices. The customer expects that each employee has two or sometimes three devices. The company has a few wireless devices such as printers too.

Users run applications such as Web browsing, print, email, SAP Online, voice, and file sharing.

The company also has a few clients who occasionally visit to the site, so it needs a guest wireless network. The customer indicates that a maximum of 1150 devices will connect to the wireless network at the same time on any given day.

The customer wants to use ClearPass to authenticate wireless devices with 802.1X EAP-TLS. The customer wants to simplify deploying certificates to all of the user devices and to run health checks only on the company-issued devices. The customer does not want wired authentication at this time.

The customer also wants to enhance monitoring for their network beyond their open source SNMP monitor.

The floor plan is shown in Figure 13-1. All three floors have a similar plan.

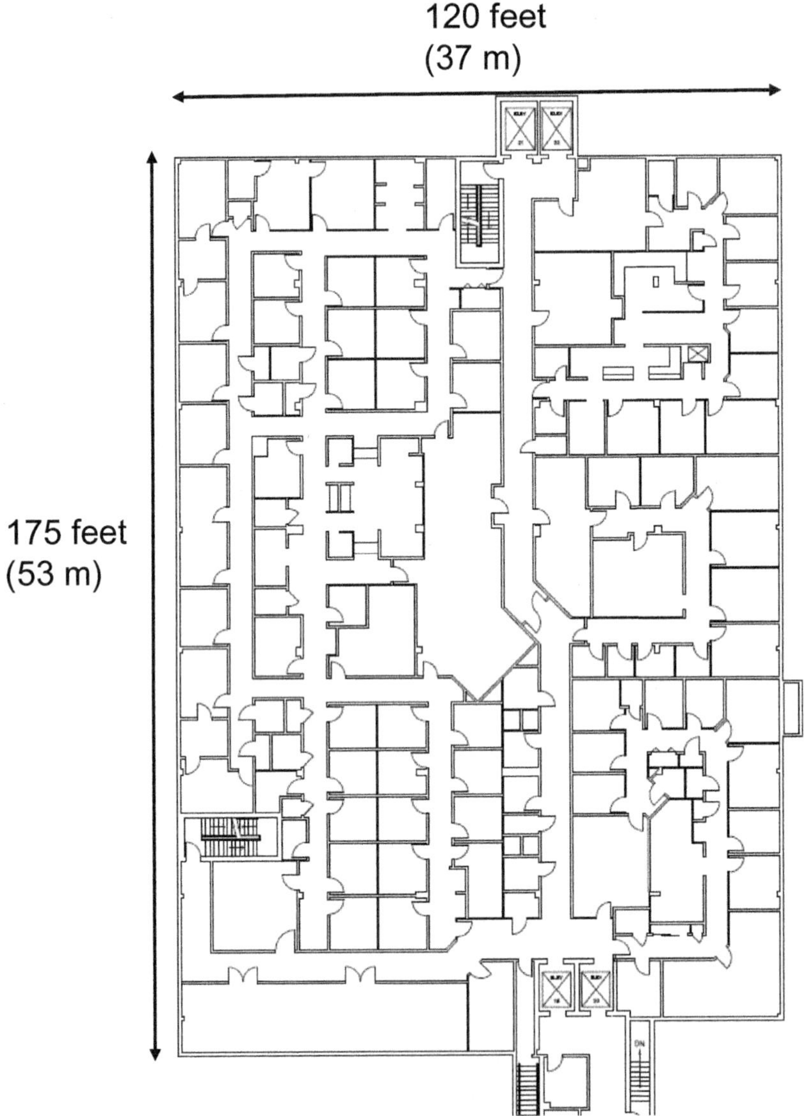

Figure 13-1 Scenario exhibit

Each floor has four switches in a wiring closet, which all connect to two core switches in a server room on floor 1. The customer has indicated that the access layer switches support:

- PoE (802.3af) on 8 ports
- BASE-T 10/100/1000 Mbps on 24 ports
- SFP on two ports

1. Refer to the scenario.

   What is some additional information that the architect collect to plan for the AP deployment? (Select two.)

   a. Location of the wiring closets
   b. Year that the cable was installed
   c. Type of cable extended across the floor
   d. Whether the existing switches support LLDP-MED
   e. Whether the existing switches support 802.1X

2. What is one key reason should the architect should recommend a wired network upgrade?

   a. Need to support SNMP-based network security
   b. Need to support 10 GbE connectivity on the edge ports
   c. Need to provide PoE+ for the APs
   d. Need to provide access layer routing

3. Refer to the scenario.

   What ClearPass licenses should the architect propose?

   a. 1200 Access licenses, 1100 Onboard licenses, and 500 OnGuard licenses
   b. 500 Access licenses, 1100 Onboard licenses, and 500 OnGuard licenses
   c. 1200 Access licenses, 500 Onboard licenses, and 500 OnGuard licenses
   d. 500 Access licenses, 500 Onboard licenses, and 1100 OnGuard licenses

4. Refer to the scenario.

   The architect plans to include AirWave in the proposal. What does the architect need to clarify to determine the number of Management Platform (AMP) licenses to propose?

   a. The total number of devices with an IP address on the network, including endpoints and infrastructure devices
   b. The total number of network infrastructure devices to be monitored or managed
   c. The total number of network infrastructure devices as well as the features that are required
   d. The number of AirWave servers, as well as the features that are required

   You have finished the scenario. Answer the rest of the items individually.

5. A network architect has learned that a customer site must support wireless smartphones. What should the architect keep in mind during the RF design?

   a. Most smartphones currently support only 2.4 GHz, which experiences more inference than 5 GHz.

   b. Smartphones tend to be highly mobile and run applications such as voice and video.

   c. Many smartphones support weaker encryption, which adds to the need for dedicated air monitors (AMs).

   d. Smartphones tend to support more spatial streams than laptops, but narrower channels for 5 GHz.

6. A network architect proposes two 7205 Mobility Controllers (MCs), which will together support about 400 APs. The customer environment will have a maximum of about 3400 wireless clients. The customer wants a virtual MM and requires redundancy for the MM services. Which LIC-MM-VA licenses should the architect recommend?

   a. One LIC-MM-VA-500 license

   b. Two LIC-MM-VA-500 licenses

   c. One LIC-MM-VA-5k licenses

   d. Two LIC-MM-VA-5k licenses

7. A network architect has learned that a site is near a military installation. What effect does this have on the RF plan?

   a. The 2.4 GHz band should be used exclusively.

   b. The 5 GHz band should be used exclusively.

   c. Channels in the U-NII-1 band should be excluded.

   d. DFS channels should be excluded.

8. A customer had an existing network with a 7024 controller deployed on each floor. The architect plans to replace the APs with 150 AP-325s. The customer currently has an ArubaOS 6.x master-local architecture. What are two customer needs that would indicate that a MM-architecture (based on ArubaOS 8x) with centralized and clustered controllers will work better for this customer? (Select two).

   a. Stateful failover for clients across the site

   b. Role-based stateful firewall policies

   c. Application filtering and deep packet inspection (DPI)

   d. Centralized load balancing for clients across MCs

   e. Integration with a ClearPass Guest portal

9. For which scenario is an Instant AP (IAP) deployment a good choice?
   a. The customer needs to provide easy-to-use VPN access to the main site for remote users at home offices.
   b. The customer needs secure wireless services, including 802.1X authentication and a stateful firewall, for an environment that requires about 60 APs.
   c. The customer requires seamless, instant roaming across a building with 10 floors and a three-tiered wired infrastructure.
   d. The customer has a challenging RF environment and requires centralized intelligence for dynamic channel planning and client load balancing.

10. Initially, an architect planned to propose Aruba 2540 switches to a customer. Which new customer requirement should make the architect rethink this plan?
    a. Deep packet buffers
    b. 10 GbE fiber uplinks
    c. 1 GbE copper edge ports
    d. Virtual Switching Framework (VSF)

11. A network architect has planned a wireless upgrade for a customer with three closely neighboring buildings with four floors each. The architect has planned between 25 and 30 APs per floor and a couple of outdoor APs between the buildings. The customer requires seamless roaming across the campus, as well as the support for both IPv4 and IPv6 endpoints.

    The customer plans to provide an SSID for employee access and an SSID for guest access.

    Which plan should the architect recommend for wireless devices?
    a. One VLAN for employees and one VLAN for guests
    b. 12 VLANs for employees, one per floor, and 3 VLANs for guests, one per building
    c. One VLAN for all wireless devices
    d. 3 VLANs, one for all wireless devices in each building

12. Refer to the exhibit in Figure 13-2.

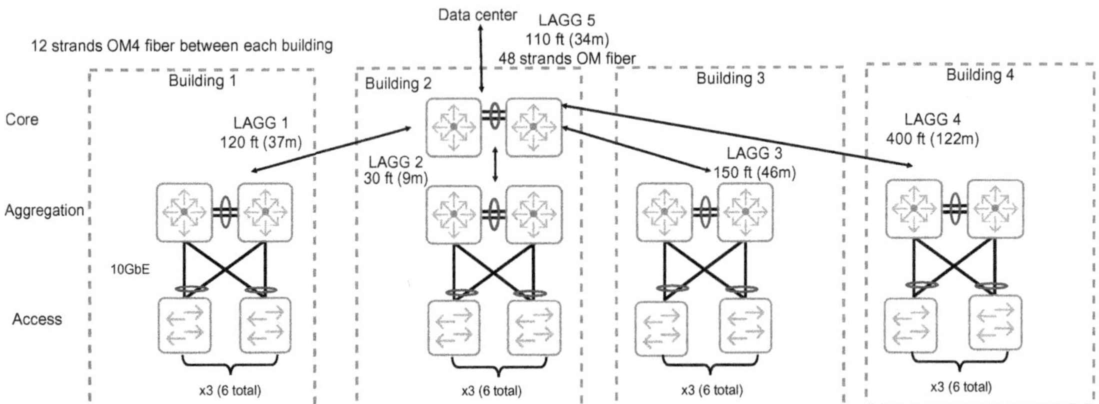

Figure 13-2 Question 12 exhibit

The network architect has selected 8320 switches for the aggregation layer and core. What is the most appropriate amount of bandwidth for the link aggregations indicated in the exhibit?

a. LAGG 1, 2, 3, and 4 = 20 GbE; LAGG 5 = 80 GbE
b. LAGG 1, 2, 3, and 4 = 40 GbE; LAGG 5 = 40 GbE
c. LAGG 1, 2, 3, and 4 = 40 GbE; LAGG 5 = 160 GbE
d. LAGG 1, 2, 3, and 4 = 60 GbE; LAGG 5 = 80 GbE

13. A network architect needs to ensure that Aruba APs will forward voice traffic in the correct queue over the air. What information is important to determine?

a. The LLDP-MED settings configured on switches that connect to the APs
b. The DSCP trust setting configured on the switches that connect to the APs
c. The IDS/WIPS capabilities that are enabled on the Aruba APs
d. The DSCP that the traffic will have when it arrives on the AP from the wired network

14. A network architect needs to plan a wireless upgrade for a building, which currently has an 802.11n solution. The building has a high density of wireless devices, including laptops, smartphones, printers, and security cameras. Which approach should the architect take to determine the AP count?

    a. Calculate the number of APs required to meet the capacity needs for the security cameras and other devices.

    b. Calculate the number of APs required to provide a strong –75 dBm signal across the complete coverage area.

    C. Divide the coverage area in square feet by 5000 or the coverage area in square meters by 465.

    d. Determine the current number of 802.11n APs and then multiply by 1.25.

15. Refer to the exhibit in Figure 13-3.

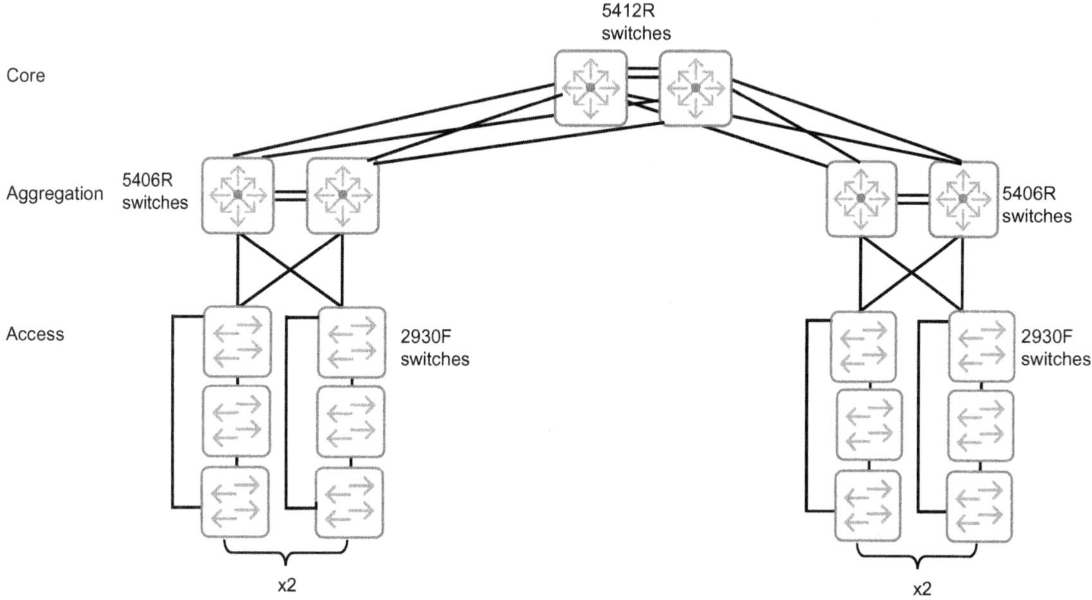

Figure 13-3 Question 15 exhibit

The exhibit shows the plan for a new network. How should the network architect implement VSF in this plan? Figure 13-4, Figure 13-5, Figure 13-6, and Figure 13-7 show four options. In the figures, each box indicates a VSF fabric. Which figure shows the correct design?

a.

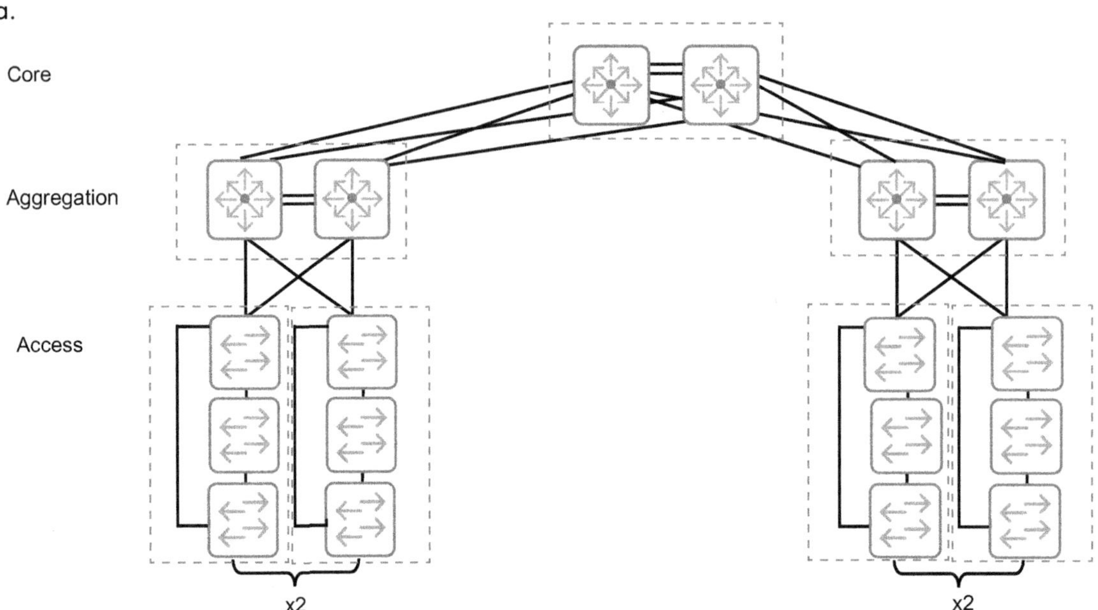

Figure 13-4 Question 15 option a

b.

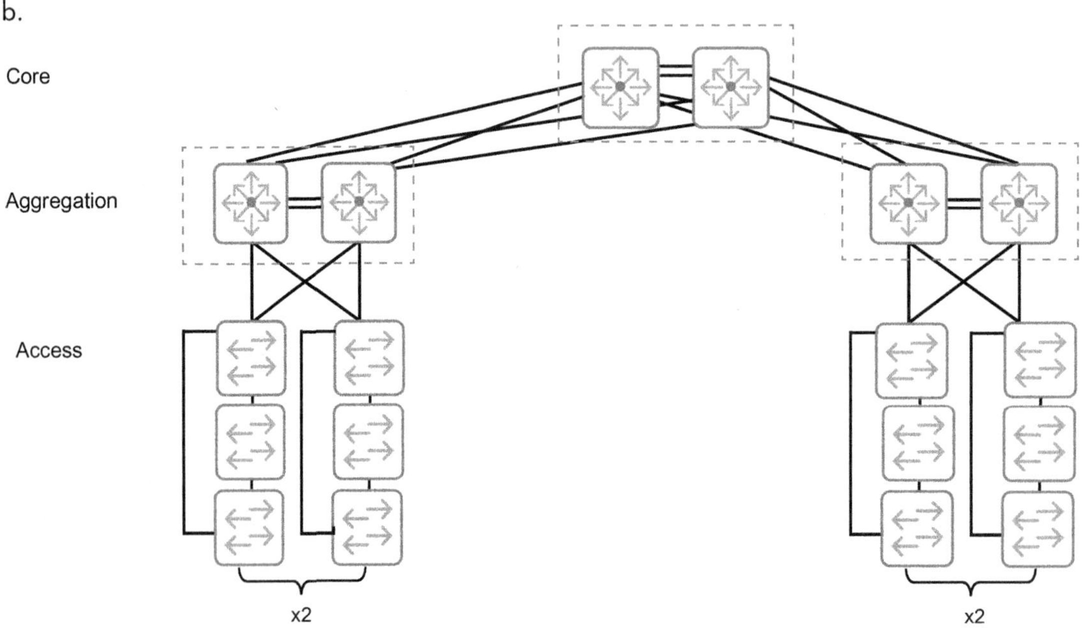

Figure 13-5 Question 15 option b

c.

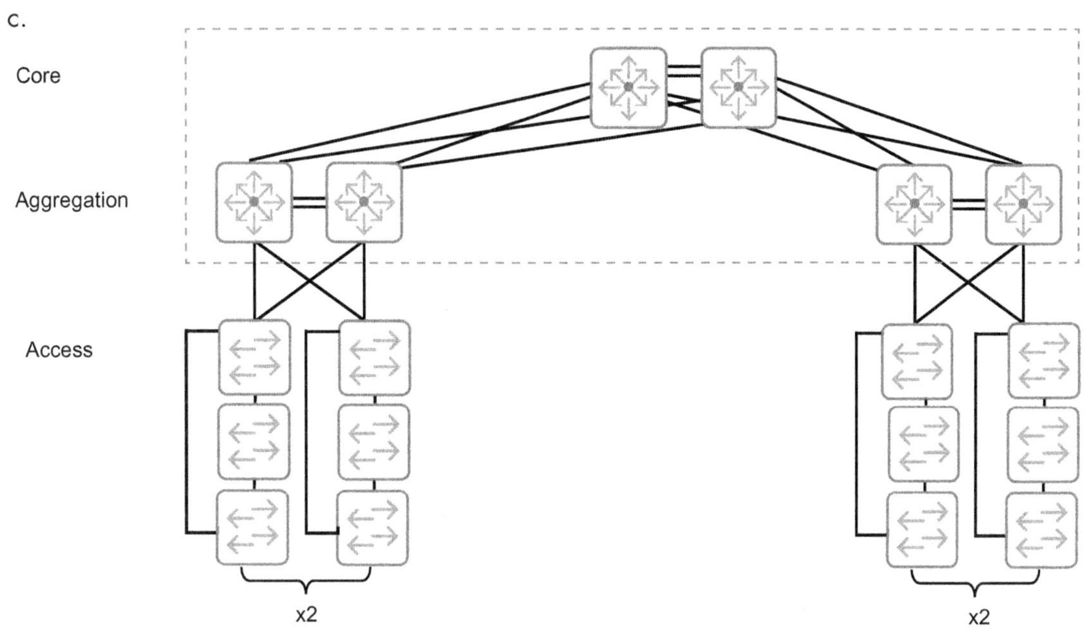

Figure 13-6 Question 15 option c

d.

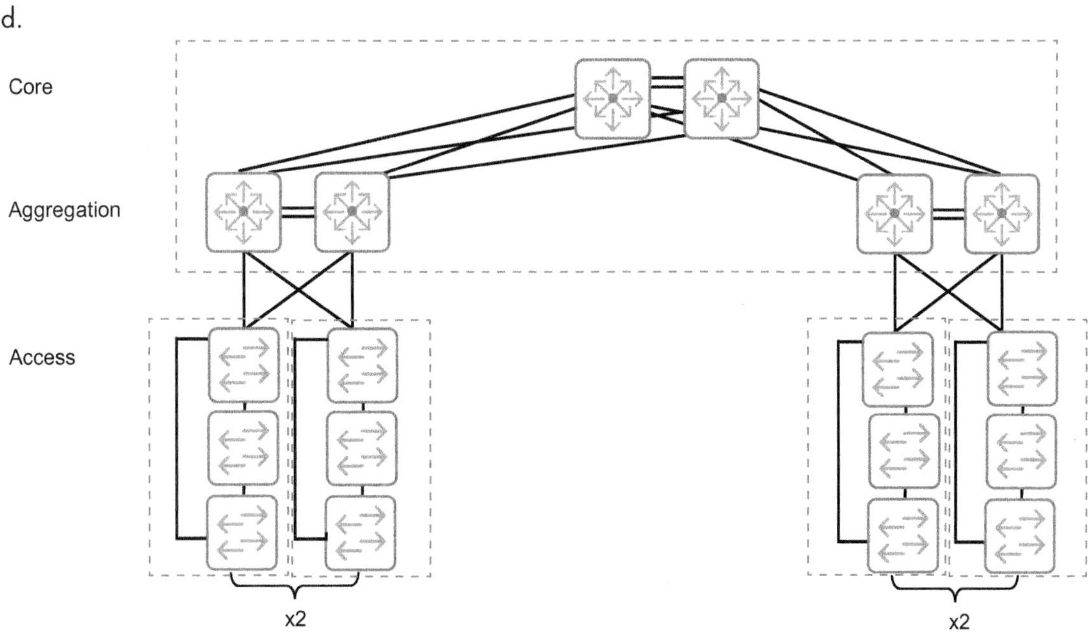

Figure 13-7 Question 15 option d

16. A customer has a RAP at a small branch office. When is bridge mode recommended?

    a. The branch office has its own Internet connection, but branch users also need to access corporate services at the main office.

    b. The customer wants the RAP to be managed by a centralized controller but forward all branch user traffic locally.

    c. The RAP needs to tunnel all branch user and control traffic to the main office, but it only needs to encrypt the control traffic.

    d. The customer wants to offload encryption and decryption to the RAP, but still tunnel branch user traffic to the main office.

17. For which scenario should a network architect recommend Aruba Central?

    a. The customer wants a simple solution that out-of-the-box RAPs can contact to determine the location of their MC.

    b. The customer wants a simple cloud-based option for managing and monitoring multiple Aruba Instant AP clusters and AOS-Switches.

    c. The customer wants a comprehensive network monitoring solution for a mix of Aruba and third-party networking devices.

    d. The customer wants a centralized licensing server to maintain licenses for multiple clusters of Aruba Mobility Controllers (MCs.)

18. A network architect needs to plan an 802.11ac deployment for an auditorium that seats 1000 people. The architect has gathered this information:

    - The expected maximum number of devices per user is 4.
    - The take rate is 80%.
    - It is expected that 75% of devices will use the 5 GHz band, and 25% will use the 2.4 GHz band.
    - The bandwidth requirements per user are about 1 Mbps per user

    What approximate number of APs should the architect propose?

    a. 8

    b. 16

    c. 32

    d. 48

19. A network architect has proposed an Aruba solution for a customer, including:
    - A Mobility Master (MM)
    - Three 7210 Mobility Controllers, deployed in a cluster
    - AP-325
    - Aruba 2930M switches at the access layer
    - 8320R switches at the core

    The customer wants to pass both wired and wireless traffic through Aruba Mobility Controllers' (MCs') firewalls. What is one implication of this requirement?

    a. The MCs will need to route all campus traffic and should be configured with a dynamic routing protocol.
    b. The firewall and other access control features built into the 8320 switches must be disabled.
    c. The access layer switches will need to implement a form of authentication on the edge ports.
    d. The access layer switches will need to implement port-based tunneled-node.

20. A network architect has proposed a single VLAN for a wireless deployment in a building with about 3000 wireless devices. The customer is concerned that this VLAN design will compromise performance. What should the architect explain about the Aruba solution?

    a. The Aruba solution supports a large broadcast domain with intelligent broadcast and multicast suppression and optimization.
    b. The Aruba Mobility Controllers (MCs) use a VLAN pool to translate the single VLAN ID into multiple logical VLANs.
    c. All wireless traffic is passed through the MM, so segmenting the wireless devices into different VLANs has no effect.
    d. A single VLAN design for wireless users is required in order for MCs to operate in a cluster.

21. A customer has wireless devices that use Microsoft Skype for Business. Which feature should the network architect recommend to ensure that MCs can recognize all voice traffic and reclassify it with the correct priority?

    a. AirGroup
    b. Wi-Fi Multimedia (WMM)
    c. AppRF
    d. Unified Communications Module (UCM)

22. What is one way to reduce co-channel interference in an environment with a high density of Aruba APs?

    a. Allow the use of only 20 MHz channels or 40 MHz channels in the 5 GHZ band.

    b. Deploy dedicated air monitors (AMs) to help detect RF interference.

    c. Make sure to install omnidirectional antennas on APs and mount the APs on the ceiling.

    d. Make sure that airtime fairness is enabled in the MCs' traffic shaping profile.

23. A network architect has planned to propose an Aruba 2930F 48G PoE+ 4SFP+ switch. The switch must power these devices:

    - 20 AP-315s, which are PoE+ devices and require a maximum of 14.4 W
    - 20 IP phones, which are PoE devices and require a maximum of 7 W

    The switch PoE+ reserve is 370 W. What can the architect do to resolve the power issue?

    a. Add a redundant power supply to the switch.

    b. Disable PoE+ on 20 of the ports and both PoE and PoE+ on eight of the ports.

    c. Select a power supply that delivers a higher wattage.

    d. Select the 740 W model for this switch.

24. Refer to the exhibits in Figure 13-8 and Figure 13-9.

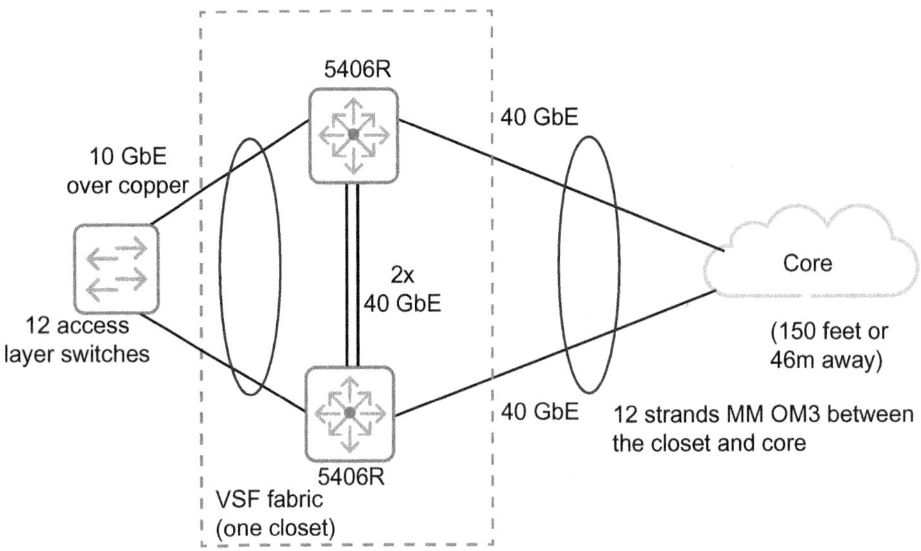

Figure 13-8 Question 24 exhibit 1

Line #	Part Number	Description	Manufacturer	Unit Price	Quantity	Total	Price List
1.00	J9821A	Aruba 5406R zl2 Switch	Hewlett Packard Enter...	$2,419.00	2	$4,838.00	USA Price List (USD)
1.01	H1MT0E	HPE 3Y FC 24x7 Aruba 5406R zl2 Switc SVC [for J9821A]	Hewlett Packard Enter...	$4,094.00	2	$8,188.00	USA Price List (USD)
1.02	U4832E	HPE Networks 54xx/82xx zl Startup SVC [for J9821A]	Hewlett Packard Enter...	$2,325.00	2	$4,650.00	USA Price List (USD)
1.03	J9828A	Aruba 5400R 700W PoE+ zl2 PSU	Hewlett Packard Enter...	$799.00	4	$3,196.00	USA Price List (USD)
1.04	J9828A ABA	INCLUDED: Power Cord - U.S. localization	Hewlett Packard Enter...	incl.	4		
1.05	J9995A	Aruba 8p 1/2.5/5/XGT PoE+ v3 zl2 Mod	Hewlett Packard Enter...	$4,799.00	6	$28,794.00	USA Price List (USD)
1.06	J9996A	Aruba 2p 40GbE QSFP+ v3 zl2 Mod	Hewlett Packard Enter...	$6,799.00	4	$27,196.00	USA Price List (USD)
1.07	JH231A	HPE X142 40G QSFP+ MPO SR4 Transceiver	Hewlett Packard Enter...	$3,299.00	2	$6,598.00	USA Price List (USD)
2.00	JH234A	HPE X242 40G QSFP+ to QSFP+ 1m DAC Cable	Hewlett Packard Enter...	$419.00	2	$838.00	USA Price List (USD)
		Quote Total				**$84,298.00**	

Figure 13-9 Scenario exhibit

Figure 13-8 shows the plan for a wiring closet, and Figure 13-9 shows the BOM. What should the architect change about the BOM to meet the requirements of the customer environment?

a. Change the QSFP+ DACs to QSFP+ MPO SR4 transceivers.

b. Change the QSFP+ DACs to stacking cables.

c. Change the QSFP+ MPO SR4 transceivers to QSFP+ BiDi transceivers.

d. Change the QSFP+ MPO SR4 transceivers to QSFP+ SR4e transceivers.

25. A customer has a deployment with two 7205 MCs deployed in a cluster and controlled with an MM. The customer has 100 AP-305s. The customer now wants to add 40 RAPs, which will connect telecommuters to the company's single data center. Redundancy is required for the RAPs' controller. The customer would like the most cost-effective option that complies with best practices. What should the architect recommend?

a. Adding the 40 RAPs to the current cluster

b. Deploying two 7030 MCs with VRRP-based redundancy

c. Deploying two 7030 MCs in a cluster

d. Adding an 7205 MC to the cluster

26. An architect needs to recommend an MC for a new 802.11ac deployment. The customer environment has about 3500 wireless devices, and the architect plans to recommend 200 APs. The customer requires full N+1 redundancy for the solution and prefers hardware controllers. Which MCs should the architect recommend for the most cost-effective solution that meets the requirements?

    a. Three 7030 MCs

    b. Two 7210 MCs

    c. Two 7205 MCs

    d. Three 7205 MCs

27. What value does a passive site survey offer beyond a virtual, or predictive, survey alone?

    a. The passive survey collects information about non-802.11 interference to help eliminate that interference from the site.

    b. The passive survey collects information about existing APs' signal to help validate the correct location for APs.

    c. The passive survey results in a heat map of the signal predicted for new APs, while the virtual survey only shows AP locations.

    d. The passive survey results in heat maps of the actual tested signal for new APs, while the virtual survey creates a heat map for the existing APs.

28. A network architect has proposed Aruba 802.11ac APs and AOS-Switches for a network upgrade. For wireless security, the architect has proposed PEAP-MSCHAPv2 for the EAP method used with WPA2-Enterprise. ClearPass will act as the RADIUS server. What is one best practice for the solution?

    a. Use ClearPass OnGuard to deploy certificates to wireless devices.

    b. Enable tunneled-node on the switch ports that connect to APs.

    c. Add IPsec IKE as the authentication method in the ClearPass policy.

    d. Install a RADIUS certificate signed by a trusted CA on ClearPass.

29. Refer to the exhibit in Figure 13-10.

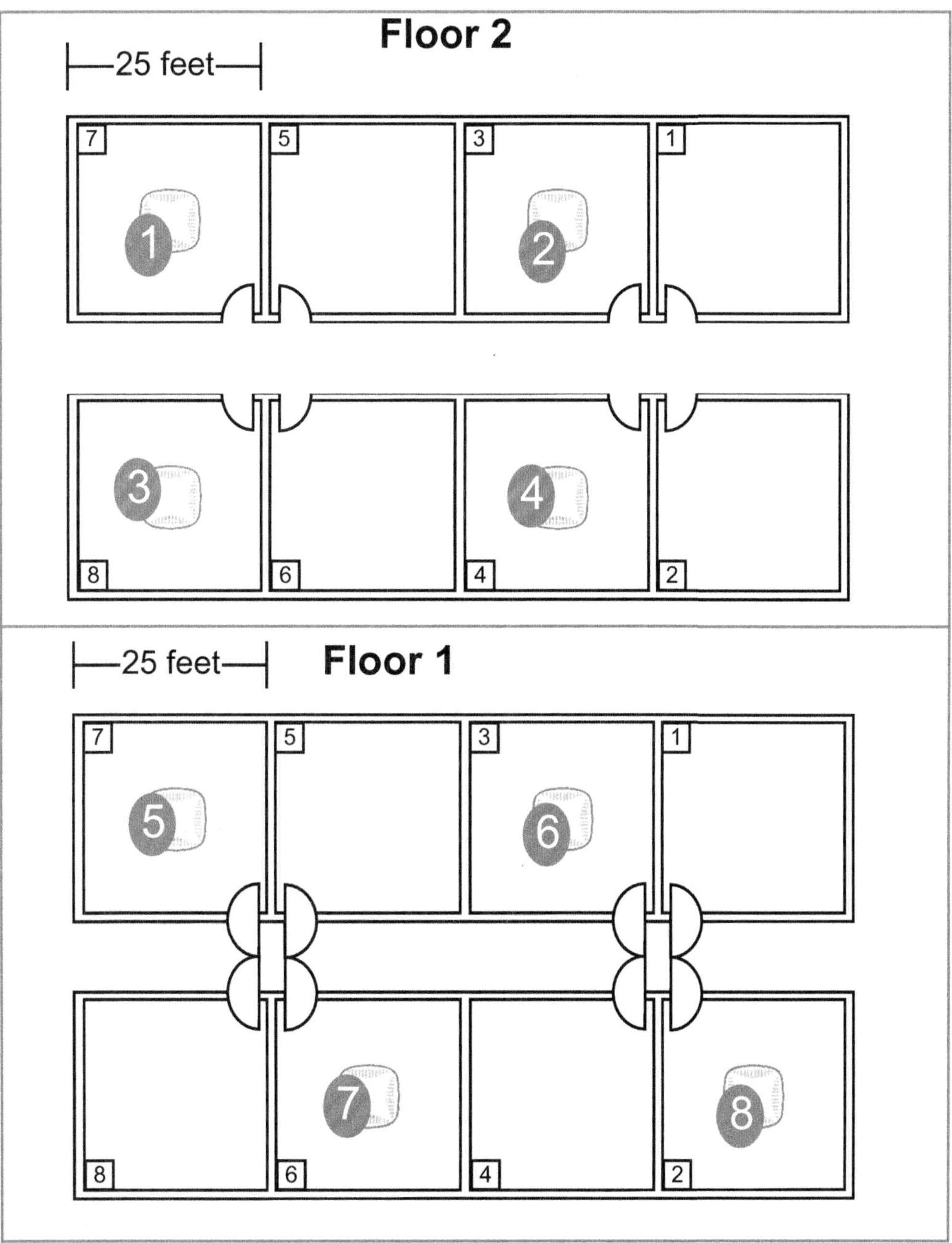

Figure 13-10 Question 29 exhibit

A university needs an 802.11ac upgrade for an office building. The exhibit shows the plan for the AP placement on one section of the building. What correction to this plan should the architect make to better align with Aruba best practices?

a. Move APs 1, 2, 5, and 6 into the hallway.

b. Move AP 1 into room 5 and AP 2 into room 1.

c. Move AP 7 into room 8 and AP 8 into room 4.

d. Remove APs 1, 2, 5, and 6 from the plan.

30. A customer has a small site with Instant APs, which tunnel traffic to an MC at a main site. The network architect needs to plan the IP addressing for wireless devices at the small site. For which IAP DHCP modes should the wireless devices receive IP addresses in the main site scopes?

    a. Centralized L2 and Distributed L2 only

    b. Centralized L2 and Centralized L3 only

    c. Centralized L2, Centralized L3, Distributed L2, and Distributed L3

    d. Local L3, Centralized L2, and Centralized L3

Answers

1. **A** and **C** are correct. The architect needs to know where wiring closets are located in order to plan the AP deployment. It is important to know both the cable type and length for the connection particularly when the connection must support PoE+ and 1GbE (or higher) speeds.

   B is incorrect. If the cable is very old, you might anticipate some issues, but this building is relatively new. The precise year that the cable was installed is less important than the cable type. **D** and **E** are incorrect. From the scenario, you know that the existing switches do not support PoE+, which is required for the 802.11ac APs. Therefore, the existing switches' other features do not matter for planning the AP deployment.

   For more information on this topic, refer to Chapter 1.

2. **C** is correct. 802.11ac APs require PoE+ to operate with their full capabilities. The existing switches only support PoE.

   A is incorrect. SNMP-based network security is a ClearPass feature for implementing authentication for legacy wired networks; it is not a reason for an upgrade. If the customer were interested in wired authentication, you would recommend an AOS-Switch, all of which support 802.1X. **B** is incorrect; in this scenario, the customer might require 10 GbE on the uplinks, but not to the edge ports. **D** is incorrect; nothing in the scenario indicates that access layer routing is a driving need. (Routing at the core is more typical.)

   For more information on this topic, refer to Chapter 1.

3. **C** is correct. Access licenses are calculated based on concurrent authentications. The customer needs 1200, rounding up from 1150. Onboard licenses are calculated by user count for users who need to onboard devices. This is 350. You could round up to 400, but one 500-count licenses is more cost-effective than four 100-count licenses. The OnGuard licenses are calculated by the number of devices that need health checks, which is 350. Again you can round up to 500.

   A, **B**, and **D** are incorrect because they do not recommend the correct number of licenses (see above).

   For more information on this topic, refer to Chapter 5.

4. **B** is correct. AMP licenses are based on monitored network infrastructure devices.

   A is incorrect because the customer does not need licenses for the clients. **C** and **D** are incorrect because licenses do not depend on features.

   For more information on this topic, refer to Chapter 11.

5. **B** is correct. Smartphones tend to be highly mobile and run time-sensitive applications such as voice and video. These characteristics affect the RF design. For example, APs must be deployed densely enough to allow users with phones to roam from one AP to another without too much of a drop in signal.

   A is incorrect; most smartphones currently support both 2.4 GHz and 5 GHz. **C** is incorrect. Most smartphones support WPA2 AES encryption. **D** is incorrect; smartphones usually support the same number of spatial streams as laptops or fewer because more spatial streams require antennas, for which smartphones do not have room.

   For more information on this topic, refer to Chapter 2.

6. **A** is correct. The VMM is licensed by the managed device count, which includes MCs and APs. An LIC-MM-VA-500 license will cover the approximately 402 devices in this environment. The customer can deploy a standby VMM which uses the same licenses.

   B is incorrect. The standby VMM does not require separate licenses. **C** and **D** are incorrect; the wireless clients are not included in the count for VMM licenses.

   For more information on this topic, refer to Chapter 3.

7. **D** is correct. The military installation indicates that APs will probably need to use DFS to move out of DFS channels if radar needs to use them. Aruba recommends against using DFS channels in these circumstances because it can lead to disrupted connections, especially since client support for DFS can be uneven.

   A and **B** are incorrect. Both bands can be used in proximity of a military installation. **C** is incorrect. The U-NII-1 band does not require DFS.

   For more information on this topic, refer to Chapters 2 and 8.

8. **A** is correct. Moving the MCs to a central location and making them a cluster managed by MM enables better failover, as well as more seamless roaming, across the site. **D** is also correct. Clustering automatically load-balances clients across MCs.

   B, **C**, and **E** are incorrect. Standalone MCs and MCs in a master-local architecture support all of these features.

   For more information on this topic, refer to Chapters 3 and 7.

9. **B** is correct. Instant APs are easy to deploy and cost-effective. In this case, the customer wants only 60 APs, and Instant APs can work well at numbers well about that. The IAPs support many of the same features as controllers, including 802.1X and a stateful firewall.

   A is incorrect. A Remote AP would better meet the needs for providing remote home office users VPN access to the main site. **C** is incorrect. An IAP deployment could work in this environment, but it would require several clusters due to the three-tiered architecture (each IAP cluster must be in the same VLAN). A controller-based design could establish a cluster at the center of the network and provide more seamless roaming. **D** is incorrect. A MM, which management MCs, not IAPs, is required to provide centralized intelligence for dynamic channel planning (AirMatch) and client load balancing (ClientMatch).

   For more information on this topic, refer to Chapter 3.

10. **D** is correct. The Aruba 2540 Series does not support VSF.

    A, **B**, and **C** are incorrect. The Aruba 2540 Series provides deep packet buffers, 10 GbE fiber uplinks, and 1 GbE copper edge ports.

    For more information on this topic, refer to Chapters 4 and 8.

11. **A** is correct. For seamless roaming and best support for IPv6, as well as overall simplicity, Aruba recommends making VLANs as large as possible within certain constraints. The VLAN should extend as far as the contiguous RF coverage area; in this case, that is the entire campus because outdoor APs extend coverage between buildings. Each different SSID should be associated with a different VLAN. The customer has a guest SSID and employee SSID, so the architect should recommend one VLAN for each SSID.

    B is incorrect because it indicates multiple VLANs within the same SSID and coverage area. The customer requirements point to a single VLAN per SSID and coverage area being the better design. **C** is incorrect because different SSIDs should be associated with different VLANs; guests and employees should be in different VLANs. **D** is incorrect because guests and employees should be in different VLANs and because one RF coverage area extends across the buildings. Employees should be in the same VLAN in all buildings, and guests should be in the same VLAN in all buildings.

    For more information on this topic, refer to Chapter 6.

12. **C** is correct. The Aruba 8320 switches support M-LAG, so the pairs of 8320 switches connected together in the exhibit use a shared link aggregation to connect to the access layer and core. The exhibit indicates that each access layer VSF fabric has two 10 GbE links and six VSF fabrics connect to the 8320 switches. The highest appropriate oversubscription at the aggregation layer is 4:1. 120/4 = 30 Gbps. However, 30 Gbps would require an unbalanced M-LAG. It is better to round up to 40 GbE, and establish two 10 GbE links on each 8320 switch (one to each core switch). The exhibit indicates enough fiber for these connections (eight strands are required). In a three-tier network no additional oversubscription should be introduced at the network core. Four aggregation layer pairs of switches have 40 GbE connections for 160 GbE total; the core should have the same with four 40 GbE links.

    A is incorrect. Planning 20 GbE at the aggregation layer would provide 6:1 oversubscription, which is higher than recommended. **B** is incorrect. The bandwidth for LAGGs 1–4 is appropriate, but, because the core connects to four aggregation layer pairs, it could receive up to 160 Gbps of traffic heading to the data center. A 40 GbE link would introduce 4:1 oversubscription, and the core should not introduce more oversubscription in a three-tier topology. **D** is incorrect. The oversubscription at the aggregation layer (2:1) is okay, although six 10 GbE links between the aggregation layer and core would create a less balanced configuration. In addition, 80 GbE bandwidth would not be enough for the core.

    For more information, refer to Chapter 4.

13. **D** is correct. The Aruba AP uses the incoming DSCP and the DSCP mappings configured on the MC to determine in which WMM queue to forward the traffic.

    A is incorrect. A switch can use LLDP-MED to configure a connected device, such as an IP phone, to send traffic with the correct VLAN ID and priority. It can also use LLDP-MED to negotiate power settings with an AP. The LLDP-MED settings on the switch do not have a direct effect on the priority passed to the AP. **B** is incorrect. The DSCP trust settings on an AOS-Switch port have an important effect on how the switch prioritizes traffic that it receives from the AP with a DSCP; therefore, it is an important part of the overall QoS considerations. However, this question is asking about how to ensure that the AP itself forwards traffic over the air, and this setting does not alter the DSCP that the switch itself sends on traffic forwarded on this port. **C** is incorrect. Whether or not the AP implements IDS/WIPS has no effect on how it queues traffic. (In addition, an AP that both serves clients and implements IDS/WIPS is voice aware and does not conduct scans when it is communicating with a voice client.)

    For more information on this topic, refer to Chapter 8.

14. **A** is correct. This environment requires a capacity-based design in which the architect determines that the cameras and other devices can receive enough bandwidth from APs, not simply that they can receive a good signal.

    B is incorrect. A strong signal is important, but the architect needs to think more in terms of how many security cameras each AP can support and still provide the required bandwidth. In addition, a –65 dBm signal is recommended for smartphones that might use voice so that they can roam properly. **C** is incorrect; this is the rule of thumb for a coverage-based design and might not meet the needs for this customer. **D** is incorrect; this is not a valid rule.

    For more information on this topic, refer to Chapter 2.

15. **A** is correct. VSF can play a valuable role in simplifying the topology at each layer of the network. The first exhibit shows that multiple access layer switches are ringed together, and then two of the switches are connected to the aggregation layer. With this physical configuration, it is best to use VSF (or backplane stacking on 2930M and 3810M switches) to eliminate a complex spanning topology. The aggregation layer and core should also use VSF to provide better resiliency and a simpler design.

    B and **C** are incorrect because they do not show VSF at the access layer. **C** and **D** are incorrect because they show the aggregation layer and core switches combined in a single fabric. However, a VSF fabric with 5400R switches does not support this many switches.

    For more information, refer to Chapter 4 and 7.

16. **B** is correct. In bridge mode, the RAP forwards all traffic locally.

    A is incorrect. This would be a use case for split-tunnel mode. **C** is incorrect. This would be a use case for tunnel mode with the default behavior. **D** is incorrect; this is not how an RAP deployment works.

    For more information, refer to Chapter 10.

17. **B** is correct. Aruba Central is a cloud-based management solution, which can manage Instant AP (IAP) clusters and AOS-Switches.

    A is incorrect. Aruba Activate supplies provisioning rules for out-of-the-box devices and is the solution for this scenario. **C** is incorrect. Aruba Central does not support third-party device management; AirWave should be used for this scenario. **D** is incorrect; the MM provides centralized licenses for Aruba MCs.

    For more information, refer to Chapter 11.

18. **B** is correct. Aruba recommends this formula to calculate the number of APs for a deployment such as this:

    Users x Devices per user x Take rate x percentage devices in the 5GHz band / 150

    In this case, the formula yields:

    1000 x 4 x .8 x .75 / 150 = 16

    A, **C**, and **D** are incorrect because they do not follow this formula.

    For more information, refer to Chapter 9.

19. **C** is correct. This scenario calls for tunneled-node on the AOS-Switches so that they can tunnel wired traffic to the MC and the MC can apply firewall rules to it. The preferred form of tunneled-node is per-user tunneled-node, and this is the only mode that works with MC clustering. To implement per-user tunneled-node, the switch must also implement authentication on edge ports.

    A is incorrect. The MCs can forward both wired and wireless traffic that is tunneled to it at either Layer 2 or Layer 3. They can route the traffic, but they are not required to do so.

    B is incorrect. The 8320 switches do not have a built-in firewall.

    D is incorrect. Port-based tunneled-node does tunnel traffic to MCs. However, it does not work with MC clustering.

    For more information, refer to Chapter 5.

20. **A** is correct. The traditional concern with a single VLAN for a wireless deployment is that a VLAN defines the broadcast domain. If every broadcast sent by one wireless device must be sent over the air on every AP, the wireless capacity is quickly used up. However, the Aruba solution has intelligent measures to suppress unnecessary broadcasts and multicasts. This allows the solution to provide the benefits of a single VLAN—such as simpler roaming—without performance drawbacks.

    B is incorrect. A single VLAN design does not use VLAN pooling. **C** is incorrect. Wireless user traffic does not pass through the MM. **D** is incorrect. While a single VLAN design is often recommended with an MC cluster, because it enables the full benefits of seamless roaming, this design is not required. The cluster can put wireless devices in multiple different VLANs. For example, it might support wireless devices in different RF contiguous zones, in which case different VLANs are recommended.

    For more information, refer to Chapter 6.

21. **D** is correct. The most effective version of UCM runs on the Mobility Master (MM). UCM integrates with Microsoft Skype for Business to enable the Aruba solution to identify voice sessions. The MC can then recognize the voice traffic and reclassify it with the proper priority, as necessary. UCM also provides central monitoring for call quality.

    A is incorrect. The AirGroup feature makes network that run protocols such as Bonjour and Apple AirPlay more efficient. **B** is incorrect. It *is* recommended that WMM be enabled when wireless devices send voice and other UC traffic. However, this feature does not itself cause MCs to reclassify traffic. Instead, it enables wireless clients and APs to send classified traffic in different priority queues. **C** is incorrect. AppRF provides deep packet inspection (DPI) for traffic flowing through the MC, but is not specifically related to reclassifying voice traffic.

    For more information, refer to Chapter 8.

22. **A** is correct. The narrower channels support lower data rates, but they allow for more nonoverlapping channels. The 5 GHz band has many 20 MHz and 40 MHz channels, but fewer 80 MHz channels particularly if DFS channels cannot be used. Using 20 MHz or 40 MHz channels reduces co-channel interference in high-density environments. (Only 20 MHz channels are recommended in the 2.4 GHz band, which is much narrower.)

    B is incorrect. AMs detect and mitigate against rogue APs and other wireless threats. Spectrum Monitors (SMs) detect RF interference. **C** is incorrect. While a properly planned high-density deployment can use omnidirectional antennas and ceiling mounting, these characteristics are not required. In fact, in very dense environments other options can reduce co-channel interference. **D** is incorrect. Airtime fairness helps to divide airtime evenly among legacy clients that only support slow data rates and higher speed clients so that the slow clients do not monopolize all of the airtime. It does not address co-channel interference.

    For more information, refer to Chapter 9.

23. **D** is correct. The question indicates that the switch must support more power than its reserve. (It supports at least 446.6 W to be precise, taking into account that the switch must have a bit more power than the requirements in reserve to bring up the last PoE+ device). The 2930F switch has an integrated power supply, so the only way to provide more power is to select the model that provides more power.

    A and **C** are incorrect because the 2930F switches have an integrated power supply. **B** is incorrect. The switch only supplies power on a port if the connected device requests it, and disabling some ports from supplying power will not solve the issues.

    For more information, refer to Chapter 4.

24. **C** is correct. The SR4 transceivers require 12 strands of fiber each, and the buildings do not have enough fiber between them to support two. A QSFP+ BiDi transceiver requires only two strands, and it can extend over the distance indicated here.

    A is incorrect. QSFP+ DACs are appropriate for connections between switches in the same closet. **B** is incorrect. Switches that use backplane stacking, such as the 3810M and 2930M, require stacking cables; however, switches that use VSF such as the 5406R can use DACs (as well as other kinds of cable such as SPF+ transceivers with fiber). **D** is incorrect. Both the SR4 and SR4e can extend the required distance, but they both require too much fiber for this environment.

    For more information, refer to Chapter 4.

25. **B** is correct. It is best practice to place RAPs on a dedicated controller. One 7030 MC can handle 40 RAPs with some room for expansion; the second 7030 MC provides redundancy. Clustering is not recommended for MCs that terminate RAP connections because clustering would require each cluster member to have a public IP address to provide to the RAPs. VRRP-based redundancy will serve the needs for a single site (LMS backup would serve the needs if the customer had a second data center connected at Layer 3 and wanted to deploy the backup controller there).

    A is incorrect. The 7205 MCs have sufficient capacity to handle the APs, even in a failover situation. However, a cluster does not usually meet the needs for RAP termination, as discussed above. And it is often best practice to separate the MCs that terminate campus APs and RAPs. **C** is incorrect because it specifies a cluster, and clustering is not recommended for the reasons listed above. **D** is incorrect. Adding an MC to the cluster does not address the needs since the customer does not need more AP capacity but rather nonclustered controllers to terminate the RAPs' tunnels.

    For more information, refer to Chapter 7.

26. **C** is correct. A 7205 MC can support 256 APs and about 8000 devices. Each MC can handle this customer's environment, so two MCs will meet the needs.

    A is incorrect. Each 7030 MC can support 64 APs and about 4000 devices. Three MCs can nearly support the required number of APs, but two (in a failover situation) cannot. **B** is incorrect. The 7210 MCs can support up to 512 APs and about 16,000 users, and two of these controllers are more than the customer requires. **D** is incorrect because the customer can meet the redundancy needs with two MCs.

    For more information, refer to Chapter 3.

27. **B** is correct. A passive survey collects information about existing APs' signal. This can help to validate the correct location for APs beyond what a virtual survey, which uses predictive software, can do.

    A is incorrect. A passive survey does not include spectrum analysis of non-RF interference. A spectrum clearing survey provides this information. **C** is incorrect. A virtual survey provides a heat map of coverage predicted for new APs. **D** is incorrect. In a passive survey, an architect creates a heat map of the coverage provide by existing APs. In an active site survey, the architect installs a test AP in proposed locations and maps actual coverage for new APs.

    For more information, refer to Chapter 2.

28. **D** is correct. With PEAP-MSCHAPv2, the RADIUS server (here ClearPass) authenticates with a certificate. It is best practice to have this certificate be signed by a trusted CA so that clients can validate it.

    A is incorrect. ClearPass Onboard, not OnGuard, deploys certificates. In addition, wireless devices do not require client certificates for PEAP-MSCHAPv2; they use usernames and passwords to authenticate. **B** is incorrect. Tunneled-node on the switches is for tunneling wired device traffic to an MC. It is against best practice to add tunneled-node on the switch ports that connect to APs since the APs already tunnel traffic to the MC. **C** is incorrect. The ClearPass policy will use 802.1X to authenticate clients, which is the correct authentication method for WPA2-Enterprise.

    For more information, refer to Chapter 5.

29. **B** is correct. The site does not have a thick floor, so Aruba recommends staggering the placement of APs on different floors such that APs are not directly above and below each other.

    A is incorrect. Aruba recommends deploying APs in rooms, not hallways. **C** is incorrect. Moving these APs would result in all of the APs on the two floors being directly above and below each other. **D** is incorrect. The current plan shows a reasonable distance between APs for an office.

    For more information, refer to Chapters 2 and 9.

30. **C** is correct. Wireless clients connected to IAPs receive IP addresses in the main site's scope when the IAPs operate in Centralized L2, Centralized L3, Distributed L2, or Distributed L3 mode—even though the IAPs act as DHCP server in Distributed L2 and Distributed L3 mode.

    A and **B** are incorrect because, even though they list correct modes that use main site scopes, they do not list all of the correct modes. **D** is incorrect because in Local mode, whether L2 or L3, the IAPs assign wireless devices IP addresses in a local scope.

    For more information, refer to Chapter 10.

Appendix: Activities and Learning Check Answers

This appendix includes answers for learning checks and activities.

Chapter 1: Information Gathering

Activity 1.1: Determine the customer's existing network

1. Briefly describe the scope of your project. For which sites and buildings do you need to propose a network solution?

   The project focuses on Phase 1 of Corp1's network upgrade. In Phase 1, Corp1 wants to completely upgrade its Wi-Fi and wired network at the main corporate campus, with includes Building 1 and Building 2. Each building has three floors.

2. Determine the customer's present Wi-Fi network structure at this site:
   a. Existing Wi-Fi network: Y
   b. Wi-Fi vendor: IBEWIFI
   c. Radio capabilities: 802.11a/bg
   d. Give a brief explanation of the customer's current Wi-Fi issues:

      The existing Wi-Fi coverage is sparse and does not cover every corner of the building. Performance is poor even though employees are not using demanding applications.

3. Determine the customer's existing wired network(s) at this site:
   a. Wired network to be replaced: Y
   b. If yes, which segments: Campus
   c. Datacenter network vendor: HPE
   d. Campus network vendor: IBESwitch
   e. Number of wired closets: 12
   f. Rack space in each closet: 20U
   g. Number of access layer switches per closet: 3 in 10 and 1 in 2
   h. Number and speed of ports on access layer switches: 48 10/100/1000 Mbps and 2 1 GbE (SPF)

i. Current access layer switch capabilities such as PoE and PoE+: PoE, 8 ports only; Layer 2

j. Is every desk wired (Y/N): Y

k. Wired network diagram supplied (Y/N): Y

l. Give a brief explanation of the customer's present wired issues. Also note features that might be required that the current switches do not support:

Users are experiencing performance issues. The current switches support only up to 1 GbE. They do not support PoE+, which the new APs require. They also lack authentication capabilities. The aggregation layer switches also lack support for 10 GbE and 40 GbE and have relatively small MAC and ARP tables.

m. Does the existing wired network support enough drops for an upgraded Wi-Fi network?: No

Activity 1.2: Determine site specifics

1. You have determined that the customer needs a Wi-Fi upgrade for two buildings at the main campus. Fill in more details in the table below. If the information is the same for both buildings, you can simply draw an arrow to the other cell.

Table 1-1 Details of Buildings 1 and 2 at the Main Campus

	Building 1	**Building 2**
Dimensions: Length Width Area per floor	310 x 173 feet (94 x 53 m) 53,630 square feet (4,982 sq m)	310 x 173 feet (94 x 53 m) 53,630 square feet (4982 sq m)
Number of floors	3	3
Ceiling height	15 feet high (4.5 m)	15 feet high (4.5 m)
Drop-ceiling height	10 feet (3 m)	10 feet (3 m)
Obstruction within drop-ceiling	Y	Y
Floor plans supplied (Y/N)	Y	Y
General description of the physical environment and list of features relevant to RF	This is an office building with mostly closed offices. Center corridor on each floor has restrooms, elevators, stairs and many closets. Floor 1 has wide open spaces, including a data center.	This is an office building with mostly closed offices. Center corridor on each floor has restrooms, elevators, stairs, and many closets.

2. What is the distance between building 1 and building 2: 50 feet (15.2 m)

Activity 1.3: Collect information on users and devices

1. Determine the customer's wireless user base:

a. Number of users in main campus: 900–920, including guests

b. User types: Employees and a few guests

c. Device types: HP Elitebook Folio G1 laptops, printers, smartphones (BYOD), and tablets (BYOD)

d. Types of applications in use: Email, Web, print, accessing files on servers

2. What are the requirements for roaming on these devices?

   Users want to be mobile with their wireless devices and want seamless roaming.

3. Use the Internet to research the wireless capabilities of some smartphones. You can use this site **http://clients.mikealbano.com/**. Or you can search for the official specifications on the providers' sites. Note features such as the Wi-Fi bands supported, 802.11ac support, and number of spatial streams. Look for IPhone 7, Samsung Galaxy S8 plus and the Sony Z4. (If you want to look up the EliteBook Folio G1 specifications, search for the HP QuickSpecs because the **clients.mikealbano.com** website does not list this specific model.)

   The iPhone 7 and Samsung Galaxy S8 plus, as well as the EliteBook Folio G1, have these capabilities:

   – 802.11a/bg/n/ac

   – 2 spatial streams

   The iPhone 7 does not support MU-MIMO, the Galaxy S8 plus does, and the EliteBook Folio G1 does not list whether it does or does not.

   The Sony Z4 is only 802.11a/bg/n capable.

4. Determine the customer's wired user and device base:

   a. Excluding APs, how many drops per closet must the new switches support?

      i. Building 1

         1. Floor 1 Closet 1:35
         2. Floor 1 Closet 2: 35
         3. Floor 2 Closet 1: 125
         4. Floor 2 Closet 2: 125
         5. Floor 3 Closet 1: 125
         6. Floor 3 Closet 2: 125

      ii. Building 2

         1. Floor 1 Closet 1: 125
         2. Floor 1 Closet 2: 125
         3. Floor 2 Closet 1: 125
         4. Floor 2 Closet 2: 125
         5. Floor 3 Closet 1: 125
         6. Floor 3 Closet 2: 125

b. User types: Employees

c. Device types: HP EliteBook Folio G1 laptops in docking stations, HD printers on Floor 1 in both buildings, video conference in conference rooms on Floor 2 of Building 1, board rooms on Floor 3 of Building 1, and conference rooms on all floors in Building 2

d. Types of applications in use: Web, Email, print, access to shared files, access to inventory software, video conferencing on special equipment

Learning check

1. What information do you need to gather about devices to create an RF plan? (Select two.)
   a. Type of applications running on the devices
   b. Size of the hard drive and available RAM
   c. Security features supported
   d. Degree of mobility
   e. Whether multiple users share laptops or workstations

2. What challenges do highly mobile devices such as voice handsets pose for architects planning a WLAN?
   a. These devices are in use while roaming, and users expect roaming transitions to be seamless.
   b. These devices require APs operating in the 5.0 GHz range, which limits the ability to implement 40 MHz or wider channels.
   c. These devices typically do not support Quality of Service, making it more difficult to ensure traffic is handled appropriately across the network.
   d. These devices don't support 802.11ac, forcing network designers to deploy APs that support 802.11 a/g.

Chapter: RF planning

Activity 2: RF planning

Activity 2.1: Select APs and begin a BOM in IRIS

1. Based on the customer requirements, which AP model will you propose? More than one option can be correct, but give your reasons for your selection.

   As of the publication of this study guide, the most typical selections would be APs in the 300 Series. The AP-305 and AP-315 could be acceptable choices because the user density is not particularly high and most clients support only two spatial streams and no MU-MIMO. But you could also propose the AP-325 or AP335 for future proofing.

2. Make a rough estimate for the number of APs required per floor. For a deployment such as this, divide the square footage by 2500 (or the area in square meters by 232). This is only an estimate; make adjustments for conference rooms and the central area.

   The square footage for each floor is 53,630. Dividing this by 2500 yields about 21 or 22 APs. However, the user density is very high (about 270 square feet per user), so you could maybe plan a bit fewer APs per floor. Floor 1 has an open area, so you should plan fewer APs on that floor.

Activity 2.6: Document your results and expand the BOM

1. How many APs on Building 1 Floor 1: Around 14–20
2. How many APs on Building 1 Floor 2: Around 16–22
3. How many APs on Building 1 Floor 3: Around 16–22
4. How many APs on Building 2 Floor 1: Around 16–22
5. How many APs on Building 2 Floor 2: Around 16–22
6. How many APs on Building 2 Floor 3: Around 16–22
7. What is your total count of APs: 94–130

Learning check

1. You see these specs for an Aruba AP-335: 4x4:4:4:3. How many spatial streams does the AP support and how many MU-MIMO spatial streams?
   a. 3 spatial streams and 4 MU-MIMO spatial streams
   b. 4 spatial streams and 3 MU-MIMO spatial streams
   c. 3 spatial streams and 3 MU-MIMO spatial streams
   d. 4 spatial streams and 4 MU-MIMO spatial streams

2. For which deployment would you suggest an active survey?
   a. A hospital that wants to upgrade its wireless network
   b. A branch site with 20 users
   c. A hospital that does not wants to upgrade its wireless network
   d. A company that has less complex environment

Chapter 3: Aruba Campus Design

Activity 3: Campus design

Activity 3.1: Document your results

1. What was your AP count?
   a. Total number of APs: About 120 (your number might range from 94 to 130)
   b. Total number of APs in building 1: About 60 (your number might range from 46 to 64)
   c. Total number of APs in building 2: About 60 (your number might range from 48 to 66)

2. What is your total number of wireless client devices?

   The site has 900 employees. You should expect up to three devices per users for 2700. In addition, the site has some guests and wireless printers, so you should add about 100 more devices. You should expect approximately 2800 devices.

3. How many controllers are you recommending?

   You could recommend one or two controllers for the campus. Two controllers could provide redundancy although you have not yet considered the customer's requirements for redundancy.

4. Which models of controller(s) are you recommending and why?

   You could recommend one 7205 Mobility Controller (MC) or a VMC, or two 7030 MCs. Two 7205 MCs could provide redundancy although you have not yet considered the customer's requirements for redundancy.

5. What is your recommended location of each controller and why?

   For this scenario, most traffic is destined to the data center and Internet. You would probably locate the controller or controllers in the Building 1 data center, where the network core switches reside (including the data center core and the campus aggregation for Building 1. You will add campus core switches here too).

6. What is your recommended setup for the controllers?
   a. Controllers in L2 mode (Y/N): Y
   b. Controllers in L3 setup (Y/N): N
   c. Controllers in Cluster setup (Y/N): Y
   d. Justify your recommendation:

   You should usually deploy the controllers in L2 mode unless you have a reason to route on the controllers. (A routing switch can typically support many more ARP entries and routes.) A cluster setup is best for redundancy. If you did not recommend a cluster setup at this point, you might after you learn more about redundancy in a later model.

7. Will you recommend an MM? Explain.

   You might recommend an MM because of the many advantages it will provide the customer. Although you are only recommending one or two controllers at this point, the customer might need more controllers later. The MM permits controller clustering. It also simplifies configuration and management because all devices are set up from a centralized dashboard.

   The MM also provides a better user experience. It provides AirMatch, an improvement over ARM for automatically optimizing RF for high-density environments.

Learning Check

1. What APs can be converted to a spectrum analyzer?
   a. Any AP
   b. Only 300 series APs
   c. RAP only
   d. Only 200 series APs
   e. Only dual radio APs

2. You have calculated that you need 500 APs in a building. What are two options for a controller can handle this many APs?
   a. 7210
   b. 7205
   c. VMC 250 + 250
   d. MM 500
   e. 7030

3. Which statements are applicable to a Mobility Master (MM)?
   a. It cannot be a centralized licensing server.
   b. It can terminate APs.
   c. It can do configuration validation.
   d. It can push a full configuration to managed node (MD/MC).
   e. It can only be deployed on a physical server.

4. You have calculated that you need an MM license of 550. What license(s) do you purchase?
   a. MM 1K
   b. MM 10K
   c. MM 500 + 50
   d. Only AP licenses are needed

5. What features do IAPs support?
    a. ARM
    b. Mesh
    c. Spectrum analyzer
    d. IDS/IPS
    e. Stateful firewall

Chapter 4: Wired Network Design

Activity 4: Plan the access layer

Activity 4.1: Access layer

1. The customer needs new cable drops to the new APs. What cable type do you recommend?

   CAT6a

2. Calculate the number of edge ports per closet with the exception of Building 1. Assume that you will connect half of the APs per floor to each closet.

   125 + ½ APs per floor = About 135 (might range from 133 to 136) number of edge ports per closet

3. Calculate the number of edge ports per closet on B1F1.

   35 + ½ APs per floor = About 43 (might range from 42 to 45) number of edge ports per B1F1 closet

4. Begin to plan the switches for one closet. These questions do not have a single correct answer, but be prepared to justify your answers.
    - Will you place APs on their own switches or on the same switches that support wired ports?
    - Will you use fixed-port (24-port or 48-port) or modular switches?
    - If you choose fixed-port switches, do you want to combine them with backplane stacking, VSF, or neither?

   The most cost-effective option for the customer is to support APs on the same switches that connect to wired devices. This customer does not have particularly intensive application needs, so three fixed port switches per closet, combined in a VSF fabric, could provide a good basic solution. The three switches can share a couple of 10 GbE links, which lets you expand the bandwidth in a simple and cost-effective way (fewer transceivers).

   You can also suggest fixed port switches using backplane stacking or even a modular switch, which might be a slightly more expensive, but perhaps higher performance, option. For the modular switch to provide the same redundancy for power supply and management/control plane as a VSF fabric or backplane stack, you would have to add extra power supply and management modules to the switch.

5. Based on your answers to the questions above, which switch series will you use?

   You might have recommended Aruba 2930F Series switches if you plan to use VSF or Aruba 2930M Series switches if you want to use backplane stacking. If you want to use modular switches, the correct option is the Aruba 5400R zl2 Series switches.

   If you recommended AP-330 Series, you should recommend a switch that supports Smart Rate ports to future proof. Therefore, you should recommend Aruba 2930M Series switches with backplane stacking. Alternatively, you could recommend two 5400R zl2 Series switches using VSF, but the backplane stacking solution is probably better for these port requirements. The 2930F Series switches do not support Smart Rate ports.

   You could also recommend 3810M switches in a backplane stack as a higher capability switch.

6. Based on the scenario, which devices require PoE or PoE+? What additional questions should you ask your customer to make sure that you plan the correct number of PoE/PoE+ ports?

   You should have identified that the APs require PoE+. You should also pay attention to the fact the customer is planning to gradually repurpose Ethernet jacks for IoT devices. These devices might require PoE. Assume that you have had further discussions with the customer about this need, and the customer wants to support PoE on all switch ports.

7. Do your APs have special requirements for their switch port?

   If you have selected 330 Series APs, you should select switches that have Smart Rate ports to take advantage of that 330 feature.

14. Take notes on your plan below, including the exact switch models and modules you selected.

    If you recommended AP 330s in earlier activities, you should select either two Aruba 2930M-40G-8SR PoE+ switches and one 2930M-48G PoE+ switch or three 2930M-40G-8SR PoE+ switches.

    If you recommended other APs, you might select three 2930F-48G PoE+ switches or three 2930M-48G PoE+ switches.

    A 5406R switch with one 20-port PoE+/4SFP+ module and five 24-port 1000BASE-T PoE+ modules would also meet the customer's requirements.

15. For how many closets can you use this plan (hint: all closets except the two on Building 1 Floor 1).

    10

17. How will you adjust your plan for the B1F1 closets? Again, think about how you will distribute APs across the switches and the requirements for the AP ports, as well as other wired devices.

    Building 1 Floor 1 closets require fewer switches each because that floor has fewer Ethernet jacks. You can meet the needs with one 48-port switch per closet, but you could also use two 24-port switches so that you can distribute APs across more switches.

Activity 4.2: Access layer uplinks and VSF/Stacking links

1. For Building 1 Floor 3 Closet B, how many uplinks will you plan and of what speed?

   Two 10 GbE links

2. Based on the scenario, what media do these links need to use?

   Fiber

3. What level of oversubscription does your plan give per-switch, VSF fabric, or backplane stack?

   Approximately 7:1 (134 x 1 Gbps/2 x 10 Gbps) = 6.7)

4. Based on the scenario, is this an appropriate level of oversubscription?

   Six 1 GbE links is probably too little bandwidth for this customer. However, two 10 GbE links will meet the needs. The customer expects moderate wireless usage and light wired usage. The fabric or backplane stack only needs to support about 9-12 APs. (An individual switch needs to support even fewer). Even if the APs were handling heavy wireless traffic, 5 or so Gbps should be adequate, leaving ample bandwidth for the light wired usage. (15 G/125 non-AP ports = about 8:1 oversubscription, which is quite generous for the access layer.)

5. Will the same plan be appropriate all other areas of the campus? Adjust your plan if necessary.

   Based on the scenario, Building 1 Floor 3 Closet B might need to support a few ports with higher utilization (APs + ports running video conferencing). Even so, the 20 Gbps uplink should be ample, given light usage on the other ports.

   Other places where you should pay attention to bandwidth requirements include:

   - Ports for the two HD printers on Floor 1 in each building
   - Ports each in the video conferencing rooms in Building 1 Floor 2

   Your plan of two 10 GbE links should work well for these areas, as well. You might monitor the uplinks for congestion on the switches supporting the areas indicated above, and add uplinks just to that closet if you detect issues.

6. How many fibers does your plan take? Make sure that your plan fits with the current cabling.

   It takes four fibers and does work with the cabling.

7. If you are using a VSF fabric or backplane stack, plan how you will distribute the uplinks across the members.

   You should place an uplink on two different switches for better redundancy. Because you expect most traffic to go to the APs, you should place the uplinks on switches that support APs, if only two of the three switches are supporting APs. If your APs don't require Smart Rate ports though, you be planning to distribute APs evenly across all three switches, though, in which case you can place the uplinks on any two switches.

12. Which type of transceiver did you select?

    An LC SR transceiver. This transceiver supports 10 GbE over multimode OM3 fiber up to 300 m, which is much further than required.

16. What topology will you plan for your fabric or stack?

    You should plan a ring for redundancy.

17. If you are using VSF, you must choose ports to dedicate to the VSF links. How many ports of what speed will you use? Do you already have ports available on the module that you chose for the uplinks, or do you need to add a module?

    For the 2930F switches, you should use two of the four provided SFP+ ports for the VSF links.

Activity 4.3: PoE budget

1. What are the maximum PoE requirements for the APs that you are proposing?

   The answer varies based on the AP model chosen in previous activities. The AP 325 consumes up to 20W with PoE+, and the AP 335 consumes up to 25.3W with PoE+.

2. For all closets except those on Floor 1, how many APs are you planning to connect to per closet? (Remember to divide the number per floor by two for the number per closet.) And how do you plan to distribute the APs across the switches?

   Approximately 8–11 APs need to connect to each closet. If you are using fixed-port switches, you should have three 48-port switches. Typically, you would place three or four APs on each switch to minimize the lost coverage if a switch fails. However, if you are connecting the APs to Smart Rate ports, you might be proposing just two switches with Smart Rate ports, in which case you would connect four to six APs on each switch. (In either case, you would connect the APs such that nearby APs connect to different switches to keep coverage gaps as small as possible if a switch fails.)

3. The Building 1 Floor 1 closets probably have a slightly different design. How many APs are you planning to connect to each switch in these closets?

   If you have one switch in each closet, you will connect between 7 and 10 APs to the switch.

4. Use the answers above to calculate how much power each representative switch in your topology requires. For example, if you are planning to connect four APs per switch on most switches but nine APs on a couple switches, calculate the requirements for four APs and for nine APs.

   The answer various based on the plan in the previous activities. As one example, if you are planning three AP 325s on one switch, the total is 73W (2x20W + 33W for the final PD). If you are planning 11 AP 325s on one modular switch, the total is 233W.

As another example, if you are planning four AP 335s on one switch, the total is 108.9W. If you are planning up to 11 AP-335s, the total 286W.

You could add 5W or so to these totals to account for cable loss.

6. Does the switch meet the PoE budget needs for the APs?

   Yes, for fixed-port switches. For a 5400R switch, it probably does; you need to have remembered to add a power supply.

7. If you need more power on a 5400R switch, you can select the Power Options tab and change power supply type or set the number to 2. What supplies have you selected?

   You should have determined that a 700W power supply should meet the needs.

10. As a final step, make sure that switches will continue to meet the needs as the customer introduces IoT devices that use PoE. The devices that the customer is considering are PoE class 1 devices. How many class 1 devices could each of your representative switches support in addition to its APs?

    The answer depends on the previous answers. In most cases, all of the switch ports that do not connect to APs could provide sufficient power to a Class 1 device.

    You might have found that a 5406R switch with a 700W supply could support fewer Class 1 devices than it has remaining ports. If the customer wants a future proof solution, you could propose an 1100W supply instead. This activity does not specifically address redundancy, but you could later decide to propose two supplies if the customer needs redundancy.

Activity 4.4: Need for aggregation layer

1. Refer back to the number of access layer uplinks that you planned for Building 1. Also refer to the cabling information in the scenario. Will you plan an aggregation layer for this building? Why or why not?

   No. You are creating a campus core, which can reside where the current Building 1 aggregation switches are. The closets in Building 1 can connect directly to this core.

   The only reason to create an aggregation layer for this building would to be consist with Building 2, but that is not required.

2. Refer back to the number of access layer uplinks that you planned for Building 2. Also refer to the cabling information in the scenario. Will you plan an aggregation layer for this building? Why or why not?

   You can make a case for planning an aggregation layer for this building or not. Only 12 strands of inter-building fiber is available, and you would need 24 strands to connect every wiring closet with two connections. So you could connect all of the closets to an aggregation layer and then connect the aggregation switches to the core using the limited inter-building fiber.

Alternatively, you could plan a VSF fabric across closets. In this case, you could connect each switch to the other building on one link while maintaining link redundancy through the VSF link to the other closet.

Activity 4.5: Aggregation layer design

1. Refer back to the number of downlinks planned in the building for which you are planning the aggregation layer. How many uplinks to the core will you plan and of what speed? Explain the reasoning for your plan.

   You might plan four 10 GbE links (two on each aggregation layer switch) for 3:1 oversubscription. Or, you might choose to consolidate to two 40 GbE links (one on each aggregation layer switch) for just 1.5 oversubscription.

2. If you are planning to connect an MC or MCs at the aggregation layer, add these ports to your plan as well. If you are not planning to connect MCs here, move on to step 3.

   c. How many ports of which type do you plan to use? Note that you will need to use transceivers because you cannot use DACs between the AOS-Switches and the controllers.

   You did not need to answer this question unless you are planning to connect a controller in Building 1. The number of ports depends on the choice made in Activity 3. For this scenario, you should typically choose to connect a controller on two SFP+ ports.

3. Determine whether to deploy one aggregation layer switch or a pair based on the customer scenario.

   It is best practice to deploy a pair of aggregation switches and use backplane stacking or VSF to combine them into one virtual switch.

4. Select a switch series.

   As you learned in the chapter, two 3810-16SFP+ switches can support up to eight or 16 access layer switches/stacks/fabrics, depending on the oversubscription requirements. Because you need to support just six, two 3810-16SFP+ switches deployed in a backplane stack should meet the requirements.

10 Record the type of transceivers that you chose.

   For 10 GbE links, you should select LC SR transceivers. For 40 GbE links, you must select LC BiDi transceivers to support the required distance over the available fiber strands. (MPO connectors require 12 strands for one connection. The inter-building cable only has 12 strands available, and you need two connections). The BiDi transceivers only require two strands each.

Activity 4.6: Core Design

1. Record the total number of transceivers required on the core switches to support the aggregation layer. Also note the transceiver type.

   If you are planning an aggregation layer: Two SPF+ LC SR transceivers on each core switch (four total) or one QSFP+ BiDi on each core switch (two total).

2. Record the total number of ports required on the core switches to support directly connected access layer switches, if any. Also note the transceiver type.

   If you are planning an aggregation layer in Building 2: Six SFP+ LC SR transceivers on each core switch (12 total)

   If you are planning no aggregation layer: 9 SFP+ LC SR transceivers on each core switch (18 total)

   If you are planning 40 GbE links to the access layer, the transceivers would be QSFP+ transceivers instead.

3. Add the answers to 1 and 2 to obtain the total number of downstream ports that your core switches must support.

   Your answers could vary but might be one of these:

   - 8 SPF+ LC SR transceivers on each core switch (16 total)
   - 9 SPF+ LC SR transceivers on each core switch (18 total)
   - 6 SPF+ LC SR transceivers on each core switch (12 total) and 1 QSFP+ BiDi transceiver (2 total)

4. How many links of what bandwidth will you plan to connect the campus core to the data center core? As a reminder, ports available on the data center core are:

   - 15 10 GbE SFP+ ports per-switch (30 total)
   - 4 40 GbE QSFP+ ports (8 total)

   You want no more than 4:1 oversubscription from the access layer and nonblocking from an aggregation layer. This means you should plan one or two 40 GbE links on each core switch for two or four total (depending on your choices for the downlinks).

5. If you are planning to connect an MC or MCs at the campus core, add these ports to your plan, as well.

   c. How many ports of what type do you plan to use? Note that you will need to use transceivers because you cannot use DACs between the AOS-Switches and the controllers.

   The number of ports depends on the choice made in Activity 3. For this scenario, if you are connecting MCs here, you should typically choose to connect the controllers on two 10 GbE SFP+ SRs to each MC.

6. Determine whether to deploy one core switch or a pair based on the customer scenario.

   It is best practice to deploy a pair of core switches and use backplane stacking or VSF to combine them into one virtual switch.

7. Choose a switch series for the campus core.

   You could have selected one of these options for the core:

   - 2x 5406R switches, each with 2x 8-port SFP+ modules and 2x 2-port QSFP+ modules
   - 2x 3810M-16 SFP+ switches (less desirable option, but possible if you are planning fewer 40GbE links)

Show your final plan

Figure A-1 and Figure A-2 show a couple of example plans. Your plan might differ a bit but should follow the same general design.

Example Design 1

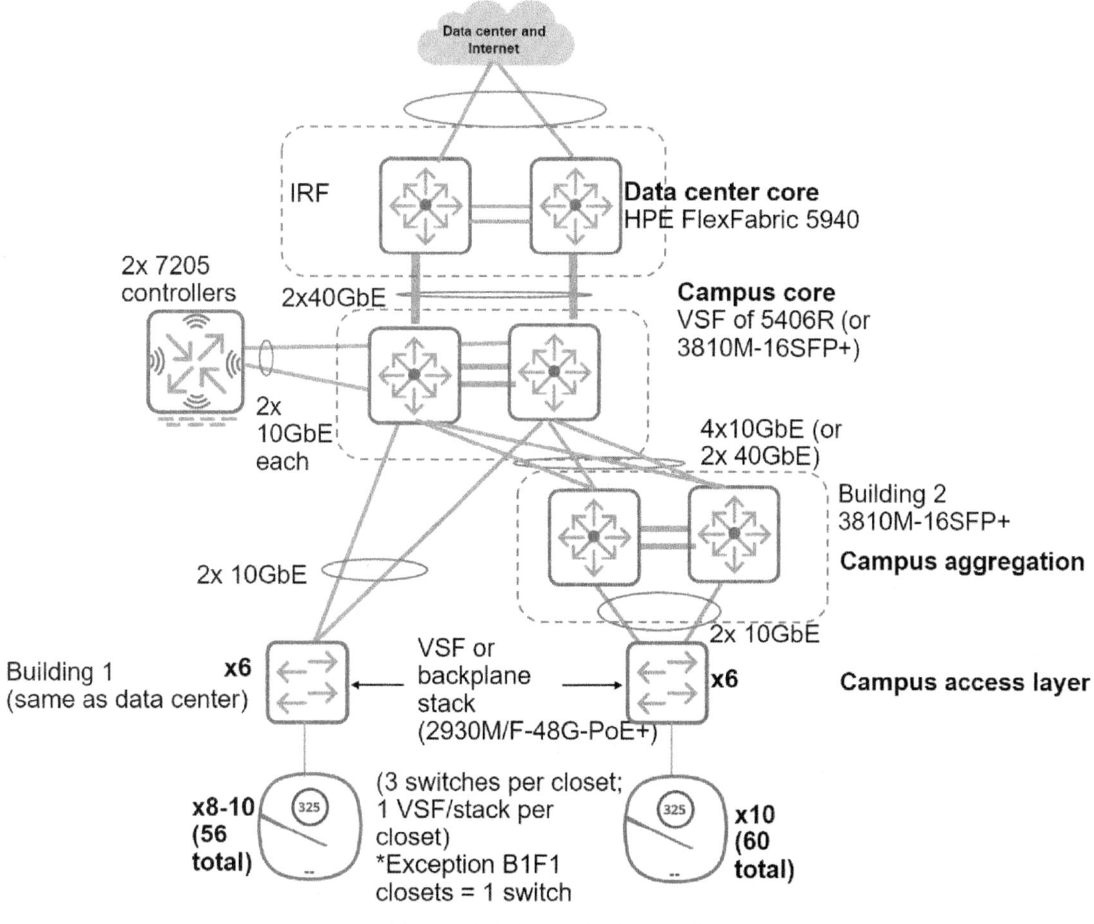

Figure A-1 Example Design 1

Example design 2

An alternate plan with no aggregation in Building 1 is shown in Figure A-2:

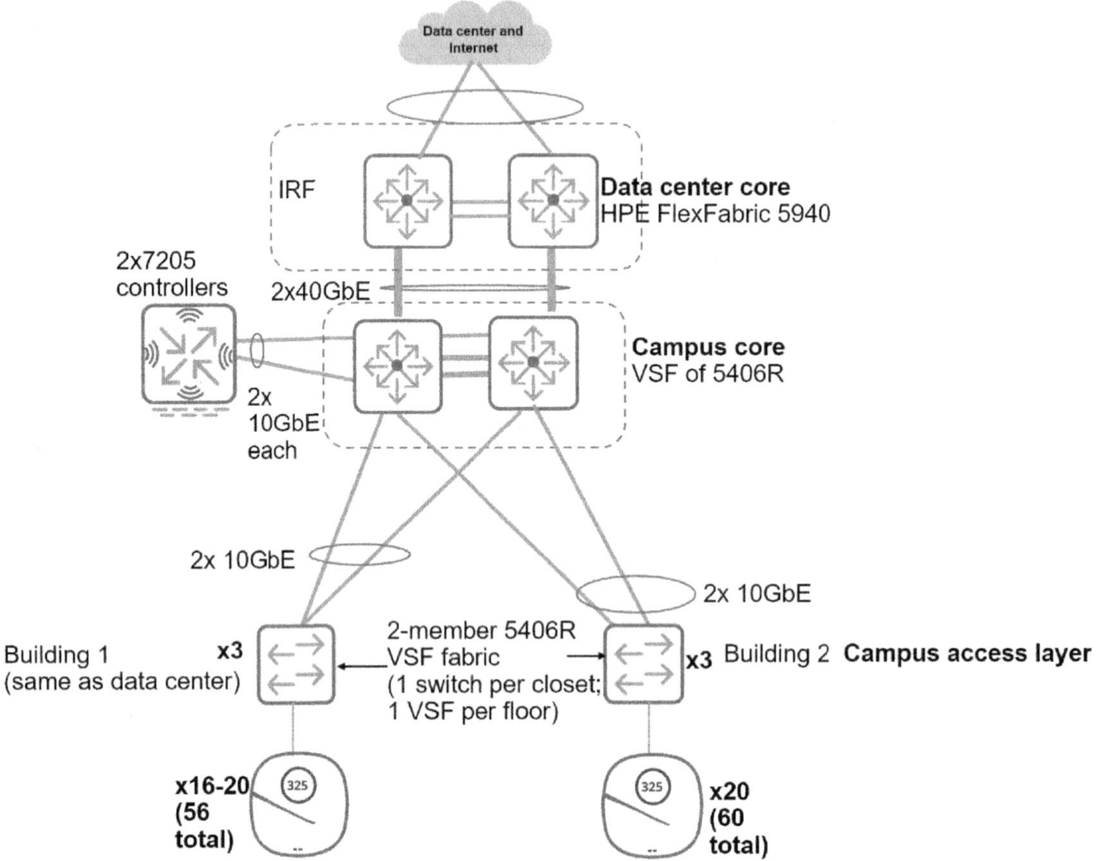

Figure A-2 Example design 2

Learning check

1. What is one advantage of deploying VSF on redundant core switches?
   a. VSF makes it easier to configure and deploy technologies such as VRRP.
   b. The topology can use link aggregations rather than spanning tree to handle redundant links.
   c. VSF uses specialized backplane stacking modules and high bandwidth stacking cables.
   d. VSF makes it easier to implement VRRP and MSTP together.

2. Which switch is most likely to require the highest bandwidth on its uplinks?
   a. A switch that connects to employees who browse the Internet and access Word documents and spreadsheets on centralized servers.
   b. A switch that connects to printers and HVAC IoT devices.
   c. A switch that connects to a high density of 802.11ac APs.
   d. A switch that connects to employees who occasionally run video conferencing softwareWhat are advantages of designing an OSPF AS to use multiple areas?

3. What is the most appropriate oversubscription ratio between the aggregation layer and core?
   a. 4:1
   b. 12:1
   c. 20:1
   d. 24:1

Chapter 5: Access Control and Security

Activity 5.1: Design authentication and access control

Activity 5.1: Design authentication and access control

1. Refer to the scenario for this activity. Make a case for why this customer should implement Aruba ClearPass. What advantages and business benefits does it provide?

   Aruba ClearPass provides more sophisticated and granular control over users' wired and wireless access. It allows organizations to identify all the devices trying to access the networking, enforce authentication, and ensure that only authenticated, authorized, or "healthy" (noninfected) devices are given access.

   In addition, ClearPass supports an enterprise guest solution with a variety of advanced features such as highly customizable portal pages, self-registration, SMS-integration, and more. ClearPass also helps IT staff to monitor users and their network usage. ClearPass also integrates with other security solutions such as IntroSpect, as well as third-party solutions.

   You should state these advantages in a way that highlights what they bring to this specific customer. For example, the customer wants to grant financial users special rights. ClearPass can recognize financial users and assign them to a role, enabling the MC to apply different policies to those users from the policies applied to other users. Similarly, ClearPass's guest, BYOD, and endpoint integrity capabilities specifically address customer initiatives.

2. In addition to ClearPass Policy Manager, which components will you recommend for the customer to use (ClearPass Guest, OnGuard, and OnBoard) and why?

   For this customer, in addition to Policy Manager, you might recommend:

   - Guest to provide a captive portal for guest authentication.
   - OnBoard to make it easier to roll out an EAP-TLS solution.
   - OnGuard to help ensure that infected devices cannot connect to the network.

3. Make a high-level plan for the wireless authentication and access control, including:

   - The number of WLANs and which users and devices will connect to each WLAN.
   - The type of authentication and encryption implemented on the WLAN.
   - The general strategy for applying access control.

   You should plan one WLAN for employees. The company is already using WPA2 with 802.1X encryption, and you should continue to recommend this level of security. The company likes the security of EAP-TLS, but is concerned about certificate management. You can recommend EAP-TLS with ClearPass Onboard to easily manage the certificates.

   You should also recommend a WLAN for guests. This WLAN will use captive portal and no encryption. As a high-level access plan, ClearPass is responsible for authenticating users and devices and assigning them a role. MCs apply the firewall policies that correspond to that role.

4. Will you recommend implementing authentication and access control at the wired edge? Explain why or why not.

   Although the customer did not request authentication on wired ports, you could recommend 802.1X authentication across the site for greater security and consistency of user experience.

5. If you have recommended implementing access control at the wired edge, make a high-level plan, including:

   - Whether to use per-user tunneled-node
   - The type or types of authentication to implement
   - The general strategy for applying access control

   You may recommend per-user tunneled-node because it enables the MC to apply the security features that the customer was interested in to all devices.

   If you recommend authentication for wired devices, you must implement authentication on switch ports connected in meeting rooms. In addition to 802.1X, you should implement MAC-Auth for the guest devices. The switch can implement MAC-Auth with captive portal to ClearPass Guest. Or if you decide to use tunneled-node, the switch and ClearPass can use MAC-Auth to assign new guests to a tunneled role and the MC can handle their traffic.

If the switches are using tunneled node, they must use role-based authorization, which activates tunneling for the role ClearPass sends for employees and guests. The switch also communicates a role name for these users to the MC, which can then apply role-based policies to them. If you have not opted for tunneled-node, the switches could enforce simpler access controls.

For example, ClearPass could use RADIUS attributes to communicate different VLANs for employees and guests, as well as filter rules to restrict guest access. Or ClearPass could send a role name, and the switch could use role-based authorization. The roles configured on it would place employees and guests in different VLANs and restrict guest access.

Applying authentication to wired connectivity does add to the number of licenses required if users are logged in on wired and wireless devices at the same time. However, the consistency of experience and security makes this solution worthwhile for many customers. You could justify implementing authentication across all wired ports to a customer who is highly interested in using ClearPass to classify employees with different roles and control the users' access based on those rules. Similarly, because the customer is excited about the benefits of the MC's DPI features, you can use authentication and tunneled-node to extend those benefits to the wired traffic.

Activity 5.2: Design ClearPass server sizing and licensing

1. Do you have enough information to plan the Access licenses? What should you clarify with the customer?

   You need to know how many devices are connected per day, and if this number will increase once users realize they can access the wireless network everywhere and performance is greatly improved.

2. How many of the following licenses will you recommend (if you are not planning to recommend a feature, you do not need to plan any licenses for it):
   - Access
   - You discussed usage patterns with the customer and determined that the site must support up to 3000 concurrently connected devices. You should recommend 3000 licenses.
   - Onboard
   - If you are recommending ClearPass Onboard, you need one license per-user to onboard any number of devices. Up to 945 different users could come to the site (although at different times) and licenses are sold in 100-license increments, so the company needs 1000 licenses.
   - OnGuard
   - The OnGuard licenses are per-device, but the company only requires them for the corporate devices. Again, you should plan 1000 licenses.

3. What would you discuss with the customer to determine whether to propose a ClearPass appliance or VM?

   You might explain the advantages that an appliance provides, such as a turnkey solution and simple deployment. The appliance can also be a better option when the network and server team are separate, and the network team is responsible for obtaining its own equipment. A VM, on the other hand, is a good option for customers who have limited space and the ability to deploy the VM on existing server hardware.

Activity 5.3: Plan Certificates

1. Which devices require certificates?
   - If you are implementing 802.1X, consider the EAP method.
   - If you are implementing Captive Portal authentication, also consider the need for HTTPS certificates.
   - Also consider the need for HTTPS certificates for the MC and ClearPass UIs.

   ClearPass requires a RADIUS certificate so that it can implement various EAP methods.

   If you are planning, EAP-TLS, endpoints require certificates, which ClearPass OnBoard can distribute.

   You should also plan HTTPS certificates for AOS-Switches and MCs, which permits these devices to redirect HTPPS traffic unauthenticated guests send to the ClearPass captive portal. Aruba also recommends an HTPS certificate on ClearPass. This certificate is mandatory when guests must log in through the captive portal. An HTTPS certificate for ClearPass Policy Manager also enhances security for the ClearPass UI. An HTTPS certificate for MM does the same for its UI.

2. Will you recommend that the customer use CA-signed certificates rather than self-signed certificates? What are the advantages of your recommendation?

   You can choose either CA-signed or self-signed certificates, but you should be able to explain the choice you make.

   CA-signed certificates offer several advantages. More devices trust CA-signed certificates, which means that users accessing the captive portal, will not see errors warning them away from the page. With 802.1X, it is also important that clients trust the server's RADIUS certificate. If ClearPass uses a self-signed certificate for this purpose, more setup is required to make the clients trust this certificate.

   The main advantage of self-signed certificates are that they are simpler to install and most cost-effective. In many cases, though, the company can obtain CA-signed certificates for not much money, when compared with the costs of IT handling the uses that self-signed certificates can cause.

If a service has little public visibility—for example, the MM UI—you may choose to use a self-signed certificate.

Note also that, in addition to permitting the switches and MCs to redirect HTTPS traffic to the captive portal, HTTPS certificates on those devices permit HTTPS access to the devices' UI.

Activity 5.4: Determine needs for AMs

1. Refer to the scenario for this activity. Will you recommend the deployment of dedicated AMs or not?

   Although the customer is interested in WIPS, the customer does not want to deploy extra equipment. You should probably recommend using the IDS features on the APs that are currently included in the plan, rather than adding APs deployed as dedicated AMs.

2. If you have not already recommended RPF licenses, will you do so?

   Yes. Although some IDS/WIPS features are available about the RFP license, this license is recommended for the advanced features.

Learning check

1. What defines the number of ClearPass Access licenses that a customer requires?
   a. Number of employee user accounts in ClearPass
   b. Number of employee and guest user accounts in ClearPass
   c. Total number of devices that ever authenticate to ClearPass
   d. Number of devices concurrently authenticated to ClearPass

2. What is one best practice for implementing 802.1X with PEAP-MSCHAPv2?
   a. Install a certificate on all client devices
   b. Configure client devices to validate the RADIUS certificate
   c. Create a DHCP scope for devices that have not yet authenticated
   d. Supplement 802.1X with WEP encryption

3. You are planning to implement per-user tunneled node on an AOS-Switch and authenticate tunneled-node endpoints with 802.1X. What device will enforce 802.1X?
   a. The AOS-Switch that implements tunneled node
   b. The core AOS-Switch
   c. The MC to which the switch tunnels traffic
   d. The MM

Chapter 6: VLAN Design

Activity 6: Recommended VLANs for new network

Activity 6.1: Wired VLAN recommendations

1. What recommendations do you have for the wired infrastructure? Consider these questions:
   - Will you change any VLANs? If so, explain your changes.
   - Will you add any VLANs? If so, plan the VLAN ID and associated subnet.
   - Which devices will act as the default routers for each VLAN? And where do VLANs need to extend? (Keep in mind your security choices from Activity 5.)
   - What VLAN or VLANs will you use between the aggregation layer, if present, and campus core? What VLAN or VLANs will you use between the campus core and data center core?

   Possible recommendations if you have decided not to use tunneled node.

   You should plan to keep the current VLANs extended to the wiring closets for the employees because there is no pressing reason to change. You might have suggested adding a VLAN or VLAN to separate video conferencing devices from employee laptops.

   If you are not using switch OOBM ports, you should add a management VLAN for the IP addresses on access layer switches. When planning to route between an aggregation layer in Building 2 and the core, you should plan a different management VLAN in each building. You can choose any VLAN IDs and subnets that make sense to you and the administrators for these VLANs. You will also need to request a subnet or subnets from the customer, explaining that it is best practice to manage switches on IP addresses separate from users' subnet.

   If you are planning to use switch OOBM ports instead, you still need to ask the customer for a subnet for those IP addresses. You also need to think through the implications. For example, to create complete isolation for the OOBM network, you should add a switch for connecting the OOBM ports. This switch might connect to a terminal server that can grant managers access to the network. If you have planned a two-tier topology between Building 2 and the core, you might not have sufficient fiber to connect those ports; a logically separated management VLAN might better meet your needs.

   Also keep in mind that if you are using tunneled-node, the switches must have an IP address on a VLAN that they can use to reach the MC.

   The core VSF fabric or backplane stack will be the default router for all VLANs in Building 1. You have a choice for the default router in Building 2. The campus core supports enough MAC addresses and ARP entries to route for both buildings; you might select this design for simplicity. However, you might recommend routing at the aggregation layer VSF fabric to create smaller fault domains. If the aggregation layer and core will route traffic between each other, you should plan have a dedicated VLAN between them for this purpose. You can choose any available VLAN ID and subnet in the 10.2.0.0/16 subnet.

You should connect the campus core and data center core on VLAN 1000, 10.1.0.0/24, which the data center core is already using to connect to the campus.

In addition to adding the management VLAN, you might suggest reserving some VLAN IDs for future IoT devices. However, you do not need to plan VLANs for these devices at this time. You might have also thought of placing video conferencing equipment in its own VLAN, although the customer did not specially request that.

Adjustments if you are using tunneled-node:

If you are planning to use tunneled-node, you can keep the current subnets for employees, but the VLANs need to be on the access layer switches, the MCs, and the MCs' link to the campus core. They do not extend on the links from the access switches to the core.) You can use the existing VLAN IDs or plan new ones with the customer. The campus core will route for these VLANs, regardless of whether you are planning an aggregation layer or not.

You will also need to ask the customer for a subnet for the APs and plan a VLAN ID for it.

If you are planning to add a VLAN for video conferencing devices, you would also need to make the choice about whether access switches are switching that traffic locally or tunneling it to the MCs.

2. Draw out the logical design using the figures.
   - In each box, indicate the VLANs configured on switches deployed in that closet or at that layer.
   - For the links between layers, indicate the VLANs carried on the links between switches and whether those VLANs are tagged or untagged.

   Figures A-3 to A-6 show some examples of designs that you might have made, including the wireless component.

No tunneled node

Figure A-3 shows a Building 1 design when you are not using tunneled-node. VLAN 100 is a management VLAN for the switches, which you will use if the switches do not have OOBM ports. The core is routing the Building 1 wired VLANs and a single campus wireless VLAN.

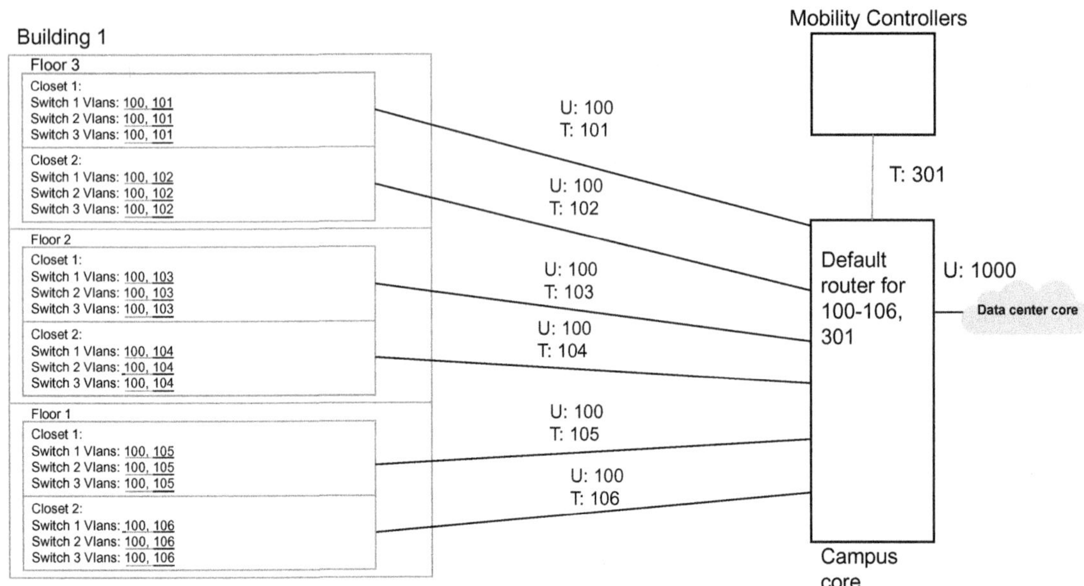

Figure A-3 Building 1

Figure A-4 shows a Building 2 design when you are not using tunneled-node and Building 2 uses aggregation. If Building 2 did not use aggregation, the VLANs shown on the links between the closets and the aggregation switch would be on the links from the closets to the core.

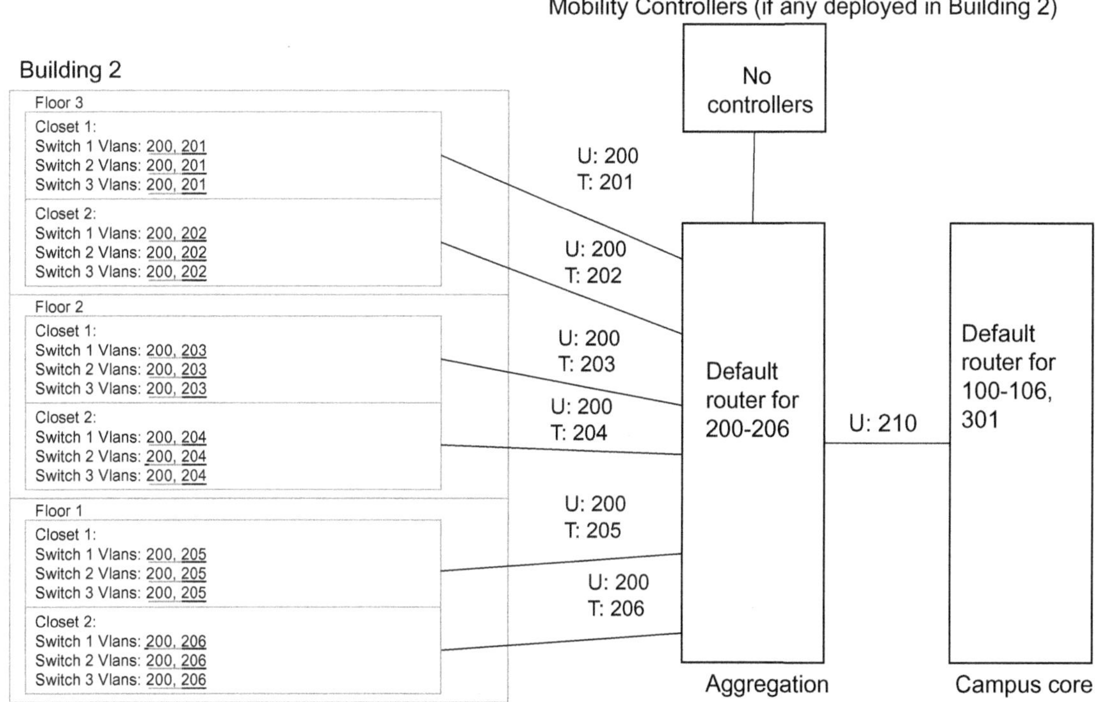

Figure A-4 Building 2

Tunneled-node

Figure A-5 shows a Building 1 design when you are using tunneled-node, and tunneled-node users are assigned to the same VLAN IDs that the customer was using. Note that the VLANs must exist on the switches in the closets, but traffic is not forwarded in those VLANs until the traffic reaches the MC. The only VLAN necessary on the link between the closets and the core is a VLAN for the APs and switches to have their IP addresses. (This example shows one VLAN for both switches and APs.) The core routes for all wired VLANs in Buildings 1 and 2, as well as a single campus wireless VLAN.

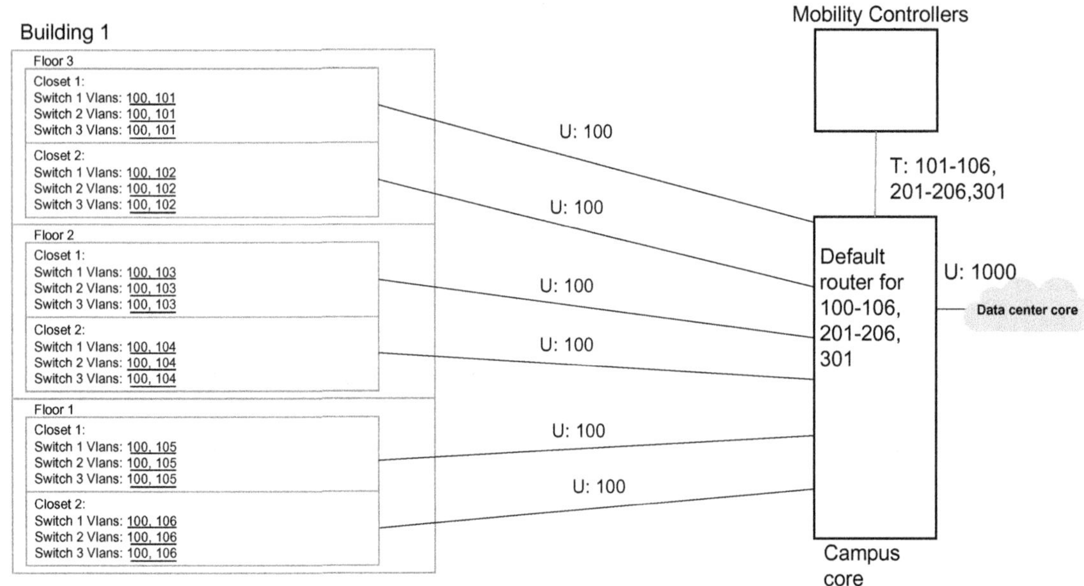

Figure A-5 Building 1

Figure A-6 shows a Building 2 design when you are using tunneled-node. Again, the APs and switches have IP addresses on a subnet that you should request from the customer (VLAN 200 in this example). The aggregation layer routes for this subnet, but the core routes for all user VLANs, which are tunneled there. If you are not using aggregation in Building 2, you would extend the AP and switch VLANs to the core.

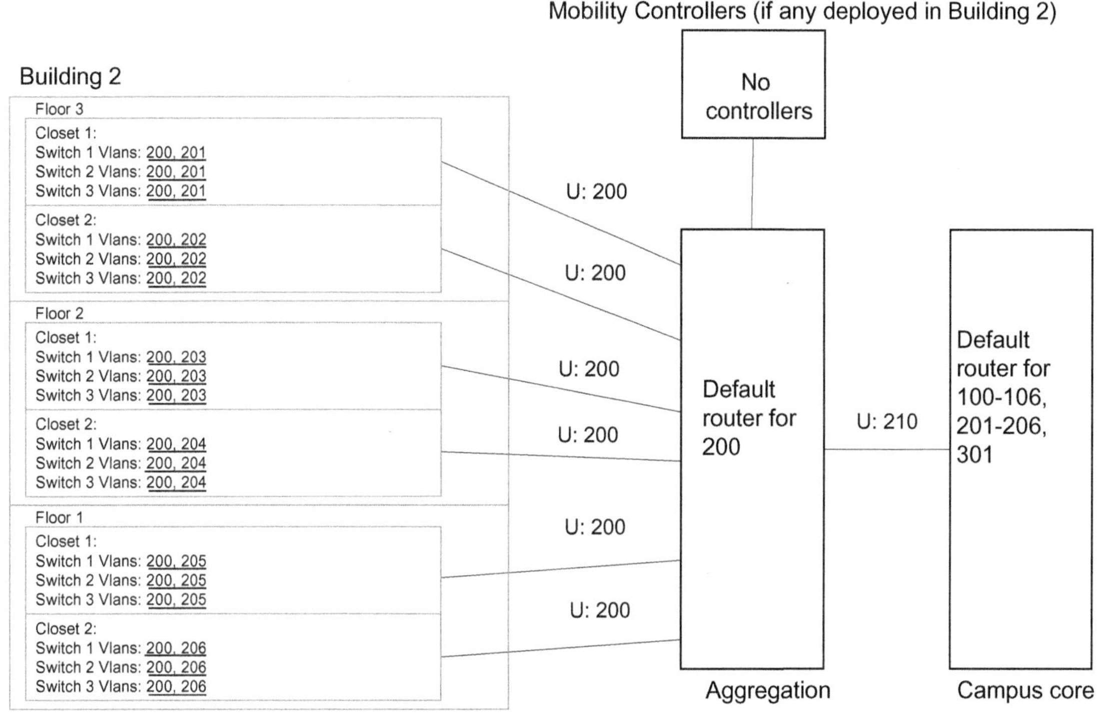

Figure A-6 Building 2

Activity 6.2: Wireless VLAN recommendations

1. What VLANs have been assigned for the Wi-Fi structure:
    a. Building 1 Floor 1: 10.3.1.0/24 VLAN 301
    b. Building 1 Floor 2: 10.3.2.0/24 VLAN 302
    c. Building 1 Floor 3: 10.3.3.0/24 VLAN 303
    d. Building 2 Floor 1: 10.3.4.0/24 VLAN 304
    e. Building 2 Floor 2: 10.3.5.0/24 VLAN 305
    f. Building 2 Floor 3: 10.3.6.0/24 VLAN 306

2. Do the VLANs and subnets assigned meet your requirements (Y/N)? N

   You should plan at least three devices per user, as well as other devices such as printers. So the site will have about 1570 to 1600 devices. The customer only gave you enough address space for 1524. Also depending on how you plan to the VLANs, you might need larger subnets which will extend into the 10.3.0.0-10.3.0.255 and 10.3.7.0-10.3.7.255 ranges.

3. What recommendation do you have for the Wi-Fi infrastructure?

   You might have suggested one of three approaches: one VLAN per floor, one VLAN per building, or one VLAN per campus. The one VLAN per floor approach is not recommended because it interferes with roaming and adds complexity. If you did take this approach, the /24 subnets are not large enough for the requirements on every floor. You would need to ask for /23 subnets.

   The one VLAN per building approach would work if you considered each building its own RF domain. It provides roaming within a building and, if you are using a single cluster at the data center, between buildings. If you deployed two MCs or clusters, one at each building, though, GRE tunneling would be required to support roaming between buildings. If you took this approach, you would need to one /22 subnets for each building, so you would need to ask for the 10.3.0.0-10.3.0.255 and 10.3.7.0-10.3.7.255 space, as well as the subnets already provided. You can then use 10.3.0.0/22 and 10.3.4.0/24.

   Because the customer wants roaming between buildings, you can consider both buildings one RF domain, and the one VLAN for the campus approach is probably best. You would need one /21 subnet to meet the device requirements. Again, you should ask for the 10.3.0.0-10.3.0.255 and 10.3.7.0-10.3.7.255 space, as well as the subnets already provided. You can then use 10.3.0.0/21.

4. List the AP group(s) that you will recommend, along with what VLANs will be used in each AP group VAP.

   For the one VLAN per floor design (not recommended), you would create six AP groups (one per floor) and assign a different VLAN to each one (301-306).

   For the one VLAN per building design, you would create two AP groups and assign a different VLAN to each (such as 301 and 302).

   For the one VLAN for the campus design, you would create one AP group with one VLAN (such as 301).

Learning check

1. Which best practice should you follow when planning VLANs for wired user devices?
   a. Place wired devices in the same VLANs used by wireless devices.
   b. Extend the wired VLANs as far as possible up to a /16 subnet.
   c. Make sure that VoIP phones and the computers connected to them use the same VLAN.
   d. Keep the wired VLANs at between about 250 and 1000 devices.

2. What is an AP group?
   a. A group of MCs
   b. APs of the same type
   c. A group of APs with the same configuration
   d. AP grouped in a location

3. What are the key considerations when planning a single VLAN in a campus?
   a. Contiguous RF
   b. APs of the same type
   c. MAC and ARP tables size
   d. DHCP server capabilities
   e. Third-party firewall

Chapter 7: Redundancy

Activity 7: Recommended Wi-Fi design and redundancy strategy for the new wireless and wired network

Activity 7.1: Controller Wi-Fi redundancy

What are your design recommendations for controller failure?

Because you have planned two controllers that are co-located, clustering is the best solution. Clustering supports seamless roaming, AP load balancing, client load balancing, and client sync state. Also, with the proper RF environment, seamless upgrades of the network.

The L3 connected cluster is not fully redundant; APs are fully replicated, but users are not synced between active and standby UAC nodes. Users are de-authorized on a failover, even from the active to standby UAC. Users must then complete a full 802,1X authentication.

The L2 connected cluster is fully redundant with APs and clients fully replicated. Users are fully synced between the active and standby UAC nodes. Because high-value sessions are synced, users are not de-authorized on a failover, even from the active to standby UAC. As the user roams from AP to AP in an L2 cluster, the user traffic is always returned to the UAC. As a result, roaming is simplified.

If MCs are in same building, you should recommend clustering the 7205 Controllers as L2.

If one MC is located in each building, you should recommend clustering the 7205 Controllers as L3

Activity 7.2: AP Wi-Fi redundancy

What are your design recommendations for an AP or POE switch failure affecting APs?

If APs are placed in high-density deployment, the loss of one AP will not affect the Wi-Fi accessibly of the network. Also, high-density deployments are recommended for voice networks.

The loss of a switch is more critical. If all your APs on one floor are physically connected to one switch, and it goes down, all the APs are lost on that floor.

Dual homing APs guarantees that a single switch failure will not cause an AP outage. However, this can be costly because it requires two drops per AP to different switches or even a different network closet.

For this customer, the best solution is a high-density AP deployment and staggering the connections for APs on a floor between two or more switches.

Activity 7.3: Wired redundancy

1. Assess the plan that you have created for the updated wired infrastructure. Does it meet the customer redundancy requirements?

   You should have designed the wired topology to provide a pair of redundant aggregation layer switches in Building 2 and a pair of redundant switches at the core. You should also have followed best practices for balancing redundant links across both members of the pair to ensure that connectivity continues if any one link or switch fails.

2. Will you recommend redundant components for any of the proposed modular switches? Will you recommend any spare components?

   Although you could recommend redundant components such as redundant power supplies and management modules for 5400R switches deployed at the core, keep in mind that you are planning switch-level redundancy with VSF. These redundant components are less important, and you cannot use redundant management modules with VSF.

   If you are proposing 5406R switches at the access layer, you might propose redundant power supplies to enhance resiliency for the APs being powered.

3. What technologies will you recommend to ensure Layer 2 and Layer 3 resiliency, and how will you implement these technologies? How will you explain the benefits of your plan to the customer?

   You should propose backplane stacking and VSF to replace MSTP and VRRP. Links between all layers will use LACP-based link aggregations, including links that connect to different physical switches within the backplane stack or VSF fabric. Make sure that you think about link redundancy for every critical device that connects to the network. For example, if the MCs connect to the core, you should connect to both members of the core VSF.

   When planning the VSF and backplane stacking implementation, you should also make sure to follow best practices. For example, plan resiliency for the links between the members. On 5400R switches, connect the redundant physical links for the VSF link on different modules. Also make sure that admins implement a form of MAD to protect against split brain. Finally, enable OSPF graceful restart to ensure quick failover if the commander fails since the customer is using OSPF.

   You could explain the benefits of VSF and backplane stacking in more detail. Emphasize that from the point of view of management, design, and functionality, the fabric or stack acts like a single switch. However, the network still has the redundancy of two physical chassis. Therefore, the failure of a member, from the point of view of connected switches, has much the same effect of the failure of a module on a modular switch. Half of the links in the link aggregation fail, but traffic quickly fails over to the other links in the link aggregation. Even if the commander fails, the standby member, which has been synchronizing the control plane, can become commander in less than a second. The customer benefits from the simpler design. Not only does IT have less to configure and manage, but the simpler the design, the less likely for issues to arise. And users and applications benefit from the fast failover.

Learning check

1. What are the two modes of HA redundancy?
   a. Active Backup
   b. Active Active
   c. Active Standby
   d. Active Primary
   e. Active Secondary

2. A user is associated to an AP. The User anchor controller must be the same as the AP's anchor controller.
   a. True
   b. False

3. You plan to deploy a pair of Aruba 5406R switches at the core of a two-tier topology. Which technology should you plan to provide the simplest setup and fastest failover for default router services?
   a. VRRP
   b. MSTP
   c. Backplane stacking
   d. VSF

Chapter 8: Quality of Service (QoS)

Activity 8: Design QoS

Activity 8.1: Classify applications

1. Refer to the scenario for this activity. Which applications might need high priority in order to provide a good experience to users running those applications or protect critical services?

   The main applications you must prioritize are voice and video in Microsoft Skype for Business. You could also identify management applications such as Telnet, SSH, and SNMP. (Network protocols run by the infrastructure should also be prioritized, but devices will automatically prioritize these protocols.)

2. What else would you discuss with the customer to further classify applications and identify high-priority ones?

   You might want to encourage the customer to make a more formal survey to ensure that you have identified all important applications in the network.

   You could also discuss whether the customer wants to provide certain groups with higher or lower priority based on their identity as opposed to the application. You could also discuss the customers' preferences for prioritizing traffic.

   You could also work with the IT staff to obtain information about current congestion points. (However, in this scenario, you are already planning to increase capacity.)

Activity 8.2: Plan QoS Measures

1. What do you need to plan to ensure that wired traffic receives the correct priority?

   Make sure that AOS-Switches trust DSCP on ports that connect to trusted employee clients so that the switches can honor the priorities.

   If you want, the switches and MCs can also use policies to classify management traffic for priority delivery. Management traffic could have a priority below video but above best effort such as 24 or 32. To create QoS policies for network management traffic, you could create traffic classes that select the TCP and UDP ports for this traffic and then create a policy that sets your selected DSCP for each class. You can look up the well-known ports for each application or use the keyword in the switch CLI.

   You can then apply that policy to the appropriate VLANs (such as the IT user VLAN and any VLAN on which you are managing the infrastructure devices).

2. What do you need to plan to ensure that wireless traffic receives the correct priority? Include ensuring that the wired infrastructure supports the plan.

   The customer said that the admins use the recommended QoS priorities for Microsoft Skype for Business. You should recommend using 56 for voice and possibly 40 for video.

   The Aruba solution should also have WMM enabled and use the recommended DSCP mappings. The voice applications should mark their own traffic. It is also best practice to have the voice firewall policies place voice traffic in the high-priority queue, too.

   Make sure that AOS-Switches trust DSCP on ports that connect to APs so that they can honor the priorities.

3. What additional technologies do Aruba solution provide to help ensure a good QoS for critical and time-sensitive traffic?

   You need to plan bandwidth contracts for the guests.

   To ensure good voice quality over wireless connections, the wireless solutions should support fast roaming and rekeying. The Aruba solutions can also implement technologies such as AirMatch, Airtime Fairness, and Client Match to optimize capacity, which improves performance in the shared medium.

Aruba MCs can apply bandwidth limits to the guest role as a per-role or per-user limit. The AOS-Switches are authenticating guests to ClearPass. ClearPass can send a rate limit for guests dynamically. Or if the AOS-Switches are using role-based authentication, admins could manually apply the rate limit to the ports that guests will use in meeting rooms.

Activity 8.3: Describe Aruba benefits

1. Plan a brief (1 or 2 minute) presentation on the benefits that an Aruba solution provides for the customer's sensitive applications such as voice. How will the solution enhance users' experience?

   Your presentation should include the following benefits:

   - High-speed wireless connectivity (802.11ac and Smart Rate)
   - Innovative technologies to optimize wireless capacity (AirMatch, Client Match) without a lot of configuration
   - SDN-based QoS to remove complexity and select the right traffic every time
   - Insights into call quality with UCM

Learning check

1. What describes the effect jitter can have on voice and video applications?
   a. Jitter generally has fewer negative effects on these applications than latency.
   b. Jitter triggers the UDP transport protocol to slow down transmissions.
   c. Jitter can cause a choppy quality, worse than that caused by a consistent delay.
   d. Jitter, which refers to the percentage of packets dropped, causes ill effects at about 5%.

2. What is one advantage that an MM provides for customers with Microsoft Skype for Business over a standalone MC?
   a. Role- and firewall-based prioritization
   b. Bandwidth contracts
   c. Support for WMM
   d. Better scalability for SDN integration

3. What is one way that an MM-based architecture helps to optimize wireless capacity better than a legacy controller architecture?
   a. MM delivers ARM, a technology that optimizes broadcast and multicast traffic.
   b. MM offers AirMatch, which can make channel plans based on centralized intelligence.
   c. MM enables support for eight queues within WMM rather than just four.
   d. MM replaces legacy ClientMatch with AirMatch, which helps match clients to the best APs on the fly.

Chapter 9: High-Density Design

Activity 9: Meet customers' high-density requirements

Activity 9.1: Plan a solution for the meeting room

1. What is your new AP count for the meeting area: 6-10

2. What is your recommendation for the layout of these APs in the meeting room?

   Possible answer: You could recommend moving APs from other areas on this floor. Other areas of this floor have low-density requirements.

Activity 9.2: Plan higher capacity across the site

4. After you make your adjustments, how many APs are you planning for the business user floor? Explain your plan.

   You should plan about 22–24 APs for this floor. A higher capacity design requires about 1 AP per 1500–2500 sq ft, which would be 22–36 APs. But the floors have a relatively open area in the center with few users, so you can plan near the lower end.

5. Multiply this number by 4 for the total number of APs on the business floors (Building 1 Floor 2 and all floors in Building 2).

   You should plan between 88 and 96 APs for these floors.

6. Similarly assess your plan for Building 1 Floor 3, the executive floor. How many APs will you deploy on this floor to support voice and video? Explain your plan.

   You should plan about 18–24 APs for this floor, which has fewer users than the others.

7. How many APs will you plan for Building 1 Floor 1, including the ones that you planned for the meeting area?

   You should plan about 14–20 APs for this floor, which has much fewer users and large open spaces. However, it does require a high density of APs in the meeting area.

8. How many APs are you now recommending for the site? (Add the numbers for steps 4, 5, and 6).

   You should be recommending between 120 and 140 APs.

9. What implications does this change have on the rest of your plan? Think about as many of the potential effects as you can, going over all the topics that you have learned about up until now.

   You will need to add licenses for the additional APs. You also need to look at the plan for the switches. If you have no more than 24 APs per floor, you can almost certainly meet the needs with the current switches.

In addition, you might have been planning to connect switches in a VSF fabric per floor and provide each floor two 10 GbE links total. You should reassess if this oversubscription level meets the new needs. If you were planning 40 GbE links or two 10 GbE links per closet, the bandwidth should be sufficient.

Learning Check

1. What is true of side coverage established by APs with omnidirectional antennas?
   a. It is required for very high density deployments in large venues.
   b. Picocells are required, increasing the number of APs in the design.
   c. Spatial reuse increases, distorting the signal.
   d. 50% of AP signals are lost to the next room.

2. As a general guideline how many associations should you plan for a high-density design?
   a. 150 per radio
   b. 170 per radio
   c. 210 per radio
   d. 255 per radio

Chapter 10: Branch Deployments

Activity 10: Remote access design

Activity 10.1: Sale employees

1. What remote solution do you recommend? VIA

2. Describe the advantages to your solution:

   VIA supports androids and IOS and works with cellular data. VIA can also use split tunnel for local and corporate access. Finally, VIA is always on.

3. Does your solution have any network requirements?

   VIA user licenses required (alternative PEFV)

   Recommend 7000 series controller in DMZ

Activity 10.2: Work from home employees

1. What remote solution do you recommend? 203 RAP for each employee; 100 RAPs total
2. Describe the advantages to your solution:

   The Aruba 203 RAP supports 802.11ac and can be easily deployed, using the same SSID implemented at Corp1. Remote employees then have the same experience as they do when they are onsite. In addition, the 203 RAP provides:

   - 1 uplink port, 1 10/100/1000 port, 1 10/100/1000 port with POE +
   - POE+ port can power VOIP phone if required
3. Does your solution have any network requirements?

   7200 series controller in DMZ

Learning check

1. What APs can be deployed as RAPs
   a. AP 300 series
   b. RAP 100 series
   c. AP 200 series
   d. 205H series
2. VIA can be installed on what devices?
   a. Linux
   b. MacOS
   c. IOS
   d. Windows
   e. Androids
3. What is an IAP's DHCP Distributed L3 mode?
   a. Default gateway on IAP and DHCP at corporate
   b. Default gateway and DHCP with corporate scopes on IAP
   c. Default gateway and DHCP with local scopes on IAP
   d. Default gateway at corporate and DHCP on IAP

Chapter 11: Network Management

Activity 11: Management design

Activity 11.1: Management design

1. What management platform do you recommend? Aruba AirWave

2. Justify your recommendation:

   The customers wants a management solution that can be used to manage devices in the datacenter. Aruba AirWave can manage HPE FlexFabric switches, but Aruba Central cannot manage HPE FlexFabric.

Activity 11.2: Licenses

1. Based on your chosen management platform list the required license(s):

   AWS Licenses for ~300 devices

   No redundancy required

   No Glass license needed

2. Explain why you need these licenses and what they are used for:

   You need an AirWave license to handle all the MMs, MCs, APs, and switches you have planned for the customer. You also need enough licenses for the HPE FlexFabric switches in the datacenter.

Learning check

1. What, in AirWave, is affected by SNMP Poll periods ?
   a. CPU and Disk capacity
   b. CPU and Memory
   c. CPU and IOPS
   d. CPU only

2. Central can manage what devices?
   a. Aruba Controllers
   b. Aruba IAPs
   c. Aruba Switches
   d. Junipers routers

Chapter 12: New Network Design

Example solution

The sections below describe an example solution for CorpXYZ.

Access Points (APs)

Square footage for each floor is 36,150 square feet (3358 sq m), excluding the central area. In the example shown in Figure A-7, you are choosing to support high speeds across the floor with 24 APs per floor.

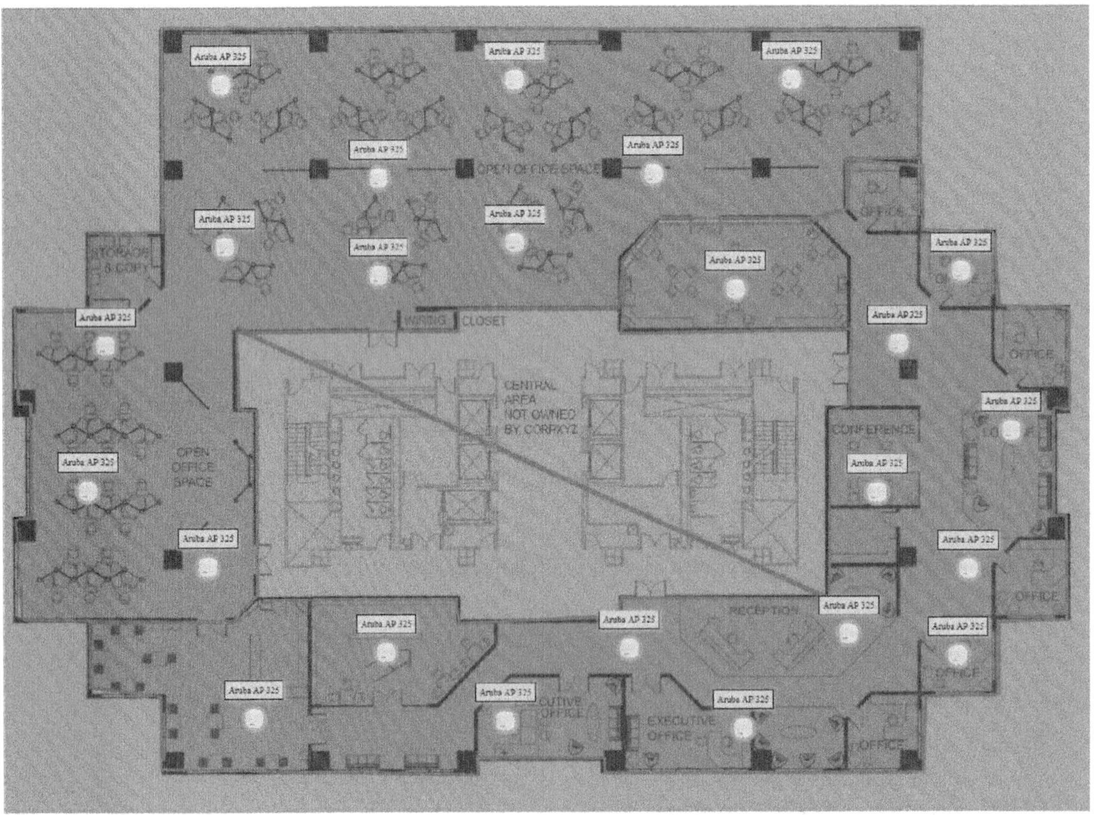

Figure A-7 Example heatmap

You should choose an 802.11ac Wave 2 AP. The example BOM shows AP-315s; the relatively high density of APs means that each AP will be generating less traffic, and they do not need dual ports or Smart Rate ports for this deployment. But you could make a case for a different AP.

For the home office employees, you should recommend at least 214 RAPs. The 203RP provides PoE ports, which accommodates the employees' VoIP phones. You could propose an even 220 RAPs.

Because the customer is planning to add about five home office employees a year, the extra six RAPs would enable CorpXYZ to have those RAPs on hand when they hire new employees.

Separate AMs are not required for this customer, but you could add them if you want to. The APs can provide IDS/WIPS.

MCs and MM

The customer wants a highly available network. You should recommend two 7205 MCs in a cluster to support about 72 APs; two 7205s works better than two 7030s because a single 7205 can support the APs in a failover situation. The MCs could be deployed in the data center, but here they are connected locally to the campus core switches on the first floor. They should be in the same VLAN so that they can cluster most effectively.

You should propose an additional two 7205 MCs to support the RAPs with high availability (but without clustering). It is better to use different MCs from the campus MCs to support the RAPs. These MCs should be deployed in a data center DMZ.

You should recommend an MM to manage the MCs and provide features such as AirMatch, ClientMatch, and UCM for integration with Microsoft Skype for Business. A virtual MM with a free standby MM would be a good choice for this customer as long as the customer has a server that meets the specifications. To support the devices in this plan, MM requires about 296 licenses (about 72 APs + 220 RAPs + 4 MCs). One 500-device license will be more cost-effective than six 50-device licenses.

Also add 292 AP, PEFNG, and RFP licenses.

Switches

You could suggest three 2930M switches per floor, deployed in a backplane stack. These switches can support SmartRate port models and modules, but you do not need them for this example design. (You could also add a SmartRate module to one of the switches later.) A 5406R switch could also work.

The backplane stack could have two 10GbE uplinks.

At the core, propose two 3810-16SFP+ switches in a backplane stack. Each switch provides enough ports to connect to the access layer (three 10 GbE links) and the MCs (one 10 GbE link to two 7205 MCs). Each switch will have one 10 GbE link to the data center, which provides 3:1 oversubscription. The switches also should connect to each other on two stacking cables. If you proposed 5406R switches at the access layer, you could propose those at the core to for consistency.

An OOBM network is recommended with backplane stacking; it also provides management access without staff having to access the closets. You could create an OOBM network with a 2530 24G switch on the first floor. This might require pulling Ethernet cable through the floors for those connections.

Aruba ClearPass

The customer only has a basic freeware RADIUS server, but is interested in increasing security. To give the access control solution more power, you should probably recommend Aruba ClearPass. You should recommend WPA2-Enterprise for the wireless network, of course, and 802.1X on the wired ports. You should probably propose tunneled node because the customer is interested in the stateful firewall and DPI controller features. Both the wireless and wired solution should implement captive portal authentication to ClearPass Guest.

To calculate the number of ClearPass Access licenses, you must determine the maximum number of devices that might be concurrently logged in through ClearPass. Because RAPs authenticate devices to ClearPass, you must take into account both devices at the office and home. You know that the customer will have at least one laptop and VoIP phone per-user, concurrently authenticated through 802.1X, which makes a total of 928 devices. But most users will probably connect one or two additional devices, so you might double that number for 1856 licenses. And you should plan another 20–30 devices for the 10 guests per day. Propose 1900 or 2000 licenses. The extra licenses will accommodate printers and other nonuser devices.

You could make a case for 500 Onboard licenses for the company's 464 users although the scenario did not indicate that those are required. The example BOM includes these.

You are proposing a virtual ClearPass appliance. High availability is important for ClearPass, but the customer will use existing processes for providing high availability for VMs.

AirWave or central

CorpXYZ needs better manageability and insight into the network. You should propose AirWave with about 300 licenses to monitor the network infrastructure (308 in the example solution).

Support and training

You could accept the default support options or adjust the support period to three years, for example. You could add training for both the wireless and wired solutions.

Bill of Materials (BOM)

A picture of the topology in shown in Figure A-8 and the BOM is shown in Figure A-9.

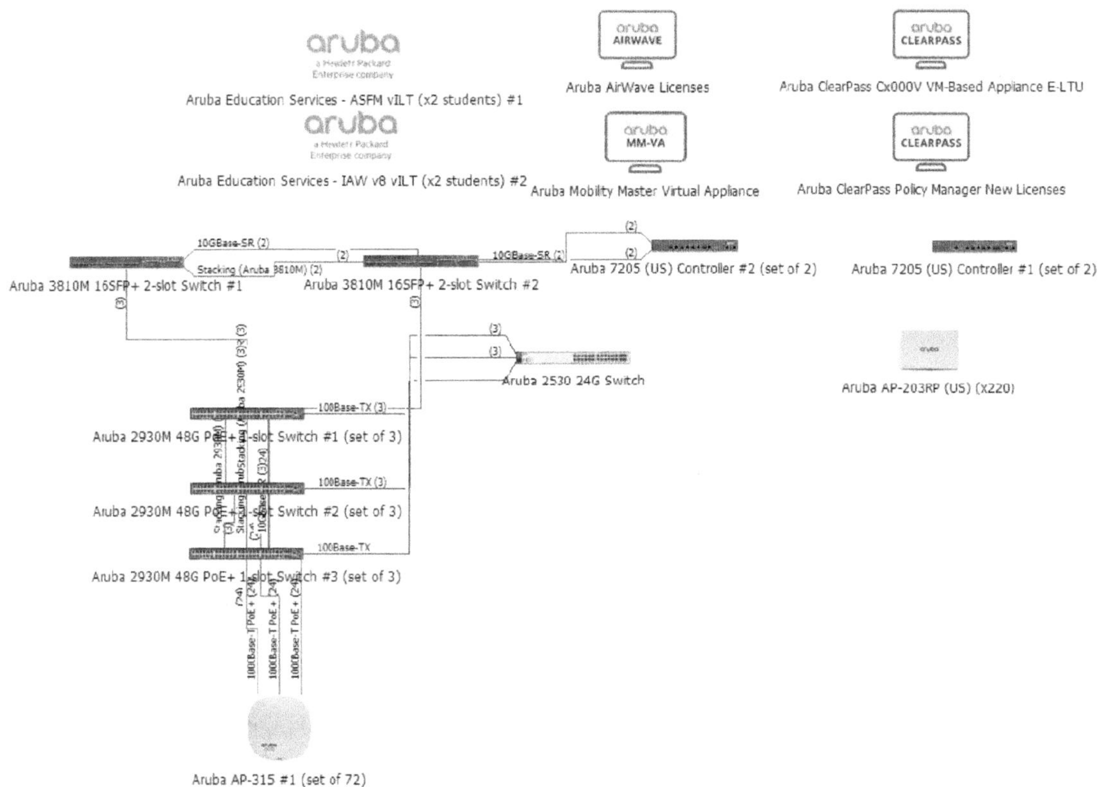

Figure A-8 Example Topology 1

Line#	Part Number	Description	Manufacturer	Unit Price	Quantity	Total	Price List
1.00	JL322A	Aruba 2930M 48G PoE+ 1-slot Switch	Hewlett Packard Enter...	$6,339.00	9	$57,051.00	USA Price List (USD)
1.01	H2CA5E	HPE 3Y FC NBD Exch A 2930M 48G P Swt SVC [for JL322A]	Hewlett Packard Enter...	$1,216.00	9	$10,944.00	USA Price List (USD)
1.02	JL086A	Aruba X372 54VDC 680W Power Supply	Hewlett Packard Enter...	$639.00	9	$5,751.00	USA Price List (USD)
1.03	JL086A ABA	INCLUDED: Power Cord - U.S. localization	Hewlett Packard Enter...	incl.	9		
1.04	JL325A	Aruba 2930 2-port Stacking Module	Hewlett Packard Enter...	$1,019.00	9	$9,171.00	USA Price List (USD)
1.05	JL083A	Aruba 3810M/2930M 4SFP+ MACsec Module	Hewlett Packard Enter...	$1,259.00	6	$7,554.00	USA Price List (USD)
1.06	J9150D	Aruba 10G SFP+ LC SR 300m MMF Transceiver	Hewlett Packard Enter...	$1,040.00	6	$6,240.00	USA Price List (USD)
2.00	JY722A	Aruba AP-203RP (US) PoE Unified RAP	Hewlett Packard Enter...	$345.00	220	$75,900.00	USA Price List (USD)
2.01	H6PZ3E	Aruba 3Y FC NBD Exch AP-203R POE SVC [for JY722A]	Hewlett Packard Enter...	$41.00	220	$9,020.00	USA Price List (USD)
2.02	J9880A	HPE 1.8m C7 to NEMA 1-15P Power Cord (North America)	Hewlett Packard Enter...	$25.00	220	$5,500.00	USA Price List (USD)
3.00	JW736A	Aruba 7205 (US) 2-port 10GBASE-X (SFP+) Controller	Hewlett Packard Enter...	$12,995.00	4	$51,980.00	USA Price List (USD)
3.01	H3CX1E	Aruba 3Y FC NBD Exch 7205 Controller SVC [for JW736A]	Hewlett Packard Enter...	$5,405.00	4	$21,620.00	USA Price List (USD)
3.02	JW088A	SFP-SX 1000BASE-SX SFP 850nm LC Connector Pluggable ...	Hewlett Packard Enter...	$395.00	4	$1,580.00	USA Price List (USD)
3.03	JW091A	SFP-10GE-SR 10GBASE-SR SFP+ 850nm Pluggable LC Conn...	Hewlett Packard Enter...	$1,245.00	4	$4,980.00	USA Price List (USD)
3.04	H6RY4E	Aruba 3Y FC NBD Exch SFP-10GE-SR SVC [for JW091A]	Hewlett Packard Enter...	$135.00	4	$540.00	USA Price List (USD)
4.00	JY895AAE	Aruba MM-VA-500 Mobility Master SW E-LTU	Hewlett Packard Enter...	$10,495.00	1	$10,495.00	USA Price List (USD)
4.01	H5UE0E	Aruba 3Y FC 24x7 MM-VA-500 ELTU SVC [for JY895AAE]	Hewlett Packard Enter...	$4,363.00	1	$4,363.00	USA Price List (USD)
4.02	JW472AAE	Aruba LIC-AP Controller per AP Capacity License E-LTU	Hewlett Packard Enter...	$75.00	292	$21,900.00	USA Price List (USD)
4.03	H2YU4E	Aruba 3Y FC 24x7 Ctrl perAP Cap ELTU SVC [for JW472A...	Hewlett Packard Enter...	$32.00	292	$9,344.00	USA Price List (USD)
4.04	JW473AAE	Aruba LIC-PEF Controller Policy Enforcement Firewall Per AP...	Hewlett Packard Enter...	$75.00	292	$21,900.00	USA Price List (USD)
4.05	H2XX4E	Aruba 3Y FC 24x7 License PEF Cn SVC [for JW473AAE]	Hewlett Packard Enter...	$32.00	292	$9,344.00	USA Price List (USD)
4.06	JW474AAE	Aruba LIC-RFP Controller RFProtect Per AP License E-LTU	Hewlett Packard Enter...	$75.00	292	$21,900.00	USA Price List (USD)
4.07	H2XV4E	Aruba 3Y FC 24x7 AP RFProtectE-LTU SVC [for JW474AAE]	Hewlett Packard Enter...	$32.00	292	$9,344.00	USA Price List (USD)
5.00	JZ402AAE	Aruba ClearPass NL AC 1K CE E-LTU	Hewlett Packard Enter...	$21,000.00	2	$42,000.00	USA Price List (USD)
5.01	H9XH3E	Aruba 3Y FC 24x7 ClearPass NL AC 1KCESVC [for JZ402AAE]	Hewlett Packard Enter...	$4,535.00	2	$9,070.00	USA Price List (USD)
5.02	JZ437AAE	Aruba ClearPass NL OB 500 USR E-LTU	Hewlett Packard Enter...	$17,000.00	1	$17,000.00	USA Price List (USD)
5.03	H9XD3E	Aruba 3Y FC 24x7 ClearPass NLOB500USRSVC [for JZ437A...	Hewlett Packard Enter...	$3,699.00	1	$3,699.00	USA Price List (USD)
6.00	JZ399AAE	Aruba ClearPass Cx000V VM Appliance E-LTU	Hewlett Packard Enter...	$4,000.00	1	$4,000.00	USA Price List (USD)
6.01	H9WX3E	Aruba 3Y FC 24x7 ClearPass Cx000V VM SVC [for JZ399AAE]	Hewlett Packard Enter...	$869.00	1	$869.00	USA Price List (USD)
7.00	JW546AAE	Aruba LIC-AW Aruba Airwave with RAPIDS and VisualRF 1 ...	Hewlett Packard Enter...	$75.00	308	$23,100.00	USA Price List (USD)
7.01	H2YV4E	Aruba 3Y FC 24x7 Airwave 1 Dev E-LTU SVC [for JW546A...	Hewlett Packard Enter...	$32.00	308	$9,856.00	USA Price List (USD)
8.00	H1EJ9E	HPE Aruba WW Education Tech Training SVC	Hewlett Packard Enter...	$200.00	36	$7,200.00	USA Price List (USD)
8.01	01077931_VILT	INCLUDED: Aruba Switching Fundamentals for Mobility (A...	Hewlett Packard Enter...	incl.	18		
8.02	01089598_VILT	INCLUDED: Implementing Aruba WLANs v8 vILT	Hewlett Packard Enter...	incl.	18		
9.00	J9776A	Aruba 2530 24G Switch	Hewlett Packard Enter...	$1,099.00	1	$1,099.00	USA Price List (USD)
9.01	J9776A ABA	INCLUDED: Power Cord - U.S. localization	Hewlett Packard Enter...	incl.	1		
9.02	H1GT6E	HPE 3Y FC NBD Exch Aruba 2530 24G Sw SVC [for J9776A]	Hewlett Packard Enter...	$170.00	1	$170.00	USA Price List (USD)
10.00	JL075A	Aruba 3810M 16SFP+ 2-slot Switch	Hewlett Packard Enter...	$11,549.00	2	$23,098.00	USA Price List (USD)
10.01	U6TC7E	HPE 3Y FC NBD Exch Aruba 3810M 16SFP SVC [for JL075A]	Hewlett Packard Enter...	$2,432.00	2	$4,864.00	USA Price List (USD)
10.02	JL085A	Aruba X371 12VDC 250W Power Supply	Hewlett Packard Enter...	$439.00	2	$878.00	USA Price List (USD)
10.03	JL085A ABA	INCLUDED: Power Cord - U.S. localization	Hewlett Packard Enter...	incl.	2		
10.04	JL084A	Aruba 3810M 4-port Stacking Module	Hewlett Packard Enter...	$1,099.00	2	$2,198.00	USA Price List (USD)
10.05	J9150D	Aruba 10G SFP+ LC SR 300m MMF Transceiver	Hewlett Packard Enter...	$1,040.00	12	$12,480.00	USA Price List (USD)
11.00	JW797A	Aruba AP-315 802.11n/ac 2x2:2/4x4:4 MU-MIMO Dual Radi...	Hewlett Packard Enter...	$995.00	72	$71,640.00	USA Price List (USD)
11.01	H4RH3E	Aruba 3Y FC NBD Exch AP-315 SVC [for JW797A]	Hewlett Packard Enter...	$119.00	72	$8,568.00	USA Price List (USD)
12.00	J9734A	Aruba 2920/2930M 0.5m Stacking Cable	Hewlett Packard Enter...	$149.00	9	$1,341.00	USA Price List (USD)
13.00	J9578A	Aruba 3800/3810M 0.5m Stacking Cable	Hewlett Packard Enter...	$249.00	2	$498.00	USA Price List (USD)
		Quote Total				**$620,049...**	

Figure A-9 Example BOM

Other design considerations

That completes the BOM. Additional components of your design would include these elements:

- Create one WLAN for employees and one WLAN for guests.
- Use a different VLAN/subnet for wireless devices and wired devices.
- For the wireless devices, the 10.62.0.0/22 subnet will provide addresses for about 1000 devices, which is sufficient for the main office but not remote users.
  - For the main office, users cannot roam between floors, so each floor is a separate RF domain. Therefore, you can use a per-floor VLAN design rather than per-campus.
  - You could divide up the 10.62.0.0/22 subnet into four /24 subnets and assign one per floor. But you should keep in mind that you cannot really count on only having only 85 users and 250 devices per floor. Users might move around the office and receive an IP address in the

floor 1 VLAN, the floor 2 VLAN, and the floor VLAN 3. The lease might persist even when the device disconnects. So you might want to ask the customer for additional IP addressing space so that you can create /23 or /22 subnets for each floor.

- The users connecting through RAPs will also need a VLAN, as will their VoIP phones. Because these devices cannot roam to the main office, you might plan some VLAN segmentation here and ask for /24 subnets.
- You can use the existing 10.50.0.0/22 subnet for wired devices. But you should recommend dividing off VoIP phones into their own VLAN, which will need a subnet address (/23 or two /24 to accommodate future growth; 250 phones is near the limit of a /24 subnet).
- Plan to use LLDP-MED to assign a high priority (such as 802.1p 6 and DSCP 48 or 802.1p 7 and DSCP 56) to VoIP phones; both switches and RAPs support LLDP-MED
- Discuss with the customer if employees also use Microsoft Skype for Business softphone on their laptops and other devices.
  - If so, recommend Windows policies to assign DSCP 56 or 48 to voice and 40 for video.
  - Make sure WMM is enabled with the recommended DSCP maps on MCs.
  - Make sure MM UCM is set up for SDN integration with Microsoft Skype for Business.
  - Make sure DSCP trust is enabled on switch ports that connect to APs and any corporate laptops.

Index

A

E

F

G

H

I

M

N

R

S

Z